THE COMPLETE OFFICIAL AUSTIN-HEALEY 100-SIX AND 3000

Model Years 1956-1968

Selected Titles From Bentley Publishers

Engineering

Robert Bentley's Repair Manual for British Cars *John Organ*
ISBN 978-0-8376-0041-3

Vintage Racing British Sports Cars *Terry Jackson*
ISBN 978-0-8376-0153-3

Scientific Design of Exhaust & Intake Systems *Philip H. Smith and John C. Morrison*
ISBN 978-0-8376-0309-4

The Design and Tuning of Competition Engines *Philip H. Smith* ISBN 978-0-8376-0140-3

New Directions in Suspension Design: Making the Car Faster
Colin Campbell
ISBN 978-0-8376-0150-2

Racing Car Design and Development *Len Terry with Alan Baker* ISBN 978-0-8376-0080-2

Austin-Healey

Austin-Healey 100/6 NS 3000 Workshop Manual: 1956-1968
British Leyland Motors
ISBN 978-0-8376-1743-5
Hardcover, 416 pages

The Complete Official 1275 cc Sprite/Midget: 1967-1974
British Leyland Motors
ISBN 978-0-8376-1748-0
Hardcover, 400 pages

The Complete Official 948 cc & 1098 cc Sprite/Midget
British Leyland Motors
ISBN 978-0-8376-1747-3
Hardcover, 256 pages

MG

The Complete Official MGB: 1962-1974 *British Leyland Motors*
ISBN 978-0-8376-1761-9
Hardcover, 480 pages
Includes: Driver's Handbook, Workshop Manual, Special Tuning Manual

The Complete Official MGB: 1975-1980 *British Leyland Motors*
ISBN 978-0-8376-1760-2
Hardcover, 304 pages

The Complete Official MG Midget 1500: 1975-1979
British Leyland Motors
ISBN 978-0-8376-1744-2
Hardcover, 384 pages

The MG Workshop Manual: 1929-1955 *W.E. Blower*
ISBN 978-0-8376-1746-6
Hardcover, 602 pages

Triumph

The Complete Official Triumph GT6, GT6+ & GT6 MK III: 1967-1973 *British Leyland Motors*
ISBN 978-0-8376-1749-7
Hardcover, 480 pages
Includes: Driver's Handbook, Workshop Manual

The Complete Official Triumph TR6 & TR250: 1967-1976
British Leyland Motors
ISBN 978-0-8376-1762-6
Hardcover, 608 pages
Includes: Driver's Handbook, Workshop Manual

The Complete Official Triumph Spitfire MK III, MK IV & 1500: 1968-1974 *British Leyland Motors*
ISBN 978-0-8376-1763-3
Hardcover, 480 pages
Includes: Driver's Handbook, Workshop Manual

The Complete Official Triumph Spitfire 1500: 1975-1980
British Leyland Motors
ISBN 978-0-8376-1745-9
Hardcover, 536 pages
Includes: Driver's Handbook, Workshop Manual

The Complete Official Triumph TR7: 1975-1981
British Leyland Motors
ISBN 978-0-8376-1742-8
Hardcover, 544 pages
Includes: Driving Manual, Workshop Manual

The Complete Official Triumph TR4 & TR4A: 1961-1968
British Leyland Motors
ISBN 978-0-8376-1741-1
Hardcover, 404 pages
Includes: Driving Manual, Workshop Manual, Competition Preparation Manual by R.W. "Kas" Kastner

The Complete Official Triumph TR2 & TR3: 1953-1961
British Leyland Motors
ISBN 978-0-8376-1750-3
Hardcover, 456 pages
Includes: Driver's Instruction Book, Service Instruction Manual

Jaguar

The Complete Official Jaguar "E" Series 1 & Series 2: 1961-1971
British Leyland Motors
ISBN 978-0-8376-1764-0
Hardcover, 790 pages
Includes: Driver's Handbook, Workshop Manual, Special Tuning Manual

B BentleyPublishers®
Automotive Reference

Bentley Publishers has published service manuals and automobile books since 1950. For more information, please contact Bentley Publishers at 1734 Massachusetts Avenue, Cambridge, MA 02138 USA, or visit our web site at `BentleyPublishers.com`

THE COMPLETE OFFICIAL AUSTIN-HEALEY 100-SIX AND 3000

Model Years 1956 –1968

BentleyPublishers®
.com

BENTLEY PUBLISHERS® | Automotive Reference™

Bentley Publishers, a division of Robert Bentley, Inc.
1734 Massachusetts Avenue
Cambridge, MA 02138 USA
800-423-4595 / 617-547-4170

Information that makes the difference®

BentleyPublishers®.com

Technical contact information

We welcome your feedback. Please submit corrections and additions to our technical discussion forum at:

`http://www.BentleyPublishers.com`

Updates and corrections

We will evaluate submissions and post appropriate editorial changes online as updates or tech discussion. Appropriate updates and corrections will be added to the book in future printings. Check for updates and corrections for this book before beginning work on your vehicle. See the following web address for additional information:

`http://www.BentleyPublishers.com/updates/`

WARNING—Important Safety Notice

Do not use this manual for repairs unless you are familiar with basic automotive repair procedures and safe workshop practices. This manual illustrates the workshop procedures for some maintenance and service work. It is not a substitute for full and up-to-date information from the vehicle manufacturer or for proper training as an automotive technician. Note that it is not possible for us to anticipate all of the ways or conditions under which vehicles may be serviced or to provide cautions as to all of the possible hazards that may result.

We have endeavored to ensure the accuracy of the information in this manual. Please note, however, that considering the vast quantity and the complexity of the service information involved, we cannot warrant the accuracy or completeness of the information contained in this manual.

FOR THESE REASONS, NEITHER THE PUBLISHER NOR THE AUTHOR MAKES ANY WARRANTIES, EXPRESS OR IMPLIED, THAT THE INFORMATION IN THIS MANUAL IS FREE OF ERRORS OR OMISSIONS, AND WE EXPRESSLY DISCLAIM THE IMPLIED WARRANTIES OF MERCHANTABILITY AND OF FITNESS FOR A PARTICULAR PURPOSE, EVEN IF THE PUBLISHER OR AUTHOR HAVE BEEN ADVISED OF A PARTICULAR PURPOSE, AND EVEN IF A PARTICULAR PURPOSE IS INDICATED IN THE MANUAL. THE PUBLISHER AND AUTHOR ALSO DISCLAIM ALL LIABILITY FOR DIRECT, INDIRECT, INCIDENTAL OR CONSEQUENTIAL DAMAGES THAT RESULT FROM ANY USE OF THE EXAMPLES, INSTRUCTIONS OR OTHER INFORMATION IN THIS MANUAL. IN NO EVENT SHALL OUR LIABILITY WHETHER IN TORT, CONTRACT OR OTHERWISE EXCEED THE COST OF THIS MANUAL.

Before attempting any work on your Austin-Healey, read **vi Warnings and Cautions** and any **WARNING** or **CAUTION** that accompanies a procedure in the manual. Review the **WARNINGS** and **CAUTIONS** each time you prepare to work on your Austin-Healey.

Your common sense and good judgment are crucial to safe and successful service work. Read procedures through before starting them. Think about whether the condition of your car, your level of mechanical skill, or your level of reading comprehension might result in or contribute in some way to an occurrence which might cause you injury, damage your car, or result in an unsafe repair. If you have doubts for these or other reasons about your ability to perform safe repair work on your car, have the work done at an authorized Austin-Healey dealer or other qualified shop.

Part numbers listed in this manual are for identification purposes only, not for ordering. Always check with your authorized Austin-Healey dealer to verify part numbers and availability before beginning service work that may require new parts.

Special tools required to perform certain service operations are identified in the manual and are recommended for use. Use of improper tools may be detrimental to the car's safe operation as well as the safety of the person servicing the car.

The vehicle manufacturer will continue to issue service information updates and parts retrofits after the editorial closing of this manual. Some of these updates and retrofits will apply to procedures and specifications in this manual. We regret that we cannot supply updates to purchasers of this manual.

This manual is prepared, published and distributed by Bentley Publishers, 1734 Massachusetts Avenue, Cambridge, Massachusetts 02138. All information contained in this manual is based on the information available to the publisher at the time of editorial closing. Austin-Healey has not reviewed and does not vouch for the accuracy or completeness of the technical specifications and work procedures described and given in this manual.

ISBN 978-0-8376-1743-5 Job code: X133-09

Library of Congress Catalog Card No. 77-72588

The paper used in this publication is acid free and meets the requirements of the National Standard for Information Sciences-Permanence of Paper for Printed Library Materials.∞

Manufactured in the United States of America.

Preface

The Austin-Healey 100-Six and 3000 models represent a highly successful range of cars that did much to bring excitement to the American sports car movement. They were, in fact, the best-selling medium-sized sports cars of their day. Beginning with the introduction of the 100-Six in the summer of 1956, the "big" six-cylinder Austin-Healeys remained in production for eleven years. Countless sports car enthusiasts, initiated by the MG, Singer, and other of the slower "British traditional" roadsters, moved up to genuine speed and power with the purchase of a six-cylinder Healey.

Unfortunately, the Austin-Healey was killed by the 1968 U.S. Federal safety regulations; too many modifications would have been necessary to make continued production possible.

Today, thanks to timeless styling, ample power, and genuine sports car feel, the six-cylinder Austin-Healeys are as eagerly sought after as they were some twenty or more years ago, when they were still in production. Needless to say, these machines remained active and successful in SCCA racing for a number of years after the last new model had rolled out of the Abingdon factory, and wise collectors began to acquire the 100-Six and 3000 models even before their long and successful histories had been written on the race track.

This is as it should be; the six-cylinder Austin-Healeys are landmarks in the evolution of the sports car. Before the "big sixes" arrived on the automotive scene, enthusiasts of modest means had very few *fast* sports cars from which to choose. The "big" Healey, while not quite so fast as the Jaguar, was at the last an "almost 120-mph" sports car with great mechanical durability and a considerably smaller price tag than the "Jag".

Because the value of Austin-Healey 100-Six and 3000 sports cars is increasing, it is impossible to overstate the importance of correct maintenance and repair. This manual has been published in order to supply complete, accurate, and comprehensive maintenance and repair data to both car owners and professional mechanics; it is the official factory manual that was originally intended for use by dealer service departments. Owners of Austin-Healey 100-Six and 3000 sports cars willl find it both informative and indispensable to the continued maintenance of their cars.

Assembly work on small, highly-tuned machines such as the Austin-Healey sports cars must be carried out with greater precision than is commonly practiced on large Amercan cars. Particular emphasis must be given to the proper use of torque wrenches and to strict adherence to the tighhtening torque specifications that are given this manual. A fastener that is too tight can be worse than one that is too loose—especially on a lightweight sprots car. Stretched or broken bolts and distorted parts, which result from overtightening by musclebound mechanics, become a serious concern where the precision fitting of light alloy and thin-wall iron castings is involved.

The importance of cleanliness cannot be overemphasized. Under no circumstances should an engine or gearbox be repaired on the ground or on a garage floor. Thoroughly clean the exteriors of major components prior to disassembly in order to keep road dirt and other grime out of the working parts. No more than a pinch of abrasive dust in a gearbox can cause rapid failure of the synchronizers and bearings.

During the final assembly of an engine or gearbox, the cleaned parts should be laid out on a clean workbench that has been covered with clean sheets of new cardboard or wrapping paper. The engine or gearbox itself, if not mounted on a special stand, should likewise be placed on a clean workbench. Sandpapering, valve grinding, or the use of bench grinders should not be permitted near the area where final assembly is taking place. If assembly cannot be completed in a day, enclose the partially-assembled engine or gearbox in a large plastic bag—such as a new trash bag or a dry cleaner's garment return bag—so that dust and dirt will be excluded until assembly work resumes.

By observing these precautions during the maintenance, repair, or restoration of your Austin-Healey, you will be preserving the value and the life of a car that is rapidly becoming one of the classics among post-World War II sports cars. If at any time you lack the skills, special equipment, tools, or workshop facilities for making repairs as they are described in this manual, we suggest you leave these repairs to and authorized dealer or other qualified shop.

Please read these warnings and cautions before proceeding with maintenance and repair work.

WARNING –

- Never work under a lifted car unless it is solidly supported on stands intended for the purpose. Do not support a car on cinder blocks, hollow tiles, or other props that may crumble under continuous load. Do not work under a car that is supported solely by a jack.

- If you are going to work under a car on the ground, make sure that the ground is level. Block the wheels to keep the car from rolling. Disconnect the battery ground strap to prevent others from starting the car while you are under it.

- Never run the engine unless the work area is well ventilated. Carbon monoxide kills.

- Friction materials such as brake and clutch linings or brake pads may contain asbestos fibers. Do not create dust by grinding, sanding or by cleaning with compressed air. Avoid breathing asbestos fibers and asbestos dust. Breathing asbestos may result in serious diseases, such as asbestosis or cancer, and cause severe injury and death.

- Tie long hair behind your head. Do not wear a necktie, scarf, loose clothing, or necklace when you work near machine tools or running engines. If your hair, clothing, or jewelry were to get caught in the machinery, severe injury could result.

- Disconnect the battery ground strap whenever you work on the fuel system or the electrical system. When you work around fuel, do not smoke or work near heaters or other fire hazards. Keep an approved fire extinguisher handy.

- Illuminate your work area adequately but safely. Use a portable safety light for working inside or under the car. Make sure its bulb is enclosed by a wire cage. The hot filament of an accidentally broken bulb can ignite spilled fuel or oil.

- Catch draining fuel, oil, or brake fluid in suitable containers. Do not use food or beverage containers that might mislead someone into drinking from them. Store flammable fluids away from fire hazards. Wipe up spills at once, but do not store the oily rags, which can ignite and burn spontaneously.

- Finger rings should be removed so that they cannot cause electrical shorts, get caught in running machinery, or be crushed by heavy parts.

- Keep sparks, lighted matches, and open flame away from the top of the battery. If hydrogen gas escaping from the cap vents is ignited, it will ignite gas trapped in the cells and cause the battery to explode.

- Always observe good workshop practices. Wear goggles when you operate machine tools or work with battery acid. Gloves or other protective clothing should be worn whenever the job requires it.

CAUTION –

- If you lack the skills, tools, and equipment, or a suitable workshop for any procedure described in this manual, we suggest you leave such repairs to an authorized dealer or other qualified shop. We especially urge you to consult your authorized dealer before attempting any repairs on a car still covered by the new-car warranty.

- British Leyland Motors is constantly improving its cars and sometimes these changes, both in parts and specifications, are made applicable to earlier models. Therefore part numbers listed in this manual are for reference only. Always check with your authorized parts department for the latest information.

- Before starting a job, make certain that you have all necessary tools and parts on hand. Read all instructions thoroughly; do not attempt shortcuts. Use tools appropriate to the work and use only replacement parts meeting specifications. Makeshift tools, parts, and procedures will not make good repairs.

- Use pneumatic and electrical tools only to loosen threaded parts and fasteners. Never use such tools to tighten fasteners, especially on light alloy parts.

- Be mindful of the environment and ecology. Before you drain the crankcase, find out the proper way to dispose of the oil. Do not pour oil onto the ground, down a drain, or into a stream, pond, or lake. Consult local ordinances that govern the disposal of wastes.

Contents

AUSTIN-HEALEY WORKSHOP MANUAL 1956-1968

INTRODUCTION

THIS workshop manual gives comprehensive and authentic information for the assistance of the Austin Distributors and Dealer organisations in maintaining and repairing the Austin-Healey '100-Six', and '3000', Mk. I, II and III.

The manual is divided into sections each of which deals with an assembly or major component of the vehicle and carries a reference letter. Where necessary, a section is divided into two or three parts. In such cases, the single letter section refers to Series BN4 cars, the double letter section to Series BN6 cars, and the treble letter section to '3000' Mk. I and II (Series BN7 and BT7) cars and Mk. II and III Sports Convertible (Series BJ7 and BJ8) cars. Double and treble letter sections should always be used in conjunction with the corresponding single letter section. Where there are no double or treble letter sections the information contained in the single letter section refers to all models.

Maintenance items contained in a section should be carried out at the intervals specified in the Passport to Service or Drivers Handbook.

Full illustration of both major and sub-assemblies accompanies the detailed instructions for the dismantling, assembling and inspection of component parts. It is emphasised that only BMC Genuine Parts are to be used as replacements, and the operator should at all times make use of his "Service Parts List".

Use of the correct Service Tools contributes to an efficient, economic, and profitable repair. References to such tools have, therefore, been made throughout the manual.

CONTENTS

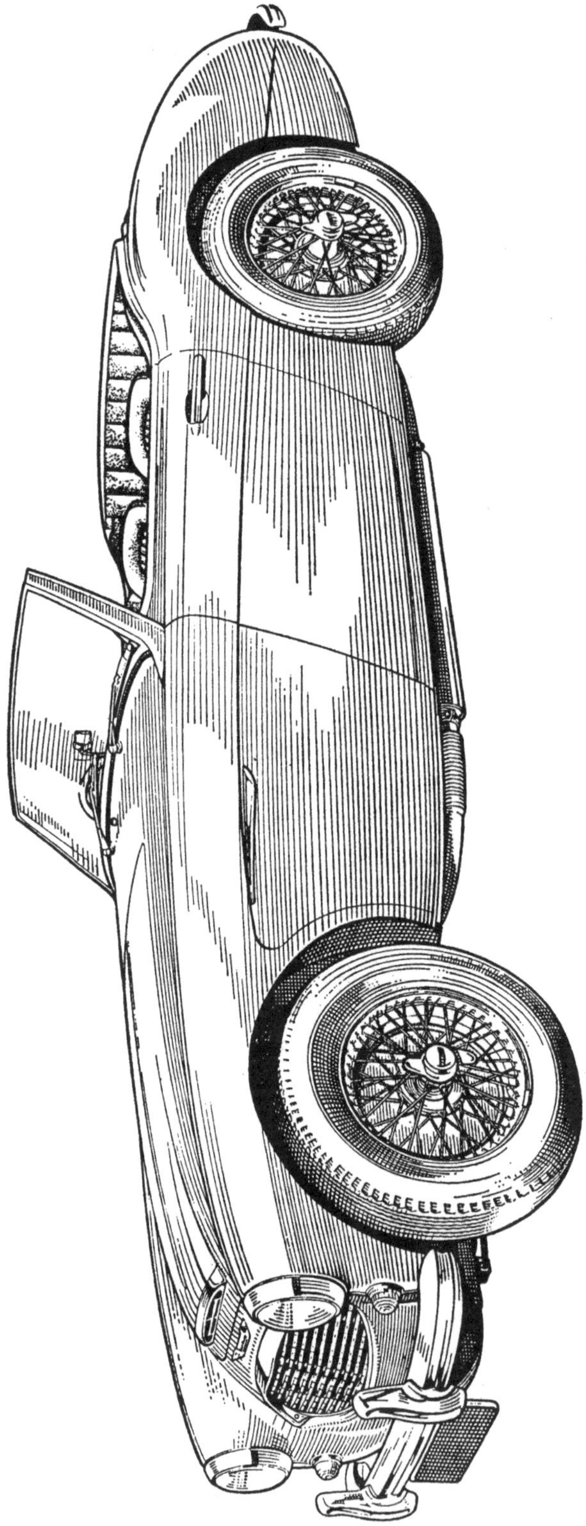

Austin-Healey '100-Six' (Series BN4) with hood lowered.

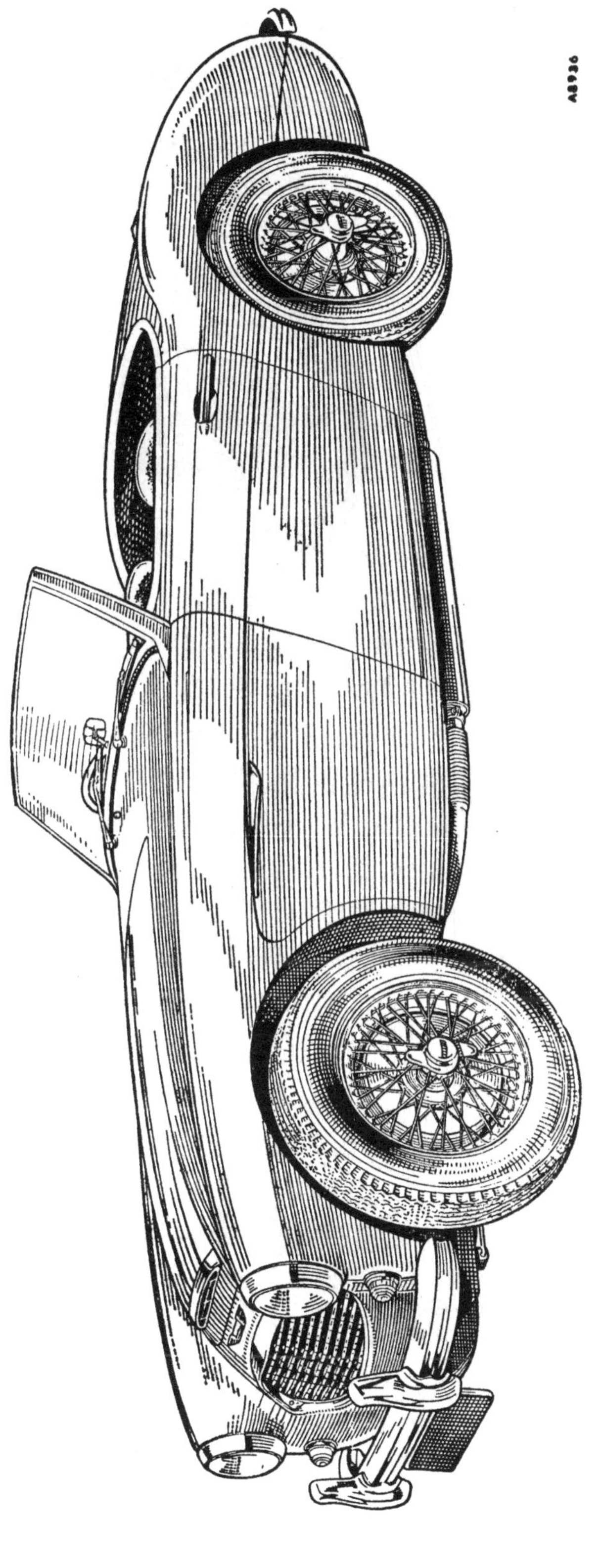

Austin-Healey '100-Six' (Series BN6) with hood lowered.

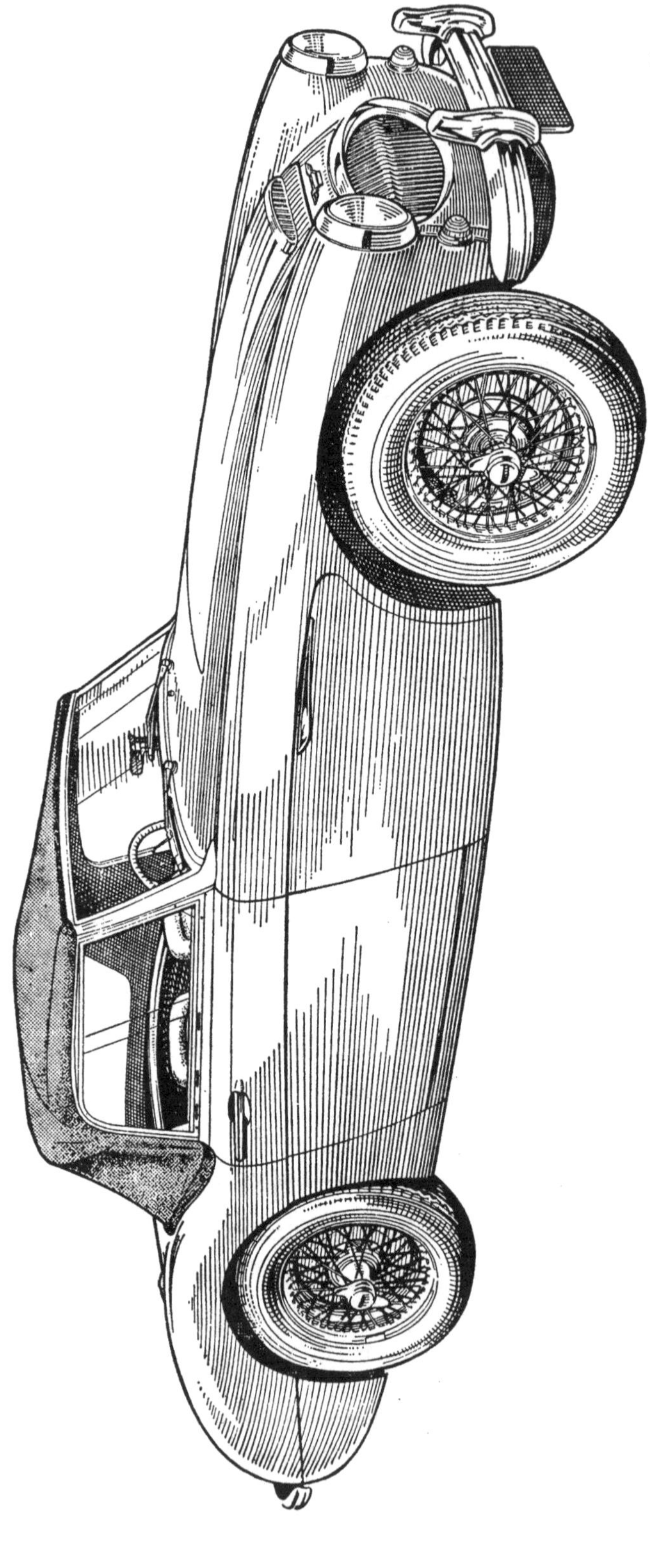

Austin-Healey '3000' (Series BNT) with the hood raised.

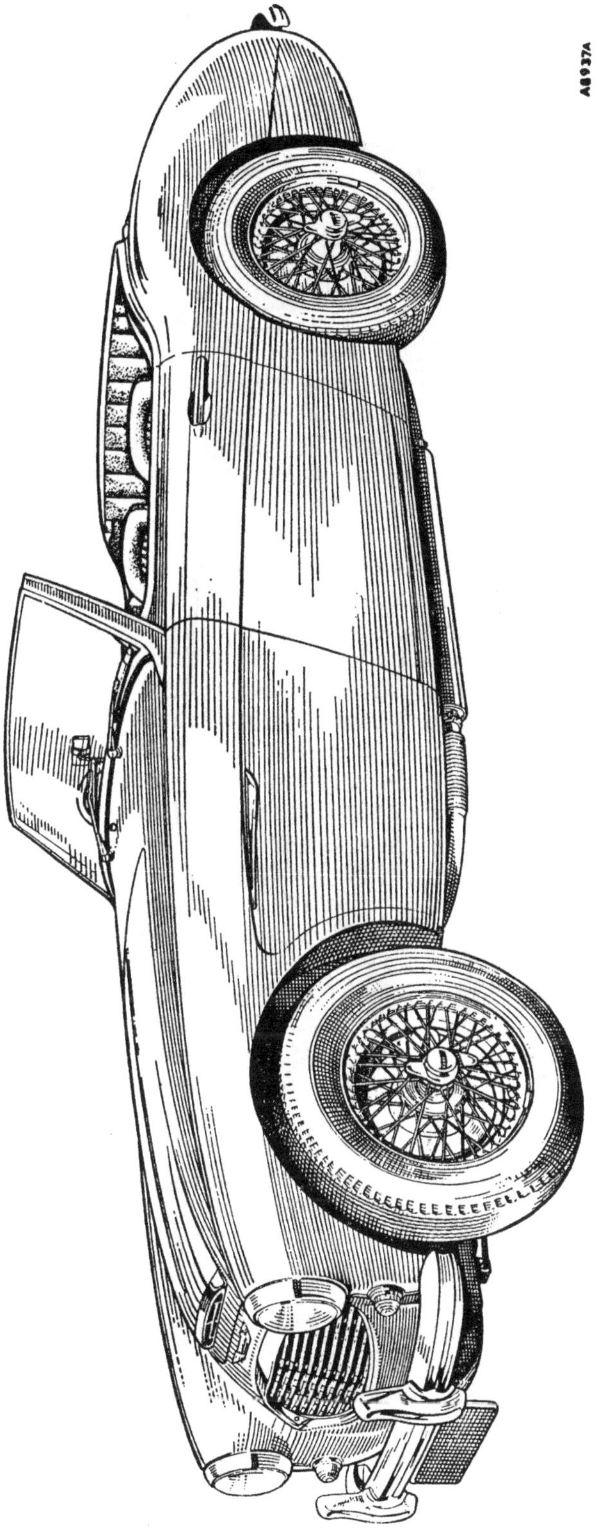

Austin-Healey '3000' (Series BT7) with the hood lowered.

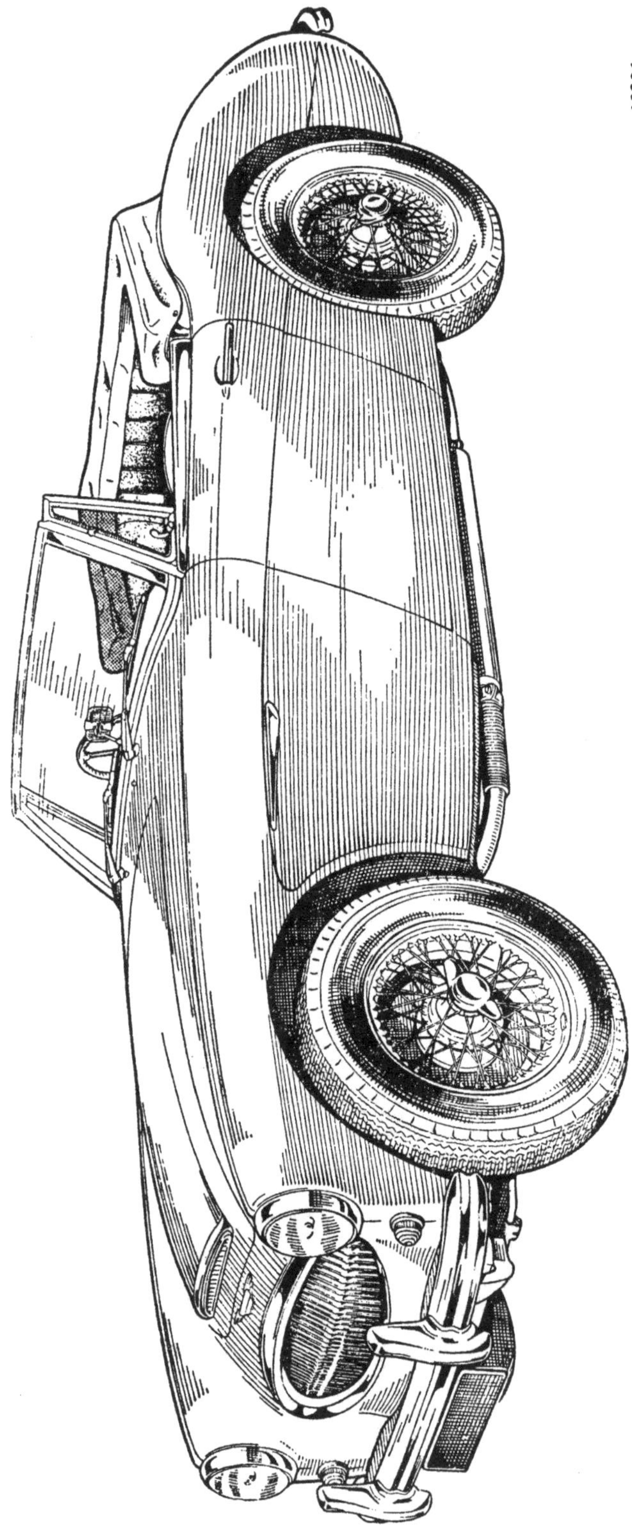

A8924

Austin-Healey '3000' Mk. II and early III Sports Convertible (Series BJ7 and BJ8) with hood lowered.

GENERAL DATA

Data variations for cars fitted with the 6 port cylinder head are given on General Data page 8 and for Series BN6 and later Series BN4 cars on page 9. Data for Series BN7, BT7, BJ7 and BJ8 cars will be found on General Data page 10.

ENGINE

Type	BN4
Number of cylinders	6.
Bore	3·125 in. (79·375 mm.).
Stroke	3·5 in. (88·9 mm.).
Capacity	2639 c.c. (161 cu. in.).
Compression ratio	8·25 : 1.
Bore, 1st Oversize	+·010 in. (·254 mm.).
Bore, 2nd Oversize	+·020 in. (·508 mm.).
Bore, 3rd Oversize	+·030 in. (·762 mm.).
Bore, 4th Oversize	+·040 in. (1·016 mm.).
Firing Order	1, 5, 3, 6, 2, 4.
Cooling	Thermo-siphon; pump, fan, and thermostat.
Torque	142 lb./ft. (19·77 kg./m.) at 2,400 r.p.m.
B.M.E.P.	139 lb./sq. in. (9·77 kg./cm.^{2}).

Valves

Position	Overhead, push-rod operated.
Lift	·3145 in. (8·054 mm.).
Diameter: Head: Inlet	1·693 to 1·683 in. (42·99 to 42·75 mm.).
Exhaust	1·420 to 1·415 in. (36·07 to 35·94 mm.).
Stem: Inlet	·34175 to ·34225 in. (8·68 to 8·69 mm.).
Exhaust	·34175 to ·34225 in. (8·68 to ·869 mm.).
Stem/guide clearance: Inlet	·0025 to ·0015 in. (·063 to ·038 mm.).
Exhaust	·002 to ·001 in. (·051 to ·025 mm.).
Valve rocker clearance	·012 in. (·3 mm.) hot.
Seat angle: Inlet	30°.
Exhaust	45°.
Seat face width: Inlet	·091 to ·097 in. (2·31 to 2·46 mm.).
Exhaust	·198 to ·217 in. (5·03 to 5·51 mm.).

Valve guides

Length: Inlet	2·266 in. (57·55 mm.).
Exhaust	2·578 in. (65·49 mm.).

Valve springs

Free length: Inner	1·969 in. (50 mm.).
Outer	2·047 in. (51·99 mm.).
Fitted length: Inner	1·517 in. (38·53 mm.) load 25·3 lb. (11·476 kg.).
Outer	1·607 in. (40·82 mm.) load 54·2 lb. (24·58 kg.).

Tappets

Type	Cylindrical, spherical foot.
Diameter	·93725 in. (23·81 mm.).
Length	2·548 in. (64·72 mm.).

Rockers

Bushes	Steel and white metal.
Outside diameter (before fitting)	·913 in. (23·17 mm.).

Engine—continued

Inside diameter (reamed in position)	·8115 to ·8125 in. (20·62 to 20·65 mm.).
Clearance	·0025 to ·0005 in. (·063 to ·012 mm.).
Bore of arm	·909 to ·910 in. (23·076 to 23·101 mm.).

Piston

Material	Low expansion aluminium alloy.
Clearance at skirt (right angles to gudgeon pin)	·0226 to ·0008 in. (·066 to ·020 mm.).
Width of ring groove: Compression	·0952 to ·0962 in. (2·410 to 2·436 mm.).
Oil	·189 to ·190 in. (4·81 to 4·83 mm.).
Oversizes	+·010 in.; +·020 in.; +·030 in.; +·040 in. (+·25 mm.; +·50 mm.; +·76 mm.; +1·02 mm.).

Piston rings

Number	3 compression (2 taper), 1 oil control.
Width: Compression	·0938 to ·0928 in. (2·383 to 2·357 mm.).
Oil	·1865 to ·1875 in. (4·737 to 4·762 mm.).
Clearance in groove: Compression	·0034 to ·0014 in. (·086 to ·036 mm.).
Oil	·0015 to ·0035 in. (·038 to ·088 mm.).
Ring gap (compression and oil)	·009 to ·014 in. (·23 to ·35 mm.).

Gudgeon pins

Type	Clamped in rod. Fully floating: From Engine No. 40501.
Fit	Selective; push in piston.
Diameter	·8748 to ·8750 in. (22·215 to 22·220 mm.).

Crankshaft

Journal diameter	2·3742 to 2·3747 in. (60·305 to 60·317 mm.).
Crankpin diameter	2·0000 to 2·0005 in. (50·80 to 50·8127 mm.).
Undersizes (journals and crankpins)	—·010 in.; —·020 in.; —·030 in.; —·040 in. (—·254mm.; —·508 mm.; —·762 mm.; —1·016 mm.).
End-float	Taken on thrust washer at front middle (No. 2) main bearing; ·0025 to ·0055 in. (·063 to ·140 mm.).
Thrust washer: Standard	·091 to ·093 in. (2·315 to 2·366 mm.).
+·0025 in. (+·063 mm.)	·0935 to ·0955 in. (2·378 to 2·429 mm.).
+·005 in. (+·127 mm.)	·0960 to ·0980 in. (2·442 to 2·493 mm.).
+·0075 in. (+·190 mm.)	·0985 to ·1005 in. (2·505 to 2·556 mm.).
+·010 in. (+·254 mm.)	·1010 to ·1030 in. (2·569 to 2·620 mm.).

Main bearings

Number	4.
Type	White-metalled steel shell.
Length	1·495 to 1·505 in. (37·973 to 38·227 mm.).
Running clearance	·0013 to ·0028 in. (·033 to ·071 mm.).
Sizes for reground journals	·010 in. U/S; ·020 in. U/S; ·030 in. U/S; ·040 in. U/S. (·254 mm.; ·508 mm.; ·762 mm.; 1·016 mm.).

Connecting rods

Length (centres)	6·601 to 6·605 in. (167·66 to 167·76 mm.).
Side clearance	·007 to ·004 in. (·18 to ·10 mm.).
Big-ends: Type	White-metalled steel shells.
Diametrical clearance	·0005 to ·002 in. (·0127 to ·051 mm.).
Small-end bush	·8750 to ·8755 in. (22·225 to 22·328 mm.).
Sizes for reground crankpins	·010 in. U/S; ·020 in. U/S; ·030 in. U/S; ·040 in. U/S. (·254 mm.; ·508 mm.; ·762 mm.; 1·016 mm.).

Camshaft

Journal diameters: Front	1·78875 to 1·78925 in. (45·434 to 45·447 mm.).
Middle front	1·76875 to 1·76925 in. (44·926 to 44·939 mm.).
Journal diameters: Middle rear	1·74875 to 1·74925 in. (44·418 to 44·431 mm.).
Rear	1·72875 to 1·72925 in. (43·910 to 43·923 mm.).
End-float	Taken on thrust plate at front end: ·003 to ·006 in. (·076 to ·152 mm.).

Camshaft bearings

Number and type	4 thin-wall rolled bush.
Outside diameter (before fitting): Front	1·9205 in. (48·780 mm.).
Middle front	1·9005 in. (48·272 mm.).
Middle rear	1·8805 in. (47·762 mm.).
Rear	1·8605 in. (47·252 mm.).
Inside diameter (reamed in position): Front	1·79025 to 1·79075 in. (45.472 to 45·485 mm.).
Middle front	1·77025 to 1·77075 in. (44·964 to 44·977 mm.).
Middle rear	1·75025 to 1·75075 in. (44·456 to 44·469 mm.).
Rear	1·73025 to 1·73075 in. (43·946 to 43·959 mm.).
Clearance	·002 to ·001 in. (·051 to ·025 mm.).

Valve timing

Marking	Adjoining gear teeth are marked.
Chain pitch and number of pitches	·375 in. (9·525 mm.). 62.
Rocker clearance for valve	·0234 in. (·610 mm.).
Inlet valve: Opens	5° B.T.D.C.
Closes	45° A.B.D.C.
Exhaust valve: Opens	40° B.B.D.C.
Closes	10° A.T.D.C.

Lubrication

System	Pressure.
Pump type	Rotor.
External filter	Full flow; Tecalemit.
Oil pressure: Running	55 to 60 lb/sq. in. (3·9 to 4·2 kg/cm²).
Idling	25 to 30 lb/sq. in. (1·758 to 2·109 kg/cm²).
Release valve spring, free length	2·562 in. (65·09 mm.).
Release valve: Number of coils	13.
Diameter	·484 in. $^{+\cdot000\text{ in.}}_{-\cdot015\text{ in.}}$ (12·30 mm. $^{+\cdot000\text{ mm.}}_{-\cdot381\text{ mm.}}$).

Flywheel

Diameter	12·8125 in. (325·45 mm.).
Number of teeth on starter ring	106.

Torque wrench settings

Cylinder head studs	400 lb. in. (4·6 kg. m.).
Cylinder head nuts	900 lb. in. (10·4 kg. m.).
Main bearing nuts	900 lb. in. (10·4 kg. m.).
Connecting rod set screws	600 lb. in. (6·91 kg. m.).
Front cover screws	$\frac{7}{16}$ in. (11·11 mm.) less than 150 lb. in. (1·73 kg. m.) $\frac{1}{2}$ in. (12·70 mm.) less than 150 lb. in. (1·73 kg. m.).
Front mounting plate screws	200 lb. in. (2·30 kg. m.).
Rear mounting plate screws	600 lb. in. (6·91 kg. m.).
Flywheel bolts	600 lb. in. (6·91 kg. m.).
Rocker shaft bracket nuts	300/324 lb. in. (3·45/3·72 kg. m.).

GENERAL DATA

IGNITION

Type	Lucas 12-volt coil
Distributor type	Lucas DM6A
Direction of rotation	Anti-clockwise at rotor arm
Contact breaker gap	·014 in. to ·016 in. (·356 to ·406 mm.)
Static setting	6° (Crankshaft) B.T.D.C.
Maximum advance	35° (Crankshaft) B.T.D.C.
Coil type	Lucas type HA 12
Sparking plug type	Champion UN12Y
Sparking plug gap	·024 in. (·60 mm.) to ·026 in. (·70 mm.)

COOLING SYSTEM

Capacity	20 Imp. pints (24 U.S. pints, 11 litres)
Circulation	Pump and thermostat

FUEL SYSTEM

Fuel delivery	S.U. electric, type H.P.
Carburetter type	Twin S.U. horizontal H.4
Needle (normal)	A.J.
Jet size	·090 in. (2·29 mm.)
Tank capacity	12 Imp., 14·4 U.S. gallons (54·5 litres)
Air cleaner	Twin "Pancake" type

CLUTCH

Make	Borg and Beck
Type	Single dry plate
Diameter	9 in. (229 mm.)
Total friction area	36·5 sq. in. (235 cm.^{2})$\times$2
Thickness of friction linings	·150 (3·8 mm.)
Release bearings	Special carbon graphite or copper carbon graphite
Number of springs	9
Total axial spring pressure	1215 to 1305 lb. (551 to 592 kg.)
Distance thrust race to thrust plate	·10 in. (2·54 mm.)
Thrust plate travel to fully released position	·42 to ·47 in. (10·66 to 11·93 mm.)

GEARBOX

Type	Synchromesh on 2nd, 3rd and top
Type of gear	Helical constant mesh
Type overdrive (optional extra)	Laycock de Normanville electrically controlled
Gear ratios :	
First	3·076 : 1
Second	1·913 : 1
Third	1·333 : 1
Overdrive, third	1·037 : 1
Fourth	Direct
Overdrive, fourth	·778 : 1
Reverse	4·16 : 1
Overall Gear Ratios—	
Standard Box :	
First	12·027 : 1
Second	7·48 : 1
Third	5·212 : 1
Fourth	3·91 : 1
Reverse	16·4 : 1

Including overdrive :	
First	12·6 : 1.
Second	7·84 : 1.
Third	5·47 : 1.
Third and overdrive	4·24 : 1.
Fourth	4·1 : 1.
Fourth and overdrive	3·19 : 1.
Reverse	17·1 : 1.
Layshaft bearing :	
Type	Needle roller.
Number of rollers	46.
Length of roller	1·551 in. (39·6 mm.).
Diameter of roller	3·118 in. (3 mm.).
Mainshaft bearing :	
Make	R. & M.
Type	MJ.35.
Size	1·39 × 3·15 × ·827 in. (35 × 80 × 21 mm.).
First motion shaft bearing :	
Make	R. & M.
Type	IM J40G.
Size	1·58 × 3·55 × ·905 in. (40 × 90 × 23 mm.).
Oil capacity (standard box)	5 pints (6 U.S. pints, 2·84 litres).
Oil capacity (overdrive fitted)	6¼ Imp. pints (7·5 U.S. pints, 3·55 litres)

REAR AXLE

Type	¾ floating.
Ratio :	
Standard	3·91 : 1.
With overdrive	4·1 : 1.
Final drive	Hypoid bevel.
Teeth on crown wheel :	
Standard	43
With overdrive	41
Teeth on pinion :	
Standard	11
With overdrive	10
Crown wheel/pinion backlash	Marked on crown wheel.
Oil capacity	3 Imp. pints (1·704 litres, 3·6 U S. pints).

REAR SPRINGS

Type	Semi-elliptic.
Number of leaves	7.
Thickness of leaves	6 at $\frac{3}{16}$ in. (4·76 mm.) and 1 at $\frac{5}{32}$ in. (3·97 mm.).
Width	1¾ in. (44·45 mm.).
Deflection	4 in. (101·6 mm.) ± ¼ in. (6·35 mm.).
Loaded camber	½ in. (12·70 mm.) ± ⅛ in. (3·18 mm.).
Number of zinc leaves	3 ($\frac{1}{32}$ in. (·80 mm) thickness).

STEERING

Make	Cam gears.
Ratio	14 : 1. 15 : 1. from Car Nos. BN4 68960 and BN6 1995
Track toe-in	$\frac{1}{16}$ in. to ⅛ in. (1·58 to 3·17 mm.).
Torque wrench settings:	
Steering wheel nut	41 lb/ft. (5·76 kg.m.).

SUSPENSION

Front :	
Type	Independent by coil spring and wishbones.
Castor angle	2°.
Camber angle	1°.
Swivel pin inclination	6½°.
Rear :	
Type	Semi-elliptic underslung leaf-springs with Panhard rod.
Shock absorbers :	
Type	Lever hydraulic.

BRAKES

Type	Girling hydraulic two leading shoes front.
Drum diameter	11 in. (280 mm.).
Total frictional area	188 sq. in. (1213 sq. cm.).
Shoe lining width	2¼ in. (57 mm.).
Shoe lining length :	
Front	10·4 in. (265·6 mm.).
Rear	10·4 in. (265·6 mm.).
Shoe lining thickness	·167 to ·174 in. (4·24 to 4·42 mm.).
Pedal free movement	⅛ in. (3·175 mm.).
Handbrake	Mechanical, rear wheels only.

ELECTRICAL

Battery	
Type :	
Home (standard)	GTW 9A.
Dry charged (export)	GTZ 9A.
Voltage	12.
Capacity :	
10 hour rate	51 amp. hr.
20 hour rate	58 amp. hr.
Electrolyte to fill one cell	1 pint (·57 litres).
Initial charging current	3·5 amp.
Normal recharge current	5 amp.
Master switch	Lucas type ST330.
Generator	
Type	Lucas C45 PV-5
Cutting-in speed	1,100 to 1,250 generator r.p.m.
Maximum output	22 amps., 13·5 volts at 1700 to 1900 generator r.p.m.
Field resistance	6 ohms.
Starting motor :	
Type	Lucas M418 G.
Lock torque	17 lb. ft. (1·2858 kg.m.) at 440 to 460 amps. and 7·0 to 7·4 volts.
Light running current	45 amps. at 7,400 to 8,500 r.p.m.
Solenoid switch	Lucas type, ST 950.
Overdrive (optional extra) :	
Control switch	Lucas type 2TS.
Transmission gear solenoid	Lucas type TGS1.
Relay—overdrive	Lucas type SB 40—1.
Interrupter switch	Lucas type 5510—1.
Rotary throttle switch	Lucas type RTS1.

Control box :

Type	Lucas RB 106/2.
Cut-out :	
Cut-in voltage	12·7 to 13·3 volts.
Drop off voltage	8·5 to 11 volts.
Reverse current	3·5 to 5 amps.

Regulator :

Setting on open circuit at 68°F. (20°C.) ... 16·0 to 16·6 volts at 3,000 generator r.p.m.

Note : For circuit temperature other than 20°C. the following allowances should be made to the above setting.

For every 10°C. (18°F.) above 20° subtract 0·1 volt.

For every 10°C. (18°F.) below 20° add 0°1 volt.

Windscreen wiper :

Type	Lucas DR2.
Normal running current	2·3 to 3·1 amp. at 12 volts.
Stall current (motor hot)	8 amp.
Stall current (motor cold)	14 amp.
Armature resistance (adjacent commutator segments)	0·34 to 0·41 ohms.
Field resistance	12·8 to 14·00 ohms.

N.B.—On some high output motors usually identified by a red insulating piece above the terminals, the field resistance is 8·0 to 11·5 ohms.

Fuse Unit :

Type (two live and two spare fuses)	Lucas SF6.

Fuses :

A1—A2	50 amp.
A3—A4	35 amp.

Sidelamps :

Type	Lucas model 594.

Headlamps :

Type	Lucas model F700

Stop-tail lamps :

Type	Lucas model 594.

Number plate illumination :

Type	Lucas model 467/2.

Flashing indicator unit :

Type	Lucas FL.5.

TYRE SIZES AND INFLATION PRESSURES

Tyre sizes	5·90—15 tubeless or 5·90—15 road speed (optional alternative).
Pressures :	
Front	20 lb./sq. in. (1·41 kg./cm.^{2}).
Rear	23 lb./sq. in. (1·62 kg./cm.^{2}).
Rear (full load)	26 lb./sq. in. (1·83 kg./cm.^{2}).
Sustained speeds over 85 m.p.h. (137 km.p.h.)	Increase all pressures by 6 lb./sq. in. (·42 kg.cm.^{2}).

WHEELS

Type	15 × 4J ventilated steel disc or 15 × 4J wire (optional alternative).

DIMENSIONS

Wheelbase	7 ft. 8 in. (2·34 m.).
Overall length	13 ft. 1½ in. (4·00 m.).
Overall height (hood raised)	4 ft. 1 in. (1·24 m.).
Overall height (hood lowered)	3 ft. 10 in. (1·17 m.).
Overall width	5 ft. 0½ in. (1·54 m.).
Height over scuttle	2 ft. 11⅞ in. (0·91 m.).
Ground clearance	5½ in. (0·14 m.).
Track, front	4 ft. 0¾ in. (1·24 m.).
Track, rear	4 ft. 2 in. (1·27 m.).
Turning circle	35 ft. 0 in. (10·67 m.).
Approximate weight (with overdrive and wire wheels)	2,436 lb. (1105 kg.).

The following are the differing details of vehicles fitted with the 6 port cylinder head engine and should therefore be used in conjunction with the preceding specification.

ENGINE

Engine type	BN6.
Torque	150 at 3,000 r.p.m
Compression ratio	8·7 : 1.
Piston rings :	
1st ring	Taper
Connecting rod, type of bearing	Steel backed lead indium.
Standard journal diameter	2·3742 in. (60·305 mm.) to 2·3747 in. (60·32 mm.).
Exhaust valve :	
Throat diameter	1·3125 in. (33·34 mm.)
Head diameter	1·5625 in. (39·69 mm.) to 1·5575 in. (39·56 mm.).
Inlet valve :	
Throat diameter	1·5 in. (38·1 mm.).
Head diameter	1·750 in. (44·45 mm.) to 1·745 in. (44·32 mm.).
Valve seat angles :	
Inlet	45°.
Exhaust	45°.

Engine—continued

Valve spring:
- Outer:
  - Fitted length ... ... 1·594 in. (40·5 mm.) load, 26 lb. (11·8 kg.).
- Inner:
  - Fitted length ... ... 1·504 in. (38·20 mm.) load, 55·7 lb. (25·27 kg.).

IGNITION

Distributor	Lucas DM6.
Static setting	6° (crankshaft) B.T.D.C.
Maximum advance	36° (crankshaft) B.T.D.C.

CARBURATION

Carburetter type	Twin S.U. H.D.6.
Needle (normal)	C.V.
Jet size	·100 in. (2·54 mm.).
Angle of fitting	30° semi-downdraught.
Spring colour code	Yellow.

The following information is applicable to the Austin-Healey Series BN6 and later BN4 cars, and should be used in conjunction with the preceding specifications.

FUEL SYSTEM

Fuel delivery	S.U. Electric, type LCS.
Fuel pipe	Outside diameter $\frac{3}{8}$ in. (9·52 mm.).
Maximum flow	12½ gallons (15 U.S. gallons, 56 litres) per hour
Suction head	33 in. (81 cm.)
Delivery head	4 ft. (121·9 cm.)
Cut off pressure	3·8 lb./sq. in. (27 kg.cm^2)

The following information is applicable to the Austin-Healey BN6 and should be used in conjunction with the preceding specifications.

ELECTRICAL

Battery
- Type:
  - Home (standard) ... ... SLG 11E.
  - Export (dry charged) ... ... SLGZ 11E.
- Voltage ... ... (2) 6 volt.
- Capacity:
  - 10 hour rate ... ... 50 amp. hr.
  - 20 hour rate ... ... 58 amp. hr.
- Electrolyte to fill one cell ... ... 1 pint (·57 litres).

The following information is applicable to the Austin-Healey 3000 Mk. I and II (Series BN7 and BT7), Mk. II (Series BJ7) and Mk. III (Series BJ8).

ENGINE

Type	*Mk. I* 29D. *Mk. II* 29E. BJ7 29F. *Mk. III* 29K.
Number of cylinders	6.
Bore	3·281 in. (83·34 mm.).
Stroke	3·5 in. (88·9 mm.).
Capacity	2912 c.c. (177·7 cu. in.).
B.M.E.P.	142 lb./sq. in. (9·98 kg./cm^2.) at 2,700 r.p.m.
Torque	167 lb. ft. (23·09 kg. m.) at 2,700 r.p.m
Compression ratio	9 : 1.
Cranking pressure	175 lb./sq. in. (12·25 kg. cm.^{2})
Engine idle speed (approx.)	500 r.p.m.
Firing order	1, 5, 3, 6, 2, 4.
Bore, 1st Oversize	+·010 in. (·25 mm.).
Bore, 2nd Oversize	+·020 in. (·50 mm.).
Bore, 3rd Oversize	+·030 in. (·76 mm.).
Bore, 4th Oversize	+·040 in. (1·02 mm.).

Valves

Position	Overhead, push-rod operated.
Lift	·3145 in. (8·054 mm.). *Mk. II and III* ·368 in. (9·36 mm.).
Diameter: Head: Inlet	1·750 to 1·745 in. (44·45 to 44·32 mm.).
Exhaust	1·5625 to 1·5575 in. (39·69 to 39·56 mm.).
Stem: Inlet	·34175 to ·34225 in. (8·68 to 8·69 mm.).
Exhaust	·34175 to ·34225 in. (8·68 to 8·69 mm.).
Stem/guide clearance: Inlet	·0025 to ·0015 in. (·063 to ·038 mm.).
Exhaust	·002 to ·001 in. (·051 to ·025 mm.).
Valve rocker clearance	·012 in. (·30 mm.) cold.
Competition (*Mk. II and III*)	·015 in. (·34 mm.) cold.
Seat angle: Inlet	45°.
Exhaust	45°.
Seat face width: Inlet	·091 to ·097 in. (2·31 to 2·46 mm.).
Exhaust	·198 to ·217 in. (5·03 to 5·51 mm.).

Valve guides

Length: Inlet	2·266 in. (57·55 mm.).
Exhaust	2·578 in. (65·49 mm.).
Fitted height above head: Inlet	1·348 in. (34·23 mm.).
Exhaust	1·036 in. (26·32 mm.).

Valve springs

Free length: Inner	1·969 in. (50 mm.).
Outer	2·047 in. (51·99 mm.). *Mk. II and III:* 2·031 in. (51·5 mm.).
Fitted length: Inner	1·504 in. (38·2 mm.) load 26 lb. (11·8 kg.).
Outer	1·594 in. (40·49 mm.) load 55·7 lb. (25·2 kg.). *Mk. II and III:* 67·5 lb. (30·63 kg.).

Tappets

Type	Cylindrical, spherical foot.
Diameter	·937 in. (23·81 mm.).
Length	2·548 in. (64·72 mm.).

Engine—continued

Rockers

Bushes	Steel and white metal.
Outside diameter (before fitting)	·913 in. (23·17 mm.).
Inside diameter (reamed in position)	·8115 to ·8125 in. (20·62 to 20·65 mm.).
Clearance	·0025 to ·0005 in. (·063 to ·012 mm.).
Bore of arm	·909 to ·910 in. (23·076 to 23·101 mm.).

Pistons

Material	Low expansion aluminium alloy.
Clearance at skirt: Top	·0032 to ·0043 in. (·081 to ·109 mm.).
Bottom	·0010 to ·0016 in. (·025 to ·040 mm.).
Width of ring groove: Compression	·1417 to ·1482 in. (3·599 to 3·764 mm.).
Oil	·1567 to ·1632 in. (3·98 to 4·137 mm.).
Oversizes	+·010 in.; +·020 in.; +·030 in.; +·040 in. (+ ·25 mm.; +·50 mm.; +·76 mm.; +1·02 mm.).

Piston rings

Number	3 compression (2 taper), 1 oil control.
Width: Compression and oil	3·2055 to 3·3832 mm.
Clearance in groove: Compression and oil	·0015 to ·0035 in. (·038 to ·088 mm.).
Ring gap: Compression and oil	·013 to ·018 in. (·33 to ·46 mm.).

Gudgeon pins

Type	Fully floating.
Fit	Selective; push in piston.
Diameter	·8748 to ·8750 in. (22·219 to 22·225 mm.).
Location	Circlips in piston.

Crankshaft

Journal diameter	2·3742 to 2·3747 in. (60·305 to 60·317 mm.).
Crankpin diameter	2·0000 to 2·0005 in. (50·80 to 50·813 mm.).
Undersizes (journals and crankpins)	—·010 in.; —·020 in. (—·254 mm.; —·508 mm.).
End-float	Taken on thrust washer at front middle (No. 2) main bearing; ·0025 to ·0055 in. (·063 to ·140 mm.).
Thrust washer: Standard	·091 to ·093 in. (2·315 to 2·366 mm.).
+·0025 in. (+·063 mm.)	·0935 to ·0955 in. (2·378 to 2·429 mm.).
+·005 in. (+·127 mm.)	·0960 to ·0980 in. (2·442 to 2·493 mm.).
+·0075 in. (+·190 mm.)	·0985 to ·1005 in. (2·505 to 2·556 mm.).
+·010 in. (+·254 mm.)	·1010 to ·1030 in. (2·569 to 2·620 mm.).

Main bearings

Number	4.
Type	White metalled steel shell, lead indium plated.
Length	1·495 to 1·505 in. (37·973 to 38·227 mm.).
Running clearance	·0013 to ·0028 in. (·033 to ·071 mm.).
Sizes for reground journals	·010 in. U/S; ·020 in. U/S (·254 mm.; ·508 mm. U/S).

Engine—continued

Connecting rods

Length (centres)	6·601 to 6·60 in. (167·66 to 167·76 mm.).
Side clearance	·005 to ·009 in. (·13 to ·23 mm.).
Big end bearings: Type	White metalled steel shells, lead indium plated.
Diametrical clearance	·002 to ·0035 in. (·051 to ·089 mm.).
Small-end bush	·8749 to 8751 in. (22·22 to 22·23 mm.).
Sizes for reground crankpins	·010 in. U/S; ·020 in. U/S (·254 mm.; ·508 mm. U/S).

Camshaft

Journal diameters: Front	1·78875 to 1·78925 in. (45·434 to 45·447 mm.).
Middle front	1·76875 to 1·76925 in. (44·926 to 44·939 mm.).
Journal diameters: Middle rear	1·74875 to 1·74925 in. (44·418 to 44·431 mm.).
Rear	1·72875 to 1·72925 in. (43·910 to 43·923 mm.).
End-float	Taken on thrust plate at front end: ·003 to ·006 in. (·076 to ·152 mm.).

Camshaft bearings

Number and type	4 thin-wall rolled bush.
Outside diameter (before fitting): Front	1·9205 in. (48·780 mm.).
Middle front	1·9005 in. (48·272 mm.).
Middle rear	1·8805 in. (47·762 mm.).
Rear	1·8605 in. (47·252 mm.).
Inside diameter (reamed in position): Front	1·79025 to 1·79075 in. (45·472 to 45·485 mm.).
Middle front	1·77025 to 1·77075 in. (44·964 to 44·977 mm.).
Middle rear	1·75025 to 1·75075 in. (44·456 to 44·469 mm.)
Rear	1·73025 to 1·73075 in. (43·946 to 43·959 mm.).
Clearance	·002 to ·001 in. (·051 to ·025 mm.).

Valve timing

Marking	Adjoining gear teeth are marked.
Chain pitch and number of pitches	·375 in. (9·525 mm.). 62.
Rocker clearance for valve timing	·030 in. (·76 mm.).

	Mk. I	*Mk. II To 29F/2285*	*Mk. II BJ7 From 29F/2286*	*Mk. III*
Inlet valve: Opens	5° B.T.D.C.	5° B.T.D.C.	10° B.T.D.C.	16° B.T.D.C.
Closes	45° A.B.D.C.	45° A.B.D.C.	50° A.B.D.C.	56° A.B.D.C.
Exhaust valve: Opens	40° B.B.D.C.	51° B.B.D.C.	45° B.B.D.C.	51° B.B.D.C.
Closes	10° A.T.D.C.	21° A.T.D.C.	15° A.T.D.C.	21° A.T.D.C.

Lubrication

Capacity (including filter)	12¾ pints, 15·3 U.S. pints, (7·25 litres)
System	Pressure.
Pump type	Gear.
External filter	Full flow; Tecalemit or Purolator.
Oil pressure: Running	50 lb./sq. in. (3·52 kg./cm²) at 40 m.p.h.
Idling	20 lb./sq. in. (1·4 kg./cm²) at 600 r.p.m.
Release valve spring, free length	2·687 in. (68·26 mm.).
Release valve: Number of coils	13.
Diameter	·484 in. $^{+\cdot 000}_{-\cdot 015}$ in. (12·30 mm. $^{+\cdot 000}_{-\cdot 318}$ mm.).

Flywheel

Diameter	12·8125 in. (325·45 mm.).
Number of teeth on starter ring	106.

IGNITION

Type	Lucas 12-volt coil.
Distributor type	Lucas DM6A. *Mk. I and II* Lucas 25D6 *Mk. II and III (from engine No. 29F/3563)* *Mk. III:* Serial No. 40966B.
Direction of rotation	Anti-clockwise at rotor arm.
Contact breaker gap	·014 in. to ·016 in. (·356 to ·406 mm.).
Dwell angle	35±3°
Condenser capacity	*Mk. I and II:* ·2 mF. *Mk. II (Series BJ7):* ·1 mF. *Mk. III:* ·18 to ·23 mF.
Static setting	5°(Crankshaft) B.T.D.C. *Mk. II and III*—10° B.T.D.C.
Strobscapic ignition timing	15° B.T.D.C. at 600 r.p.m.
Maximum advance	35° (Crankshaft) B.T.D.C.
Coil type	Lucas type HA 12.
Resistance	3·1 to 3·5 ohms.
Sparking plug type	Champion UN12Y
Sparking plug gap	·024 in. to ·026 in. (·6 mm. to ·66 mm.).
Mk. III	
Maximum advance	34—38° at 6,400 r.p.m. (Crankshaft)
Vacuum advance: starts	5 in. (12·7 cm.) Hg.
Ends	16° at 12 in. (30·4 cm.) Hg.
Decelerating check	28° to 32° at 4400 r.p.m. (Crankshaft). 22° to 26° at 3400 r.p.m. (Crankshaft). 15° to 19° at 2000 r.p.m. (Crankshaft). 10° to 16° at 1500 r.p.m. (Crankshaft). 2° to 8° at 1100 r.p.m. (Crankshaft.)

COOLING SYSTEM

Type	Pressurised thermo-syphon, thermostatically controlled, pump and fan assisted.
Capacity (without heater)	18 Imp. pints (21·6 U.S. pints, 10·2 litres).
Thermostat setting	70°C. (158°F.). 83°C. (182°F.). *Mk. II and III (from engine No. 29F/2592.)*

FUEL SYSTEM

Fuel pump	S.U. electric, type LCS. S.U., type LCS (*Mk. II and early Mk. III*) S.U., type AUF 300 (*later Mk. III*)
Maximum flow	15 gallons (18 U.S. gallons 68·25 litres) per hour.
Suction head	18 in. (45·7 cm.)
Delivery head	4 ft. (121·9 cm.)
Cut off pressure	3·8 lb./sq. in. (27 kg./cm.^{2}).
Carburetter type	Twin S.U. type HD6, semi-downdraught. *Mk. II:* Triple S.U. type HS4 *Mk. II (Series BJ7):* Twin S.U. type HS6. *Mk. III:* Twin S.U. type HD8
Needle	CV. *Mk. II:* DJ (normal), DK (rich), DH (weak) *Mk II (Series BJ7):* BC (normal), RD (rich). TZ (weak). *Mk. III* UH (normal), UN (rich), UL (weak).
Jet size	·100 in. (2·54 mm.). *Mk. II:* ·090 in. (2·3 mm.) *Mk. II (Series BJ7):* ·100 in. (2·54 mm.). *Mk. III:* ·125 in. (3·18 mm.).

Fuel system—continued

Carburetter spring	*Mk. II:* Red. *Mk. III:* Red and green.
Float setting	*Mk. III:* ⅛ in. (3·17 mm.)
Venturi	*Mk. II (Series BJ7):* 1¾ in. (44·45 mm.) *Mk. III:* 2 in. (50·80 mm.)
Tank capacity	12 Imp., 14·4 U.S. gallons (54·5 litres).
Air cleaners	Oil wetted.

CLUTCH

(*Mk. I and II*)

Make	Borg and Beck.
Type	Single dry plate.
Diameter	10 in. (25·4 cm.).
Total friction area	39 sq. in. (241·5 cm^2)×2.
Thickness of friction linings	·150 (3·8 mm.).
Release bearing	Special carbon graphite or copper carbon graphite.
Number of springs	12.
Colour of springs	Yellow/Light Green
Total axial spring pressure	1,620 to 1,740 lb. (735 to 789 kg.).
Distance thrust race to thrust plate	·10 in. (2·54 mm.).
Thrust plate travel to fully released position	·42 to ·47 in. (10·66 to 11·93 mm.).

(*Mk. II and III from engine No. 29F/4898*)

Make and type	Borg & Beck 9 in. DS.G diaphragm spring.
Clutch plate diameter	9·63 in. (20·32 cm.).
Facing material	Wound yarn.
Number of damper springs	6.
Damper spring load	110 to 120 lb. (49·8 to 54·3 kg.).
Damper spring colour	Dark grey/light green.
Clutch release bearing	Graphite (MY3D).
Clutch fluid	Castrol Girling Amber Brake fluid.

GEARBOX

Type	Synchromesh on 2nd, 3rd and top.
Type of gear	Helical constant mesh.
Type overdrive (optional extra)	Laycock de Normanville electrically controlled

Overall Gear Ratios—

Gear ratios:		*From Engine No. 10897 with overdrive and Engine No. 11342 without overdrive*	*Mk. III*
First	2·93:1	2·83:1	2·637:1
Second	2·053:1.	2·06:1	2·071:1
Third	1·309:1.	1·31:1	1·306:1
Fourth	Direct.		Direct
Overdrive	·822:1.		8·20:1
Reverse	3·78:1.	3·72:1	3·391:1

Standard Box: (3·545:1 axle)		*From Engine No. 11342*	*Mk. III*
First	10·386:1.	10·209:1	9·348:1
Second	7·877:1	7·302:1	7·341:1
Third	4·640:1.	4·743:1	4·629:1
Fourth	3·545:1.		3·545:1
Reverse	13·400:1.	13·127:1	12·021:1

Gearbox—continued

		From Engine No. 10897	Mk. III
Including overdrive (3·909:1 axle):			
First	11·453:1	11·257:1	10·308:1
Second	8·025:1.	8·052:1	8·095:1
Third	5·116:1.	5·120:1	5·105:1
Third and overdrive	4·195:1.	4·198:1	4·188:1
Fourth	3·909:1.	3·909:1	3·909:1
Fourth and overdrive	3·205:1.	3·205:1	3·207:1
Reverse	14·776:1.	14·541:1	13·255:1

Layshaft bearing :	
Type	Needle roller.
Number of rollers	46.
Length of roller	1·551 in. (39·6 mm.).
Diameter of roller	3·118 in. (3 mm.).
End float	·012 in. (·30 mm.).
Mainshaft bearing:	
Make	R. & M.
Type	MJ.35.
Size	1·39 × 3·15 × ·827 in. (35 × 80 × 21 mm.)
End float	Nil
First motion shaft bearing:	
Make	R. & M.
Type	IM J40G.
Size	1·58 × 3·55 × ·905 in. (40 × 90 × 23 mm.).
Oil capacity (standard box)	5 pints (6 U.S. pints, 2·8 litres).
Oil capacity (overdrive fitted)	6¼ Imp. pints (7·5 U.S. pints, 3·6 litres).
Synchromesh hubs:	
End float	2nd speed, ·007 in. (·18 mm.). 3rd and 4th speeds, ·031 in. (·8 mm.).

REAR AXLE

Type	¾ floating.
Ratio:	
Standard	3·545 : 1.
With overdrive	3·909 : 1.
Final drive	Hypoid bevel.
Teeth on crown wheel and pinion:	
Standard	39/11.
With overdrive	43/11.
Crown wheel/pinion backlash	Marked on crown wheel.
Adjustment	Shims.
Oil capacity	3 Imp. pints, 3·6 U.S. pints, (1·70 litres).

REAR SPRINGS

Mk. I, II and III

Type	Semi-elliptic leaf.
Number of leaves	7.
Thickness of leaves	6 at $\frac{3}{16}$ in. (4·8 mm.) and 1 at $\frac{5}{32}$ in. (4 mm.).
Width	1¾ in. (44·5 mm.).
Deflection	4 in. (101·6 mm.) ± ¼ in. (6·4 mm.).
Loaded camber	½ in. (12·70 mm.) ± ⅛ in. (3·18 mm.) negative.
Number of zinc leaves	3 ($\frac{1}{32}$ in. [·80 mm.] thickness).

Rear Springs—continued

Mk. III from chassis No. 26705

Type	Semi-elliptic leaf
Number of leaves	6
Thickness of leaves	4 at $\frac{3}{16}$ in. (4·8 mm.) 2 at $\frac{5}{32}$ in. (4 mm.)
Width	$1\frac{3}{4}$ in. (44·5 mm.)
Deflection	$5\frac{1}{4}$ in. (133·35 mm.)
Loaded camber	1 in. (25·4 mm.)

STEERING

Make and type	Cam Gears: cam and peg.
Ratio	15 : 1.
Track toe-in	$\frac{1}{16}$ in. to $\frac{1}{8}$ in. (1·58 to 3·17 mm.).
Steering lock angle : Outer wheel turned 20° inner wheel should be	21°

SUSPENSION

Front :

Type	Independent by coil spring and wishbones.
Castor angle	2°.
Camber angle	1°.
Swivel pin inclination	$6\frac{1}{2}$°.

Rear :

Type	Semi-elliptic underslung leaf-springs with Panhard rod.

Shock absorbers:

Make	Armstrong.
Type	Lever hydraulic.

BRAKES

Type	Girling hydraulic.
Front	Disc.
Rear	Drum. One leading, one trailing shoe.
Disc diameter	11 in. (28·57 cm.).
Drum diameter	11 in. (28.0 cm.).
Drum width	$2\frac{7}{16}$ in. (61·91 mm.).
Total frictional area (rear)	95 sq. in. (612·75. cm^2.).
Shoe lining width	$2\frac{1}{4}$ in (57 mm.).
Shoe lining length	10·53 in. (267·4 mm.).
Shoe lining thickness	·187 in. (4·76 mm.).
Disc pad area	4·25 sq. in. (10·8 cm^2.) × 4
Disc pad thickness	$\frac{13}{32}$ in. (10·32 mm.).
Pedal free movement	$\frac{1}{8}$ in. (3·175 mm.).
Handbrake	Mechanical, rear wheels only.

ELECTRICAL

Battery

Type (Series BN7—two seater) :	
Standard	SLG11E (two).
Dry-charged (export only).	SLGZ11E (two).
Type (Series BT7, BJ7 and BJ8—four seater) :	
Standard	BT9A
Dry-charged (export only)	BTZ9A.
Voltage	12 volt positive earth.
Capacity, 20 hour rate :	58 amp. hr.

Electrical System—continued

Electrolyte to fill one cell	1 pint (·57 litres).
Initial charging current	3·5 amp.
Normal recharge current	5 amp.
Master switch	Lucas type ST330.
Dynamo	
Mk. I and II	
Type	Lucas C45 PV-6.
Cutting-in speed	1,100 to 1,250 generator r.p.m.
Maximum output	25 amps., 13·5 volts at 1,700 to 1,900 generator r.p.m.
Field resistance	6 ohms.
Brush length (minimum)	$\frac{7}{16}$ in. (11·11 mm.).
Brush spring tension (maximum)	34-44 oz. (965-1248 gm.).
Mk. III	
Type	Lucas C42
Cutting-in speed	1,250 generator r.p.m.
Maximum output	30 amps., 13·5 volts at 1,250 generator r.p.m.
Field resistance	4·5 ohms.
Starting Motor	
Type	Lucas M418 G.
Lock torque	17 lb. ft. (1·2858 kg.m.) at 440 to 460 amps. at 7·0 to 7·4 volts.
Lock current draw	430 to 450 amps. at 7·0 to 7·4 volts.
Light running current	45 amps. at 7,400 to 8,500 r.p.m.
Solenoid switch	Lucas type, ST 950.
Brush length (minimum)	$\frac{5}{16}$ in. (7·94 mm.).
Brush spring tension (maximum)	30-40 oz. (850-1133 gm.).
Overdrive (optional extra)	
Control switch	Lucas type 2TS.
Transmission gear solenoid	Lucas type 11 S.
Relay—overdrive	Lucas type SB 40—1.
Interrupter switch	Lucas type SS 10.
Rotary throttle switch	Lucas type RTS1.
Control Box	
Mk. I and II	
Type	Lucas RB 106/2.
Cut-out relay:	
Cut-in voltage	12·7 to 13·3 volts.
Drop off voltage	8·5 to 11 volts.
Reverse current	3·5 to 5 amps.
Setting on open circuit at 20° C (68°F.).	16·0 to 16·6 volts at 3,000 generator r.p.m.

Note: For circuit temperature other than 20·C. the following allowances should be made to the above setting.
For every 10°C. (18°F.) above 20° subtract 0·1 volt.
For every 10°C. (18°F.) below 20° add 0·1 volt.

Mk. III	
Type	Lucas RB 340
Cut-out relay:	
Cut-in voltage	12·7 to 13·3 volts.
Drop-off voltage	9·5 to 11·0 volts.
Voltage setting at 4,500 r.p.m.: 10° C. (50° F.)	14·9 to 15·5 volts.
20° C. (68° F.)	14·7 to 15·5 volts.
30° C. (86° F.)	14·5 to 15·1 volts.
40° C. (104° F.)	14·3 to 14·9 volts.

Electrical System—continued

Reverse current	3·0 tp 5·0 amps.
Current regulator	30±1 amp.

Windscreen Wiper

Mk. I and II

Type	Lucas DR2.
Normal running current	2·3 to 3·1 amp. at 12 volts.
Stall current (motor hot)	8 amp.
Stall current (motor cold)	14 amp.
Armature resistance (adjacent commutator segments)	0·34 to 0·41 ohms.
Field resistance	8·0 to 9·5 ohms.

N.B.—On some high output motors usually identified by a red insulating piece above the terminals, the field resistance is 8·0 to 11·5 ohms.

Mk. II and III (from body No. 60792 Mk. II)

Type	Lucas DR3A
Drive to wheelboxes	Rack and cable.
Armature end-float	·008 to ·012 in. (·20 to ·30 mm.).
Running current	2·7 to 3·4 amps.
Wiping speed	45 to 50 cycles per minute.

Fuses

A1—A2	50 amp.
A3—A4	35 amp.

TYRE SIZES AND INFLATION PRESSURES

Tyre sizes	5·90—15 road speed (with tubes).

Pressures:

Front	20 lb./sq. in. (1·41 kg/cm²). } for speeds up to 110m.p.h.(177km.p.h.)
Rear	25 lb./sq. in. (1·76 kg/cm²). } for speeds up to 110m.p.h.(177km.p.h.)
Maximum performance speeds	increase all pressures by 5 lb/sq. in. (·35 kg./cm²).

WHEELS

Type	4J×15 ventilated steel disc or 4J×15 wire (optional alternative).

DIMENSIONS

Wheelbase	7 ft. 7$\frac{23}{32}$ in. (2·33 m.).
Overall length	13 ft. 1½ in. (4·00 m.).
Overall height (hood raised)	4 ft. 2¾ in. (1·29 m.).
Overall height (hood lowered)	3 ft. 10 in. (1·17 m.).
Overall width	5 ft. 0 in. (1·52 m.).
Height over scuttle	2 ft. 11⅞ in. (0·91 m.).
Ground clearance	4⅝ in. (0·17 m.).
Track, front	4 ft. 0¾ in. (1·24 m.).
Track, rear	4 ft. 2 in. (1·27 m.).
Turning circle	35 ft. 7 in. (10·84 m.).
Unladen weight	2,380 lb. (1080 kg.).
Kerbside weight	2,460 lb. (1115 kg.).

WEIGHTS OF COMPONENTS

Engine	611 lbs. (277 kg.).
Gearbox, standard	79·5 lbs. (36·027 kg.).
Gearbox, with overdrive	117 lbs. (52·907 kg.).
Rear axle	166 lbs. (75·454 kg.).

TORQUE WRENCH SETTINGS

Cylinder head studs	400 lb. in. (4·6 kg. m.).
Cylinder head nuts	900 lb. in. (10·4 kg. m.).
Main bearing nuts	900 lb. in. (10·4 kg. m.).
Connecting rod set screws	600 lb. in. (6·91 kg. m.).
Front cover screws	less than 150 lb. in. (1·73 kg. m.).
Front mounting plate screws	200 lb. in. (2·3 kg. m.).
Rear mounting plate screws	600 lb. in. (6·91 kg. m.).
Flywheel bolts	600 lb. in. (6·91 kg. m.).
Rocker shaft bracket nuts	300 to 324 lb. in. (3·45 to 3·72 kg. m.).
Bell housing bolts	420 lb. in. (4·83 kg. m.).
Clutch to flywheel	300 lb. in. (3·45 kg. m.)
Differential bearing cap nuts	780 lb. in. (8·99 kg. m.).
Crown wheel bolts	680 lb. in. (7·83 kg. m.).
Pinion bearing nut	1,680 lb. in. (19·4 kg. m.).
Wheel nuts (pressed steel wheels)	720 to 750 lb. in. (8·3 to 8·64 kg. m.).
Steering wheel nut	492 lb. in. (5.76 kg. m.).

SECTION A

ENGINE

SERIES BN4

Section A.1

GENERAL DESCRIPTION

The engine is a six cylinder unit with the bores cast integrally with the crankcase. Adequate cooling under the most arduous operating conditions is ensured by the provision of large area water circulating passages, and full length water jackets.

The detachable cylinder head is of cast iron and carries the overhead push rod operated valve gear.

Forged steel is used for the counter balanced crankshaft which is supported by four large diameter main bearings of the preformed "Thinwall" type. The same type of bearings are used for the connecting rod big end assemblies.

Particular attention has been paid to the design of the lubrication system to ensure that the moving parts of the engine are adequately supplied with oil at all times. The choice of oils is of great importance and those recommended on page Q.1 have been tested under various running conditions and should be used in accordance with the schedule of regular attentions.

Section A.2

VISUAL INSPECTION

Examine the engine for any signs of oil leakage, with particular attention to the sump drain plug, the joint between the oil filter bowl and its head casting, and the rocker cover to cylinder head joint.

The connections to the distributor should be checked occasionally for tightness, and any perished or cracked high tension leads renewed.

Section A.3

ADJUSTMENTS IN THE VEHICLE

The purpose of the following adjustments is to maintain the performance of the engine at its maximum, and consists of a series of cleaning, inspecting and adjusting operations. A compression test of each cylinder should first be made to determine the general condition of the engine before proceeding with any adjustments. If a compression gauge is not available, a simple method to test the compression is to remove all the sparking plugs with the exception of the one in the cylinder being tested, and then rotate the engine with the starting handle through at least two complete revolutions. If the cylinder compression is satisfactory, proceed as detailed below.

(1) Clean the engine generally and lubricate as recommended.

(2) Adjust the fan belt tension in accordance with the instructions given in Section C.

(3) Remove the valve gear cover and test the cylinder head studs for tightness, using a torque spanner set to the figures quoted under "General Data".

(4) Check and adjust the valve and rocker clearances as outlined in Section A.11.

(5) Check for evidence of cracked valve springs or scored or worn stems.

(6) Replace the valve gear cover, using a new gasket if necessary.

(7) Disconnect the high-tension cables and remove the sparking plugs.

(8) Check to make sure that the correct type of sparking plug is being used.

(9) Clean the sparking plugs and examine the insulation for breaks or cracks.

(10) Adjust the sparking plug gaps as specified in Section B.

(11) Test the sparking plugs and renew any found to be unfit for further service.

(12) Refit the sparking plugs, using new copper washers.

(13) Check the high tension cables for wear and deterioration before refitting.

(14) Remove the distributor head cover and clean it inside and out. Examine it for cracks and burned contacts and renew it if necessary.

(15) Inspect the contact breaker points to determine whether new points are needed. Follow the procedure given in Section B to clean and adjust the points.

(16) Check the distributor rotor arm, making sure the carbon brush makes contact. Check the capacitor terminal to make sure it is clean and tight.

(17) Check the ignition timing as outlined in Section B.

(18) Clean the air cleaners in accordance with the instructions in Section D.

(19) Make sure the fuel system is operating properly and clean all filters in the system as detailed in Section D.

(20) Check the carburetter flange gaskets for evidence of leakage.

(21) Adjust the carburetters if necessary, in accordance with the procedure given in Section D.

Section A.4

ENGINE ASSEMBLY

To remove (without gearbox)

(1) The battery master switch, which is situated inside the luggage compartment, should be turned to the "off" position.

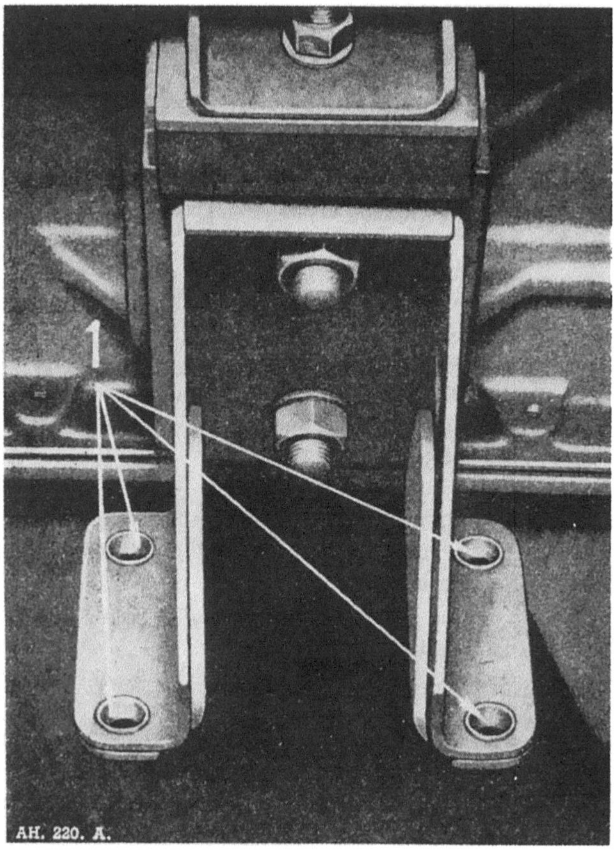

Fig. A.1. The engine left-hand front mounting bracket showing the four setpin holes at 1.

(2) Remove the radiator as described in Section C and detach the fan by unscrewing the four securing setpins.
(3) Disconnect the throttle linkage and choke control cable. The throttle linkage is freed by unclamping the throttle control rod at its projection from the bulkhead.
(4) Unscrew the setpins securing the air cleaners to the carburetter inlets and remove the air cleaners.
(5) Disconnect the petrol feed pipe at its carburetter union.
(6) Remove the high tension cables from their connections at the coil and the sparking plugs.
(7) Release the dynamo, distributor and coil low tension cables, and place the complete harness to one side.
(8) Release the heater inlet and outlet rubber hoses from their connections at the rear of the cylinder head and the heater outlet pipe (when fitted).
(9) Remove the distributor as described in Section B.
(10) Remove the dynamo, complete with coil, as described in Section N.
(11) Remove the external oil filter as described in Section A.6.
(12) Release the oil pressure flexible pipe at its upper connection.
(13) Remove the starter motor as described in Section N.
(14) Withdraw the four setpins which secure each engine mounting bracket to the chassis frame. Detachment of the left-hand bracket is facilitated by a slit in the carburetter heat shield.
(15) Unscrew the six brass nuts securing the exhaust down pipe to the exhaust manifolds, and pull the down pipe away from the manifold studs.
(16) Remove the valve rocker as described in Section A.11 and secure two suitable lifting brackets.
(17) By means of lifting tackle, similar to that illustrated in fig. A.2, support the engine so that the engine mounting brackets are just clear of their chassis mountings.
(18) Unscrew the four setpins securing the right-hand engine mounting bracket to the cylinder block and withdraw the mounting.
(19) Place a suitable support underneath the gearbox bell housing and unscrew the nuts, bolts and setpins securing the bell housing to the engine backplate.

Fig. A.2. Showing the engine being removed at the correct lifting angle.

(20) Hoist the engine to give clearance between the crankshaft damper and the chassis cross member and pull the engine forward so that the clutch driven plate slides off the first motion shaft splines when the engine can be lifted through the bonnet opening and clear of the car.

To Replace

Replacing the engine is the reverse of the procedure **"To Remove"**

To Remove (with gearbox)

To avoid possible damage, either to individual components or to the car, removal of the generator, distributor and right-hand mounting bracket is advised.

(1) Follow the instructions (1) to (10) and (12) to (18) as detailed in the engine removal less gearbox.

(2) Inside the car remove the seat cushions and release the clips securing the padded arm rest to the central tunnel.

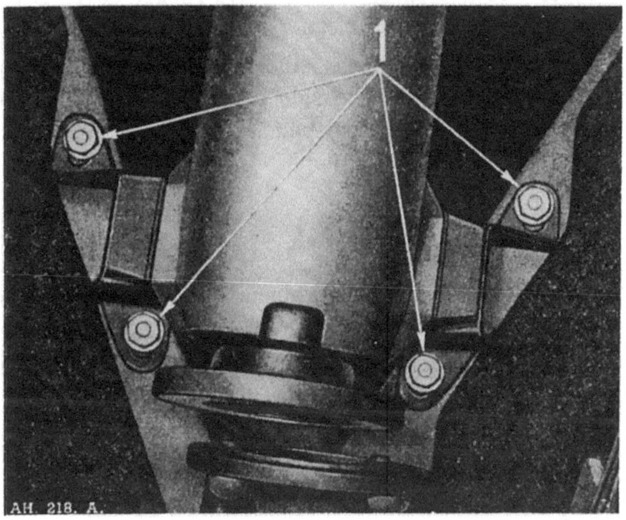

Fig. A.3. The gearbox rear upper securing bracket showing the four setpins at 1.

(3) Unclip and roll back the carpet over the short gearbox tunnel to expose the twelve screws securing the tunnel to the body of the car. Unscrew the setscrews and remove the tunnel and its carpeting.

(4) Unscrew the six setscrews, three either side, which secure the carpet covered bulkhead and remove the bulkhead.

(5) Using a suitable tool tap back the locking washers on the propeller shaft flange bolts and remove the bolts.

(6) Unscrew the four setpins from the gearbox mounting brackets (see fig. A.3), also unscrew the speedometer cable at its connection to the gearbox.

Fig. A.4. The gearbox lower securing points showing 1, Setpins; 2, Stabiliser Adjusting Nut; 3, Securing Pin.

Note.—When an overdrive gearbox is fitted it will also be necessary to unclip the cable to the gearbox switch and release it at its terminal on the switch.

(7) Working beneath the vehicle remove setpins (1) fig. A.4 and unscrew the nuts (2) and (3) to release the stabiliser bar.

(8) Detach the clutch slave cylinder from the gearbox bell housing by removing the two securing setpins. The slave cylinder push rod is released from the clutch operating lever by the removal of the securing clevis pin.

(9) Hoist the engine complete with gearbox through the bonnet opening as shown in fig. A.5, ascertain that no damage is done by the gearbox when manœuvring it through the bulkhead aperture.

Fig. A.5. Showing the engine and gearbox being removed.

Section A.5

LUBRICATION

Description

The oil supply is carried in the sump below the cylinder block and the filler cap is fitted on the forward end of the rocker cover. The dipstick is on the right-hand side of the engine and is marked to indicate the maximum and minimum levels. The eccentric rotor type oil pump driven by the camshaft is mounted below the crankcase and is partially submerged in the oil reservoir.

Oil is drawn through a gauze strainer secured to the oil pump and passes through a drilling up the right-hand side of the crankcase to the oil filter, passing the non-adjustable pressure relief valve. After leaving the full flow filter the oil-way divides, one drilling passing up the right-hand side of the cylinder block through the cylinder head to a pipe feeding oil to No. 4 rocker shaft bracket. From here, oil passes through the hollow centre of the rocker shaft to lubricate all rocker bearings, and through drillings in the rockers, to lubricate the valve gear. Oil returning to the sump from the rockers lubricates the tappets. The second oil-way from the oil filter passes round No. 3 camshaft bearing, (lubricating this bearing as it does so), to the oil gallery on the left-hand side of the engine. From the gallery, drillings in the cylinder block take oil to each main bearing and through the crankshaft to the big ends. Oil-ways from the main bearings also supply the camshaft bearings. The connecting rods have jet holes to deliver oil to the cylinder walls.

A vent pipe is attached to the rear tappet chamber cover and a breather in the valve rocker cover is connected to the rear air cleaner.

An oil pipe connects the rear end of the main oil gallery on the left-hand side of the engine with the oil gauge on the instrument panel.

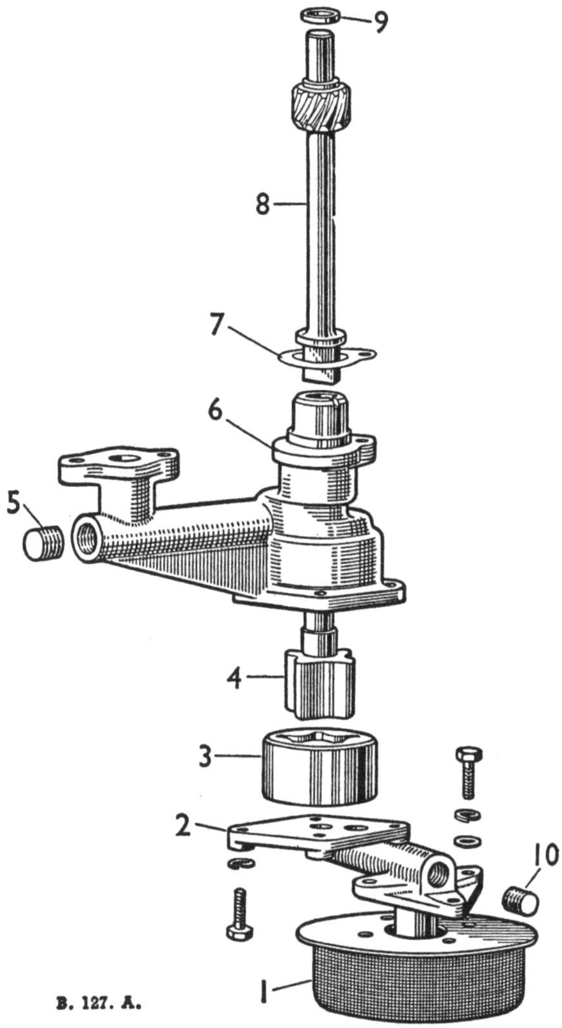

Fig. A.6. Components of the oil pump.

1. *Pick-up strainer.*
2. *Bottom lower plate.*
3. *Outer rotor.*
4. *Inner rotor.*
5. *Screw plug.*
6. *Pump body.*
7. *Joint washer.*
8. *Drive spindle.*
9. *Drive spindle thrust washer.*
10. *Screw plug.*

Draining the Sump

The sump must be drained and filled with new oil at the recommended mileage.

The hexagon-headed sump drain plug is at the rear on the right-hand side.

The sump should be allowed to drain for at least ten minutes before the drain plug is replaced. The oil will flow more readily if it is drained while the engine is hot. When the sump has been drained, approximately 10¼ pints (12·3 U.S. pints, 5·8 litres) of oil are required to fill it. The capacity of the filter is approximately 1¼ pint (1·5 U.S. pints, ·8 litres), giving a total of 11½ pints (13·8 U.S. pints, 6·5 litres). Do not forget to replace the sump drain plug.

Never use petrol or paraffin for flushing purposes. Such cleaning mediums are never completely dispersed from the engine lubrication system, and will remain to contaminate any fresh oil. This may cause premature bearing failure.

Refilling

When refilling the sump do not pour the oil in too quickly, as it may overflow from the filler orifice and

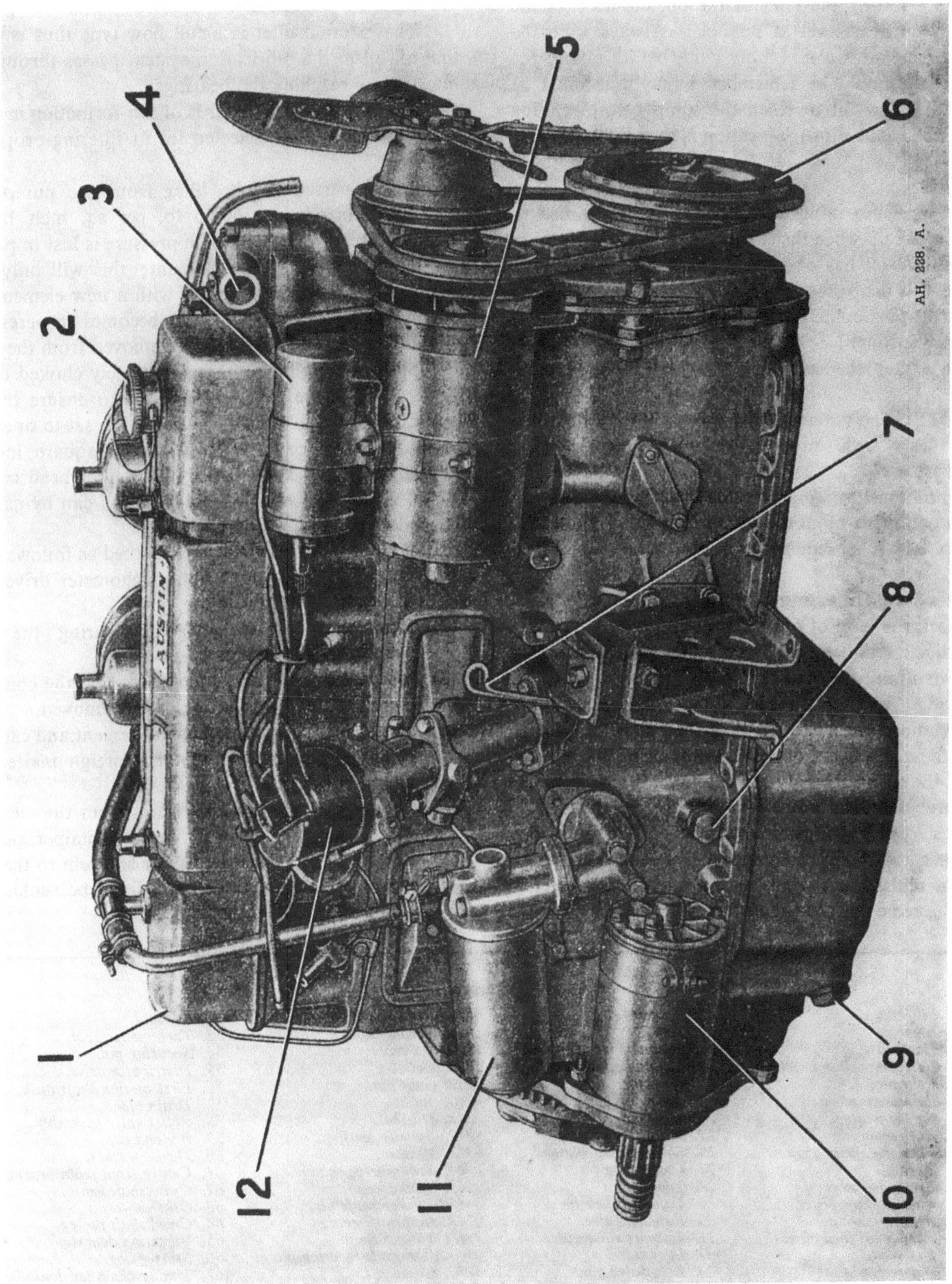

Fig. A.7. Right-hand view of engine.

1. *Valve rocker cover.*
2. *Oil filler cap.*
3. *Ignition coil.*
4. *Thermostat cover and coolant outlet.*
5. *Generator.*
6. *Crankshaft stabilizer.*
7. *Oil level dipstick.*
8. *Oil relief valve.*
9. *Oil sump drain plug.*
10. *Starter motor.*
11. *Oil filter.*
12. *Distributor cap.*

mislead the operator as to the quantity of lubricant in the engine.

Before testing the level of the oil, ensure that the vehicle is as near level as possible. Always wipe the dipstick clean with a non-fluffy cloth before taking the reading. It should be remembered that time must be allowed for new oil to reach the sump before reading the dipstick. The dipstick location is shown in fig. A.7.

Oil Pressure

The pressure indicated by the gauge may rise to 60 lb. per sq. in. when the engine is started up from cold, but after the oil has circulated and become warm, the pressure will drop to approximately 55 lb. per sq. in., with a proportionately lower idling pressure, (about 25 lb. per sq. inch). **If no oil pressure is registered by the gauge, stop the engine at once and investigate the cause.**

NOTE: The automatic relief valve in the lubrication system deals with any excessive oil pressure when starting from cold.

Continuous running with unnecessary use of the mixture control is often the cause of serious oil dilution by petrol, and a consequent drop in pressure.

Check for Low Oil Pressure

Check the level of oil in the engine sump by means of the dipstick and top up if necessary. Ascertain that the gauze strainer in the sump is clean and not choked with sludge, also that there is no air leakage at the strainer union on the suction side of the pump.

In the unlikely event of the oil pump being defective, remove the unit and rectify the fault, see Section A.8. The oil relief valve should be examined, see Section A.9.

If the engine bearings are worn the oil pressure will be reduced. A complete bearing overhaul and the fitting of replacement parts is the only remedy, necessitating the removal of the engine from the chassis.

Section A.6

OIL FILTER

The external filter is a full flow type thus ensuring that all oil in the lubrication system passes through the filter before reaching the bearings.

The element of the filter is of star formation in which a special quality felt, selected for its filtering properties, is used.

Oil is passed to the filter from the pump at a pressure controlled at 50/55 lb. per sq. inch by the engine oil relief valve. Some pressure is lost in passing the oil through the filter element; this will only be a pound or two per square inch with a new element, but will increase as the element becomes progressively contaminated by foreign matter removed from the oil.

Should the filter become completely choked due to neglect, a balance valve is provided to ensure that oil will still reach the bearings. This valve, set to open at a pressure difference of 15/20 lbs. per square inch, is non-adjustable and is located in the filter head casting. When the valve is opened, unfiltered oil can by-pass the filter element and reach the bearings.

To renew the filter element proceed as follows :—

(1) Unscrew and remove the tachometer drive from the distributor housing.

(2) Remove the two setpins securing the filter bracket to the crankcase.

(3) Unscrew the centre fixing bolt, and the container complete with element can be removed.

(4) Withdraw the contaminated element and carefully cleanse the container of all foreign matter that has been trapped.

(5) After ensuring that no fibres from the cleansing operation have been left in the container, put in a new element, prime the filter and refit to the head casting, tightening the centre fixing bolt sufficiently to make an oil-tight joint.

Fig. A.8. Engine components.

1. Filter bowl.
2. Element.
3. Head casting.
4. Filter extension bracket.
5. Split pin.
6. Plain washer.
7. Spring washer.
8. Rocker bush.
9. Rocker adjusting screw.
10. Rocker.
11. Spacing washer.
12. Rocker shaft bracket.
13. Spacing spring.
14. Rocker oil feed pipe.
15. Push rod.
16. Gudgeon pin.
17. Compression rings.
18. Oil control ring.
19. Piston.
20. Oil pump drive spindle.
21. Oil pump body.
22. Inner rotor.
23. Outer rotor.
24. Bottom cover.
25. Pick-up stainer.
26. Shell bearing big end.
27. Connecting rod.
28. Tappet.
29. Camshaft bearings.
30. Exhaust valve.
31. Exhaust valve guide.
32. Inlet valve.
33. Inlet valve guide.
34. Inner valve spring.
35. Outer valve spring.
36. Spring cap.
37. Oil seal.
38. Collets.
39. Split pin.
40. Bush.
41. Oil seal.
42. Spindle housing.
43. Washer.
44. Tachometer spindle.
45. Rocker shaft.
46. Rocker shaft plug.
47. Camshaft gear.
48. Timing chain.
49. Camshaft location plate.
50. Camshaft.
51. Plug.
52. Oil seal.
53. Tachometer gear.
54. Securing pin.
55. Distributor drive.
56. First motion shaft bush.
57. Drain pipe.
58. Relief valve assembly.
59. Big end cap.
60. Thrust washer.
61. Centre front main bearing.
62. Crankshaft gear.
63. Oil thrower.
64. Crankshaft pulley.
65. Vibration damper.
66. Starter dog.
67. Timing chain tensioner.

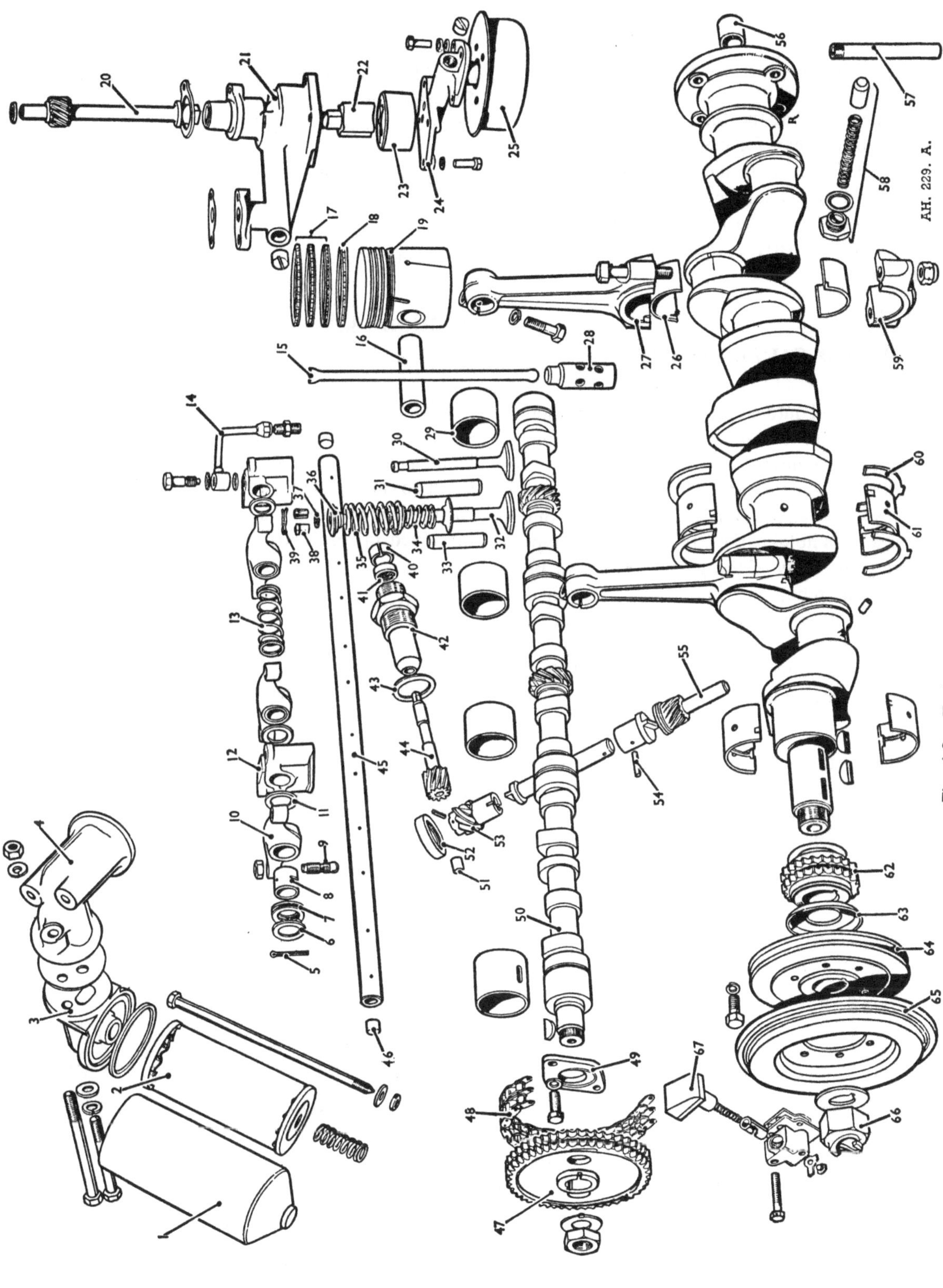

Fig. A.8. Engine components.

Note—Commencing Engine No. 40501 a fully floating small end was introduced thus deleting the clamping pin as shown in this illustration.

(6) Replace the filter and bracket complete by means of the two setpins.
(7) Refit the tachometer drive to the distributor housing.
(8) Check the level of oil in the sump by means of the dipstick.

It is recommended that the filter container should not be disturbed other than for cleaning and the fitting of a new element at the recommended mileages; to do so invites the hazard of added contamination from accumulated dirt on the outside of the filter entering the container, and thus being carried into the bearings on restarting the engine.

Section A.7

SUMP AND GAUZE STRAINER

Removing

(1) Drain off the oil into a suitable container then extract the setscrews and washers, thus enabling the sump to be removed.
(2) Detach the bottom of the strainer by removing the nut, washer and distance piece. Take out the three setpins holding the strainer to the pump, so allowing the body of the gauze strainer to be removed. The pump and strainer can be swilled out with petrol or paraffin and thoroughly dried with a non-fluffy rag.
(3) Inspect the two joint washers and renew if they are damaged in any way.

Refitting the Sump

Clean out the sump by washing it in paraffin. Take care to remove any traces of the paraffin before refitting the sump to the engine. Pay particular attention to the sump and crankcase joint faces, and remove any traces of old jointing material. Examine the joint washer and renew it if necessary. The old joint washer can be used again if it is sound, but it is advisable to fit a new one.

Smear the faces of the joint with grease and fit the joint washer. Lift the sump into position and insert the setscrews into the flange tightening them up evenly.

Section A.8

OIL PUMP

Removing the Oil Pump

(1) Remove the sump and pick up strainer.
(2) Take off the nuts and spring washers from the three studs which secure the oil pump assembly to the crankcase, when the pump can be withdrawn.
If the pump is removed with the engine still in the car, the drive shaft will be free to disengage from the camshaft, and care must be taken to prevent it falling out. Note also the thrust washer fitted on the drive shaft above the gear.

Dismantling the Oil Pump

Mark the flange and pump body to assist reassembly. Separate the body from the bottom flange. The outer rotor can then be lifted out of the body.

Replacing the Oil Pump

Insert the pump from below and push the shaft right home until the driving gear is meshed with the gear on the camshaft.

Section A.9

RELIEF VALVE

The non-adjustable oil pressure relief valve is situated at the rear of the right-hand side of the cylinder block below the oil filter and is held in position by a hexagon nut sealed by a copper washer. The relief valve spring retains a valve cup against a seating machined in the block.

Section A.10

VALVE ROCKER SHAFT

The valve rocker shaft on the cylinder head is hollow. It is supplied with oil by a pipe connection and is drilled for lubrication to each rocker bearing.

This shaft is plugged at each end, one of these being screwed in order that the shaft may be cleaned internally.

Fig. A.9. Valve adjustment.
1. Feeler gauge. 2. Rocker. 3. Lock nut. 4. Adjusting screw.

Section A.11

ADJUSTING VALVE CLEARANCE

Lift off the valve cover after removing the two flat and two dome cap nuts.

Between the rocker arm and the valve stem there must be a clearance of ·012 in. (·30 mm.) for both inlet and exhaust, clearance being set with the engine cold.

(1) If adjustment is necessary, slacken the locknut whilst continuously applying sufficient pressure to the adjusting screw with a heavy screwdriver, and raise or lower the adjusting screw in the rocker arm. Check the clearance with a feeler gauge.

(2) Tighten the locknut when the adjustment is correct, but always check it again afterwards in case the adjustment has been disturbed during the locking process.

(3) When replacing the valve cover, take care that the joint washer (using a new one if necessary) is properly in place to ensure an oil tight joint.

Section A.12

ROCKER SHAFT ASSEMBLY

Removal

(1) Disconnect the breather pipes at their rocker cover terminals.

(2) Unscrew the two flat and two dome nuts securing the rocker cover to the cylinder head, taking care not to damage the cork gasket, and remove the rocker cover.

(3) Detach the oil feed pipe at the union on the cylinder head.

(4) Unscrew and remove the twelve nuts and spring washers which hold the rocker shaft brackets to the cylinder head.

(5) Remove the rocker assembly, complete with brackets, rockers and oil feed pipe.

Dismantling

(1) Unscrew and remove the oil feed pipe banjo from its bracket noting its corresponding position on the shaft.

(2) Remove the split pins from the end of the rocker shaft to release thrust washers and double coil springs.

(3) Withdraw rocker, rocker shaft brackets, thrust washers and springs, retaining them in their original order for reassembly.

Reassembling

When reassembling the rocker gear, commence with No. 4 bracket and secure the oil feed pipe with the washers in position, ensuring that the dowel on the banjo bolt locates in the rocker shaft.

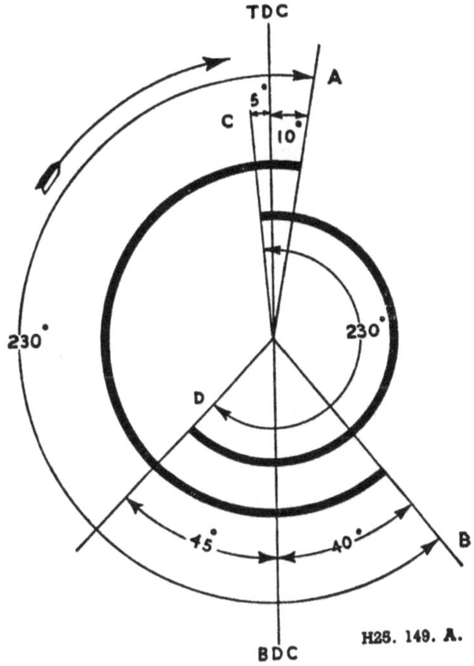

Fig. A.10. Valve timing diagram.
Exhaust closes at A, and opens at B, inlet opens at C and closes at D.

The brackets are fitted with the highest lug to the camshaft side of the engine, and the rocker shaft is fitted with the screwed in end plug to the front. The rear end tapered plug is a drive fit.

A thrust washer is fitted each side of each rocker shaft bracket, and all springs, and rockers, and the remaining brackets are interchangeable.

Section A.13

PUSH ROD REMOVAL

If the valve rocker assembly has already been removed all that remains is for the push rods to be lifted out. They may on the other hand be taken out without detaching the rocker assembly as described below :—

(1) Remove the valve rocker cover as described in Section A.12 and slacken the valve adjustment screw to its full extent.

(2) With the aid of a screwdriver supported under the rocker shaft, depress the valve and slide the rocker sideways free of the push rod.

(3) Withdraw the push rod.

(4) In the case of the rocker at each end, it is necessary to take out the split pins at the end of the shaft.

(5) The above sequence should be reversed when replacing push rods and rockers.

Section A.14

ROCKER ARM BUSHES

(1) While the rocker gear is detached from the head, check for play between the rocker shaft and the rocker arm bushes. If this is excessive new bushes should be fitted. To do this dismantle the rocker assembly as described in Section A.12.

(2) The bush is best removed by using a drift and anvil (Service Tool No. 18G 21). The anvil is recessed to retain the rocker in position while the bush is gently knocked out by the drift. File and drill out the rivet in the rocker arm oilway.

(3) The flange of the drift is also recessed to prevent the new split bush from opening when being driven into position with the joint immediately above the rocker arm oilway.

(4) Drill an oilway through the bush from the top of the rocker using a ·0785 in. diameter drill. A second oilway must be drilled through the bush via the rocker arm using a ·089 in. diameter drill.

(5) Plug the oilway in the rocker arm with a rivet and weld its head to the rocker boss. Ream the internal diameter of the bush to suit the shaft.

Section A.15

TAPPETS

Removal

(1) Remove the valve rocker shaft assembly as detailed in Section A.12.

(2) Disconnect the dynamo terminals and remove the set bolt securing the dynamo to the slotted link. Take out the bolts on which the dynamo pivots and remove the dynamo and coil.

Fig. A.11. Removing a tappet.

(3) Release the front tappet chamber cover by removing the five securing bolts. The centre and rear tappet chamber covers give access to the valves for No. 3, 4, 5 and 6 cylinders when the single retaining bolts are removed.

(4) Withdraw the push rods, keeping them in their respective positions to ensure replacement onto the same tappets. Lift out the tappets, keeping them in the same respective locations. Inspect the tappet cam contacting surfaces for wear. New tappets should be fitted by selective assembly so that they just fall into their guides under their own weight when lubricated.

Replacement

Assembly is a reversal of the above procedure, but care should be taken to see that the tappet cover joints are oil-tight and that the rockers are adjusted to give the correct valve clearance.

Section A.16

RENEWING VALVE SPRING IN POSITION

(1) In an emergency a new valve spring(s) can be fitted without lifting the cylinder head, but it is advisable first to bring the piston to top dead centre, to ensure that the valve cannot fall into the cylinder during the process.

(2) Remove the sparking plug, and by means of a length of copper tubing or similar tool inserted through the plug hole, the valve can be held on its seat whilst the spring is compressed. The valve rocker shaft can be used as a fulcrum point, by an operator using two screwdrivers to bear on the valve spring cap each side of the valve stem, whilst the cotters are removed.

Section A.17

MANIFOLD

Removal and Replacement

(1) Detach the air cleaners from the carburetters by unscrewing the four setpins and releasing the breather pipe attached to the air cleaner.

(2) Disconnect the heat shield by removing the two securing nuts and washers.

(3) Unscrew and remove the six brass nuts and plain washers which secure the exhaust manifold to the down pipes.

(4) Disconnect the throttle and choke linkages to the carburetters, together with the vacuum control pipe and petrol feed pipe.

(5) Unscrew and remove the 14 nuts and washers which secure the exhaust manifold and carburetters to the cylinder head (four on the

carburetter flanges, ten on the exhaust manifold). This will automatically release the heater outlet pipe.

(6) The exhaust manifold and carburetters can then be drawn off their respective studs and lifted clear of the engine.

(7) Reassembly is the reverse of the above procedure; always use a new joint washer for the exhaust manifold to ensure an air tight joint.

Section A.18

CYLINDER HEAD

Removing

(1) Drain all water from the cooling system, if the water contains anti-freeze mixture, it should be run into a clean container and used again.

(2) Detach the top water hose from the cylinder head.

(3) Disconnect the high tension wires from the sparking plugs and remove the plugs.

(4) Detach the exhaust manifold, complete with carburetters, as detailed in Section A.17.

(5) Remove rocker cover and breather pipes as described in Section A.12.

(6) Release the suction advance pipe clip from its securing point on the cylinder head. Also slacken the retaining clip and detach the heater inlet hose.

(7) Remove the rocker assembly as described in Section A.12.

(8) Withdraw the push rods, keeping them in order of removal taking care not to pull the tappets out of their guides in the block.

(9) Remove the sixteen cylinder head nuts together with their flat washers and lift off the cylinder head.

Replacing

(1) Replace the cylinder head joint washer with the side marked "Top" uppermost, it is not necessary to use jointing compound or grease for the gasket.

(2) Having slipped the gasket over the studs, next lower the cylinder head into position and position the cylinder head stud nut washers. Ensure that a bronze washer is fitted below the steel washer on each stud which passes through the inlet manifold on the left-hand side of the head; also ensure that the suction advance pipe clip is replaced in its original position on the cylinder head.

(3) Fit the nuts finger tight and then tighten them a turn at a time, in the order given in fig. A.12, to the recommended torque spanner readings.

(4) Insert the push rods, ensuring that the ball ends are correctly located in the tappets.

(5) Replace the rocker gear and connect the oil feed pipe, as described in Section A.12.

(6) Reset the valve clearance, and replace the rocker cover using a new joint washer if the old one is damaged in any way.

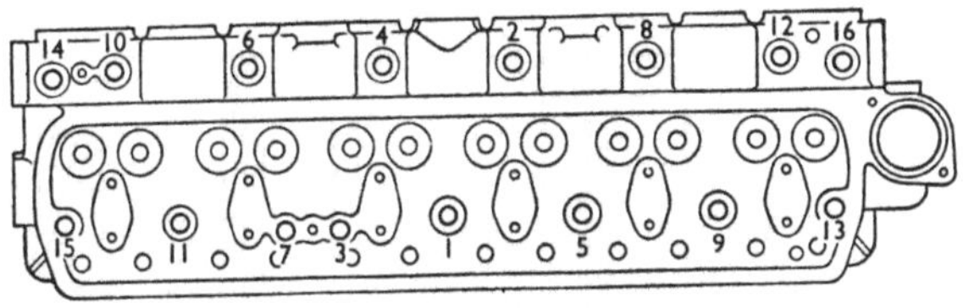

B. 40. A.

Fig. A.12. The order of tightening for the cylinder head nuts.

(7) Replace the exhaust manifold and carburetters and connect up the petrol pipe, throttle and choke controls and heater outlet pipe. Tighten the manifold nuts evenly ensuring that a good joint is made.

(8) Reconnect heater inlet pipe, water hose from the thermostat housing to the radiator, and breather pipes.

(9) Refill the cooling system, replace the sparking plugs and their washers, and the high tension wires to their respective plugs.

(10) Check the valve clearance again after the vehicle has run about 100 miles (160 km.) as the valves have a tendency to bed down. At the same time it is advisable to test the cylinder head nuts for tightness. By using the special cylinder head nut spanner (Service Tool 18G 545) it is possible to perform this operation without removing the rocker shaft assembly. Tightening the cylinder head nuts may affect valve clearances, although not usually enough to justify resetting.

Section A.19

REMOVING AND REFITTING VALVES

With the cylinder head removed, a valve lifting tool can be used to compress the springs (Service Tool No. 18G 106). Take away the circlip, split cotters, and valve stem cap, so releasing the springs and allowing the valve to be removed.

(1) When removing the valves, place them in a rack, thus enabling them to be paired up with their correct cylinders. The valve springs should be tested and their free length checked, the correct length being approx. 1·969 in. (50·03 mm.) for the inner spring and 2·047 in. (52 mm.) for the outer spring.

(2) Clean the carbon from the top and bottom of the valve heads, as well as any deposit that may have accumulated on the stems. The valve heads should, if necessary, be refaced at an angle of

45° for the exhaust valve and 30° for the inlet valve. If the valve seats show signs of excessive pitting it is advisable to reface these also.

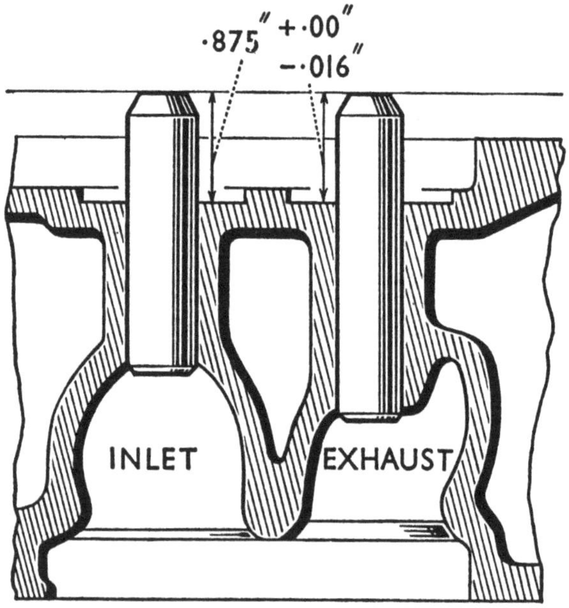

Fig. A.13. Showing the position of the valve guides after fitting.

(3) The valves are made without any indentures or slots in the head, this necessitates the use of a rubber suction valve grinding tool.

(4) Reassembly is a reversal of the operations for removal.

Section A.20

VALVE GRINDING

(1) For valve grinding a little grinding paste should be smeared evenly on the valve face, and the valve rotated backwards and forwards against its seat (using Service Tool No. 18G 29), advancing it a step at short intervals until a clean and unpitted seating is obtained. The cutting action is facilitated by allowing a light spring situated under the valve head, to periodically lift the valve from its seat. This allows the grinding compound to re-penetrate between the two faces after being squeezed out.

(2) On completion, all traces of compound must be removed from the valve and seating. It is essential that each valve is ground-in and refitted to its own seating.

(3) It is also desirable to clean the valve guides; this can be done by dipping the valve stem in petrol or paraffin, and moving it up and down in the guide until it is free.

Section A.21

VALVE GUIDES

(1) The valve guides are of a one-piece design. They are pressed into the cylinder head to allow ·859 to ·875 in. (21·8 to 22·23 mm.) of the guide to protrude above the machined face when fitted.

(2) To position each valve spring on the cylinder head, a stepped pressed steel seating collar is fitted over the part of the guide protruding from the cylinder head.

(3) Valve guides should be tested for wear whenever valves are removed, and if excessive side play is present, a close check should be made of the valve stem and the guide. In the event of wear being noticeable, the defective components should be renewed. If a valve is at fault the wear will be evident on the stem. It should be borne in mind that the valve stem and guide should be a running fit to avoid the possibility of an air leak.

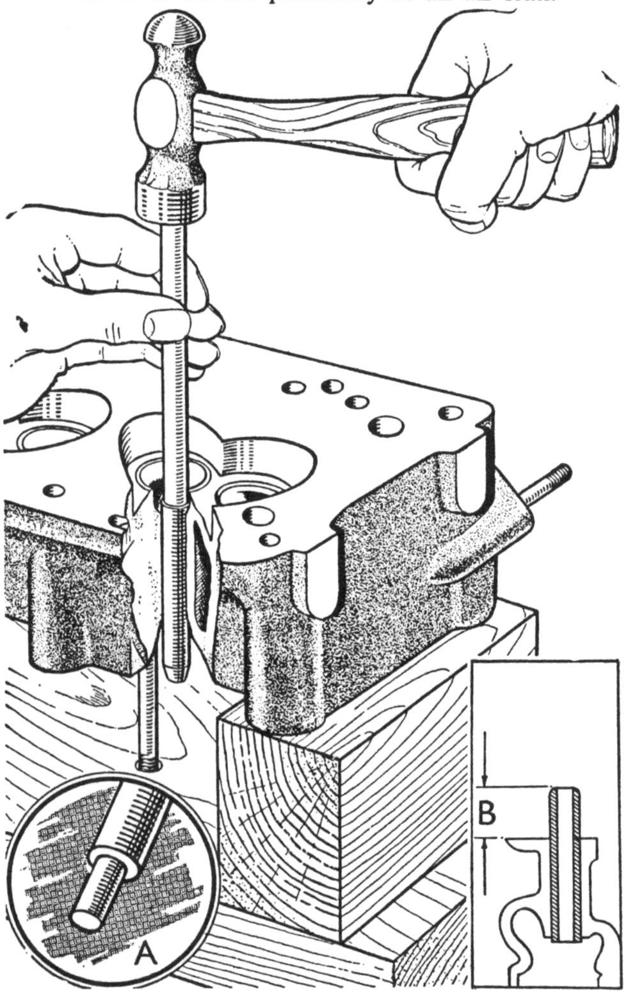

Fig. A.14. Removing a valve guide.
A shows the stepped end of the tool. B indicates the portion of the guide which must stand above the surface.

(4) If renewal is necessary due to wear, the valve guide may be driven out after removal of the valve, as shown in fig. A.14.
(5) The drift is stepped from a $\frac{1}{2}$ in. (12·7 mm.) diameter to a $\frac{5}{16}$ in. (7·9 mm.) diameter locating spigot in order to obviate it slipping off the guide and damaging the port. Knock out the guide in the direction shown.
(6) A new guide should be driven into position in the same direction, that is, inserting it through the valve seating and driving towards the top of the cylinder head.
(7) The final position of the guide is shown in fig. A.13.

Section A.22

DECARBONISING

(1) Remove the cylinder head as described in Section A.18.
(2) Scrape off all carbon deposit from the cylinder head and ports. Clean the carbon from the piston crowns, care being taken not to damage the pistons, and not to allow dirt or carbon deposit to enter the cylinder barrels or push rod compartment.
When cleaning the top of the pistons do not scrape right to the edge as a little carbon left on the chamfer assists in keeping down oil consumption; with the pistons cleaned right to the edge or new pistons, oil consumption is often slightly though temporarily increased.
(3) Blow out the oil passages and swill out the water passages using a water hose. The gasket contacting surfaces of the head should be checked for flatness with a straight-edge and the surfaces examined for scores. If the cylinder head is found to be badly out of true it should be renewed.
(4) Remove all carbon accumulation from the valves and thoroughly clean them. Inspect the valve bases and seats and if they are slightly pitted or rough, grind them in, as described in Section A.20. If the valves and seats show signs of excessive pitting, or the faces are not flat, the valves and seats should be replaced.
(5) Examine the valve guides, as described in Section A.21.
(6) Broken or weak valve springs should be renewed. The other valve springs should be tested and the results compared with figures given under "General Data".
(7) Clean the rocker shaft gear and blow out the oil passages as described in Section A.10.
(8) Inspect the rocker shaft, rockers and bushes for wear. Renew any worn rocker bushes as described in Section A.14.
(9) Reassemble and install the cylinder head assembly.

The following operations should be carried out with the engine removed, although in some cases it is possible to perform them with the engine in position.

Before removing or replacing any component it is important to ensure that all surrounding surfaces are perfectly clean, to prevent the entry of foreign matter into the engine. This can best be accomplished by the use of a paraffin bath and brush, and it is also important to note that fluffy rags should never be used, as there is danger of causing obstruction to small oil ways.

Section A.23

CONNECTING RODS AND BEARINGS

Removal

(1) Remove the cylinder head assembly as described in Section A.18.
(2) Drain and remove the sump (see Section A.5).
(3) Remove the self-locking nuts securing the caps and bearings to the connecting rods. Remove the caps and bearings.
(4) Withdraw the pistons and connecting rods upwards through the cylinder bores.

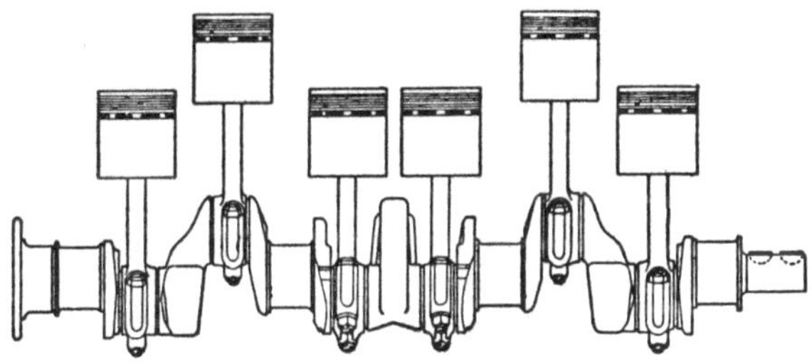

B. 125. A.

Fig. A.15. Showing the positions of connecting rod off-sets.

(5) It may be necessary to remove the carbon or ridge from the top of the bores prior to pushing the pistons upwards, to avoid piston-ring fracture.
(6) Remove the pistons from the connecting rods by unscrewing the clamp bolt from the small end of the connecting rod and pushing the gudgeon pin out.
(7) Ensure that each connecting rod, cap and bearing is marked with the cylinder number from which it was removed.
(8) The big ends are offset, and rods in numbers 1, 3 and 5 cylinders are offset towards the front, with 2, 4 and 6 cylinders offset towards the rear.

(9) The alignment of the connecting rods should be checked on an alignment fixture. On no account must the rods or caps be filed.

(10) Examine the bearing shells for wear and pits. Renew the bearing shell if necessary. Bearings are pre-finished with the correct diametrical clearance and do not require bedding in.

(11) Check the crankpins with a micrometer if they are worn oval or are scored, the crankshaft will have to be removed for regrinding, see Section A.29.

Replacing

Before installing the connecting rods and bearings it is assumed that the pistons and rings have been serviced, see Section A.24.

The pistons and connecting rods must be fitted in the same cylinder bores and the same way round as when removed.

(1) Assemble the piston and the connecting rod to the gudgeon pin, so that the split in the piston skirt is adjacent to the split in the top of the connecting rod.

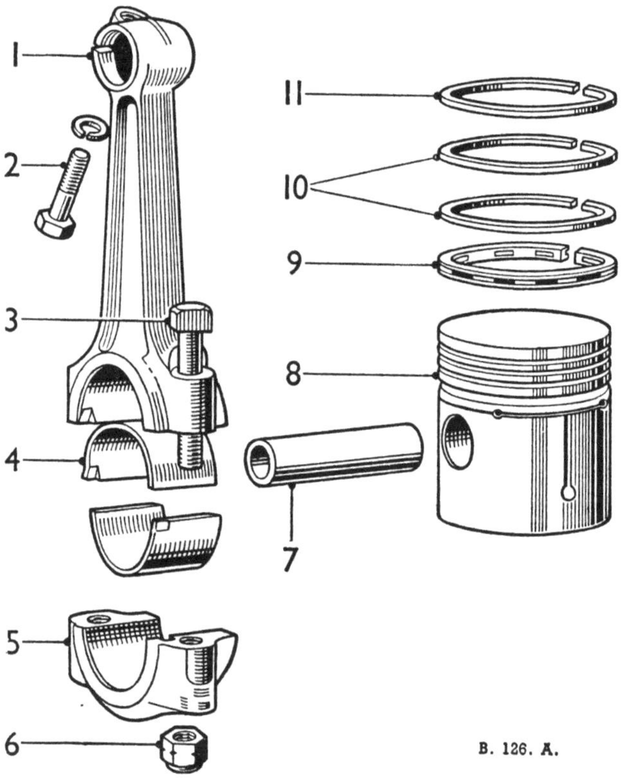

Fig. A.16. Connecting rod and piston assembly.

1. *Connecting rod.*
2. *Small end clamping screw.*
3. *Big end bolt.*
4. *Shell bearing.*
5. *Big end cap.*
6. *Big end nut.*
7. *Gudgeon pin.*
8. *Piston.*
9. *Oil control ring.*
10. *Taper compression rings.*
11. *Plain compression ring.*

(2) Refit the piston rings very carefully, make quite sure that the pistons and bores are perfectly clean and smear the bores with clean engine oil.

(3) Use a piston ring clamp, service tool No. 18G 55A, when replacing the pistons from the top of the bore, and make sure that the split in the piston faces away from the camshaft.

(4) Clean the crankpins and both sides of the shell bearings, locate the feathered ends in the connecting rod and its cap, and smear the crankpins with engine oil.

(5) Before fitting the cap, check that the number stamped on the rod is the same as that on the cap. Note that the recesses in the cap and rod must be on the same side. Tighten the nuts. Turn the crankshaft after fitting each rod, to ensure that the bearing is not binding on the crankpin. Also check the side clearance of each rod, as given under "General Data".

(6) Refit the cylinder head assembly, see Section A.18.

(7) Refit the sump and refill with recommended grade of oil, see Section Q.

Section A.24

PISTONS, RINGS AND GUDGEON PINS

Removal

The split-skirt pistons are of aluminium alloy material. Four rings are fitted above the gudgeon pin, the bottom ring being of the oil-control type. The pistons are fastened to the connecting rods by gudgeon pins which are clamped rigidly in the small ends of the connecting rods. Bushings are not needed in the gudgeon pin bosses of the pistons because the aluminium alloy material serves as a suitable bearing for the gudgeon pins, the bearing surfaces of which are lubricated by means of splash through the two holes drilled in each boss. To remove the pistons see Section A.23.

To view and overhaul

(1) Remove the rings over the tops of the pistons.

(2) Scrape all accumulation of carbon off the piston heads and, using a piston ring groove-cleaning tool or an old ring section, carefully scrape all carbon out of the ring grooves of the pistons. Clean the carbon out of the oil holes in the piston ring grooves.

(3) Thoroughly clean all the dismantled components in paraffin.

(4) Examine all parts for wear and damage, renew if necessary.

(5) If cylinder reconditioning is required (see Section A.30), determine the amount of material to be removed (refer to "General Data" concerning oversize pistons available).

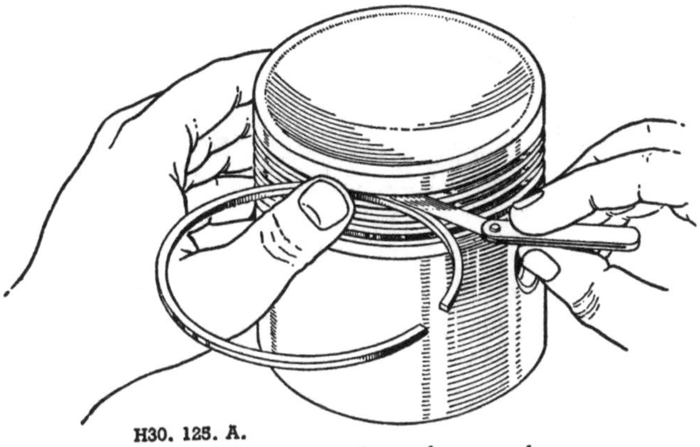

Fig. A.17. Checking piston ring and groove clearance.

(6) When fitting new or oversize pistons and rings to reconditioned cylinder bores, the clearances should be controlled within the limits given under "General Data".
The cylinder bore glazing should be removed before fitting new rings to a worn cylinder bore.

(7) Piston rings should have a gap clearance (see "General Data") when installed in the cylinder bores. If new rings are being installed, each ring should be checked in the cylinder bore to determine whether its gap clearance is within the range specified. To do this, use the bottom of a piston to insert the ring part way into the bore. The ring will be squared up in the bore to measure the gap clearance as shown in fig. A.18. To check the ring clearance in the piston grooves, install the rings on the pistons and determine the clearances with a feeler gauge. If the piston ring grooves are worn excessively, as indicated when comparing the actual clearances with those given under "General Data", renew the rings and pistons.

(8) Gudgeon pins should be a hand-push fit in the piston. The fit can be checked after the rod has been assembled by holding the piston with the connecting rod in an approximately horizontal position. The weight of the large end of the connecting rod should be just insufficient to turn the gudgeon pin in the piston. On no account must gudgeon pin piston bosses be reamed out as oversize gudgeon pins are not supplied or permitted.

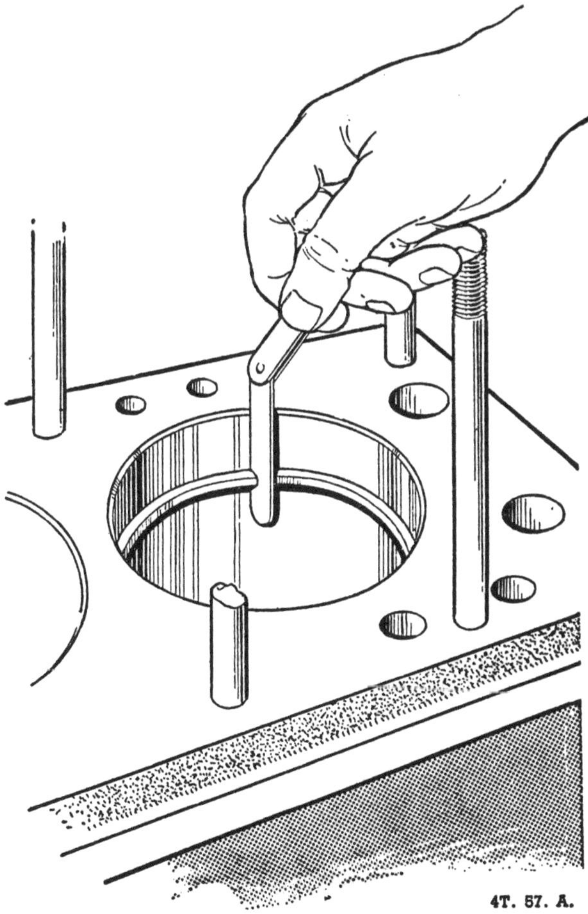

Fig. A.18. Checking piston ring gap.

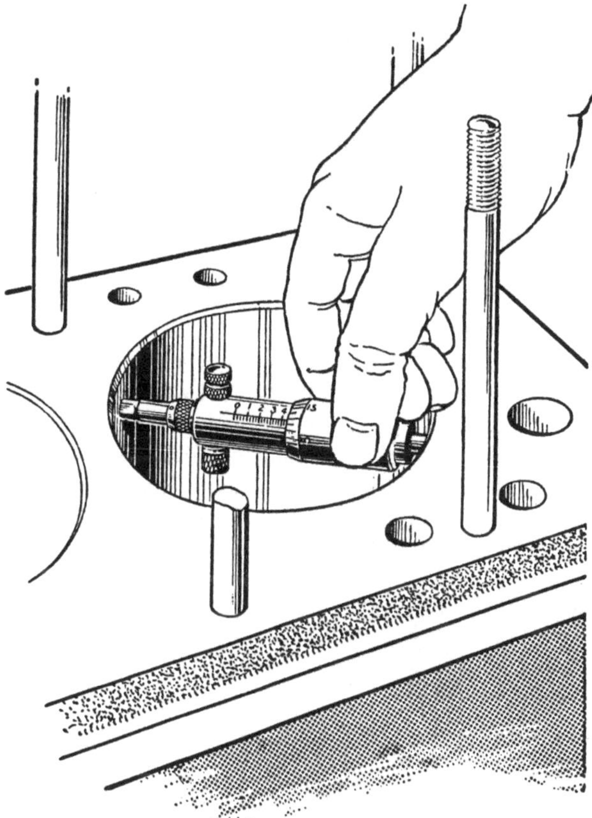

Fig. A.19. Checking bore wear.

Replacement

See Section A.23.

Section A.25

TIMING CHAIN AND WHEELS

Removal

(1) Remove the radiator as described in Section C, if the removal is to be done with the engine in position.

(2) Slacken the generator fixing bolts and take off the fan belt. Unscrew the starter dog nut using Service Tool No. 18G 391. Before doing this the tab washer must be knocked back.

(3) In some cases it may now be possible to remove the crankshaft damper and pulley complete as one unit. If, however, the pulley is tight on the crankshaft, it will be necessary to undo the six nuts securing the damper, and with this component removed, draw off the pulley with Service Tool No. 18G 2.

(4) Take out the five $\frac{1}{4}$ in. and the seven $\frac{5}{16}$ in. setpins from the timing cover flange, taking care to retrieve the special elongated washers fitted under the heads. The cover can now be removed and the joint washer separated, taking care to remove and retain the oil thrower.

(5) Remove the automatic chain tensioner, see Section A.26.

(6) Unlock and remove the camshaft chain wheel nut and remove the nut and lockwasher. Note that the locating tag on the lockwasher fits into the keyway of the camshaft chain wheel.

(7) The camshaft and crankshaft chain wheels may now be removed, together with the timing chain, by easing each wheel forward a fraction at a time, with suitable small levers or Service Tool No. 18G 58. As the crankshaft gear wheel is withdrawn care must be taken not to lose the gear packing washers immediately behind it.

Fig. A.20. Timing cover locating tool.

Fig. A.21. The timing chain showing bright links opposite spot marks on gears. 'A' shows the position of the short run of chain between the bright links.

(8) Clean and examine the joint faces of the timing cover and the front mounting plate.

(9) Examine the felt oil seal for signs of wear, hardening or damage. If the slightest wear or damage is apparent the timing cover and seal must be renewed as an assembly.

(10) Inspect chain wheels for worn or broken teeth.

(11) Inspect the chain for excessive wear or stretch.

(12) Examine the chain tensioner, see Section A.26.

Reassembling

The installation of the timing chain and wheels is the reversal of the removal procedure but for the following points :—

(1) Replace the same number of washers as was found when dismantling, unless new camshaft or crankshaft components have been fitted, which will disturb the alignment of the two gear wheels. To determine the thickness of washers required, place a straight-edge across the sides of the camshaft wheel teeth and measure with a feeler gauge the gap between the straight-edge and the crankshaft gear.

(2) When replacing the timing chain and gears, set the crankshaft and camshaft with the keyways approximately at T.D.C. when seen from the front.

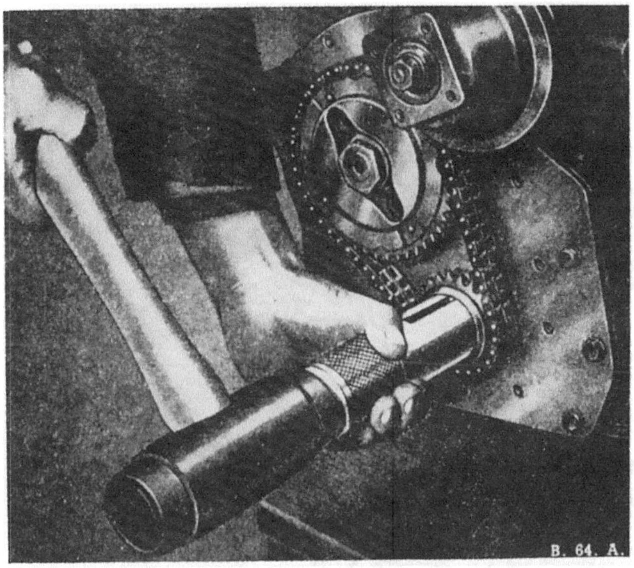

Fig. A.22. Crankshaft gear replacer tool No. 18G 16.

Double the timing chain, bringing both bright links together. This gives a long and short portion of the chain on either side of the bright links. With the shorter part of the chain on the **Right,** (the bright links facing the operator), and the longer on the **Left,** engage the marked camshaft sprocket tooth with the top bright link, and the crankshaft sprocket with the marked tooth coinciding with the other bright link.

Place the sprockets in their respective positions on the camshaft and crankshaft, complete with the chain, and push the assembly home. Carefully keep the sprockets in line with each other all the time to avoid straining the chain.

When replaced on the engine, the bright links and the marked teeth should take up the position shown in fig. A.21.

(3) Replace the camshaft chain wheel locking washer and nut.
(4) Apply engine oil to the timing chain and wheels before installing the cover.
(5) Replace the oil thrower, convex side towards the sprocket.
(6) When refitting the timing cover ensure that the seal is concentric with the crankshaft, using Service Tool 18G3. In the absence of the tool, lubricate the hub of the crankshaft pulley and push it into the seal, while turning it to avoid damaging the seal. Fill the annular groove between the lips of the rubber seal (later engines) with grease. Slide the pulley onto the crankshaft, aligning its keyway with the key on the shaft. Turn the cover to line up the setscrew holes with those in the crankcase, without straining the cover against the flexibility of the seal. A new joint washer should be fitted between the timing cover and the front mounting plate.

Section A.26

AUTOMATIC CHAIN TENSIONER

Description

The tensioner is secured to the engine front mounting plate by two bolts and a locking plate. When the engine is running, oil enters the spigot on the back face under pressure and lubricates the bearing surface through a hole in the tensioner slipper pad.

The tensioner consists of a cylinder with a helical slot which moves in a plunger by the action of a spring in the tensioner body. The helical slot has a recessed lower edge. Should the chain wear through use, the spring pushes the plunger and pad outwards against the chain. A limiting peg in the plunger, bearing on the upper edge of the helical slot, rotates the cylinder until the next recess in the lower edge engages it, and the plunger is prevented from returning to its original position and allowing the chain to become slack.

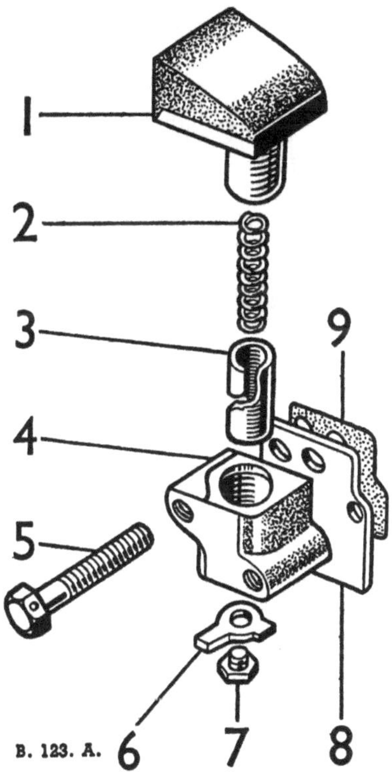

Fig. A.23. Timing chain tensioner.

1. *Slipper head.*
2. *Spring.*
3. *Locating sleeve.*
4. *Body.*
5. *Setpin.*
6. *Lockwasher.*
7. *Plug.*
8. *Backplate.*
9. *Joint washer.*

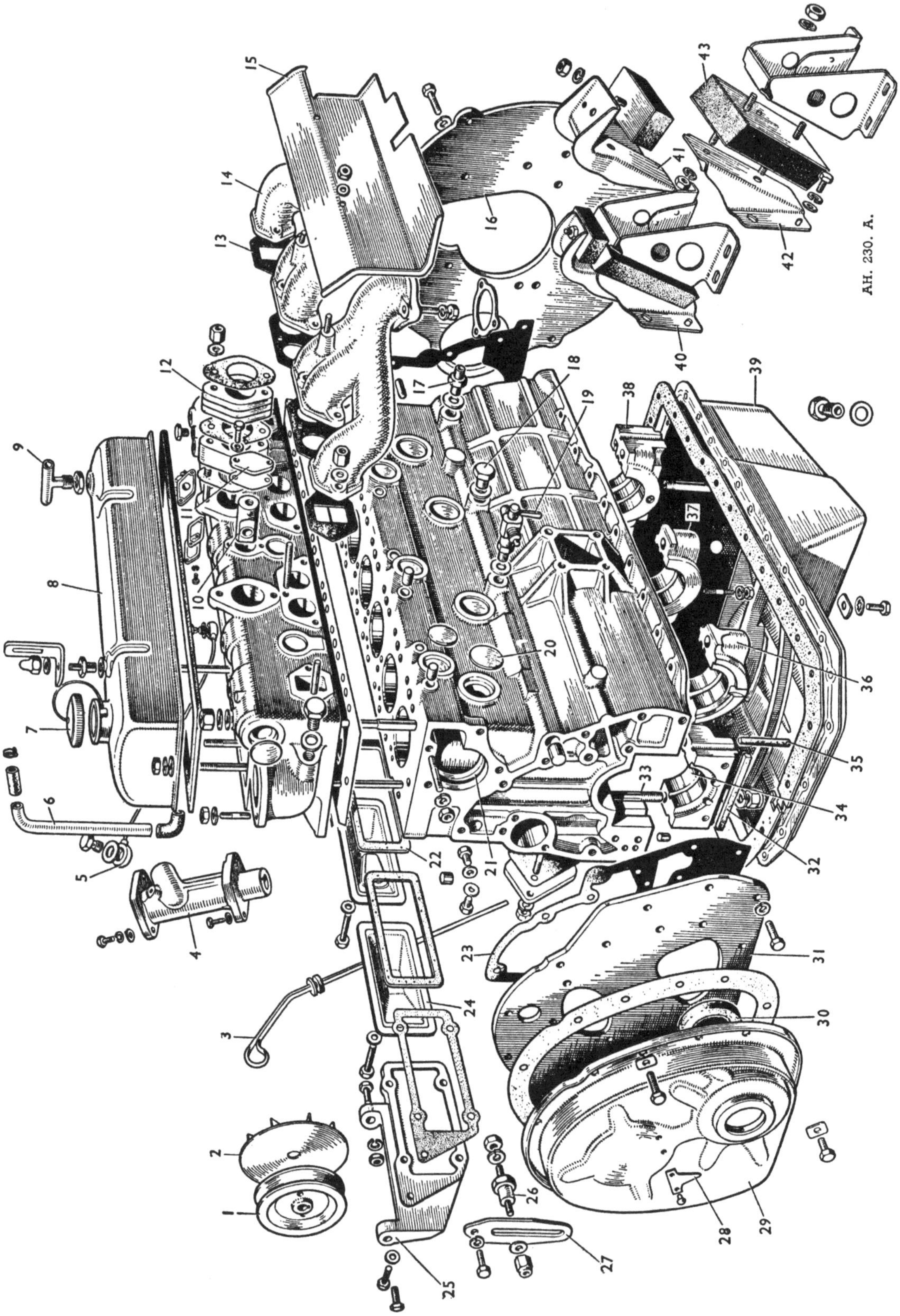

Fig. A.24. Engine exploded.

Fig. A.24. Engine exploded

1. *Generator pulley.*
2. *Generator fan.*
3. *Dipstick.*
4. *Tachometer housing.*
5. *Tachometer oil feed pipe.*
6. *Heater pipe.*
7. *Oil filler cap.*
8. *Rocker cover.*
9. *Breather pipe.*
10. *Balance plug.*
11. *Balance plug cover.*
12. *Inlet manifold joint.*
13. *Exhaust manifold joint.*
14. *Exhaust manifold.*
15. *Deflector plate.*
16. *Engine back plate.*
17. *Oil gauge pipe connection.*
18. *Plug, oil filter feed hole.*
19. *Cylinder block drain tap*
20. *Welch plug.*
21. *Cylinder block.*
22. *Tappet cover joint.*
23. *Engine front plate joint washer.*
24. *Tappet cover.*
25. *Generator mounting.*
26. *Generator mounting stud.*
27. *Generator swinging link.*
28. *Timing pointer.*
29. *Front cover.*
30. *Felt washer.*
31. *Engine front plate.*
32. *Seal for main bearing.*
33. *Front main bearing stud.*
34. *Front main bearing cap.*
35. *Sump joint washer.*
36. *Centre front main bearing cap.*
37. *Centre rear main bearing cap.*
38. *Rear main bearing cap.*
39. *Sump.*
40. *Engine mounting bracket.*
41. *Bracket carrying mounting rubber.*
42. *Bracket for mounting rubber.*
43. *Engine mounting rubber.*

To Remove

(1) Unlock the tab washer fitted to the tensioner bottom plug and remove the plug from the body.

(2) Insert a ⅛ in. (3 mm.) Allen key into the plug hole to engage the cylinder, and turn the key in a clockwise direction (viewed from the opposite end to the slipper), until the rubber slipper is completely free of spring pressure. Between a half and one full turn is necessary.

(3) Unlock and remove the bolts to release the chain tensioner assembly and back plate.

To Dismantle

(1) Withdraw the plunger and slipper assembly from the tensioner body and engage the lower end of the cylinder with the Allen key.

(2) Turn the key in an anti-clockwise direction, gripping the plunger and key securely, until the cylinder and spring are released from inside the plunger.

To Inspect

(1) Clean the components thoroughly in petrol.

(2) The oil hole in the spigot, and outlet oil hole in the slipper should be blown out by compressed air before reassembly.

(3) Check the tensioner spring and examine the slipper pad for wear. Renew as necessary. (See Section A.36).

To Reassemble

(1) Insert the spring in the plunger and place the cylinder on the other end of the spring.

(2) Compress the spring until the cylinder enters the plunger bore and engages with the peg in the bore.

(3) Hold the assembly compressed in this position and engage the Allen key. Turn the cylinder in a clockwise direction until the end of the cylinder is below the peg, thus fully compressing the spring and locking cylinder.

(4) Withdraw the key and insert the plunger assembly into the body.

To Replace

(1) Position the back plate and secure the assembly to the cylinder block.

(2) Move the timing chain into position and release the tensioner for operation by inserting the Allen key and turning it in an anti-clockwise direction as far as possible, assisting the slipper to rise initially with the finger.

(3) Secure the bolts into their locking plate, replace the bottom plug, and lock it with a tab washer.

Section A.27

CAMSHAFT AND BEARINGS

Removal

(1) Drain the sump and release it from the engine. Remove the oil pump, and then take off the rocker assembly, see Sections A.7, A.8 and A.12.

(2) Remove the push rods and take out the tappets, see Section A.13.

(3) Remove the timing cover, timing chain tensioner, chain and gears, see Sections A.24 and A.25.

(4) Remove the distributor and spindle drive, see Section B. Do not slacken the clamping plate bolt or the ignition timing setting will be lost.

(5) Take out the two setscrews which secure the camshaft locating plate to the cylinder block.

(6) Withdraw the camshaft forward rotating it slowly to assist the withdrawal.

(7) Inspect the camshaft bearing journals and cams for signs of scoring. If the journals are not within the required diameter limits (see under "General Data"), the camshaft should be renewed.

(8) Examine the camshaft bearings for scores, pits or evidence of failure. If the bearings have to be renewed it will necessitate the removal of the engine back plate as described in Section A.29. The old bearings can then be withdrawn and new ones installed, using Service Tools 18G 124A, 18G 124C, 18G 124D, 18G 124E, 18G 124F, 18G 124H, 18G 124L.

Oil holes must be lined up carefully and all bearings reamed in line to give ·001 to ·002 in. (·025 to ·051 mm.) clearance in each, using Service Tools 18G 123A, 18G 123C, 18G 123D, 18G 123E, 18G 123F, 18G 123AB, 18G 123R, 18G 123T, 18G 123AA, 18G 123L.

(9) Inspect the tappet cam contacting surfaces for wear. New tappets should be installed wherever evidence of unusual wear is found.

(10) The installation of the camshaft and tappets is a reversal of the procedure **"Removal"**. Lubricate the camshaft journals with engine oil.

Section A.28

FLYWHEEL AND ENGINE REAR PLATE

To Remove

The flywheel complete with starter ring is secured to the flange on the rear of the crankshaft by four set bolts, which are locked in position by three lockplates. The engine rear plate is secured to the crankcase by set bolts and lockwashers. To remove the flywheel and rear plate, after the engine is removed from the vehicle, proceed as follows :—

(1) Remove the gearbox from the engine (see Section F),

(2) Remove the clutch (see Section E).

(3) Knock back the tabs of the lockplates, unscrew the bolts and withdraw the flywheel.

(4) Unscrew the set bolts and withdraw the engine rear plate. Note the cork sealing strip behind the engine rear plate under the crankshaft flange.

(5) Examine the flywheel teeth and friction face for excessive wear. If the teeth on the starter ring are damaged or badly worn, a replacement flywheel and ring may be fitted.

NOTE: **For instructions on fitting a flywheel starter ring only see Section A.31.**

(6) Examine the engine rear plate for distortion and damage and clean the joint faces of the plate and crankcase and check for scores.

To Install

(1) Refit the engine rear plate to the crankcase, using a new joint washer. Tighten the securing bolts evenly and firmly.

(2) Place the flywheel over the flange and flange bolts of the crankshaft so that the timing mark "1" is at T.D.C. when the first throw of the crankshaft is at T.D.C. The joint faces should be perfectly clean. Fit the lockplates and nuts on the bolts and tighten them in diagonal sequence.

Fig. A.25. Flywheel T.D.C. mark.

Section A.29

CRANKSHAFT AND MAIN BEARINGS

To Remove

The forged-steel crankshaft is statically and dynamically balanced and is supported in the crankcase by three renewable main bearings of the sintered copper and lead steel-backed type. Crankshaft end float is controlled by thrust washers fitted on both sides of the centre main bearing.

(1) Remove the engine from the vehicle (see Section A.4) and place it upside-down in a dismantling fixture.

(2) Remove the sump and oil strainer (see Section A.7).

(3) Remove the timing chain (see Section A.25).

(4) Remove the flywheel and engine rear plate (see Section A.28).

(5) Check the crankshaft end float to determine whether the renewal of the thrust washers is necessary.

(6) Remove the connecting rod bearing caps and shells, keeping the shells with their respective caps for correct replacement, and release the connecting rods from the crankshaft. Remove the sparking plugs from the cylinder head to facilitate the turning of the crankshaft.

(7) Withdraw the main bearing caps complete with bearing bottom shells. Caps and both bearing half-shells should be kept together. The use of Service Tool 18G 42 and adaptor 18G 42B will assist in the removal of the

bearing caps. Remove the screwed plug from the rear bearing cap oil return pipe and withdraw the pipe in order to use the extractor. Note that each main bearing is stamped with a common number, which is also stamped on the centre web of the crankcase near the main bearing. The bottom halves of the two thrust washers will be removed with the centre main bearing cap.

(8) Remove the crankshaft, the two remaining halves of the thrust washers and the top half-shells of the main bearings from the crankcase.

(9) Inspect the crankshaft main journals and crankpins for wear, scores, scratches and ovality. If necessary the crankshaft may be re-ground to the minimum limits shown under "General Data". Main bearings for re-ground crankshafts are available in sizes shown under "General Data".

(10) Clean the crankshaft thoroughly, ensuring that the connecting oilways between the journals and crankpins are perfectly clear. They can be cleaned out by applying a pressure gun containing petrol or paraffin. When clean, inject a thin oil in the same manner.

(11) Thoroughly clean the bearing shells, caps and housings above the crankshaft.

(12) Examine the bearing shells for wear and pitting, and look for evidence of breaking away or picking-up. Renew the shells if necessary.

(13) Bearings are pre-finished with the correct diametrical clearance, and do not require bedding in. New bearings should be marked to match up with the marking on the cap, and **on no account should they be filed to take up wear or to reduce running clearance.**

(14) Check the thrust washers for wear on their bearing surfaces, and renew if necessary to obtain the correct end float.

To Install

The installation of the crankshaft and main bearings is a reversal of the procedure **"To Remove"**, noting the following points :—

(1) Ensure that the thrust washers are replaced the correct way round and locate the bottom half tab in the slot in the bearing cap.

(2) The bearing shells are notched to fit the recesses machined in the housing and cap.

(3) In the case of the front and rear main bearing caps, install new cork or felt sealing strips.

(4) The rear main bearing cap horizontal joint surfaces should be thoroughly cleaned and lightly covered with Hylomar jointing compound before the cap is fitted to the cylinder block. This will ensure a perfect oil joint when the cap is bolted down to the block.

(5) Lubricate the bearings freely with engine oil.

(6) Tighten the main bearing nuts, see "General Data" for torque spanner settings.

Section A.30

CYLINDER BLOCK

To Remove and Dismantle

(1) Remove and dismantle the engine (see Section A.4).

(2) Remove all studs, unions and screwed plugs, etc., if necessary.

(3) If an expansion plug has blown, drill a hole in its centre and lever it out with a screwdriver.

To View and Overhanl

(1) Scrape as much sediment as possible from the water passages and thoroughly swill out with a water hose.

(2) Clean all gasket surfaces.

(3) Inspect for cracks and scores on gasket surfaces.

(4) It may be advisable to remove the ridge above the ring travel at the top of the cylinder bores before checking the fit of the pistons.

(5) Wipe the cylinder bores clean and examine them for scores, out-of-round and taper. If the cylinders are found to be out-of-round or excessively tapered when measured with a dial test indicator, they should be reconditioned.

(6) If cylinder reconditioning is required, determine accurately the amount of material to be removed (refer to "General Data" concerning oversize pistons available).

(7) Make sure that all traces of abrasives are cleaned from all parts of the cylinder block after the cylinder reconditioning operation is completed.

(8) Check the camshaft bearings (see Section A.27).

To Reassemble and Install

(1) Install all studs, unions and screwed plugs, etc.

(2) When installing new expansion plugs, coat the edge of the plug with a sealing compound and insert the plug with the "bulge" on the outside. A carefully aimed blow at the centre of the plug with a small hammer direct or with a blunt punch will expand the plug sufficiently to make a watertight joint. If too heavy a blow is used, the plug will be useless and must be replaced by another new one.

(3) Reassemble, install and test the engine (see Section A.4).

Engine Type	Liner Part No.	Machine Bores of Cylinder Block to this Dimension before fitting Liner	Outside Diameter of Liner	Interference fit of Liner in Cylinder Block Bore	Machine Liner Bore to this Dimension after fitting
"C"	AEC 607	3·301 to 3·303 in. 83·845 to 83·896 mm.	3·305 to 3·307 in. 83·947 to 83·997 mm.	·002 to ·006 in. ·051 to ·152 mm.	3·1245 to 3·126 in. 79·362 to 79·4 mm.

Fig. A.26
Cylinder bore and liner dimensions.

Section A.31

FITTING FLYWHEEL STARTER RINGS

To remove the old starter ring from the flywheel flange split the ring gear with a cold chisel taking care not to damage the flywheel. Make certain that the bore of the new ring and its mating surface on the flywheel are free from burrs and are perfectly clean.

To fit the new ring it must be heated to a temperature of 300° to 400°C. (575° to 752°F.) indicated by a light blue surface colour. If this temperature is exceeded the temper of the teeth will be affected. The use of a thermostatically controlled furnace is recommended. Place the heated ring on the flywheel with the lead of the ring teeth next to the register on the flywheel. The expansion will allow the ring to be fitted without force by pressing or tapping it lightly until the ring is hard against its register.

This operation should be followed by natural cooling when the "shrink fit" will be permanently established and no further treatment required.

Section A.32

FITTING CYLINDER LINERS

Should the condition of the cylinder bores be such that they cannot be cleaned up to accept standard oversize pistons, dry cylinder liners can be fitted. This operation may be carried out by the use of specialized proprietary equipment, or with a power press using pilot adaptors to the dimensions shown in Fig. A.27. The press must be capable of 3 tons (3048 kg.) pressure to fit new liners and 5 to 8 tons (5080 to 8128 kg.) to remove old liners.

Remove the engine from the vehicle as detailed in Section A.4. Dismantle the engine and remove the cylinder head studs. If liners have not previously been fitted the bores must be machined and honed to the dimensions given in Fig. A.26.

To Remove Worn Liners

Place the cylinder block face downwards on suitable wooden supports on the bed of the press, making sure that there is sufficient space between the block and the bed of the press to allow the worn liner to pass down. Insert the pilot in the bottom of the liner and carefully press the liner from the bore.

To Press in New Liners

Thoroughly clean the inside of the bores and the outside of the liners. Stand the cylinder block upright on the bed of the press, insert the pilot guide in the top of the liner and position the liner with its chamfered end in the top of the bore. Make certain that the liner is square with the top of the block and that the ram or the press is over the centre of the pilot. Press the liner into the bore.

Each liner must be machined to the dimensions given in Fig. A.26 after pressing into position.

Section A.33

MODIFIED CONNECTING RODS

Connecting rods with fully floating gudgeon pins were fitted from Engine No. 40501. They are interchangeable only in sets of six. The dismantling procedure is as follows.

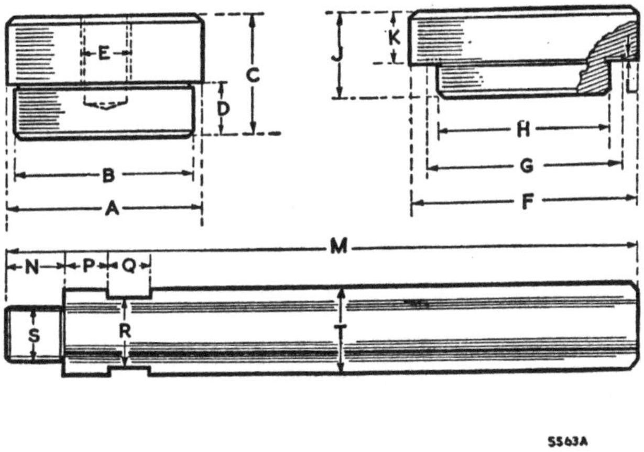

Fig. A.27

Cylinder liner pilots should be made to the above dimensions from case hardening steel, and case hardened. The pilot extension should be made from 55 ton hardening and tempering steel and hardened in oil at a temperature of 550·C. (1020·F.).

Pressing-out Pilot

A. 3$\frac{17}{64}$ + ·005 in. (82·94 + ·127 mm.
— ·000 in. — ·000 mm.).
B. 3·112 + ·000 in. (79·04 + ·000 mm.
— ·005 in. — ·127 mm.).
C. 1$\frac{3}{4}$ in. (44·45 mm.).
D. $\frac{3}{4}$ in. (19·05 mm.).
E. $\frac{3}{4}$ in B.S.W. Thread.

Pressing-in Pilot

F. 3$\frac{11}{16}$ in. (93·66 mm.).
G. 3$\frac{11}{32}$ in. (84·93 mm.).
H. 3·095 + ·000 in. (78·61 + ·000 mm.
— ·005 in. — ·127 mm.).
J. 1$\frac{1}{4}$ in. (31·75 mm.).
K. $\frac{3}{4}$ in. (19·05 mm.).
L. ·015 in. (·38 mm.).

Pilot Extensions

M. 14$\frac{1}{2}$ in. (36·83 cm.).
N. $\frac{7}{8}$ in. (22·22 mm.).
P. $\frac{5}{8}$ in. (15·87 mm.).
Q. $\frac{5}{8}$ in. (15·87 mm.).
R. 1 in. (25·4 mm.) flats.
S. $\frac{3}{4}$ B.S.W. Thread.
T. 1$\frac{1}{4}$ in. (31·75 mm.).

Remove the two circlips securing each gudgeon pin in its piston and press the gudgeon pin out. Mark the gudgeon pins and pistons for reassembly to their original positions and to their original connecting rods.

Check the gudgeon pin and the connecting rod little-end bearings for wear, with the dimension given in the General Data section. If the bush is worn it should be removed and a new bush fitted. A light press is most suitable for this operation.

When pressing in the new bush ensure that the oil hole in the bush is in line with the oil hole in the connecting rod.

Replacement bushes must be finish reamed to size after pressing into the connecting rod (see the General Data section for the correct dimensions). The piston gudgeon pin bosses must not be reamed as oversize gudgeon pins are not supplied.

Assemble the pistons to the connecting rods by inserting the gudgeon pins, which should be a hard hand-push fit at a room temperature of 20°C. (68°F.). Secure each gudgeon pin in position with the two circlips, ensuring that they fit well into their grooves.

IMPORTANT.—When assembling the piston to the connecting rod make certain that the slot in the piston will be on the opposite side to the camshaft when the assembly is fitted to the cylinder block.

Section A.34

6 PORT CYLINDER HEAD

The Austin-Healey C-Series engine has been modified by the introduction of an entirely new cylinder head, together with a detachable induction manifold.

Two S.U. HD6 carburetters with 1$\frac{3}{4}$ inch throats replace the two S.U. H4 type used in the normal "C" type engine. Details of the new carburetters are given in Section D.12.

New features of the head include modified combustion chambers, larger inlet and exhaust valve head diameters and re-shaped inlet and exhaust ports to provide an even more efficient gas-flow.

The introduction of solid skirt, flat topped pistons has increased the compression ratio from 8·25:1 to 8·71:.

Dimensional and power changes bought about by these modifications are given in the General Data.

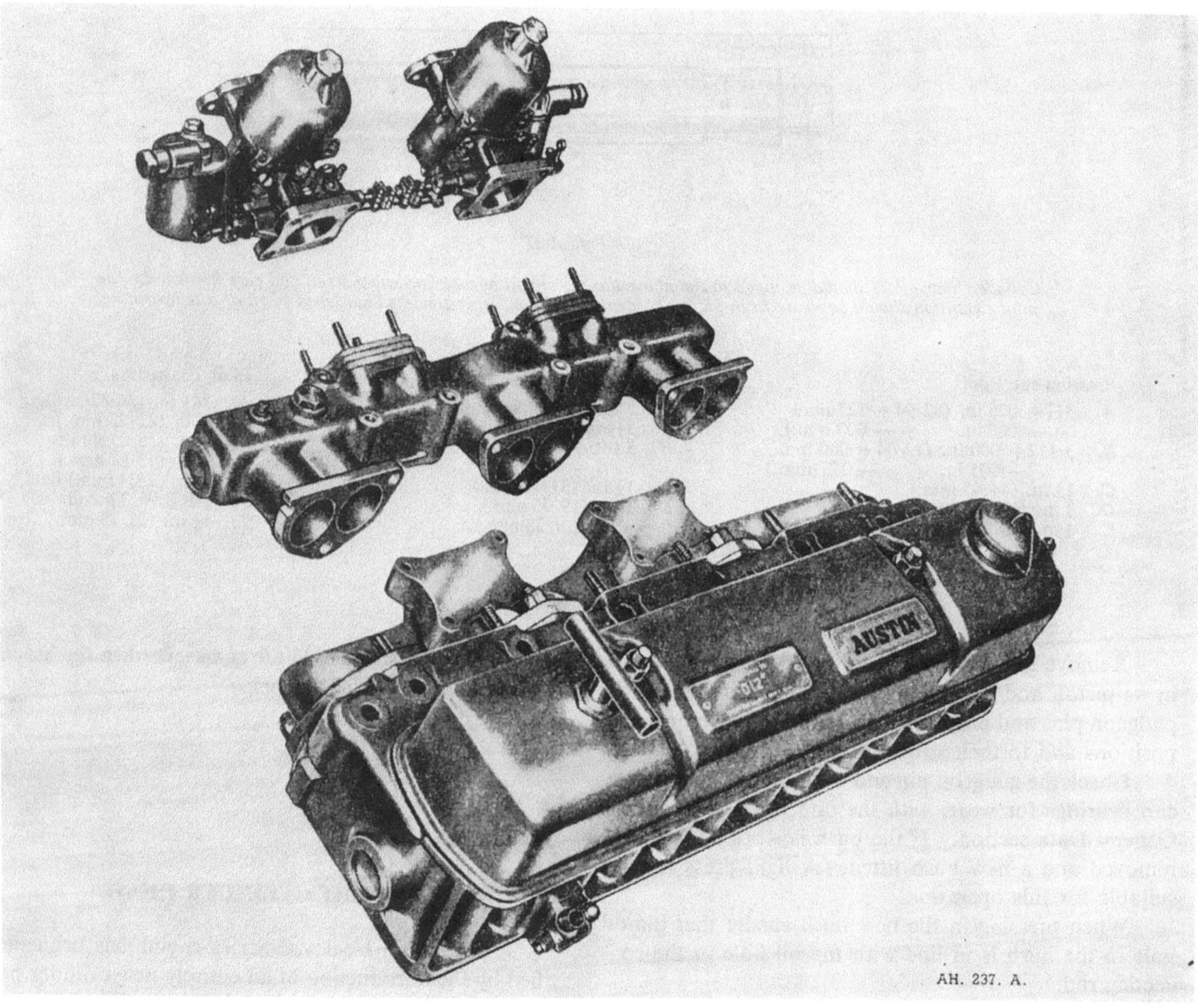

Fig. A.28 View of induction manifold showing the six inlet ports.

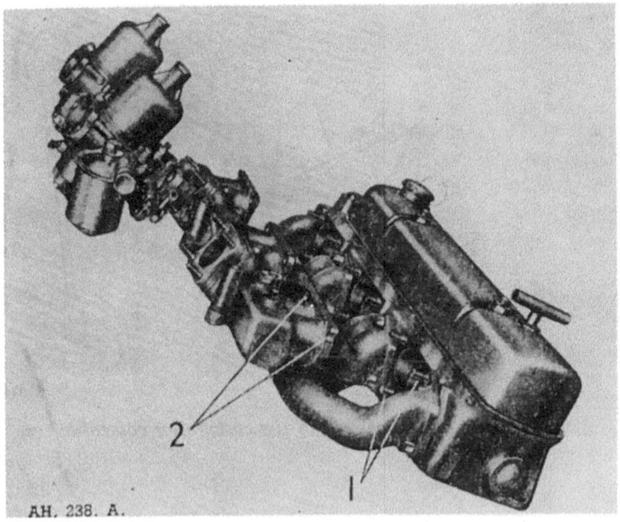

Fig. A.29 Depicting the fitting of the manifold on the cylinder head.

1. *Induction manifold mounting studs.*
2. *Locating holes on the induction manifold.*

Section A.35

6 PORT MANIFOLD

Removal and Replacement.

(1) Detach each air cleaner from the carburetters by unscrewing the four setpins and releasing the breather pipe from the rear air cleaner.

(2) Remove the carburetters as detailed in Section D.13.

(3) Withdraw the heat shield off the induction manifold flange studs.

(4) Unscrew the nine nuts securing the induction manifold to the cylinder head and remove the manifold.

(5) Unscrew and remove the six brass nuts and plain washers which secure the exhaust manifold to the down pipes.

(6) The remaining eight manifold securing nuts should now be removed and the exhaust manifold detached.

(7) Reassembly is the reversal of the above procedure but always use a new joint washer for the manifolds to ensure an air tight joint.

Section A.36

TIMING CHAIN TENSIONER OVERHAUL

If the rubber slipper head of the timing chain tensioner is found to be badly worn, then either the complete adjuster, or the slipper head and cylinder assembly must be renewed.

Remove the tensioner and dismantle as described in Section A.26. Check the bore in the adjuster body for ovality. If the degree of ovality when measured on the diameters at the mouth of the bore is greater than ·003 in. (·076 mm.) the complete adjuster assembly must be renewed. If the bore is within the acceptable limits, then fit a new slipper head and cylinder assembly in the existing body. Ensure that the bore of the body, and all components parts are scrupulously clean before reassembling the adjuster (Refer to Section A.26).

Refit the adjuster to the engine (Section A.26) and check that the slipper head does not bind on the back plate when it is moved in the body.

Section A.37

THROTTLE CONTROL LINKAGE

To prevent the throttle linkage being strained and the throttle levers working loose, the linkage may be adjusted to allow the toe board to act as a positive

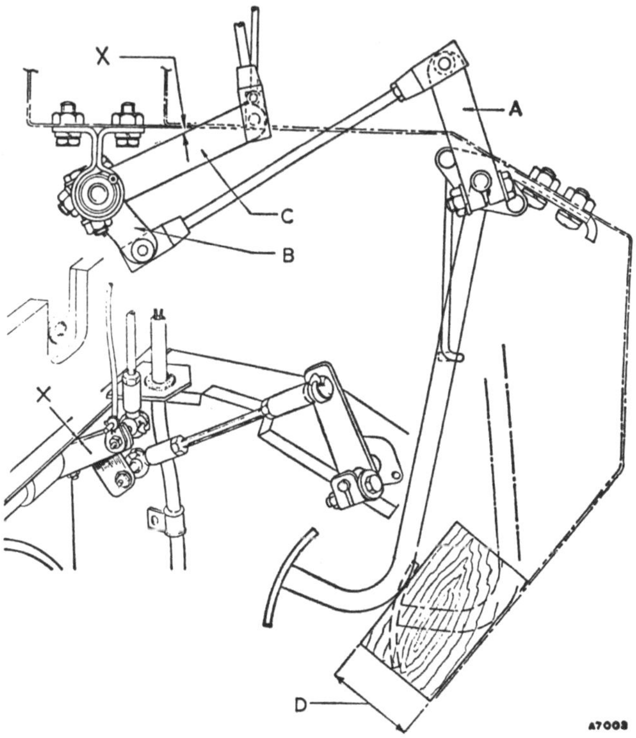

Fig. A.30 Adjustment of the throttle control linkage.

stop to the accelerator pedal when the throttles are fully open.

Slacken the pinch bolt on lever "A" (Fig. A.30). The illustration shows the linkage layout for L.H.D. cars. On R.H.D. cars lever "B" is on the right-hand side of the accelerator relay shaft. Place a wooden block "D" 2½ in. (63·5 mm.) thick between the pedal and the toe board. Push the pedal down to retain the block against the toe board. Adjust lever "A" in relation to the pedal cross-shaft to obtain a clearance of $\frac{1}{16}$ in. (1·59 mm.) at point "X" between lever "C" and the body flange (Fig. A.30). Tighten the pinch bolt on lever "A".

The carburetter control levers must then be set as follows.

Slacken the pinch bolt on levers "A" and "B" (see Fig. A.31).

Set lever "B" at approximately 45° as shown and tighten the pinch bolt, ensuring at the same time that the throttles are not being held open by the idling adjustment screws.

Adjust the length of rod "C" to bring lever "A" parallel with lever "B" (Fig. A.31).

With the pinch bolt of lever "A" still slack, press rod "D" downwards ⅛ in. (3·2 mm.) to tension the pedal return spring slightly, and then tighten the pinch bolt on lever "A".

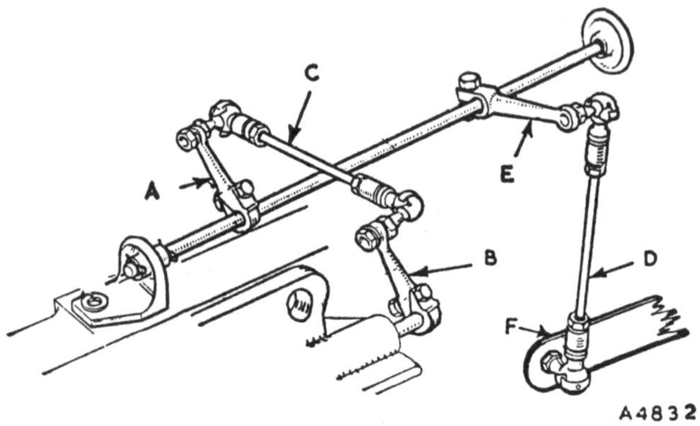

Fig. A.31 Adjustment of the carburetter control levers.

Depress the accelerator pedal fully and check the travel of lever "E". This must be such that it is at least 20° short of the vertical position when full throttle is reached on the carburetters. Adjust rod "D" as necessary to achieve this position.

When the accelerator pedal is fully depressed check that the carburetter throttles are being fully opened.

NOTE.—After the linkage has been set it will be necessary to check, and adjust if necessary, the throttle switch operation on cars fitted with overdrive (see Section G.12, "Throttle switch adjustment").

SECTION AA

ENGINE

SERIES BN6

Section No. AA1. Accelerator shaft bushes

NOTE

The engine of the BN6 is the same as that fitted to later BN4 cars, being fitted with the 6 port cylinder head. Reference should, therefore, be made to Section A and particular attention given to Sections A.35 and A.36.

Section AA.1

ACCELERATOR SHAFT BUSHES

When the felt bushes on the accelerator pedal shaft and the cross shaft have to be renewed, the following standard bushes are to be fitted:

Part No. AHB8748 Bush, Nylon, accelerator pedal shaft—engine side.

Part No. AHB8950 Bush, Compo, accelerator pedal shaft—pedal side.

Part No. AHB8950 Bush, Compo, cross shaft.

SECTION AAA

ENGINE

Mk. I and II (SERIES BN7 and BT7)
AND Mk. II and III (Series BJ7 and BJ8)

NOTE

The engine is the same as that fitted to Series BN6 cars except for the information contained in the above Sections.

Section AAA.1

GEAR TYPE OIL PUMP

The oil pump fitted to early engines is of the Hobourn Eaton rotary vane type, the service procedure for which is detailed in Section A.8. Later engines, however, are equipped with a gear type pump and instructions for servicing this type of pump are given below.

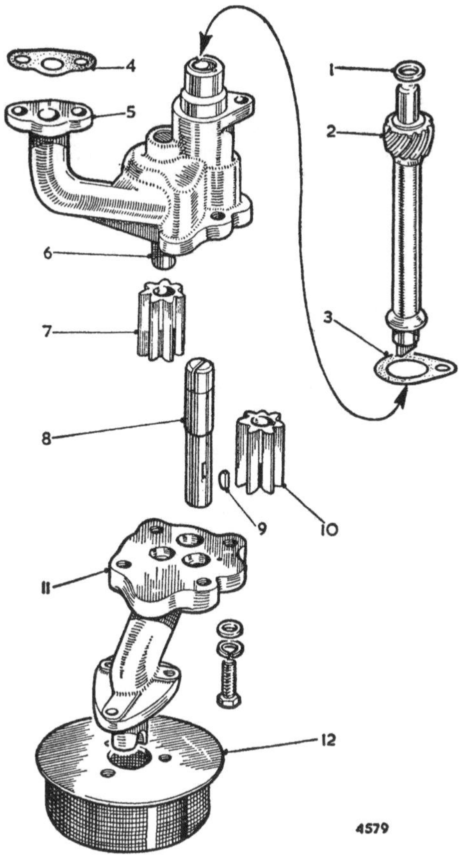

Fig. AAA.1 Components of the oil pump.

1. *Thrust washer.*
2. *Drive spindle.*
3. *Joint washer.*
4. *Joint washer.*
5. *Pump body.*
6. *Driven gear spindle.*
7. *Driven gear.*
8. *Drive gear spindle.*
9. *Key.*
10. *Drive gear.*
11. *Pick up.*
12. *Pick up strainer.*

Removing the Oil Pump

(1) Remove the sump and pick up strainer.

(2) Take off the nuts and spring washers from the three studs which secure the oil pump assembly to the crankcase, when the pump can be withdrawn.
If the pump is removed with the engine still in the car, the drive shaft will be free to disengage from the camshaft, and care must be taken to prevent it falling out. Note also the thrust washer fitted on the drive shaft above the gear.

Dismantling the Oil Pump

(1) Mark the flange and pump body to assist reassembly and remove the four retaining bolts with their spring and plain washers.

(2) Separate the pick-up from the pump body, withdraw the drive gear and spindle and the driven gear from its fixed spindle.
Thoroughly clean all parts in fuel or paraffin and dry off.

(3) Replace the gears and check for wear.

(a) The radial clearance between the gears and the pump body should not exceed ·00125 in. to ·0025 in. (·032 mm. to ·063 mm.).

(b) Check the clearance between the gears and the end cover by placing a straight edge across the pump body and checking with a feeler gauge. End float should be between ·0005 in. and ·002 in. (·013 mm. and ·051 mm.).

Renew worn parts as necessary. The pump driving gear is a press fit on its shaft and is keyed in position. The spindle should protrude ·312 in. (7·94 mm.) from the gear face.

Replacing the Oil Pump

Care should be taken to see that the abutting faces of the cylinder block and the pump are clean before replacing and that the joint washers are in good condition. Insert the pump from below ascertaining that the driving gear is in mesh with the gear on the camshaft. Secure the pump with the three nuts and spring washers. Replace the pick-up strainer and sump.

Section AAA.2

FITTING CYLINDER LINERS

The dimensions required for the fitting of cylinder liners to the 2·9 litre "C" type engine are given below.

These dimensions should be used in conjunction with the removal and fitting instructions given in Section A.33.

Pressing out Pilot

A. $3\frac{7}{16}$ + ·005 in. (87·31 + ·127 mm.
— ·000 in. — ·000 mm.).
B. 3·270 + ·000 in. (83·06 + ·000 mm.)
— ·005 in. — ·127 mm.
C. $1\frac{3}{4}$ in. (44·45 mm.).
D. $\frac{3}{4}$ in. (19·05 mm.).
E. $\frac{3}{4}$ in. B.S.W. Thread.

Pressing in Pilot

F. $3\frac{7}{8}$ in. (98·43 mm.).
G. $3\frac{1}{2}$ in. (88·9 mm.).
H. 3·258 + ·000 in. (82·75 + ·000 mm.).
— ·005 in. — ·127 mm.
J. $1\frac{1}{4}$ in. (31·75 mm.).
K. $\frac{3}{4}$ in. (19·05 mm.).
L. ·015 in. (·38 mm.).

Pilot Extensions

M. $14\frac{1}{2}$ in. (36·83 cm.).
N. $\frac{7}{8}$ in. (22·22 cm.).
P. $\frac{5}{8}$ in. (15·87 mm.).
Q. $\frac{5}{8}$ in. (15·87 mm.).
R. 1 in. (25·4 mm.) flats.
S. $\frac{3}{4}$ B.S.W. Thread
T. $1\frac{1}{4}$ in. (31·75 mm.).

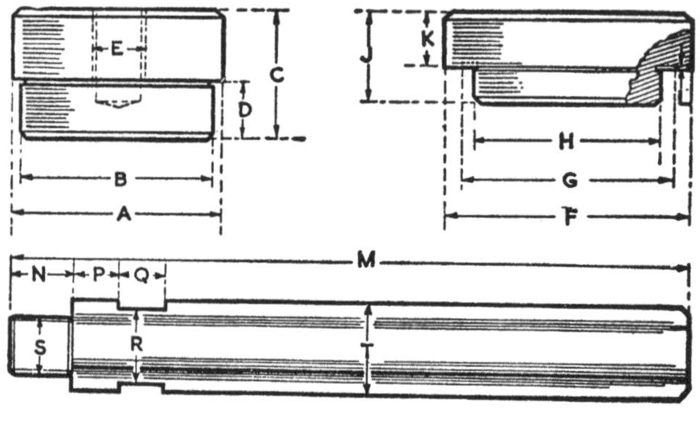

Fig. AAA.2. Cylinder liner pilots should be made to the above dimensions from case hardening steel, and case hardened. The pilot extension should be made from 55-ton hardening and tempering steel and hardened in oil at a temperature of 550° C. (1,020° F.).

Cylinder bore and liner dimensions.

Engine Type	*Liner Part No.*	*Machine Bores of Cylinder Block to this Dimension before fitting Liner*	*Outside Diameter of Liner*	*Interference fit of Liner in Cylinder Block Bore*	*Machine Liner Bore to this Dimension after fitting*
"C"	AEC 879	3·457 to 3·459 in. 87·81 to 87·86 mm.	3·461 to 3·463 in. 87·91 to 87·96 mm.	·002 to ·006 in. ·076 to ·127 mm.	3·2805 to 3·282 in. 83·32 to 83·36 mm.

Section AAA.3

TIMING CHAIN VIBRATION DAMPER

Coinciding with the introduction of the 3000 Mk. II power unit a timing chain vibration damper was fitted in addition to the automatic chain tensioner already in use (see Fig. AAA.3). The damper consists of an angled bracket bolted with a single set screw to the cylinder block and located on the engine front mounting plate by a dowel. An oil resistant rubber pad bonded to the bracket maintains light rubbing contact with the timing chain and dampens chain vibration under light load running conditions.

This modification involved the introduction of a new cylinder block, front mounting plate and gasket.

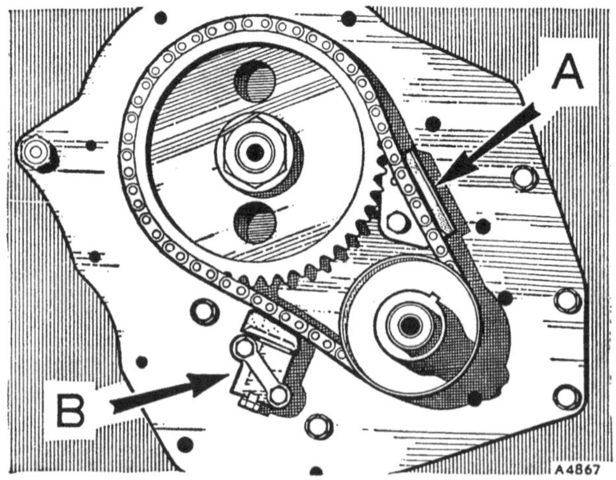

Fig. AAA.3 The engine with the timing cover removed.
A. *Vibration damper* B. *Tensioner.*

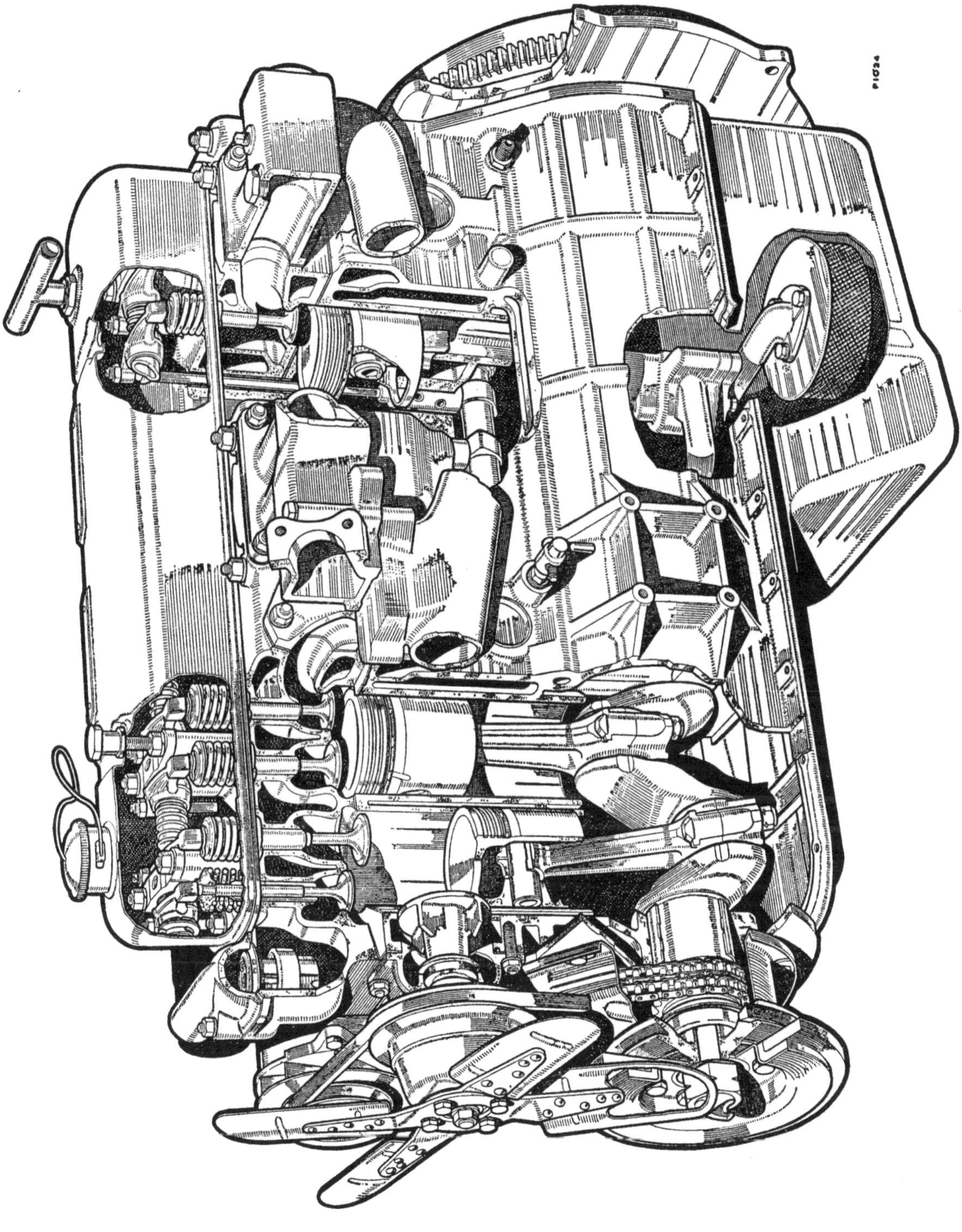

THE AUSTIN-HEALEY 3000 ENGINE

SECTION B

IGNITION

SERIES BN4

Section B.1

DESCRIPTION

The ignition system consists of two circuits—primary and secondary. The primary circuit includes the battery, ignition switch, the primary or low-tension circuit of the coil and the distributor contact breaker and capacitor. The secondary circuit includes the secondary or high-tension circuit of the coil, the distributor rotor and cover segments, the high-tension cables and the sparking plugs.

The ignition coil, which is mounted on the right-hand side of the engine, consists of a soft iron core around which is wound the primary and secondary windings. The coil carries at one end a centre high-tension terminal and two low-tension terminals marked (SW) (switch) and (CB) (contact breaker) respectively.

The ends of the primary winding are connected to the (SW) and (CB) terminals and the secondary winding to the (CB) terminal and the high-tension terminal.

The distributor is mounted on the right-hand side of the engine and is driven by a shaft and helical gear from the camshaft. Automatic timing control of the distributor is controlled by a centrifugal mechanism and a vacuum-operated unit each operating entirely independently of each other. The centrifugal mechanism regulates the ignition advance according to engine speed, while the vacuum control varies the timing according to engine load. The combined effect of the two mechanisms gives added efficiency over the full operating range of the engine. A micrometer adjuster is provided to give a fine timing adjustment to allow for the engine condition and the grade of fuel used.

A keyed moulded rotor with a metal electrode is mounted on top of the cam. Attached to the distributor body above the centrifugal advance mechanism is a contact breaker plate carrying the contact breaker points and a capacitor connected in parallel. A cover is fitted over the distributor body and retained by two spring clips attached to the body.

Inside the cover is a centre electrode and spring-loaded carbon brush which makes contact with the rotor. The brush is of composite construction, the top portion being made of a resistive compound, while the lower portion is made of softer carbon to prevent wear of the rotor electrode. Under no circumstances must a short non-resistive brush be used to replace this long resistive type. A measure of radio interference suppression is given by this brush.

spaced circumferentially around the centre electrode are the sparking plug high-tension cable segments. The distributor is secured in position on the cylinder block by a clamp plate.

The sparking plugs are located on the right-hand side of the engine and have a 14 mm. thread with a $\frac{3}{4}$ in. reach.

When the ignition is switched on, the current from the battery flows through the primary circuit, and a magnetic field is built up around the core of the coil. When the contact breaker points are opened by rotation of the distributor cam, the current flow is interrupted, causing a high voltage to be induced in the secondary winding of the coil by sudden collapse and consequent change in the magnetic field. The high-tension current thus generated in the secondary winding of the coil is conveyed by the coil high-tension cable to the centre terminal of the distributor cover. From here the current passes through the carbon brush to the rotor, where the high-tension current passes along the rotor electrode and is distributed to the segments and thence to the sparking plugs via the high-tension cables.

Section B.2

ADJUSTMENTS IN THE VEHICLE

The purpose of the following adjustments is to maintain efficient engine performance and economical running.

(1) Adjust the sparking plugs at the recommended intervals as follows :—

The gap of the plug points should be within the limits of ·024 to ·026 in. (·61 to ·66 mm.).

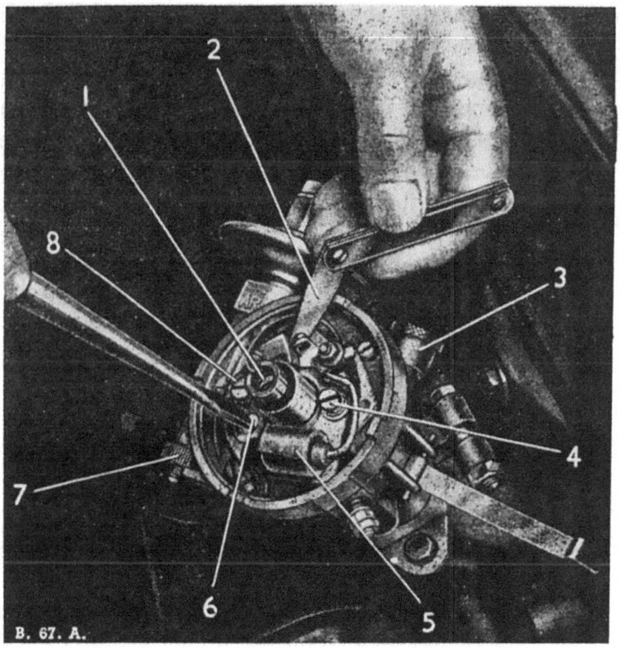

Fig. B.1. Contact breaker adjustment.

1. *Oiling point.*
2. *Feeler gauge.*
3. *Shaft lubricator.*
4. *Contact locking screws.*
5. *Condenser.*
6. *Contact adjusting screw.*
7. *Micrometer adjuster.*
8. *Contact locking screws.*

Fig. B.2. Showing the timing pointer set opposite to the notch in the crankshaft pulley. (Number 1 piston at T.D.C.)

Gap adjustment should be made by bending the side electrode only. Never bend the central electrode. If the plugs are dirty, damaged or excessively burned, see Section B.1.

(2) Adjust the contact breaker points at the recommended intervals as follows :—

Remove the distributor cover and rotor. Rotate the engine with the starting handle until the fibre heel of the rocker is on the peak of one of the cam lobes. The gap of the contact breaker points should be within ·014 to ·016 in. (·36 to ·40 mm.).

Gap adjustment should be made by slackening the fixed contact plate securing screws and moving the plate until the gap gauge is a sliding fit between the two contacts. Tighten the securing screws and recheck the gap. Replace the rotor and cover. If the points are dirty or pitted see Section B.5.

(3) Adjust the ignition timing, if the distributor has been disturbed, as follows :—

Remove the valve rocker cover so that the valve action can be observed. Rotate the engine with the starting handle until No. 1 piston is at the top of its compression stroke (*i.e.* the exhaust valve of No. 6 cylinder is just closing and the inlet valve just opening). Turn the crankshaft until the recess in the crankshaft pulley flange is in line with the pointer on the timing chain cover (see Fig. B.2). If the timing chain cover has been removed, align the bright links on the timing chain with the marked teeth on the camshaft and crankshaft sprockets (see Section A.25) when No. 6 and No. 1 pistons will be at T.D.C. Set the micrometer adjustment on the distributor to its central position. The crankshaft should now be rotated backwards to obtain its correct position before setting the distributor points, this setting is correct for premium grade fuels only. With the cover removed the distributor body must be rotated until the rotor arm is pointing to the position of No. 1 electrode in the cover. With the contact points just opening, tighten the clamp plate bolt.

Finer adjustment can be obtained under road conditions, by means of the micrometer adjustment. Note this adjustment should not be used for initial setting of the ignition; it is only altered if the main setting requires adjustment to meet the characteristics of the grades of petrol being used. There is a considerable amount of latitude for adjustment, but only extremely small movement of the adjustment knob should be made at one time.

Replace the distributor cover and cylinder head cover.

Section B.3

TO TEST IN THE VEHICLE

If the ignition system fails, or misfiring occurs, first make sure that the trouble is not due to defects in the engine, carburetter or fuel supply. Faults should be diagnosed by applying the following tests :—

(1) Examine the high-tension cables, *i.e.* the cables from the coil to the distributor, and from the distributor to the plugs. If the rubber insulation shows signs of deterioration or cracking, the cable should be renewed.

(2) Test the plugs and high-tension cables by removing the plugs in turn and allowing them to rest on the cylinder head or other convenient earthing point, and observing whether a spark occurs at the points when the engine is turned by hand. It should, however, be noted that this is only a rough test, since it is possible that a spark may not take place when the plug is under compression. If necessary, clean and test the plugs, using a plug cleaning and testing machine.

(3) To trace a fault in the low-tension circuit, release the instrument panel from the dash, switch on the ignition, and turn the engine until the distributor contacts are opened. Refer to the wiring diagram, Fig. N.25, and, with the aid of a voltmeter (0 to 20), check the circuit as follows :—

(*a*) **Cable—Battery to starter switch**

Connect the voltmeter between the supply terminal of the starter switch and an earthing point. No reading indicates a faulty cable or loose connection.

(*b*) **Cable (brown)—Starter switch to two-way fuse unit A.1 terminal**
Connect the voltmeter between the fuse unit A.1 terminal and earth. No reading indicates a faulty cable or loose connection.

(*c*) **Control box**
Connect the voltmeter between the control box terminal (A.1) and earth. No reading indicates a faulty control box.

(*d*) **Cable (brown and blue)—Control box to lighting and ignition switch**
Connect the voltmeter between the lighting switch terminal (A) and earth. No reading indicates a faulty cable or loose connection.

(*e*) **Ignition switch**
Connect the voltmeter between the ignition switch (white cable terminal) and earth. No reading indicates a faulty ignition switch.

(*f*) **Cable (white)—Ignition switch to fuse unit A.3 terminal**
Connect the voltmeter between the fuse unit A.3 terminal and earth. No reading indicates a faulty cable or loose connection.

(*g*) **Cable (white)—Fuse unit A.3 terminal to ignition coil**
Connect the voltmeter between the ignition coil terminal (SW) and earth. No reading indicates a faulty cable or loose connection.

(*h*) **Ignition coil**
Connect the voltmeter between the ignition coil terminal (CB) and earth. No reading indicates a faulty ignition coil.

(*i*) **Cable (white and black)—Ignition coil to distributor**
Connect the voltmeter between the distributor terminal and earth. No reading indicates a faulty cable or loose connection.

(*j*) **Distributor**
Connect the voltmeter across the distributor contacts. If no reading is given, remove the capacitor and test again. If a reading is given, the capacitor is faulty.

(5) If, after carrying out the foregoing tests, the fault has not been located, remove the high-tension cable from the centre terminal of the distributor. Switch on the ignition and crank the engine until the contacts close. Flick the contact breaker lever open while the high-tension cable from the ignition coil is held about $\frac{3}{16}$ in. (5 mm.) away from the cylinder block. If the ignition equipment is in order a strong spark should be obtained. If no spark is given it indicates a faulty ignition coil.

Section B.4

IGNITION COIL

To remove

(1) Disconnect the high-tension cable from the coil centre terminal.

(2) Disconnect the low-tension cables from the (SW) and (CB) terminals of the coil.

(3) Unscrew the bolts fastening the coil to the generator strap, and remove the coil.

To install

The installation of the ignition coil is a reversal of the procedure **"To remove"**.

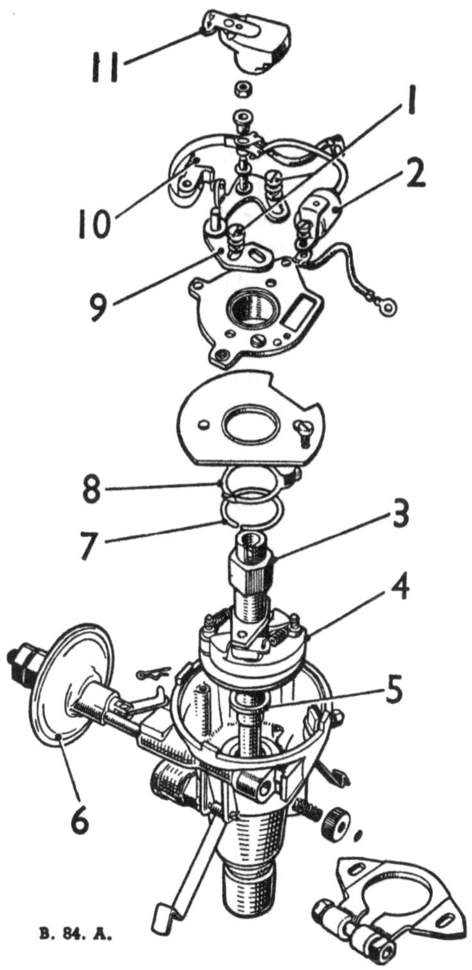

Fig. B.3. Distributor exploded.

1. *Screws for contact plate.*
2. *Condenser.*
3. *Cam.*
4. *Automatic timing contact.*
5. *Distance collar.*
6. *Vacuum control.*
7. *Felt ring.*
8. *Spring.*
9. *Fixed contact plate.*
10. *Moving contact.*
11. *Rotor.*

Section B.5

DISTRIBUTOR

To remove

(1) Disconnect the high-tension cables from the sparking plugs.
(2) Disconnect the high-tension cable from the centre terminal of the ignition coil.
(3) Disconnect the low-tension cable from the terminal on the side of the distributor.
(4) Disconnect the vacuum control pipe at the union.
(5) Unscrew and remove the tachometer drive and tachometer drive oil feed pipe.
(6) Remove the two set bolts attaching the clamp plate to the tachometer drive housing and withdraw the distributor from the tachometer housing. **Do not loosen the clamp plate.**

To dismantle

(1) Spring back the two securing clips and remove the distributor cover.
(2) Unscrew the terminal screws from the distributor cover and withdraw the high-tension cables.
(3) Remove the rotor.
(4) Take out the split pin securing the vacuum link to the sliding contact breaker plate and remove the two screws at the edge of the contact breaker base.
(5) Slacken the two nuts on the low tension terminal and pull the connection on the contact breaker away from the terminal block. Lift out the contact breaker assembly.
(6) Remove the nut, insulating piece and connections from the pillar on which the contact breaker spring is anchored, and lift off the contact breaker lever and the insulating washers beneath it.
(7) Remove the two screws securing the fixed contact plate, together with their spring and plain washers and take off the plate.
(8) Withdraw the single screw securing the capacitor and contact breaker earthing lead. The contact breaker base assembly can be dismantled by removing the circlip and star washer located under the lower plate.
(9) Remove the circlip on the end of the micrometer timing screw and turn the micrometer nut until the screw and vacuum assembly are freed, take care not to loosen the ratchet and coil springs located under the micrometer nut.
(10) Tap out the parallel pin securing the driving dog and tachometer gear to the lower end of the spindle. The driving dog and tachometer gear can then be removed.
(11) Unscrew and remove the lubricator from the distributor body together with its spring and felt washer.
(12) The distributor shaft, cam and centrifugal timing control can be pressed upwards through the distributor body.
(13) Remove the cam fixing screw from the top of the driving shaft and withdraw the cam and centrifugal timing control.

Note: A distance collar is fitted on the shaft underneath the action plate.

(14) Clean the distributor cover and examine it for signs of cracks and evidence of "tracking", *i.e.* a conducting path may have formed between adjacent segments. This is indicated by a thin black line between the segments; when this has occurred the cover should be renewed.
(15) Ensure that the carbon brush moves freely in the distributor cover.
(16) Examine the attachment of the metal electrode to the rotor moulding. If slack or abnormally burned, renew the rotor.
(17) The contact face of the contact breaker points should present a clean, greyish, frosted appearance. If burned or blackened, renew the contact set or polish the contact face of each point with a fine oil-stone, working with a rotary motion. Care should be taken to maintain the faces of the points flat and square, so that when reassembled full contact is obtained. Clean the points thoroughly in petrol.
(18) Check that the movable contact arm is free on its pivot without slackness.

Fig. B.4. Distributor drive, showing the slot in the "Twenty-to-two" position.

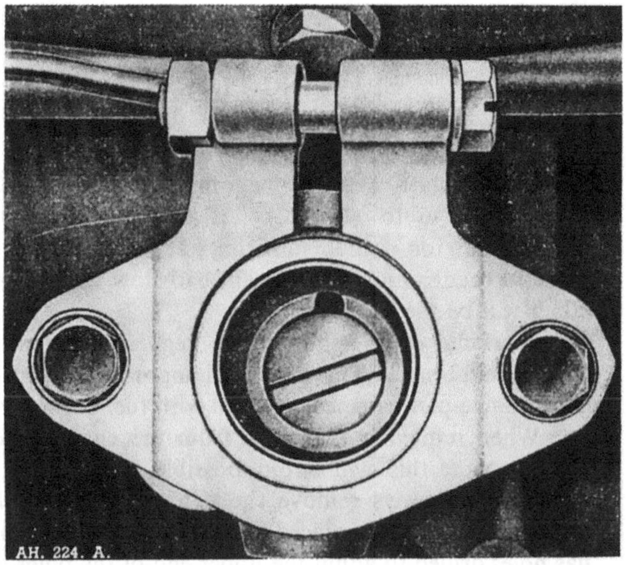

Fig. B.5. Tachometer spindle, showing the smaller segment of the offset dog in the downward position.

(19) Check the centrifugal timing control balance weights and pivot pins for wear and renew the cam assembly or weights if necessary.

(20) The cam assembly should be a free sliding fit on the driving shaft. If the clearance is excessive, or the cam face is worn, renew the cam assembly or shaft as necessary.

(21) Check the fit of the shaft in the body bearing bushes. If slack, renew the bushes and shaft, as necessary.

Press out the old bushes. The new bushes should be allowed to stand completely immersed in thin engine oil for twenty-four hours, or alternatively for two hours in oil which has been heated to 100° C., (212° F.) before pressing them into the distributor body.

To reassemble

Reassembly of the distributor is a reversal of the procedure **"To dismantle"**, noting the following points:—

(1) Apply a few drops of engine oil to the centrifugal timing control mechanism and cam bearing.

(2) Lightly smear the cam surface with engine oil.

(3) Apply a drop of engine oil to the top of the pivot on which the moving contact fibre rocker arm works.

(4) Secure the distributor driving dog to the driving spindle with a new peg, ensuring that the small offset of the driving tongue is on the right when the rotor arm driving slot is downwards.

(5) Connect the internal cables as follows:—

(*a*) Red cable, capacitor to contact breaker spring anchor post.

(*b*) Cable, low-tension terminal to contact spring anchor post.

(*c*) Cable, base plate locating screw to capacitor locating screw.

(6) Adjust the contact breaker points to the instructions given in Section B.2.

(7) Reassemble the high-tension cables to the distributor to the instructions given in Section B.7.

(8) Turn the vacuum control adjusting nut to the half-way position when refitting the unit.

To Install

Replacing the distributor to the reverse of the procedure **"To Remove"**, noting the following points :—

(1) Insert the assembled distributor, with its cap removed, into the tachometer housing and turn the rotor arm until the driving slot on the distributor engages with the dog in the housing.

(2) Turn the distributor body to align the clamping plate holes with their respective holes on the tachometer housing and not the set bolts.

(3) Check the contact breaker gap and ignition timing as described in Section B.2.

(4) When replacing the distributor cap, renew the coating of Silicone grease on the lip of the cap and the ends of the leads where they enter the holes in the moulding.

Section B.6

DISTRIBUTOR DRIVING SPINDLE

Removal

(1) Remove the distributor as described in Section B.5.

(2) Release the three setscrews which secure the tachometer housing to the cylinder block and withdraw the housing.

(3) By using a $\frac{5}{16}$ in. U.N.F. bolt approximately $3\frac{1}{4}$ in. long, screwed into the tapped end of the drive spindle, the spindle can be withdrawn.

(4) Examine the drive gear for worn teeth.

To Replace

(1) Remove the valve rocker cover (see Section A).

(2) Crank the engine until No. 1 piston is at the top of its compression stroke (*i.e.* the exhaust valve of No. 6 cylinder is just closing and the inlet valve just opening).

(3) Turn the crankshaft until the recess in the crankshaft pulley flange is in line with the (T.D.C.) indicating pointer on the timing chain cover (see Fig. B.2.).

(4) Screw the $\frac{5}{16}$ in. U.N.F. bolt into the threaded end of the distributor drive and replace the drive in the block so that the centrally cut slot takes up the position shown in Fig. B.4 (*i.e.* "twenty-to-two" position).

(5) Replace the tachometer housing, rotating the external drive dog until it mates up with the slot in the distributor driving spindle. Ascertain that the smaller segment of the offset dog, situated within the tachometer housing is in the downward position.

Note: The internal dog should now be in the "twenty-to-two" position, see Fig. B.5. Secure the tachometer housing to the block with its three setpins.

(6) Replace the distributor following the instructions given in Section B.5.

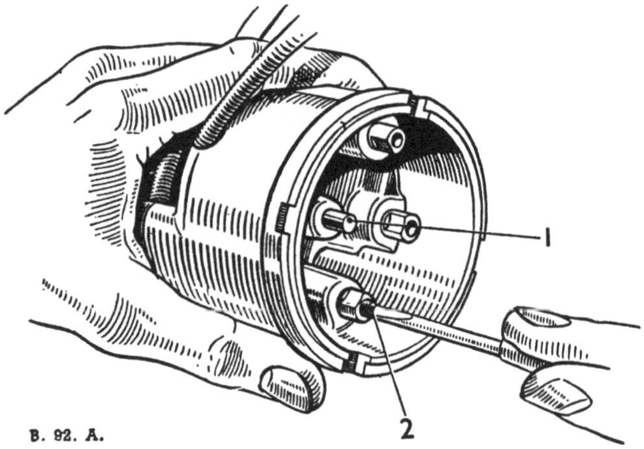

Fig. B.6. Distributor cap.
1. Carbon brush. *2. Screw securing cable.*

Section B.7

HIGH-TENSION CABLES

To remove

(1) Pull the high-tension cable off the sparking plug.
(2) Unscrew the moulded terminal to release the cable from the coil.
(3) Straighten out the bare strands of cable, remove the brass washer and withdraw the cable from the moulded terminal.
(4) Release the screw securing the cable in the distributor cover and withdraw the cable.

To Replace

(1) Thread the cable through the moulded terminal and brass washer and bend back the bare strands of the cable against the brass washer.
(2) To fit the cables to the distributor cap, fill the cable sockets in the cap with Silicone grease, push the cables completely home, and then secure with the pointed screws. Apply a smear of Silicone grease to the lip of the cap.
(3) Install the cables in the coil and distributor cover and onto the sparking plugs in the correct order. The firing order is 1, 5, 3, 6, 2, 4, following round in an anti-clockwise direction.

Section B.8

SPARKING PLUGS

The sparking plug gap (for type of plug see 'General Data') should be maintained at ·024 to ·026 in. (·6 to ·7 mm.). If the gap is allowed to become too wide, misfiring at high speeds is liable to occur; and if too small, bad slow running and idling will be the result.

Sparking plugs should be regularly inspected, cleaned and tested. This is of vital importance to ensure good engine performance, coupled with fuel economy.

When removing the plugs from the engine, use a box spanner, this will avoid possible damage to the insulators. Always remove the copper washers. The plugs should then be placed in a suitable holder which has holes drilled to admit the upper end of the plugs and marked to identify each one with the cylinder from which it was removed.

The plugs should now be carefully examined.

Oil fouling will be indicated by a wet shiny black deposit on the insulator. This condition is usually caused by worn cylinders, pistons or gummed rings. Oil vapour is forced from the crankcase, during the suction stroke of the piston which fouls the plugs.

Petrol fouling will cause a dry, fluffy, black deposit to be apparent on the plugs. This is usually caused by faulty carburation, but a faulty coil or leaking and worn out ignition leads, may have the same effect.

Under the above conditions, if the plugs otherwise appear to be sound, they should be cleaned thoroughly, adjusted, and tested.

When preparing for cleaning, the plug washers should be removed and examined. The condition of these washers is important in that a large proportion of the heat from the plug insulator is dissipated to the cylinder head by them. The washer should therefore be reasonably compressed. A loose plug can be easily overheated, thus shortening plug life. On the other hand, do not over-tighten. All that is needed is a good seal between the cylinder head and the plug. Tightening too much will cause distortion of the washer with the possibility of blow-by which will again lead to overheating and resulting danger. If there is any question of defect, replace with new washers.

The plugs should now be thoroughly cleaned of all carbon deposit, resorting to scraping if necessary, removing as much as possible from the space between the insulator and shell. An oily plug should be washed out with petrol. If a plug cleaning machine is available, 5 to 10 seconds in this will remove all remaining signs of carbon. Remember to thoroughly "blow-out" the plug after treatment under these conditions, in order to remove all traces of abrasive.

After cleaning, thoroughly examine the plug for cracked insulator or worn away insulator nose. Should either of these conditions be apparent a new plug should be installed.

Carbon deposit on the threads of the plugs, should be carefully removed by using a wire brush, or if available a wire buffing wheel. Take care not to damage the electrodes or insulator tip. Neglect of this cleaning operation will lead to tight threads and resultant loss of heat dissipation due to the carbon deposit, thereby causing overheating.

The condition of the electrodes should now be noted and any signs of corrosion removed, if it is felt that the plugs are worthy of further use. This can be carried out with the use of a small file to carefully dress the gap area. The gap should then be reset, to a clearance of ·024 to ·026 in. (·6 to ·7 mm.). When resetting bend the side electrode only.

It is advisable whilst the plugs are under pressure in the testing machine, to apply a spot of oil to the terminal end, to check for air leakage. Excessive leakage here will tend to cause compression loss, rapid deterioration of the electrode and overheating of the electrode tip. The top half of the insulator should also be carefully examined for any paint splashes or accumulation of grime and dust, which should be removed. Should there be any signs of cracks due to faulty use of the spanner, the plug should be renewed. When replacing the plug lead, make sure that it is securely attached.

At the recommended mileage the sparking plugs should be cleaned and tested or renewed.

Remember, plugs in good condition will ensure better fuel consumption and good engine performance.

SECTION BB

IGNITION

Series BN6

NOTE

For details of the ignition system fitted on BN6 cars refer to Section B.

SECTION BBB

IGNITION

Mk. I and II (Series BN7 and BT7)
AND Mk. II and III (Series BJ7 and BJ8)

NOTE

For details of the distributor fitted to Series BN7, BT7, and early BJ7 cars and all other service and repair procedures, see Section B.

THE DISTRIBUTOR COMPONENTS

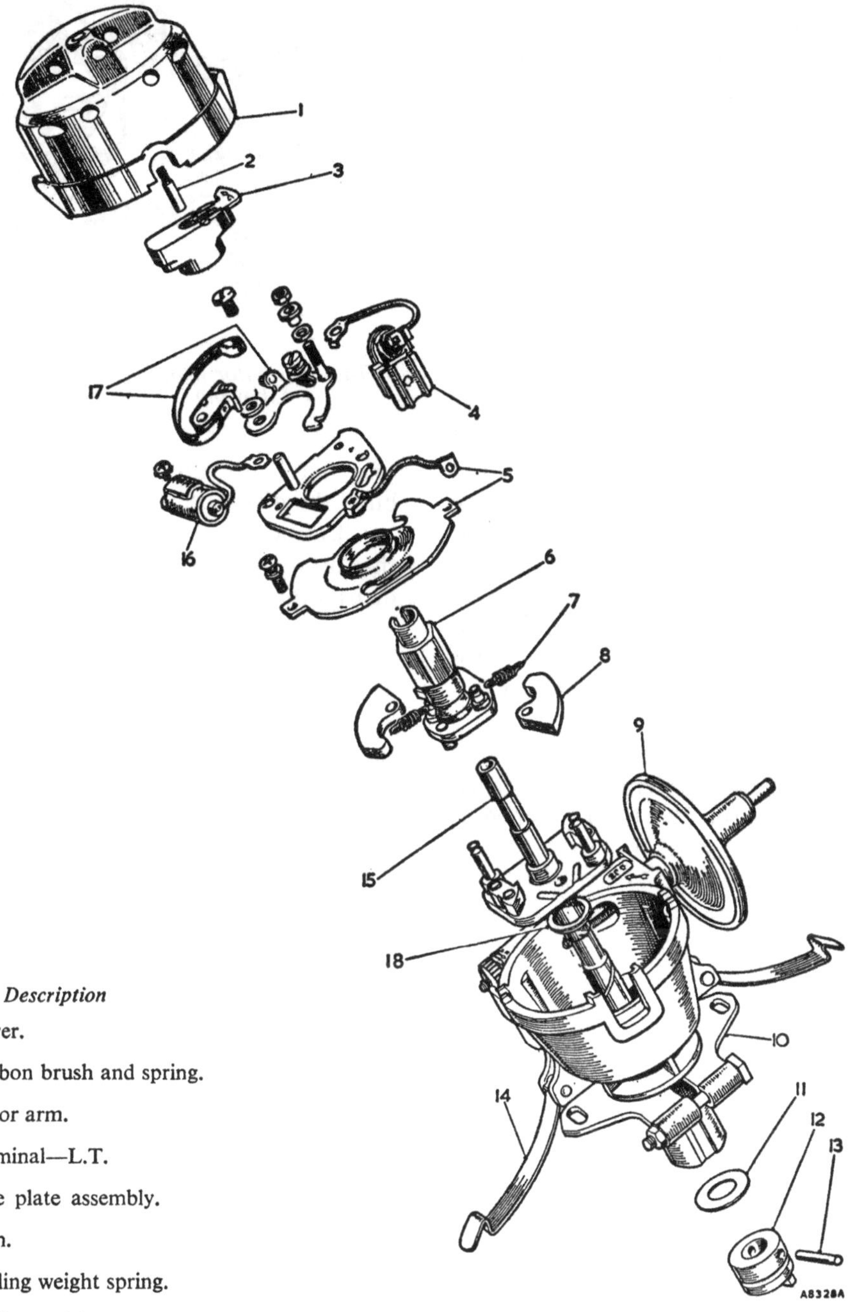

No.	Description
1.	Cover.
2.	Carbon brush and spring.
3.	Rotor arm.
4.	Terminal—L.T.
5.	Base plate assembly.
6.	Cam.
7.	Rolling weight spring.
8.	Rolling weight.
9.	Vacuum unit.
10.	Clamping plate.
11.	Thrust washer.
12.	Driving dog.
13.	Driving dog pin.
14.	Cover retaining clip.
15.	Shaft and action plate.
16.	Condenser.
17.	Contacts.
18.	Distance collar.

Section BBB.1

DISTRIBUTOR

The distributor as fitted to Mk. II and III (Series BJ7 and BJ8) from engine No. 29F/3563, has a pretilted contact breaker unit. The moving contact breaker plate is balanced on two nylon studs and the angle through which the plate may be tilted is controlled by a stud riveted to the moving contact breaker plate located in a slot in the base plate. The moving contact plate is connected to the vacuum unit by a flexible link. The plate carrying the fixed contact is secured by one screw only.

The centrifugal advance mechanism is of the rolling weight type. This consists of a shaft and action plate with two action cams and two spring pillars riveted to the plate. Two rolling weights, each located by a boss on the under side of the cam foot, roll around the action plate cam and thus alter the position of the contact breaker cam relative to the distributor driving shaft when the distributor is rotating within the speed limits of centrifugal operation. The rate of advance is controlled by springs anchored between the pillars on the action plate and two pillars on the cam foot. The maximum amount of advance is governed by the cam foot which abuts one of the spring pillars when maximum advance has been reached.

The 25D6 distributor is manufacted to reduce drive shaft and bearing diameter tolerances. After assembly to the body the sintered iron bush is honed to very fine limits, and it is therefore not possible to rebuild the distributor.

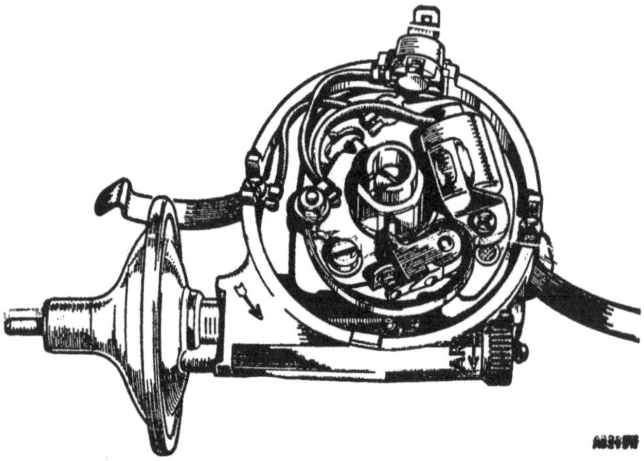

Fig. BBB.1
The distributor with the cover and rotor arm removed showing contact breaker mechanism

Lubrication is as detailed in Section B.3, except that no drive shaft grease cap is fitted to this type of distributor.

For contact breaker adjustment, see Section B.4, but note that there is one fixed contact plate securing screw in place of two screws and an eccentric adjuster.

For dismantling and reassembling the distributor, the sequence given in Sections B.5 and B.6 should be followed, with attention being given to the above-mentioned detailed differences.

SECTION C

COOLING SYSTEM

SERIES BN4

Section C.1

DESCRIPTION

The circulation of the cooling water is effected by a centrifugal pump mounted in front of the cylinder block and driven by a belt from the crankshaft pulley. A thermostat is fitted in the water outlet pipe at the front end of the engine.

When filling or topping-up the radiator, do so when the engine is cold, and if possible use rain water or clean soft water. Fill up to the filler plug orifice.

The capacity of the system is given in "General Data".

Fig. C1. Showing the radiator filler with its cap removed.

Section C.2

ADJUSTMENTS IN VEHICLE

Overheating may be caused by a slack fan belt, excessive carbon deposit in the cylinders, running with the ignition too far retarded, improper carburetter adjustment, a partially choked radiator causing failure of the water to circulate, or loss of water due to leakage or evaporation.

The belt should be just sufficiently tight to prevent slip, yet it should be possible to move it laterally about 1 in. (2·54 cm.). To make an adjustment, slacken the bolts 1 and 4, fig. C.2, which hold the dynamo in position, then raise or lower the dynamo until the desired tension of the belt is obtained. Securely lock the dynamo in position again at securing points 1 and 4, fig. C.2, when the adjustment has been made. It must be understood that there is a correct and incorrect method of fitting fan blades. The blades are not flat, but shaped, and the concave or hollow side should be the leading one, thus, when fitting to an engine the convex or arched side must always face the radiator. This convex side is further easily identified as stiffeners are pressed into the blades; they project on the concave face.

In cases of overheating, the position of the fan blades should at once be examined; make sure, after dismantling, that the fan is fitted the right way round.

Section C.3

RADIATOR

To Remove

(1) Drain the cooling system.
(2) Slacken the hose clip, on the upper water hose, at the thermostat housing and with the aid of a screwdriver ease the pipe off the housing extension.
(3) Take off the radiator bottom hose by releasing the clips on the water pump and the heater outlet pipe.
(4) Disconnect the thermometer element from the radiator header tank.
(5) Take off the six nuts (three on each side) which secure the radiator to the mounting flanges and remove the radiator.
(6) Inspect the radiator core for damage and test it for water leaks. Solder at the points where leakage occurs or renew the core if necessary.
(7) Inspect the flexible mountings for wear.
(8) Inspect the drain tap for leaks and renew it if necessary.
(9) Test the filler cap.
(10) Inspect the hose connections for deterioration and renew them if necessary.

Fig. C.2. Fan belt adjustment.
1 and 2. Dynamo securing bolts.
3. Swinging link. 4. Locknut.

To Replace

Installation is a reversal of the procedure "**To remove**".

Section C.4

THERMOSTAT

To Remove

(1) Drain the cooling system (see Section C.7).
(2) Disconnect the water outlet hose from the outlet elbow.
(3) Remove the two nuts securing the outlet elbow to the cylinder head and lift off the outlet elbow.
(4) Remove the outlet elbow to cylinder head joint.
(5) Lift out the thermostat, fig. C.3.
(6) Test the thermostat opening temperature by immersing it in water at a temperature between 158° and 167°F (70° and 75°C). If the thermostat valve does not start to open, or the valve sticks in the fully open position, the thermostat should be renewed. No attempt should be made to repair the thermostat.
(7) Clean the joint faces at the outlet elbow and at the housing in the cylinder head.

To Replace

The installation of the thermostat is a reversal of the procedure "**To remove**". Fit a new joint gasket between the outlet elbow and the cylinder head. In an emergency, the engine can be run with the thermostat removed.

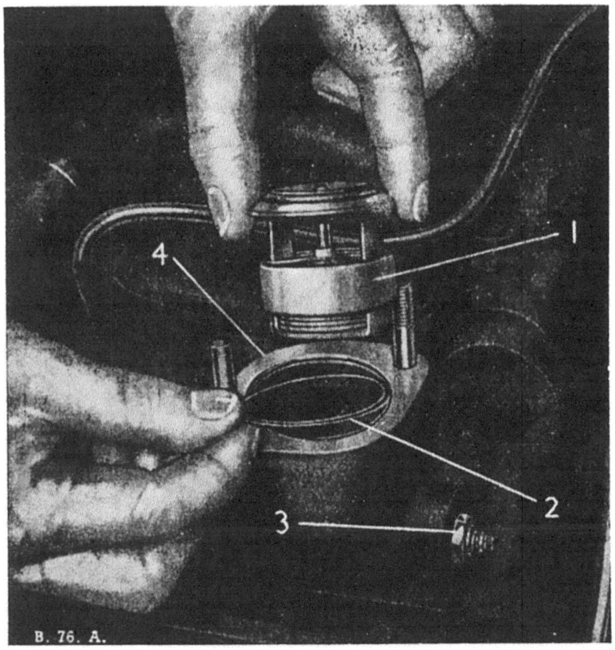

Fig. C.3. Illustrating the removal of the thermostat from its housing
1. Thermostat. 2. Joint washer.
3. Temperature gauge connection. 4. Thermostat housing.

Section C.5

TEMPERATURE GAUGE

A temperature gauge unit, consisting of a thermal element and dial indicator is fitted to the vehicle. The thermal element is held in the radiator header tank by a gland nut. On later engines the thermal capillary is fitted to the engine. (Fig. C3). The dial indicator is situated in the instrument panel and is connected to the element by a capillary tube filled with mercury.

Damage to any of the above mentioned parts will necessitate the renewal of the complete temperature gauge unit.

Section C.6

FAN AND PUMP ASSEMBLY

To Remove

(1) Remove the radiator (see Section C.3).
(2) Remove the dynamo (see Section N).
(3) Unscrew the four set bolts securing the water pump to the crankcase. Withdraw the fan and pump assembly and remove the fan belt.

To Dismantle

(1) Remove the four set bolts and withdraw the fan blades from the pulley.
(2) Remove the nut and washers from the pump spindle and withdraw the pulley and hub which is keyed to the spindle.
(3) Remove the key and tap the spindle rearwards, complete with the impeller.
(4) Remove the front bearing circlip and withdraw the dished grease retainer.
(5) Remove the rubber water seal assembly and its distance-piece from the rear of the housing.
(6) Drift out the front bearing using Tool No. 18G 61; the bearing distance-piece will follow.
(7) To remove the rear bearing, pace the dummy bearing of Tool No. 18G 61 in the pump body, and the drift, which is piloted in the rear bearing, is screwed into it. Tap the complete assembly out through the housing of the front bearing.
(8) Withdraw the felt grease seal and retainer.
(9) Clean all the dismantled pump parts.
(10) Inspect the spindle for wear.
(11) Inspect the seals for damage and wear. It is advisable to install new seals whenever the pump has been dismantled.
(12) Inspect the bearings for pits and scores. They should be renewed if evidence of excessive wear is detected. Coat the bearings with engine oil and wrap them in a clean cloth or paper until required for reassembly.

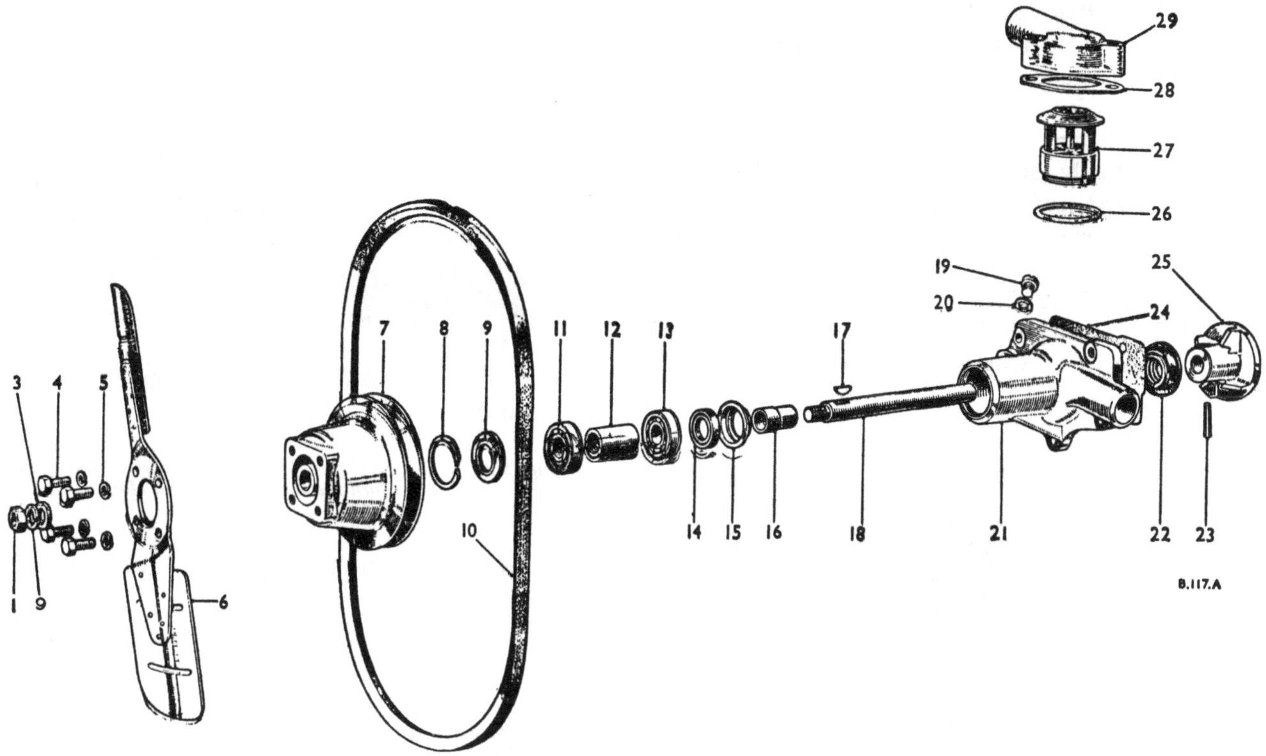

Fig. C.4. The water pump exploded.

1. Spindle nut
2. Spring washer
3. Plain washer
4. Fan blade setpins
5. Spring washer
6. Fan (some early models 2 blades)
7. Fan pulley
8. Split ring
9. Grease retainer
10. Fan belt
11. Ball race
12. Distance piece
13. Ball race
14. Rubber seal
15. Seal housing
16. Distance piece
17. Key
18. Spindle
19. Oiling plug
20. Fibre washer
21. Pump body
22. Water seal
23. Locking pin
24. Joint washer
25. Impeller
26. Joint washer
27. Thermostat
28. Joint washer
29. Thermostat cover

(13) If the bearings do not fit properly on the pump spindle or in the body, renew the parts as required.

(14) Inspect the fan belt for uneven wear or frayed fabric, and renew the belt if required.

To Reassemble

Reassembly of the fan and pump is a reversal of the procedure "**To dismantle**". Particular note should be made of the following:

(1) Pack the bearings with the recommended grade of grease during assembly.

(2) To install the bearings, assemble Tool No. 18G 61, and drive the rear bearing into its housing. The front bearing and distance piece are fitted in a similar manner.

(3) Refit the fan blades to the pulley with the radiused tips of the blades leading.

(4) Check to see that the bearings run freely without excessive end play, by spinning the fan.

To Replace

The installation of the fan and pump assembly is a reversal of the procedure "**To remove**". Particular note should be made of the following:

(1) Install a new joint gasket between the pump body and the cylinder block.

(2) Adjust the fan belt (see Section C.2).

(3) Lubricate the pump as detailed on page Q.6.

Section C.7

DRAINING AND FLUSHING THE SYSTEM

To Drain the System

When the vehicle is to be stored, the entire cooling system should be drained to protect against corrosion and, in certain instances, freezing. To drain the system proceed as follows:

(1) With the vehicle standing on level ground, remove the radiator filler cap.

Caution.—As the system is pressurised, do not remove the radiator filler cap while the engine is running and always wait until the water has cooled.

If it is necessary to remove the filler cap while the engine is hot it is essential to remove it gradually, and the filler neck is provided with a shaped cam to enable this to be done.

Unscrew the cap slowly until the retaining tongues are felt to engage the small lobes at the end of the filler neck cam, and wait until the pressure in the radiator is fully released before finally removing the cap.

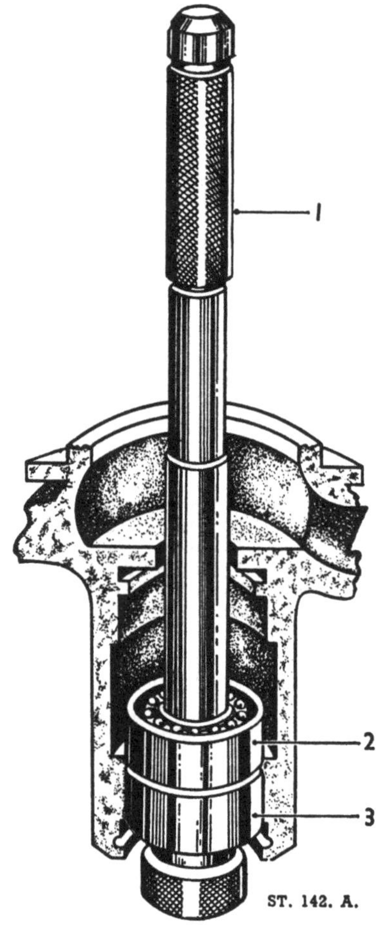

Fig. C.5. Removing rear bearing with service tool 18G 61.
1. Drift. 2. Rear bearing. 3. Dummy bearing.

(2) Open the cylinder block and radiator drain taps. If the system contains anti-freeze mixture it should be drained into clean containers, strained and preserved for re-use.

(3) Insert a length of wire into the open taps to disturb any sediment, etc., that may block the flow.

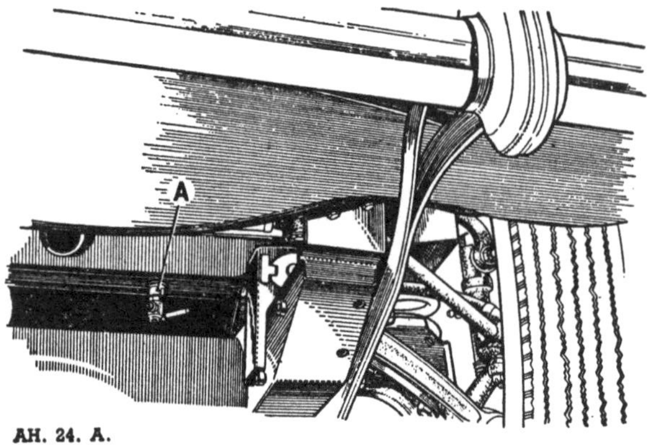

Fig. C.6. Showing the radiator drain tap A in the open position. Turn the tap lever down to close.

(4) To prevent the possibility of operating the vehicle with the system drained, make sure that a suitable notice is placed on the vehicle, or other suitable precautions taken.

To Flush the System

If an inhibitor is not used, the cooling system should be drained, cleaned and flushed at intervals depending upon the type of vehicle operation and the local water conditions. Do not use strong caustic or acid solutions for cleaning purposes because they have a detrimental effect on various parts of the system. To clean and flush the system, proceed as follows :—

(1) Drain the system completely as described above.

(2) With a hose pipe, or fresh quantities of clean water, flush the system through until water issuing from the drain taps appears to be clean.

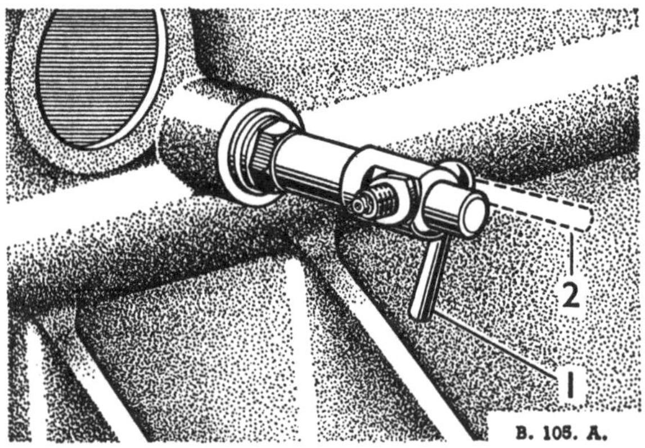

Fig. C.7. Showing the cylinder block drain tap in:—
1. Closed position. 2. Open position.

(3) Allow the system to drain completely, then close the drain taps.
(4) Fill the system with clean water (or anti-freeze solution), slowly, to allow air to escape past the thermostat valve, up to the bottom of the filler neck.
(5) Replace the filler cap by turning it approximately 90° in a clockwise direction.

Section C.8

FROST PRECAUTIONS

Care should be taken to see that the water is drained off completely, for in case of freezing it will do harm by expansion taking place, and fracture of the cylinder block may result. There are two drain taps, one of them on the right-hand side of the cylinder block, and the other at the base of the radiator. Both taps must be opened to drain the system and the vehicle must be on level ground while draining.

Freezing may occur first at the bottom of the radiator or in the lower hose connection. Ice in the hose will stop water circulation, and may cause boiling.

A muff can be used to advantage, but care must be taken not to run with the muff fully closed, or boiling will result.

If a heater is fitted to the car do not resort to draining the cooling system as an alternative to the use of anti-freeze. It is not possible to drain the heater unit completely by means of the cooling system drain taps.

Protection by Use of Anti-Freeze Mixture

When frost is expected or when the vehicle is to be used in very low temperatures, make sure that the strength of the solution is, in fact, up to the strength recommended on General Information page 9 for the conditions likely to be encountered.

Only anti-freeze solutions of the ethylene glycol type are suitable for use in the cooling system: Bluecol or any anti-freeze to B.S. 3151/2 is recommended.

The specific gravity should be checked from time to time and additional antifreeze added to ensure adequate protection against frost.

After the second winter it is recommended that the system should be drained and refilled with fresh water and the appropriate quantity of antifreeze added when required.

Fig. C.8. Showing location of the oiling plug on the water pump body.
1. Thermostat cover bolts. 2. Thermostat outlet.
3. Water pump lubricating plug.

The strength of the solution must be maintained by topping-up with anti-freeze solution as necessary. Excessive topping-up with water will reduce the degree of protection afforded.

If the cooling system has to be emptied, run the mixture into a clean container and use it again.

The correct quantities of anti-freeze are:-

Anti-freeze	*Commences to freeze*		*Frozen solid*		*Amount of anti-freeze*		
%	°*C*	°*F*	°*C*	°*F*	*Pts.*	*U.S. pts.*	*Litres*
25	—13	9	—26	—15	5¼	6¼	3
33⅓	—19	—2	—36	—33	7	8½	4
50	—36	—33	—48	—53	10½	12½	6

Section C.9

MODIFIED THERMOSTAT AND FILLER CAP

Commencing at engine no. 3099 a new thermostat (Part No. 11K399) having an opening temperature of 68°C. (154°F.) has been introduced to suit the 7 lb. pressure radiator filler cap. and to reduce the average running temperature of the engine.

SECTION CC

COOLING SYSTEM

SERIES BN6

NOTE

For details of the cooling system of BN6 car refer to Section C.

SECTION CCC

COOLING SYSTEM

Mk. I and II (SERIES BN7 and BT7)
AND Mk. II and III (SERIES BJ7 and BJ8)

NOTE

For all other repair and service procedures see Section C.

Section CCC.1

MODIFIED WATER PUMP

Later 3000 Mk. II engines are fitted with a modified water pump having a one-piece bearing and spindle. The fan pulley is an interference press fit on the spindle, instead of being keyed and secured by a nut and washers as previously. This pump was introduced at Eng. No. 29E-H-2246.

Dismantling

Remove the fan pulley with a suitable extractor.

Pull out the bearing locating wire through the hole in the pump body.

Gently tap the pump bearing assembly rearwards out of the pump body. This will release the combined bearing and spindle assembly, together with the seal and vane.

Remove the vane from the bearing assembly with a suitable extractor and remove the pump seal assembly.

Reassembling

Reassembly is a reversal of the dismantling procedure, but care must be taken to see that the seal assembly is in good condition. If there is any sign of damage the seal must be renewed. When the bearing assembly is fitted into the pump, the hole (A) in the bearing must coincide with the lubricating hole in the pump body, and a clearance of ·020 to ·030 in. (·5 to ·8 mm.) must be maintained between the vane and the pump body (see Fig. CCC.1.)

As the interference fit of the fan pulley on the spindle will have been impaired when dismantling, a new pulley must always be fitted and pressed home until its recessed face (B) is flush with the end of the spindle.

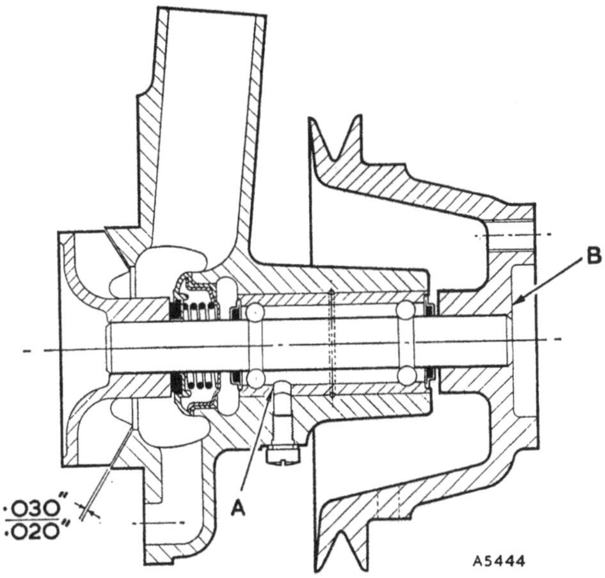

Fig. CCC.1. A section through the water pump showing the components. When assembled, the hole in the bearing (A) must coincide with the lubricating hole in the pump body and there must be a clearance of ·020 to ·030 in. (·508 to ·762 mm.) between the vane and the pump body. The recessed face of the hub (B) must be flush with the end of the spindle.

Section CCC.2

MODIFIED THERMOSTAT

A non-bellows wax element type of thermostat, interchangeable with the bellows type used previously, was fitted from Engine No. 29F/2592 to improve the effectiveness of the car heating equipment. The crack-open temperature is given in **'GENERAL DATA'.**

When refitting this type of thermostat it is essential that the threaded stem of the unit faces upwards.

Section CCC.3

FROST PRECAUTIONS

The correct quantities of anti-freeze are:—

Anti-freeze	*Commences to freeze*		*Frozen solid*		*Amount of anti-freeze*		
%	°C	°F	°C	°F	*Pts.*	*U.S. pts.*	*Litres*
25	—13	9	—26	—15	4¾	5¾	2·7
33⅓	—19	—2	—36	—33	6½	7¾	3·6
50	—36	—33	—48	—53	9½	11½	5·4

SECTION D

FUEL SYSTEM

SERIES BN4

Section D.1

THE FUEL TANK

(1) Remove the drain plug from the tank and drain the petrol into a suitable receptacle.

(2) Within the luggage compartment release and remove the spare wheel by disconnecting its securing strap.

(3) Remove the carpet which covers the floor of the luggage compartment.

(4) Remove the petrol tank feed pipe cover, situated in the top right-hand corner of the boot, by unscrewing the six securing Phillip screws. Disconnect the feed pipe from the tank.

(5) Disconnect the petrol tank filler pipe at its union with the tank. The union is made by a rubber joint hose and two securing clips.

(6) Detach the insulated lead from the petrol gauge unit terminal.

(7) Release the tank securing straps by unscrewing the nut and locknut of each tank strap stud. These nuts are visible on the underside of the luggage compartment floor just in front of the rear body panel. Pull the straps through the compartment floor and hinge them back on their clevis pin anchorages.

(8) Lift out the tank.

1. Outlet union.
2. Fibre washer (thick, orange).
3. Spring clip.
4. Delivery valve disc.
5. Valve cage.
6. Fibre washer.
7. Suction valve disc.
8. Pump body.
9. Diaphragm assembly.
10. Armature guide rollers.
11. Retaining plate.
12. Filter.
13. Fibre washer (thick, orange)
14. Filter plug.
15. Steel armature.
16. Push rod.
17. Magnet iron core.
18. Magnet coil.
19. Rocker hinge pin.
20. Terminal screw.

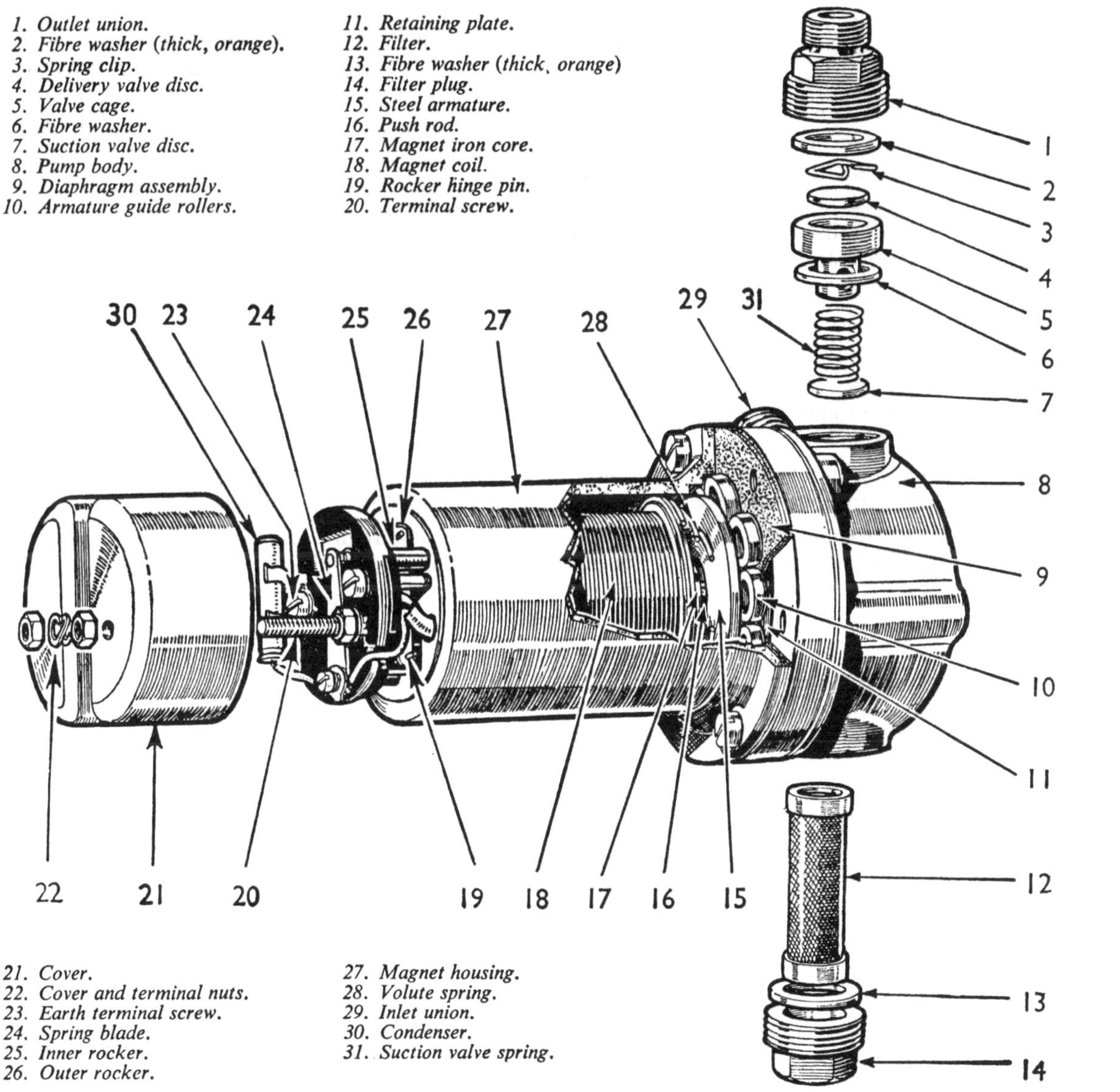

21. Cover.
22. Cover and terminal nuts.
23. Earth terminal screw.
24. Spring blade.
25. Inner rocker.
26. Outer rocker.
27. Magnet housing.
28. Volute spring.
29. Inlet union.
30. Condenser.
31. Suction valve spring.

Fig. D.1. Lift pump—components.

Section D.2

FUEL PUMP

Description

The fuel pump is an S.U. type H.P., 12-volt electric pump. (See Fig. D.1).

The pump consists of three main assemblies, the body, the magnet assembly and the contact breaker.

The body is composed of a hollow alloy die-casting (8) in two parts, into the bottom of which the filter (12) is screwed. The pump inlet union (29) is screwed in at an angle on one side. The outlet union (1) is screwed into the top and tightens down on the delivery valve cage (5) which is clamped between the two fibre washers (2) and (6). In the top of the delivery cage is the delivery valve, a thin brass disc (4) held in position by a spring clip (3). Inserted in the bottom of the cage is the suction valve (7), being a similar disc to (4) and held lightly on a seating machined in the body by a spring. Holes connect the space between the valves and the pumping chamber, a shallow depression on the forward face of the body. This space is closed by a diaphragm assembly (9) clamped at its outside edge between the magnet housing (27) and body (8) and at its centre between the retaining plate and the steel armature (15). A bronze rod to which the diaphragm is attached (16) is screwed through the centre of the armature, passes through the magnet core to the contact breaker, located at the other end. A volute spring (28) is interposed between the armature and the end plate of the coil to return the armature and diaphragm.

The magnet housing consists of a cast-iron pot containing an iron core (17), wound with a coil of copper wire to energise the magnet. Between the magnet housing and the armature are fitted eleven spherical-edged brass rollers (10). These locate the armature centrally within the magnet at all times, and allow absolute freedom of movement in a longitudinal direction. The contact breaker consists of a small bakelite moulding carrying two rockers (25) and (26), which are both hinged to the moulding at one end and are connected together at the top end by two small springs, arranged to give a "throw over" action. A trunnion is fitted into the centre of the inner rocker, and the bronze push-rod (16) connected to the armature is screwed into this. The outer rocker (26) is fitted with a tungsten point, which makes contact with a further tungsten point on a spring blade (24). This spring blade is connected to one end of the coil, and the other end of the coil is connected to the terminal (20).

A short length of flexible wire is connected to the outer rocker and to the other terminal (23) which also serves to hold the bakelite moulding on the magnet housing.

The rocker mechanism is insulated by fibre bushes Two fibre bushes are fitted to one of the spindles of the "throw over" mechanism in order to silence the operation of the contact breaker.

Action of the Fuel Pump

When the pump is at rest, the outer rocker lies in the outer position and the tungsten points are in contact. The current passes from the terminal through the coil back to the blade, through the points and to the earth return, thus energising the magnet and attracting the armature. This comes forward, bringing the diaphragm with it and sucking petrol (gasoline) through the suction valve into the pumping chamber. When the armature has advanced nearly to the end of its stroke the "throw over" mechanism operates, and the outer rocker flies back, separating the points and breaking the circuit. The spring (28) then pushes the armature and diaphragm back, forcing petrol (gasoline) through the delivery valve at a rate determined by the requirements of the engine. As soon as the armature gets near the end of this stroke the "throw over" mechanism again operates, the points again make contact, and the cycle of operations is repeated.

Section D.3

SERVICING THE PUMP

When a pump comes in for reconditioning, the first thing to do is to determine, by the sense of smell, whether the parts in contact with the fuel have become coated with gum. The gum is a substance similar to varnish and can cause the eventual destruction of the diaphragm. Its presence can be detected by smelling the outlet union: if an unpleasant stale smell is noticed, gum is present. The ordinary smell of petrol (gasoline) denotes that no gum has been formed.

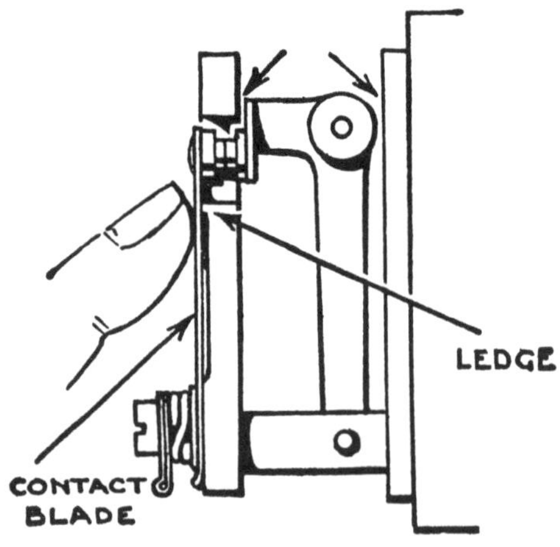

Fig. D.2.
The correct armature setting.

To Dismantle the Pump

(1) Unscrew the filter plug and remove the plug, washer and filter. The latter may be found clogged with gum.
(2) Remove the inlet union and washer.
(3) Remove the outlet union and its washer.
(4) Extract the valve cage, valve cage washer, suction valve and spring. Remove the circlip retaining the delivery valve and withdraw the valve disc.
(5) Unscrew the six screws holding the two main components of the pump together. If the presence of gum has been detected, all parts (NOT ALUMINIUM) must be boiled in 20 per cent. caustic soda solution, dipped in nitric acid and then washed in boiling water. Aluminium parts must be cleaned by thoroughly soaking in methylated spirits.
(6) If no evidence of gum formation has been found, separate the two parts of the pump and check the action of the valves. It should be possible to blow freely but not to suck air back through the inlet union, and to suck, but not blow, air through the delivery valve. If valve action is satisfactory there is no need to disturb their assembly.
(7) Clean the filter with a brush and swill out the body of the pump.
(8) Unscrew the diaphragm assembly from its trunnion in the contact breaker by rotating the whole assembly in an anti-clockwise direction. Take care not to lose the rollers fitted behind the diaphragm.
(9) Remove the contact breaker cover and the nut on the terminal acting as a seating for the cover. Cut away the lead washer squeezed on the terminal threads below the nut, and push the terminal down a short way so that the tag on the coil end is free on the terminal.
(10) Unscrew the contact blade retaining screw and the two long pedestal screws; remove the blade and the pedestal. Do not damage the coil end in disengaging the tag from the terminal.
(11) Push out the rocker hinge pin.
Do not disturb the core of the magnet: special press tools are required for its correct location.

To Reassemble the Pump

(1) Make sure that all parts are clean.
(2) Fit each valve with its smooth side downwards and ensure the correct location of the circlip in its groove.
(3) Fit the red fibre washers as follows: the thin one below the valve cage, the next thickest above the cage, and the thickest on the inlet union. The washer on the filter plug is also a thick red fibre one.

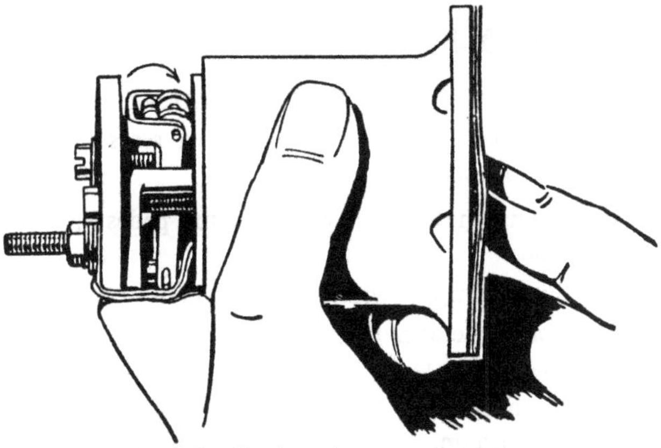

Fig. D.3. Checking the armature setting.

(4) Assemble the contact breaker on its pedestal so that the rockers are free in their mountings without appreciable sideplay. Any excessive sideplay on the outer rocker allows the points to be out of line, while excessive tightness interferes with the action of the pump through sluggish contact breaker operation.
(5) In cases of tightness it may be necessary to square up the outer rocker with a pair of thin-nosed pliers.
(6) The hinge pin is case hardened and ordinary wire must never be used as a replacement.
(7) If the contact blade has been removed, replace it underneath the tag, bearing directly against the pedestal. When the points are separated, the blade should rest against the ledge of the pedestal and must not be so stiff as to prevent the outer rocker from coming right forward when the points are in contact. The points must make contact when the rocker is in the midway position. To check, hold the blade in contact with the pedestal without pressing on the overhanging portion, and test the gap between the white rollers and the body of the pump with a ·030 in. (·76 mm.). If necessary, set the tip of the blade to give the correct clearance.

Note.—Fit the spring washer on the earth connection screw between the tag and the pedestal as the spring washer is not a reliable conductor and the tag must bear directly against the head of the screw.

Solder the coil ends to their tags and the two terminals to the earthing wire.

The assembly of components on the terminal screw holding the cover in position is as follows: spring washer, wiring tag, lead washer and recessed nut. In no circumstances omit the spring washer or shorten the assembly in any way or the pedestal may be broken when the cover retaining nut is tightened.

Fit the armature return spring with its larger diameter towards the coil and the smaller to the armature. Do not stretch the spring.

Section D.4

FUEL PUMP ADJUSTMENT

If the armature has been removed, reassemble and adjust as follows :—

(1) Swing the contact blade on the pedestal to one side.

(2) Fit the impact washer to the armature recess.

(3) Screw the armature into position.

(4) Place the eleven guide rollers in position around the armature. Use no jointing compound on the diaphragm.

(5) Hold the magnet assembly in an approximately horizontal position and push in the armature firmly and steadily. If the contact breaker throws over, screw the armature farther in until it ceases to do so; unscrew the armature one-sixth of a turn at a time until a position is found where the rocker just throws over. It is important to press steadily and not to jerk the armature. When the correct position is found unscrew the armature a further two-thirds of a turn; this is important.

When a new diaphragm is fitted it is probable that considerable pressure will be needed to push the armature right home. If there is any doubt concerning the point at which the contact breaker throws over, turn it back one-sixth of a turn.

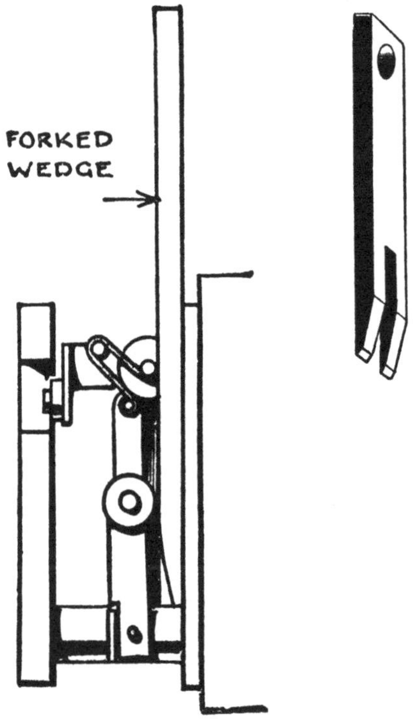

Fig. D.4. The use of a forked wedge is necessary to stretch the pump diaphragm to its outermost position.

(6) Place the magnet housing in position on the main body with the drain hole at the bottom; make sure that the rollers are still in their correct positions. If a roller drops it may get trapped between the two ports and cut a hole in the diaphragm.

Insert the coupling screws and the earth terminal screw. Do not screw up tightly before stretching the diaphragm to its outermost position. This is best accomplished by the use of a wedge as shown in Fig. D.4. Insert the wedge between the white rollers of the outer rocker and pressed under the tips of the inner rocker until it lifts the trunnion in the centre of the inner rocker as far as it will go.

If no wedge is available, insert a matchstick under one of the white rollers and pass a current through the pump. This will excite the magnet, actuate the armature and stretch the diaphragm: the screws may then be tightened down fully while the diaphragm is held in this position. The spring blade rests against a small projection on the bakelite moulding, and it must be set so that when the points are in contact it is deflected back from the moulding. The width of the gap at the points is approximately ·030 in. (·76 mm.).

(7) Now place the pump on test, using a cut-away cover to allow observation of the contact breaker and prevent the hinge pin from falling out.

A test rig of the type illustrated in Fig. D.5 is advised and either petrol (gasoline) or paraffin (kerosene) may be used for testing.

When the pump is switched on it should prime itself promptly and if there is any air leak in the pump or in its connections, bubbles will be seen coming out of the pipe projecting into the flow-meter. Bubbles normally appear when the pump is first started up but should cease to appear when the pump has been running for a minute or two.

Turn off the tap fully ; the pump should stand without repeating its action for at least 15 seconds, if not, the suction valve is not seating correctly.

Next, turn the tap off slowly and note whether the pump idles satisfactorily, and that the outer rocker comes fully forward and contacts the pedestal. While in this position, press the tip of the blade inwards to reduce the stroke of the pump gradually. However much the stroke is reduced the pump should continue to pump normally until it fails, when there is no gap left. If it buzzes instead of pumping, the cause is usually excessive flexibility in the diaphragm, and is unlikely to occur on a new one.

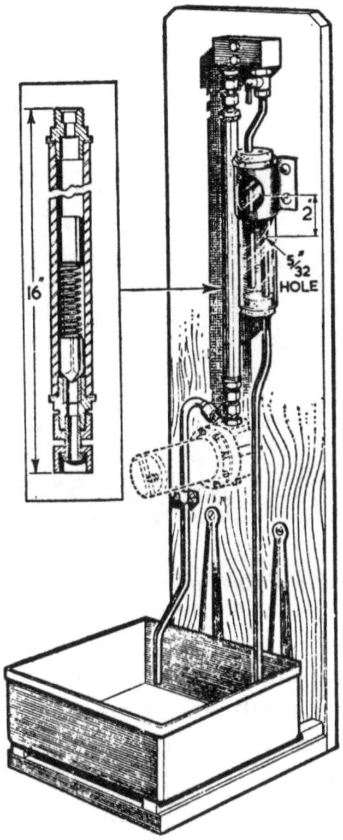

Fig. D.5.
A test rig for the fuel pump.

Finally test the pump on 9 volts, when it should work satisfactorily though probably with a somewhat reduced output.

Note.—Three important points, which will seriously affect the working of the pump if overlooked, are the following :—

(1) Keep the contact breaker blade out of contact while setting the diaphragm.
(2) Press firmly without jerking on the diaphragm.
(3) Stretch the diaphragm to its limit while tightening up the body screws.

Section D.5

TRACING PUMP TROUBLES

Should the pump cease to function, first disconnect the fuel delivery pipe from the pump. If the pump then works the most likely cause of the trouble is sticking needles in the float-chambers of the carburetters. Should the pump not work, disconnect the lead from the terminal and strike it against the body of the pump after switching on the ignition. If a spark occurs it indicates that the necessary current is available at the terminals, and that the trouble arises with the pump mechanism. If no spark can be detected, then it is an indication that the current supply has failed and that attention should be given to the wiring and battery. If current is present further investigation should be carried out by removing the bakelite cover which is retained by the terminal nut. Touch the terminal with the lead. If the pump does not operate and the contact points are in contact yet no spark can be struck off the terminal, it is very probable that the contact points are dirty and require cleaning. These may be cleaned by inserting a piece of card between them, pinching them together and sliding the card backwards and forwards.

If, when the wire is connected to the terminal and the tickler of the carburetter is depressed, the points fail to break, it is possible that there is either an obstruction in the suction pipe, which should be cleared by blowing it through with air, or some irregularity in the pump itself is preventing the correct movement. This may be due either to the diaphragm having stiffened, or to foreign matter in the roller assembly which supports the diaphragm, in which case the diaphragm should be removed and the whole assembly cleaned and reassembled in accordance with the instructions on page D.3.

On the other hand, if the points are not making contact, see that the tips of the inner rocker (25) (Fig. B.1) are in contact with the magnet housing. If they are not, it is an indication that the armature has failed to return to the end of its normal travel.

To cure this, loosen the six screws which attach the magnet housing to the pump body, and make sure that the diaphragm is not sticking to the face of the magnet housing by carefully passing a penknife between the two. The hinge pin (19) should then be removed and the six retaining screws tightened up again. The tips of the inner rockers will probably now be found to be making contact with the face of the magnet housing, but if they are not, it will be necessary to remove and dismantle the whole magnet assembly in order to ascertain if an accumulation of foreign matter has caused a jam. Remember that whenever the magnet housing is removed, care should be taken to see that the guide rollers (10) do not drop out.

Pump Noisy

If the pump becomes noisy and works rapidly, it is usually an indication that there is an air leak on the suction side of the pump. Check the level of the fuel in the tank and see that it is not too low.

The simplest way to test for air leakage is to disconnect the fuel pipe from the carburetter and place its end in a glass jar (approximately 1 pint or half a litre) and allow the pump to deliver fuel into it. If air bubbles appear when the end of the pipe has become submerged in the fuel, it is a clear indication of an air leak on the suction side of the pump in the fuel feed pipe between the tank and the pump, which should be found and cured. Check all the unions and joints, making sure that the filter union and inlet unions are all quite airtight.

Failure to Deliver Fuel

Should the pump continue beating without delivering fuel, it is probable that some dirt has become lodged under one of the valves, in which case they should be dismantled by unscrewing the top or delivery union and lifting out the valve cage, when they can be cleaned and reassembled. When replacing it, see that the thin hard red fibre washer is *below* the valve cage and the thick orange one above.

If the pump struggles to pump and becomes very hot, it is probable that the filter has become clogged or there is an obstruction on the suction side. The filter is readily removed for cleaning by unscrewing its retaining plug at the bottom of the pump.

Section D.6

FUEL PUMP MAINTENANCE

Apart from keeping the contacts clean there is no maintenance required on the fuel pump.

The filter can be removed for cleaning by unscrewing the hexagon plug at the bottom of the pump. Clean the filter in fuel with a stiff brush: do not use rag.

Many of the troubles encountered with the pump are a result of the terminals not being tight, resulting in poor connections. Make sure that the earth wire terminal, in particular, is quite tight.

Section D.7

THE CARBURETTERS

The two S.U. carburetters are of the variable jet type, fitted with air cleaners.

A damper is provided in each carburetter, consisting of a plunger and non-return valve attached to the oil cap nut, and operates in the hollow piston rod. Top up to within ½ in. (13 mm.) from the top of the hollow piston rod with oil. Its function is to give a slightly enriched mixture on acceleration by controlling the rise of the piston and to prevent piston flutter.

Section D.8

CARBURETTER ADJUSTMENT

It is first essential to run the engine until it has attained its normal running temperature before commencing any mixture or slow-running adjustments.

The slow-running is governed by the setting of the jet adjusting screws and the throttle stop screws, all of which must be correctly set and synchronised if satisfactory results are to be obtained.

Fig. D.6. Throttle linkage oiling points. With an oil can lubricate the points indicated.

The two carburetter throttles are interconnected by a coupling shaft and spring coupling clips which enable them to be correctly synchronised when adjustments take place.

Before blaming the carburetter settings for bad slow-running, make quite sure that it is not due to badly set contact points, faulty plugs, bad valve clearance setting or faulty valves and valve springs.

Good slow-running cannot be obtained if the setting for the jets is incorrect. It is therefore advisable to commence any adjustments at this point.

In order to adjust the carburetters successfully it is necessary to remove the air cleaners and intake pipe assembly from the carburetters and engine valve cover and make sure the pistons work freely and the jets are properly centred (see below).

Adjusting the Jets

(1) Slacken off the pinch-bolt of one of the spring coupling clips locating the carburetter interconnecting shaft to the carburetter throttle spindles and also release the two screws securing the choke spring to the jet levers, so that each carburetter can be operated independently.
(2) Release the throttle lever adjusting screws until both throttles are completely closed.
(3) Turn the throttle lever adjusting screw for the rear carburetter clockwise until it is just touching the web on the carburetter body and then give it one full turn. This will set the rear carburetter for fast idling and leave the front one out of action. This can be ensured further by lifting the front carburetter piston a matter of ½ in. (13 mm.).

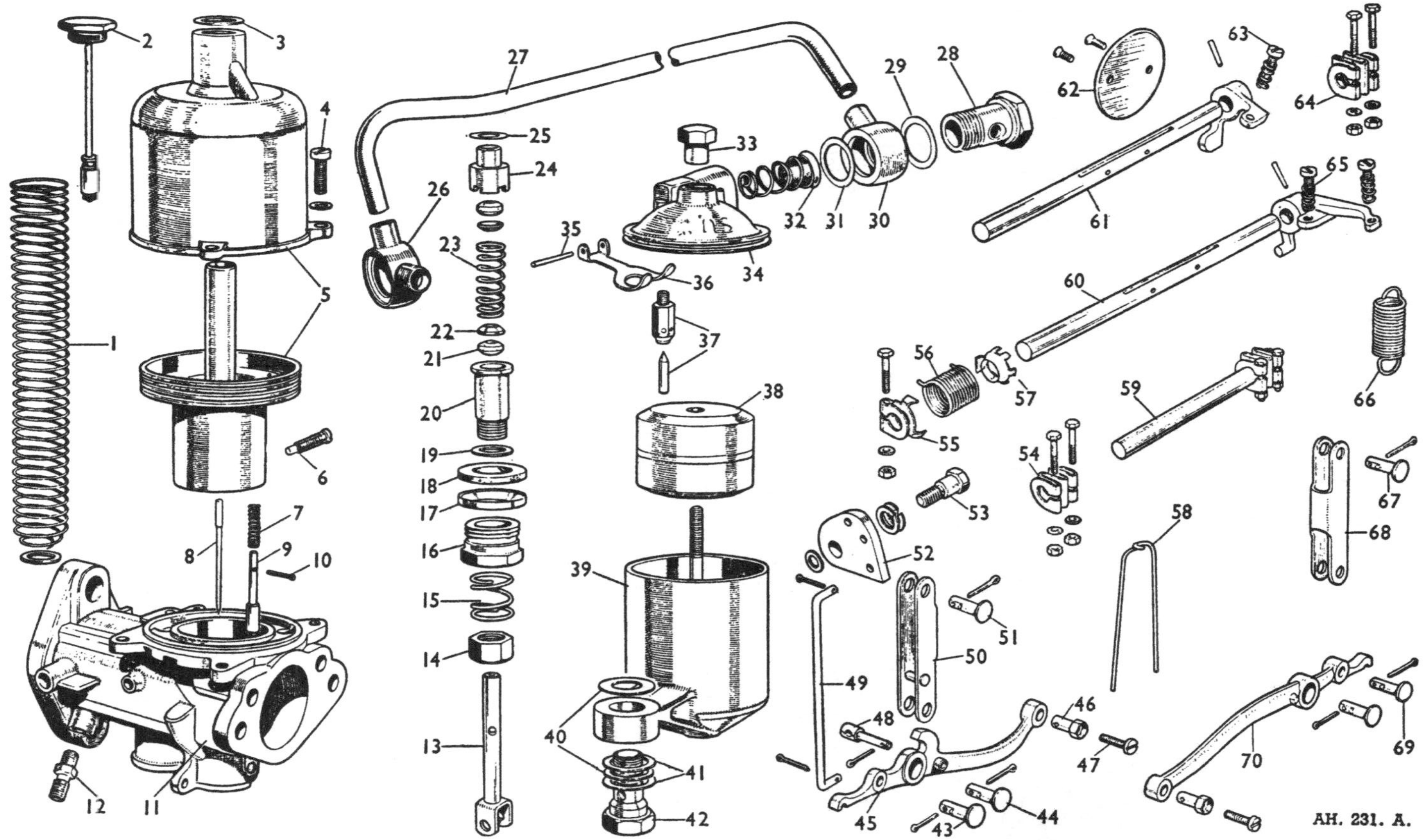

Fig. D.7. Exploded view of carburetter.

1. Piston spring.
2. Cap assembly.
3. Washer for cap.
4. Chamber securing screw.
5. Chamber and piston assembly.
6. Needle locking screw.
7. Spring for pin.
8. Needle.
9. Piston lifting pin.
10. Circlip.
11. Rear of carburetter.
12. Vacuum control union.
13. Jet assemblies.
14. Adjusting nut.
15. Adjusting spring.
16. Seating gland nut.
17. Brass sealing ring.
18. Cork sealing ring.
19. Washer.
20. Bottom bearing.
21. Gland cork washer.
22. Gland brass washer.
23. Gland spring.
24. Top bearing.
25. Copper washer.
26. Banjo union.
27. Petrol pipe.
28. Banjo bolt.
29. Fibre washer.
30. Banjo union.
31. Fibre washer.
32. Filter.
33. Lid cap.
34. Carburetter lid.
35. Lever pin.
36. Float lever.
37. Needle and seating.
38. Float.
39. Float chamber.
40. Fibre washers.
41. Brass washers.
42. Holding bolt.
43. Pin for jet lever.
44. Pin for link lever.
45. Jet lever.
46. Pin for stirrup.
47. Screw for pin.
48. Pin for connecting link.
49. Link for cam.
50. Link for jet lever.
51. Pin for link to body.
52. Slow running cam.
53. Bolt for cam.
54. Throttle coupling.
55. End clip.
56. Front return spring.
57. Anchor plate.
58. Connecting stirrup.
59. Throttle spindle connecting rod.
60. Front throttle spindle.
61. Rear throttle spindle.
62. Disc.
63. Stop screw.
64. Throttle coupling.
65. Slow running adjusting screw.
66. Spring for jet lever.
67. Link for body pin.
68. Link for lever.
69. Link for lever pin.
70. Jet lever.

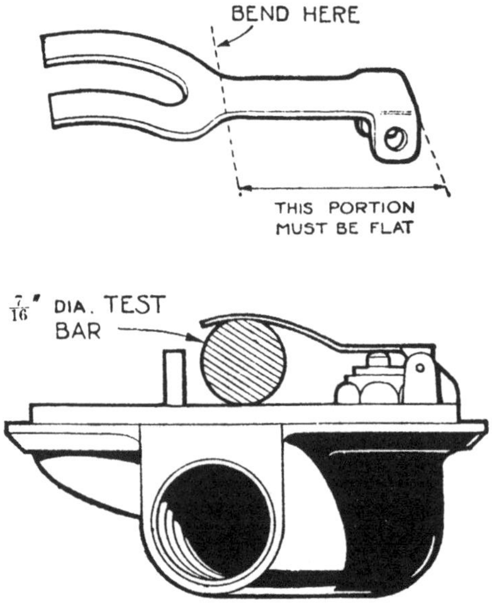

Fig. D.8. The correct setting of the float lever.

(4) With the engine running, set the jet adjusting screw for the rear carburetter so that a mixture strength is obtained which will give the best running speed for this throttle opening, taking care to see that the jet head is kept in firm contact with the adjusting nut the whole time.

(5) The correctness or otherwise of this setting can be checked by raising the suction piston with a small screwdriver, or similar instrument, to the extent of $\frac{1}{32}$ in. (1 mm.). This should cause a very slight momentary increase in the engine speed without impairing the evenness of the running in any way.

If this operation has the effect of stopping the engine it is an indication that the mixture setting is too weak.

If an appreciable speed increase occurs and continues to occur when the piston is raised as much as $\frac{1}{4}$ in. (6 mm.) it is an indication that the mixture is too rich.

(6) When the rear carburetter mixture setting has been carried out correctly release its throttle adjusting screw so that it is clear of the stop and the throttle completely closed, and lift the piston $\frac{1}{2}$ in. (13 mm.) to render it inoperative. Then repeat the jet-adjusting operations on the front carburetter.

(7) When both carburetters are correctly adjusted individually for mixture strength the throttles of each should be set so as to give the required slow-running and synchronisation.

Slow-running and Synchronisation

Screw each throttle lever adjusting screw so that its end is only just making contact with the web on the carburetter body, then give each screw one full turn exactly.

Start the engine, which will now idle on the fast side.

Unscrew each throttle lever adjusting screw an equal amount, a fraction of a turn at a time, until the desired slow-running speed is achieved.

Correct synchronisation can be checked by listening at each carburetter air intake in turn through a length of rubber tube and noticing if the noise produced by the incoming air is the same at both. Any variation in the intensity of the sound indicates that one throttle is set more widely open than the other—the louder sound indicating the throttle with the greater opening.

When the same intensity of sound is given by both carburetters the intercoupling shaft clip should be tightened up firmly to ensure that the throttles work in unison.

Since the delivery characteristics, when both carburetters are operating together, vary somewhat from those existing when each is working separately, it will be found necessary to check them again for correctness of mixture strength by lifting the pistons in turn as described in **"Adjusting the Jets,"** making such adjustments of the jet adjusting screws as are required to balance the mixture strength and to ensure that it is not too rich.

Fitting New Needles

If the road performance is not satisfactory after the above adjustments have been made, larger or smaller needles may be necessary.

To change the needles, remove the screws and lift off the suction chambers, having marked them to ensure

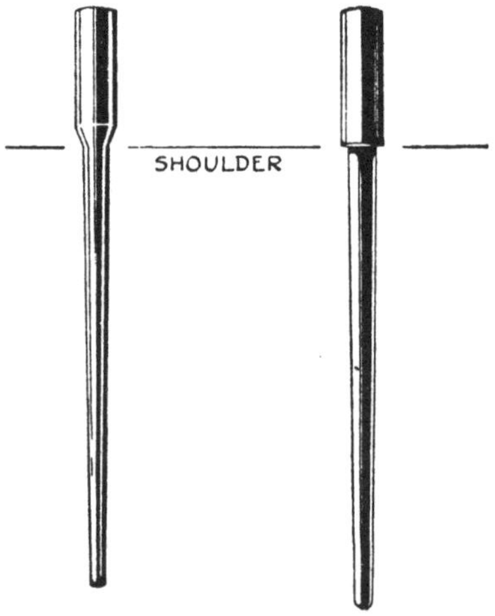

Fig. D.9. The shoulder of the needle should be flush with the under face of the piston. Two types of shoulder are in use and the correct datum point for each is shown.

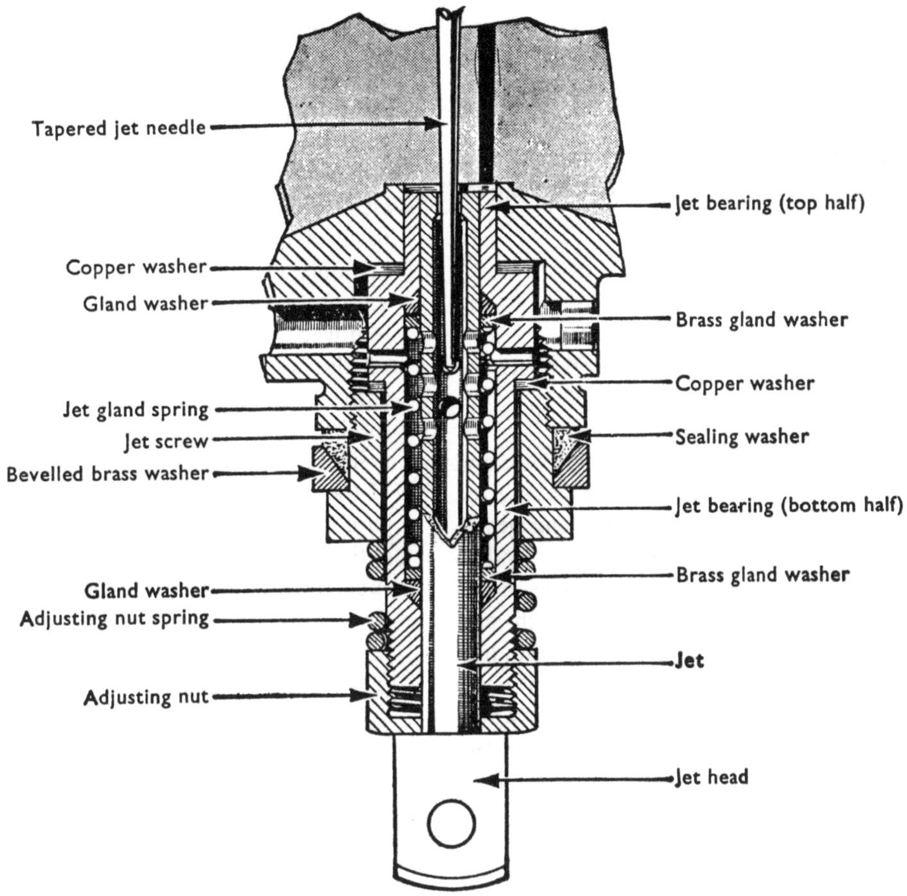

Fig. D.10. The jet assembly.

their refitting to their respective carburetters. Remove the pistons and return springs.

Unscrew the screw at the side of each piston tube and withdraw the needles.

Fit the new needles: a needle should be fitted with its shoulder flush with the face of the piston as shown in Fig. D.9.

The Float-chamber

The position of the forked lever in the float-chamber must be such that the level of the float (and therefore the height of the fuel at the jet) is correct.

This is checked by inserting a $\frac{7}{16}$ in. (11 mm.) round bar between the forked lever and the machined lip of the float-chamber lid. The prongs of the lever should just rest on the bar (see Fig. D.8) when the needle is on its seating. If this is not so, the lever should be reset at the point where the prongs meet the shank. Care must be taken not to bend the shank, which must be perfectly flat and at right angles to the needle when it is on its seating.

Centring a Jet

First remove the clevis pin at the base of the jet which attaches the jet head to the jet operating lever; withdraw the jet completely, and remove the adjusting nut and the adjusting nut spring. Replace the adjusting nut without its spring and screw it up to the highest position. Slide the jet into position until the jet head is against the base of the adjusting nut. When this has been done, feel if the piston is perfectly free by lifting it up with the finger with the dashpot piston removed. If it is not, slacken the jet holding screw and manipulate the lower part of the assembly, including the projecting part of the bottom half jet bearing, adjusting nut and jet head. Make sure that this assembly is now slightly loose. The piston should then rise and fall quite freely as the needle is now able to move the jet into the required central position. The jet holding screw should now be tightened and a check made to determine that the piston is still quite free. If it is not found to be so, the jet holding screw should be slackened again and the operation repeated. When complete freedom of the piston is achieved the jet adjusting nut should be removed, together with the jet, and the spring replaced. The adjusting nut should now be screwed back to its original position.

Experience shows that a large percentage of carburetters returned for correction have had jets removed and incorrectly centred on replacement.

Section D.9

SOURCES OF CARBURETTER TROUBLE

Piston Sticking

The piston assembly comprises the suction disc and the piston forming the choke, into which is inserted the hardened and ground piston rod which engages in a bearing in the centre of the suction chamber and in which is, in turn, inserted the jet needle. The piston rod running in the bearing is the only part which is in actual contact with any other part, the suction disc, piston, and needle all having suitable clearances to prevent sticking. If sticking does occur the whole assembly should be cleaned carefully and the piston rod lubricated with a spot of thin oil. No oil must be applied to any other part except the piston rod. A sticking piston can be ascertained by removing the dashpot piston damper, inserting a finger in the air intake and lifting the piston, which should come up quite freely and fall back smartly onto its seating when released.

Water or dirt in the Carburetter

When this is suspected, lift the piston: the jet can then be seen. Flood the carburetter and watch the jet; if the fuel does not flow through freely there is a blockage. To remedy this, start the engine, open the throttle, and block up the air inlet momentarily without shutting the throttle, keeping the throttle open until the engine starts to race. This trouble seldom arises with the S.U. carburetter owing to the size of the jet and fuel ways. When it does happen the above method will nearly always clear it. Should it not do so, the only alternative is to remove the jet.

Float-chamber Flooding

This can be seen by the fuel flowing over the float-chamber and dripping from the air inlet, and is generally caused by grit between the float-chamber needle and its guide. This can usually be cured by depressing the float depressing plunger to allow the incoming flow of fuel to wash the grit through the guide and into the float chamber.

Float Needle Sticking

If the engine stops, apparently through lack of fuel, when there is plenty in the tank and the pump is working properly, the probably cause is a sticking float needle. An easy test for this is to disconnect the pipe from the electric pump to the carburetter, switch on the ignition to check if fuel is delivered; if it is, starvation has almost certainly been caused by the float needle sticking to its seating, and the float-chamber lid should therefore be removed, the needle and seating cleaned, and refitted. At the same time it will be advisable to clean out the entire fuel feed system, as this trouble is caused by foreign matter in the fuel, and unless this is removed it is likely to recur. It is of no use whatever renewing any of the component parts of the carburetter, and the only cure is to make sure that the fuel tank and pipe lines are entirely free from any kind of foreign matter or sticky substance capable of causing this trouble.

Section D.10

THE AIR CLEANERS

Remove the units and wash the gauze in fuel at the recommended mileage.

When the gauze is clean and dry, re-oil it with engine oil and allow it to drain before refitting to the engine.

Section D.11

CARBURETTERS (Type H.D.)

General Description

S.U. H.D. carburetters are fitted to 6 port cylinder head engines.

A damper is provided in each carburetter, consisting of a plunger and non-return valve attached to the oil cap nut, and operates in the hollow piston rod which is partly filled with oil. Its function is to give a slightly enriched mixture on acceleration by controlling the rise of the piston, and to prevent piston flutter.

Remove the suction chamber cap and damper assembly and replenish the oil reservoir as necessary at the recommended mileage.

The carburetter differs from the more familiar S.U. type in so far that the jet glands are replaced by a flexible diagram, and the idling mixture is conducted along a passage-way, in which is located a metering screw, instead of being controlled by the throttle disc; the throttle and jet interconnection mechanism is also re-designed.

The constructional details are as follows (fig. D.12):—

The jet (1) which is fed through its lower end is attached to a synthetic rubber diaphragm (5) by means of the jet cup (4) and jet return spring cup (7), the centre of the diaphragm being compressed between these two parts; at its outer edge it is held between the diaphragm casing (9) and the float-chamber arm. The jet (1) is controlled by the jet return spring (8) and the jet actuating lever (10), the latter having an adjusting screw (18) which limits the upward travel of the jet (1) and thus constitutes the idler adjustment; screwing it in (clockwise) enriches the mixture, and unscrewing it weakens the mixture.

Throttle and Jet Interconnection

The throttle and jet interconnection mechanism is operated by a cam (21) mounted on the jet lever spindle (19), the whole being housed in the diaphragm casing (9). The cam (21) on being rotated by means of the jet hand control lever (22) actuates the cam shoe (20), thereby causing vertical movement of the push-rod (17). To the top of this push-rod is attached the top plate (16), which is fitted with an adjusting screw which makes contact with the throttle stop lever (15).

It will be seen that angular movement of the jet hand control lever (22) will turn the jet lever spindle (19) and, therefore, the jet actuating lever (10) controls the jet cup (4) and the jet (1). The cam controls the cam shoe (20), push-rod (17), top plate (16) and the throttle. Suitable setting of the two adjustment screws (14) and (18) will give any desired combination of mixture enrichment and throttle opening.

Vacuum Controlled Ignition and Economiser Ports

The connection to the vacuum ignition control is made at the top of the carburetter instead of underneath or at the side, as with the older type.

Throttle Spindle Glands

Provision is made for the use of throttle spindle glands consisting of the cork gland itself (23), a dished retaining washer (24), a spring (26) and a shroud (25). This assembly should not require servicing and can only be removed by dismantling the throttle spindle and disc.

Idling

The H.D. carburetter idles on the main jet, the mixture, passing under the throttle disc, is conducted along the passage-way (11) connecting the choke space to the other side of the throttle disc.

The quantity of mixture passing through the passage-way (11) and, therefore, the idling speed of the engine, is controlled by the "slow-run" valve (12), quality, or relative richness of the mixture, being determined by the jet adjusting screw (18). It follows that when idling, once the engine has reached its running

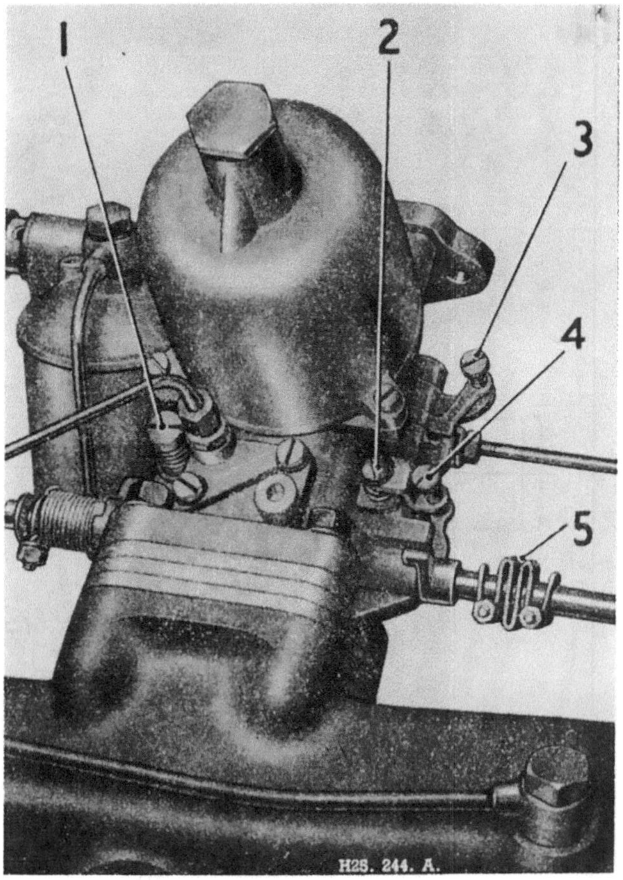

Fig. D.11. Carburetter adjustment.
1. Slow-run valve. 2. Top plate securing screw. 3. Jet adjusting screw. 4. Throttle stop lever screw. 5. Throttle shaft interconnection clip.

temperature, the throttle remains completely closed against the bore of the carburetter; for fast idle, when the engine is cold, it continues to be partially open, the mixture passing under the throttle disc as well as along the passage-way (11).

Centring the Jet

This is carried out in much the same way as on the standard type carburetter, except that the float-chambers must be removed and the jet held in the uppermost position by hand, the jet adjusting screw (18) having first been undone sufficiently to allow the jet cup (4) to make contact with the jet bearing (2), with a distinct clearance between the jet adjusting screw (18) and its abutment. It is important to keep the diaphragm and therefore the jet in the same radial position, in relation to the carburetter body and jet casing throughout this operation, as the jet orifice is not necessarily concentric with its outside diameter, and turning might cause decentralisation. The simplest way to do this is to mark one of the diaphragm and corresponding jet screw casing holes with a soft pencil.

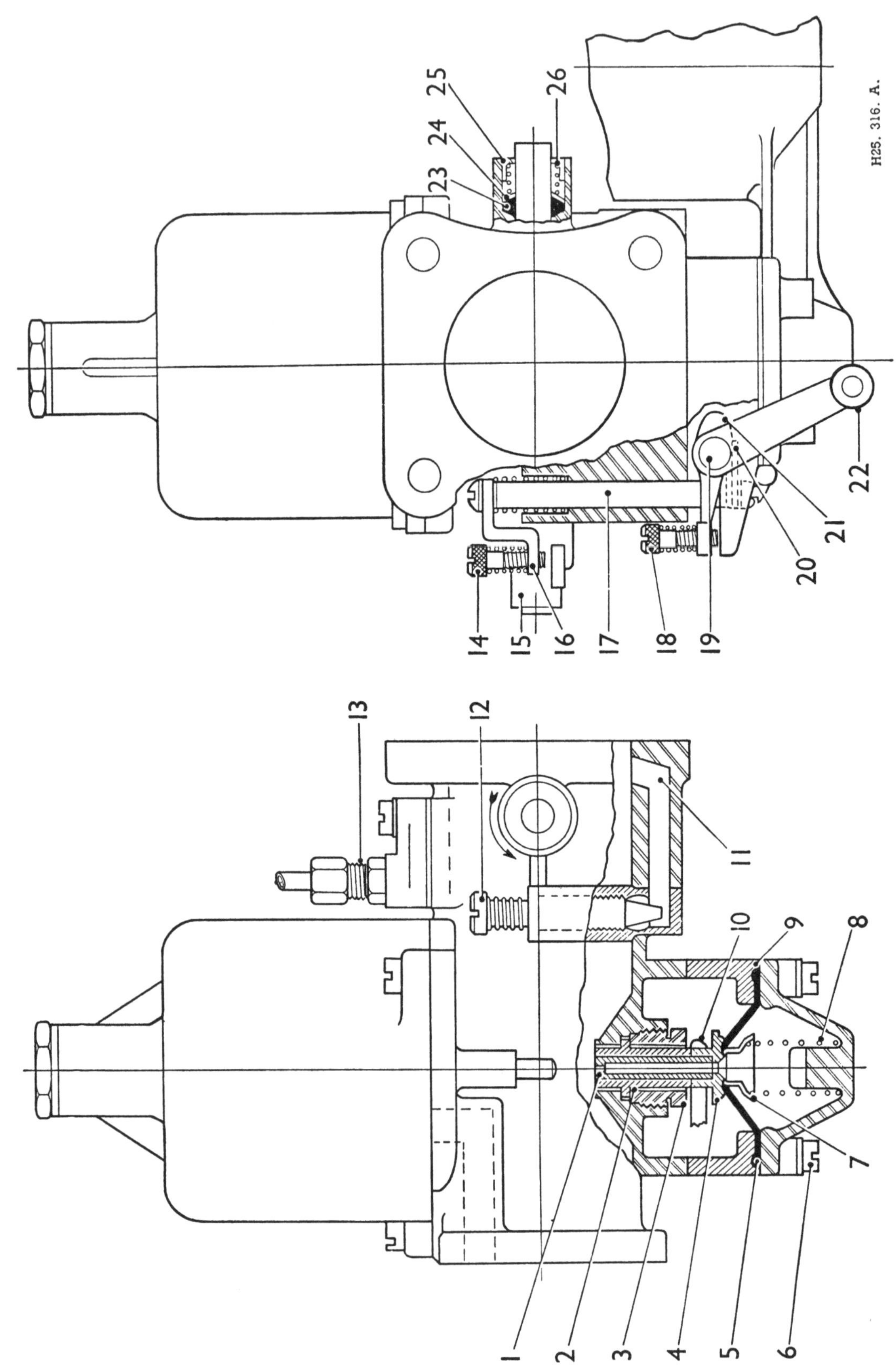

Fig. D.12. A sectional view of the carburetter.

1. *Jet.*
2. *Jet bearing.*
3. *Jet screw.*
4. *Jet cup.*
5. *Diaphragm.*
6. *Float chamber securing screw.*
7. *Jet return spring cup.*
8. *Jet return spring.*
9. *Diaphragm casing.*
10. *Jet actuating lever.*
11. *Slow-run passage.*
12. *Slow-run screw.*
13. *Vacuum advance union.*
14. *Throttle adjusting screw.*
15. *Throttle stop lever.*
16. *Top plate*
17. *Push rod.*
18. *Jet adjusting screw.*
19. *Jet lever spindle.*
20. *Cam shoe.*
21. *Cam.*
22. *Jet hand control lever.*
23. *Cork gland.*
24. *Dished washer.*
25. *Shroud.*
26. *Spring.*

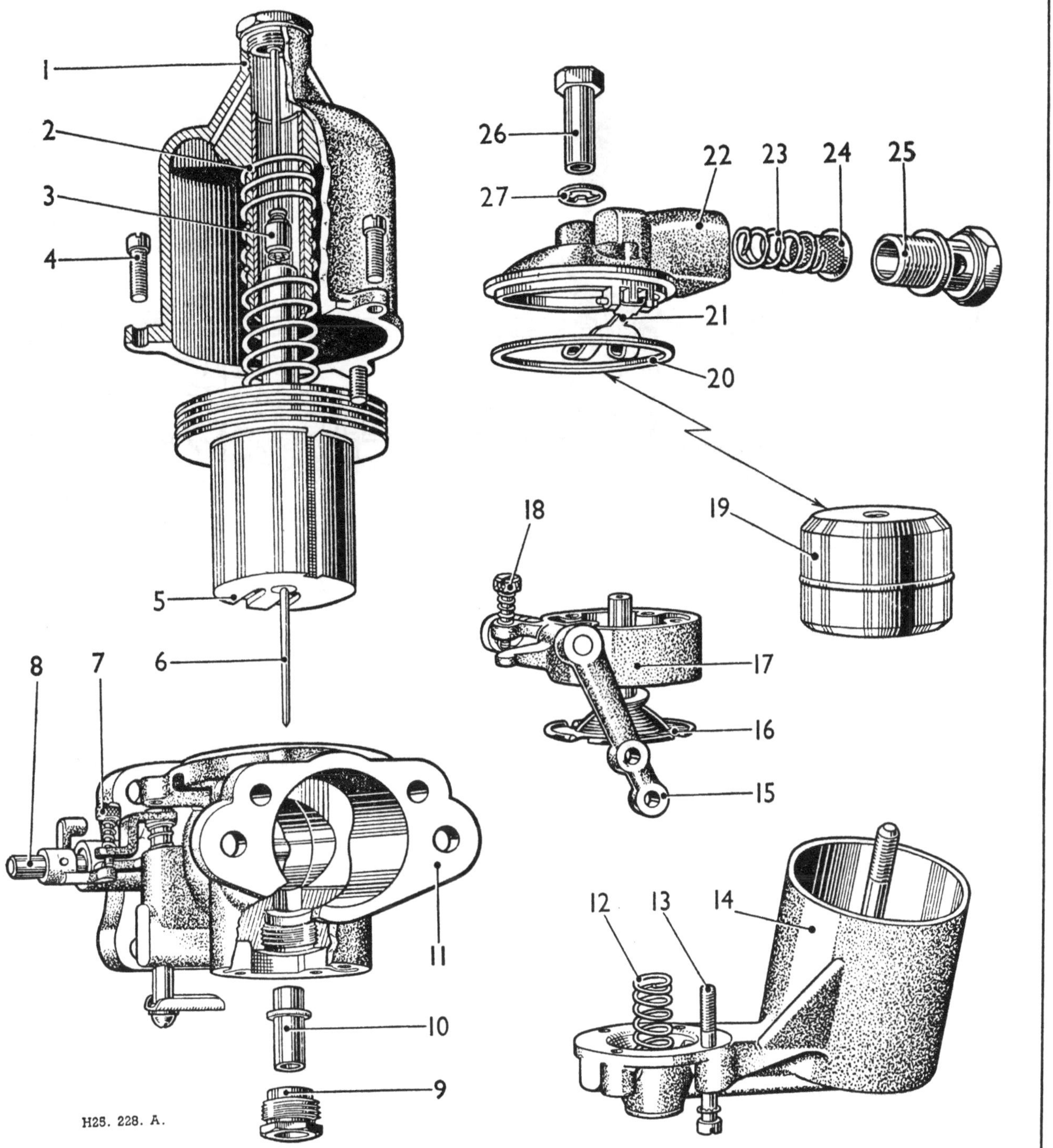

Fig. D.14. Carburetter exploded.

1. *Suction chamber.*
2. *Piston spring.*
3. *Hydraulic damper.*
4. *Suction chamber screw.*
5. *Piston.*
6. *Needle.*
7. *Throttle stop lever adjusting screw.*
8. *Throttle spindle.*
9. *Jet screw.*
10. *Jet bearing.*
11. *Carburetter body.*
12. *Jet return spring.*
13. *Float chamber securing screw.*
14. *Float chamber.*
15. *Jet hand control lever.*
16. *Jet and diaphragm.*
17. *Diaphragm casing.*
18. *Jet adjusting screw.*
19. *Float.*
20. *Cover joint washer.*
21. *Float lever.*
22. *Float chamber cover.*
23. *Filter spring.*
24. *Filter.*
25. *Inlet union.*
26. *Float chamber cover screw.*
27. *Fibre washer.*

In this installation item 15 is at the rear of the carburetter.

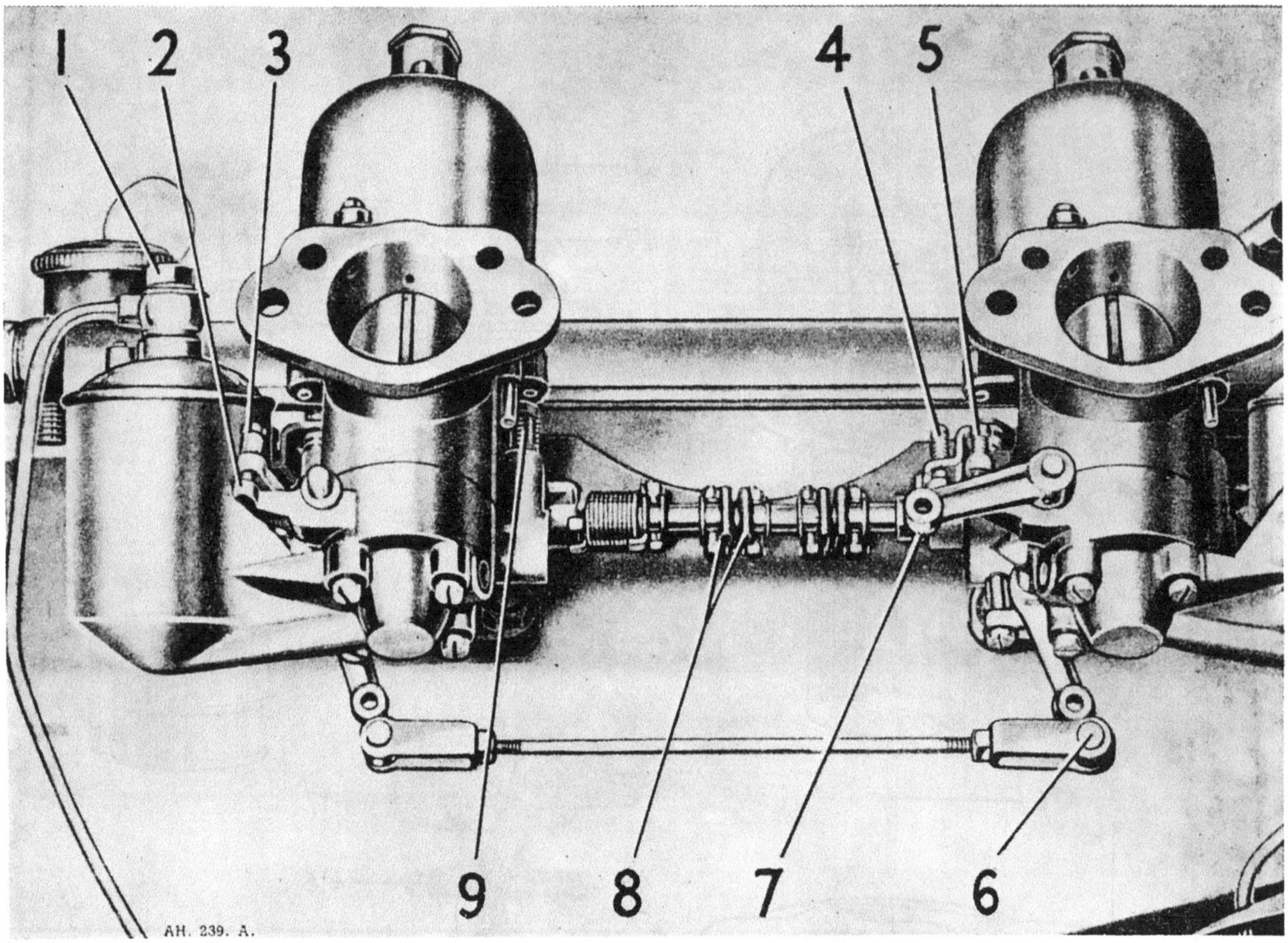

Fig. D.13 Carburetter linkage.

1. *Petrol overflow pipe union.*
2. *Jet adjusting screw stop.*
3. *Jet adjusting screw.*
4. *Throttle stop screw.*
5. *Jet adjusting screw.*
6. *Jet lever connecting yoke.*
7. *Jet lever.*
8. *Throttle shaft interconnection clip.*
9. *Slow-run valve.*

Adjustment

The adjustment of the H.D. carburetter is extremely simple. Whereas with the older type the jet was controlled by a nut, it is now set by a screw (18), and whereas the engine speed was determined by adjustment of the throttle, it is now controlled by the "slow-run" valve (12). To enrich the mixture the screw (18) should be screwed in, and to increase the idling speed the "slow-run" valve (12) should be unscrewed.

The adjustment procedure is as follows :—

(1) Run the engine until its normal operating temperature is reached.

(2) Disconnect the interconnecting rod between the jet actuating levers.

(3) The throttle stop screws on each carburetter must be undone so that they are clear of the stops. This ensures that the throttles are fully closed.

(4) Screw the slow running valve screw right down on each instrument and then unscrew them $2\frac{1}{4}$ turns.

(5) If the engine runs too fast when this has been done, screw in both slow running screws a little at a time until even idling is achieved.

(6) Set the mixture strength by means of the jet lever adjusting screws.

(7) The correctness or otherwise of this setting can be checked by raising the suction piston with a small screwdriver, or the piston lifting pin, to the extent of $\frac{1}{32}$ inch (1 mm.). This should cause a very slight momentary increase in the engine speed without impairing the evenness of the running in any way. If this operation has the effect of stopping the engine it is an indication that the mixture setting is too weak.

If an appreciable speed increase occurs and continues to occur when the piston is raised as much as $\frac{1}{4}$ inch (6 mm.) it is an indication that the mixture is too rich.

(8) The interconnecting rod should now be refitted taking care not to alter the positions of the jet actuating levers. It may be necessary to adjust its length.

(9) With the foregoing adjustments complete, it is only necessary to reset the amount of automatic throttle opening which should occur when the choke is operated. Do this by screwing down the throttle stop screw on each carburetter an equal amount until a fast idle is obtained with approximately half choke. This will give the necessary cold start throttle opening. After this is done, ensure that when the choke is fully released the throttles are closed.

Defects in Operation

Since the jet of the H.D. carburetter is fed through its centre and has no glands, leakage can only be caused by an insecure fit of the jet cup, an imperfect seal of the diaphragm, either at its outer edge, where it is compressed between the float-chamber and the diaphragm casing, or at its inner edge, where it is fitted to the jet, or by fracture of the diaphragm. Leakage at the outer edge may be cured by tightening the float-chamber securing screws (6) but fracture, or leaking at the inner edge will probably call for a new jet assembly.

The jet may also stick, either up or down, due to dirt between it and its bearing (2), or due to corrosion. The cure is to remove the parts by undoing the jet screw (3), clean and refit.

Section D.12

CARBURETTER REMOVAL

To remove the carburetters from the inlet manifold proceed as follows:

(1) Turn the battery master switch to the off position.

(2) Disconnect the petrol feed pipe from the union on the forward carburetter.

(3) Remove the air cleaners from the carburetter flanges.

(4) The float chamber overflow pipes must be removed.

(5) Release the choke wire from the jet hand control lever on the rear carburetter, and from the clamping bracket.

(6) Disconnect the throttle valve rod from the lever on the carburetter throttle shaft.

(7) Disconnect the accelerator link rod from the carburetter throttle shaft.

(8) Release the vacuum advance pipe from its union on the rear carburetter.

(9) Remove the four nuts from each carburetter flange and pull the units off the studs together, after taking the bracket, which locates the rear extension of the throttle shaft, off the car bulkhead.

If only one carburetter is to be removed the interconnecting petrol feed pipe must be released. In this case it is also necessary to split the throttle shaft at the centre connecting clip. The connecting rod for the jet hand control levers must also be released by taking out the clevis pins from the yokes.

Section D.13

FUEL PUMP (Type LCS)

The type LCS fuel pump is fitted to later Series BN.4 cars consists of three main assemblies, body, magnet assembly (coil housing assembly), and the contact breaker. This pump is consistent in description with the type HP pump given in Section D.2, with the exception of the body which is an aluminium die-casting with two identical cover plates each secured by 6, 2 BA screws. Removal of the lower cover plate gives access to the fuel filter and the top cover plate access to the valve cage and with this exception the servicing, tracing pump troubles and maintenance are the same as those given for the HP pump in Sections D.3, 4, 5 and 6.

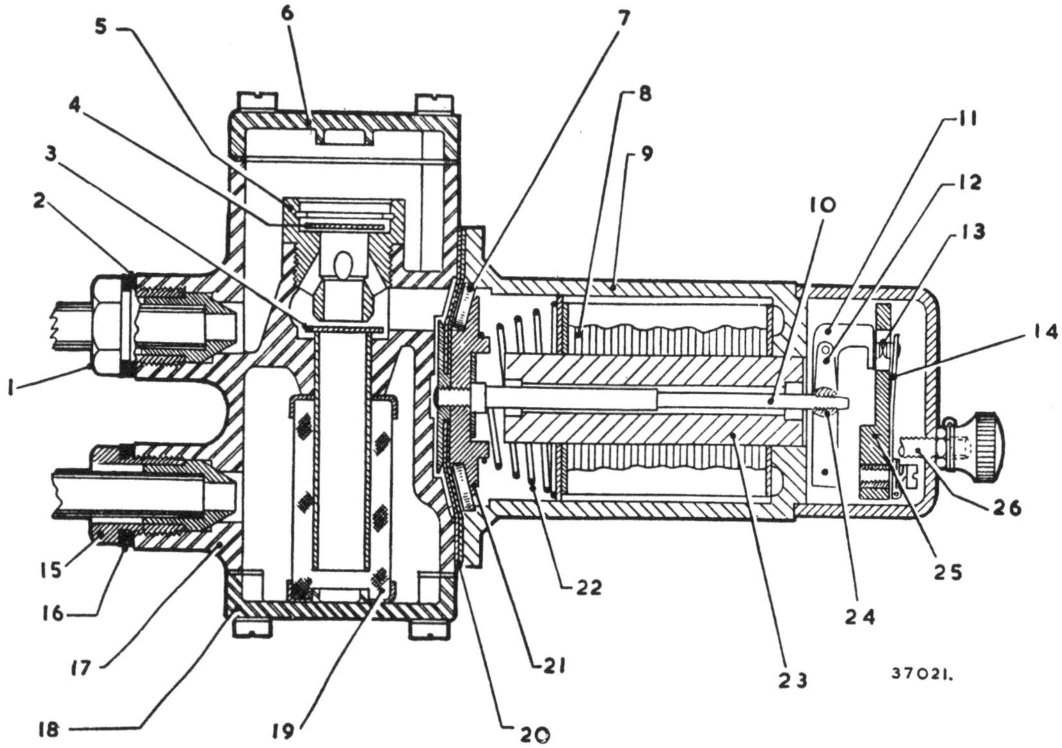

Fig. D.15.
The type LCS fuel pump with the inlet and outlet connections shown 90° out of position for clarity.

1. *Outlet union.*
2. *Rubber ring.*
3. *Inlet valve.*
4. *Outlet valve.*
5. *Outlet valve cage.*
6. *Top cover plate.*
7. *Spherical rollers.*
8. *Magnet coil.*
9. *Iron coil housing.*
10. *Bronze rod.*
11. *Outer rocker.*
12. *Inner rocker.*
13. *Tungsten points.*
14. *Spring blade.*
15. *Inlet union.*
16. *Rubber ring.*
17. *Body.*
18. *Lower cover plate.*
19. *Filter.*
20. *Diaphragm.*
21. *Armature.*
22. *Armature spring.*
23. *Magnet core.*
24. *Trunnion.*
25. *Bakelite moulding.*
26. *Terminal screw.*

SECTION DD

FUEL SYSTEM

NOTE

The fuel system of the BN6 is the same as that of later BN4 cars. All BN6 cars are fitted with H.D. type carburetters and LCS type fuel pumps. Reference should, therefore, be made to Section D and particular attention given to Sections D.12, D.13 and D.14.

SECTION DDD

FUEL SYSTEM

Mk. I and II (SERIES BN7 and BT7)
AND Mk. II and Mk. III (SERIES BJ7 and BJ8)

NOTE

Use the information given in this section in conjunction with Section D.

Sections DDD.1 to DDD.5 should be used for carburetters fitted with the auxiliary thermo carburetter. This information supersedes that given in Sections D.12 and and D.13. Full information is given in Sections DDD.6 to DDD.9 and DDD.11 for the carburetters fitted to the Austin-Healey 3000, Mk.II.

Section DDD.1

CARBURETTERS

(Type H.D. with Thermo-Carburetter fitted from Engine No. 29D/U/H2864)

General Description

The two S.U. H.D. carburetters are fitted with pancake type air cleaners.

A damper is provided in each carburetter, consisting of a plunger and non-return valve attached to the oil cap nut, and operates in the hollow piston rod which is partly filled with oil. Its function is to give a slightly enriched mixture on acceleration by controlling the rise of the piston, and to prevent piston flutter.

Remove the suction chamber cap and damper assembly and replenish the oil reservoir as necessary at the recommended mileage.

The carburetter differs from the more familiar S.U. type in so far that the jet glands are replaced by a flexible diaphragm, and the idling mixture is conducted along a passage-way, in which is located a metering screw, instead of being controlled by the throttle disc.

An auxiliary carburetter controlled by a thermo-switch, provides the enriched mixture required when starting from cold. The fuel for this carburetter is drawn from the rear main carburetter supply but its operation is completely independent.

Constructional Details of the H.D. Carburetter

The jet (1), Fig. DDD.2, which is fed through its lower end, is attached to a synthetic rubber diaphragm (5) by means of the jet cup (4) and jet return spring cup (7), the centre of the diaphragm being compressed between these two parts; at its outer edge it is held between the diaphragm casing (9) and the float-chamber arm. The jet (1) is controlled by the jet return spring (8) and the jet actuating lever (10), the latter having an adjusting screw (15) which limits the upward travel of the jet (1) and thus constitutes the idling adjustment; screwing it in (clockwise) enriches the mixture, and unscrewing it weakens the mixture.

Throttle Spindle Glands

Provision is made for the use of throttle spindle glands consisting of the cork gland itself (17), a dished retaining washer (18), a spring (20) and a shroud (19). This assembly should not require servicing and can only be removed by dismantling the throttle spindle and disc.

Vacuum-controlled Ignition and Economizer Ports

The connection (13) to the vacuum ignition control is made at the top of the carburetter instead of underneath or at the side, as with the older type.

Idling

The H.D. carburetter still idles on the main jet, but the mixture, instead of passing under the throttle disc, is conducted along the passage-way (11) connecting the air intake passage to the manifold side of the throttle disc.

The quantity of mixture passing through the passage-way (11) and, therefore, the idling speed, of the engine, are controlled by the "slow-run" valve (12) the quality or relative richness of the mixture being determined by the jet adjusting screw (15).

It follows that when idling, once the engine has reached its running temperature, the throttle remains completely closed against the bore of the carburetter

Constructional Details of the Auxiliary Enrichment Carburetter

The enrichment apparatus to assist cold starting is, in effect, an auxiliary carburetting system and is shown in Fig. DDD.3. The main body casting (36) containing a solenoid operated valve and fuel metering system is attached by means of a ducted mounting arm to the base of the main carburetter fuel inlet.

The auxiliary carburetter forms, therefore, a separate unit additional to the normal float-chamber, but drawing its fuel supply directly from it. Fuel is supplied to the base of the jet (29) which is obstructed to a greater or lesser degree by the tapered slidable needle (25).

When the device is in action air is drawn from the atmosphere through the air intake (26) and thence through the passage (28), being carburetted with fuel as it passes the jet (29). The mixture is thence carried upwards past the shank of the needle (25) through the passage (37) and so past the aperture provided between the valve (33) and its seating (35). From here it passes directly to the induction manifold through the external feed pipe shown.

The device is brought into action by energizing the winding of the solenoid (31) from the terminal screws (30). The centrally located iron core (32) is thus raised magnetically, carrying with it the ball-jointed disc valve (33) against the load of the small conical spring (34) and thus uncovering the aperture provided by the seating (35).

Considering the function of the slidable needle (25), it will be seen that this is loaded upwards in its open position by means of the slight compression spring (24) which abuts against a disc (23), attached to the shank of the needle. The needle continues upwards through the vertically adjustable stop (22) in which it is slidably mounted and it finally terminates in an enlarged head.

Depression within the space surrounding the spring (24) is directly derived from that prevailing in the induction tract, and this exerts a downward force upon the disc (23), which is provided with an adequate clearance with its surrounding bore. This tends to overcome the load of the spring (24) and to move the needle downwards, thus increasing the obstruction afforded by the tapered section which enters the jet (29).

The purpose of this device is to provide two widely different degrees of enrichment, the one corresponding to idling or light cruising conditions and the other to conditions of open throttle or full-power operation. In effect, under the former conditions the high induction depression prevailing will cause the disc (23) to be drawn downwards, drawing the tapered needle into the jet (29), while under the latter, the lower depression existing in the induction tract will permit the collar to maintain its upward position with the needle withdrawn from the jet. The only adjustment provided is the needle stop screw (22) which limits the degree of movement provided for the needle assembly.

The size and degree of taper of the lower end of the needle (25), the diameter of the disc (23), and the load provided by the spring (24) are not adjustable.

The solenoid (31) is energized by means of a thermostatically operated switch housed within the cylinder head water jacket. This is arranged to bring the apparatus into action at temperatures below 30–35°C. (86–95°F.)

Centring the Jet

This is carried out in much the same way as on the standard H-type carburetter, except that the float chamber and jet casing must be removed and the jet held in the uppermost position by hand. It is important to keep the diaphragm and, therefore, the jet in the same radial position in relation to the carburetter body and jet casing throughout this operation, as the jet orifice is not necessarily concentric with its outside diameter, and turning might cause decentralisation. The simplest way to do this is to mark one of the diaphragm and corresponding jet casing screw holes with a soft pencil.

Adjustment

The adjustment of the H.D. carburetter is extremely simple. Whereas with the older type the jet was controlled by a nut, it is now set by a screw (15), and whereas the engine speed was determined by adjustment of the throttle, it is now controlled by the "slow-run" valve (12). To enrich the mixture the screw (15) should be screwed in, and to increase the idling speed the "slow-run" valve (12) should be screwed out.

The adjustment procedure is as follows:—

(1) Run the engine until its normal operating temperature is reached.

(2) Disconnect the interconnecting rod between the jet actuating levers. This ensures that the throttles are fully closed.

(3) Screw the slow running valve screw right down on each instrument and then unscrew them $2\frac{1}{4}$ turns.

(4) If the engine runs too fast when this has been done, screw in both slow running screws a little at a time until even idling is achieved. It is essential to remember that **the adjustments made to each carburetter must be identical.**

(5) Set the mixture strength by means of the jet lever adjusting screws.

(6) The correctness or otherwise of this setting can be checked by raising the suction piston with a small screwdriver, or the piston lifting pin, to the extent of $\frac{1}{32}$-inch (1 mm.). This should cause a very slight momentary increase in the engine speed without impairing the evenness of the running in any way. If this operation has the effect of stopping the engine it is an indication that the mixture setting is too weak. If an appreciable speed increase occurs and continues to occur when the piston is raised as much as $\frac{1}{4}$-inch (6 mm.), it is an indication that the mixture is too rich.

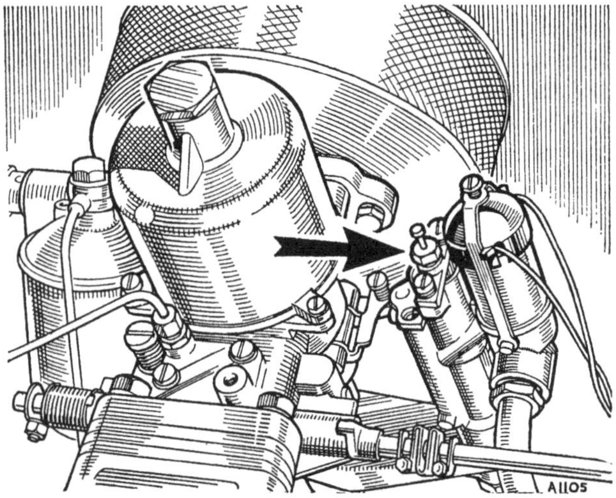

Fig. DDD.1. The needle stop screw indicated by the arrow adjusts the mixture strength of the thermo-carburetter.

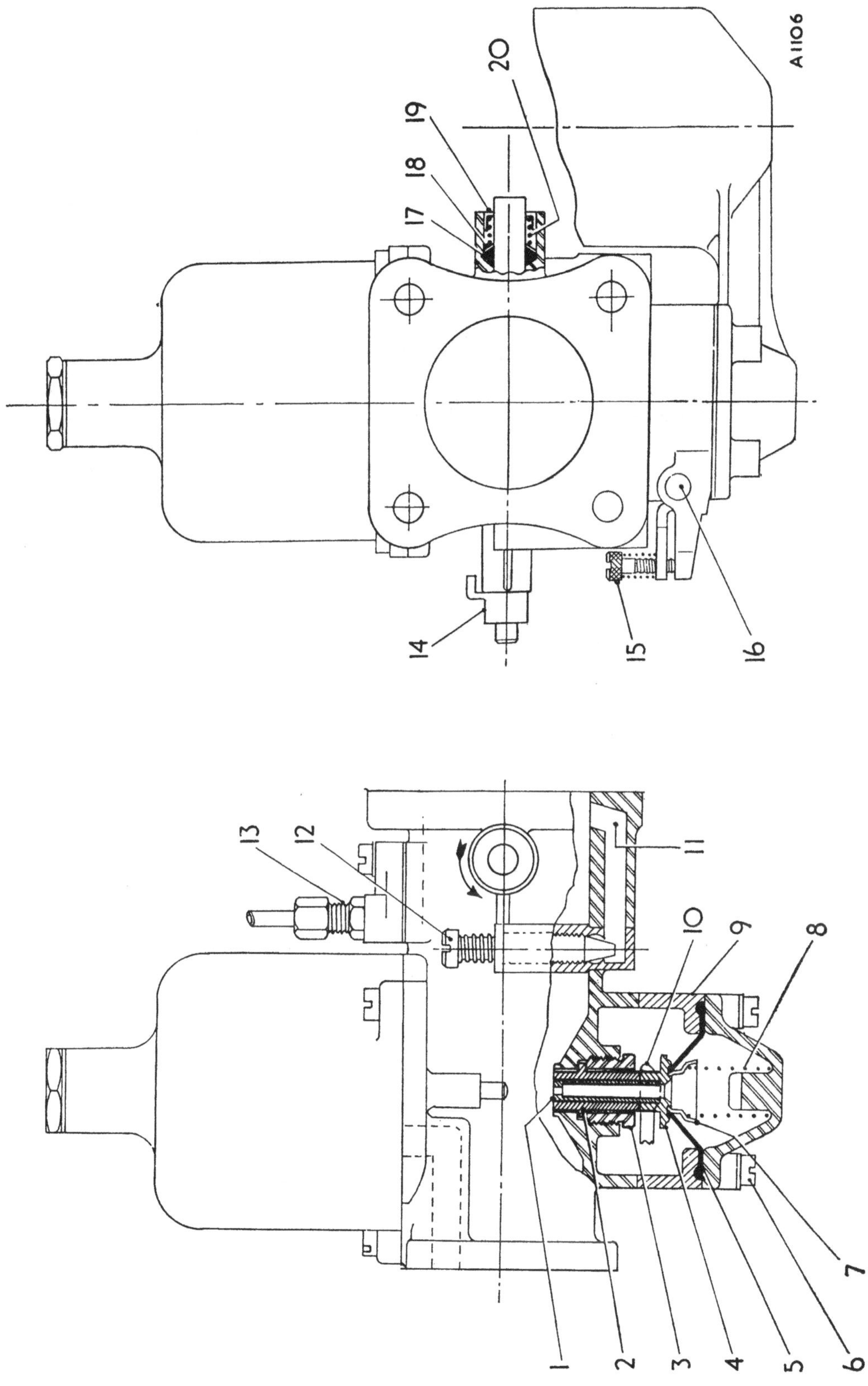

Fig. DDD.2 A section view of a carburetter

1. *Jet.*
2. *Jet bearing.*
3. *Jet screw.*
4. *Jet cup.*
5. *Diaphragm.*
6. *Float chamber securing screw.*
7. *Jet return spring cup.*
8. *Jet return spring.*
9. *Diaphragm casing.*
10. *Jet actuating lever.*
11. *Slow-run passage.*
12. *Slow-run screw.*
13. *Vacuum advance union.*
14. *Throttle stop lever.*
15. *Jet adjusting screw.*
16. *Jet lever spindle.*
17. *Cork gland.*
18. *Dished washer.*
19. *Shroud.*
20. *Spring.*

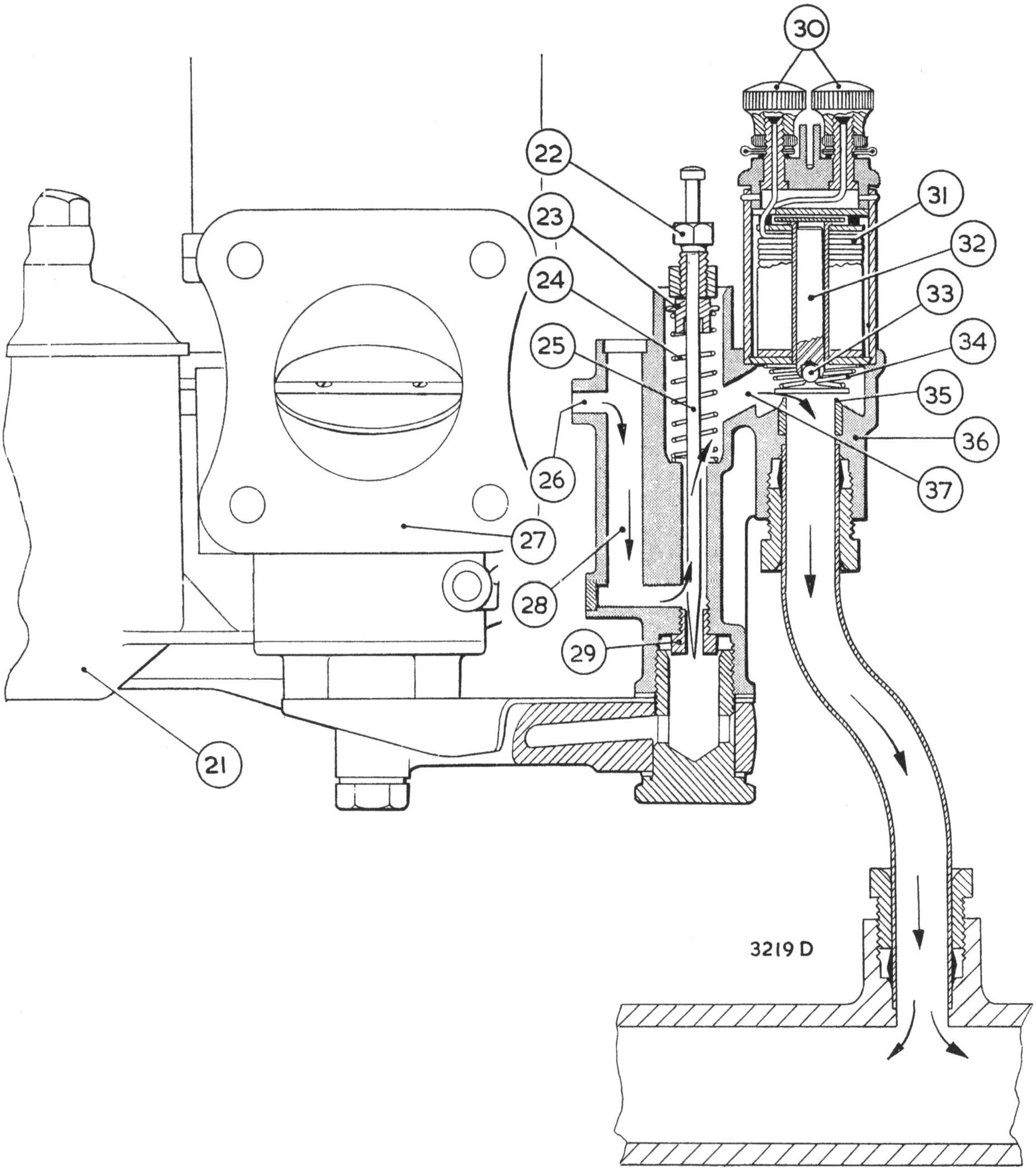

Fig. DDD.3. The auxiliary enrichment carburetter

21. *Float chamber.*
22. *Stop screw.*
23. *Disc.*
24. *Spring.*
25. *Needle.*
26. *Air intake.*
27. *Carburetter body.*
28. *Air passage.*
29. *Jet.*
30. *Terminals.*
31. *Solenoid.*
32. *Core.*
33. *Valve.*
34. *Conical spring.*
35. *Valve seating.*
36. *Body casting.*
37. *Passage.*

(7) Secure the throttle interconnecting rod and make final adjustments to the "slow running" if necessary.

(8) Adjustment of the auxiliary carburetter is confined to the stop screw which limits the downwards movement of the needle. Anti-clockwise rotation of the stop screw will raise the needle and increase the mixture strength, while rotation in the opposite direction will have the opposite effect.

(9) An approximate guide to its correct adjustment is provided by energizing the solenoid when the engine has already attained its normal running temperature. The stop screw should then be so adjusted that the mixture is distinctly although not excessively rich, that is to say, until the exhaust gases are seen to be discernibly black in colour, but just short of the point where the engine commences to run with noticeable irregularity.

(10) In order to energize the solenoid under conditions when the thermostatic switch will normally have broken the circuit, it is merely necessary to short-circuit the terminal of the thermostatic switch directly to earth or, by means of a separate wire, earth the solenoid terminal which is connected to the switch (Blue—White Wire).

NOTE.—**Should difficulty be experienced when starting from cold on the next occasion, unscrew the stop screw (22) one or two flats only.**

Defects in Operation

Since the jet of the H.D. carburetter is fed through its centre and has no glands, leakage can only be caused by an insecure fit of the jet cup, an imperfect seal of the diaphragm, either at its outer edge, where it is compressed between the float-chamber and the diaphragm casing, or at its inner edge, where it is fitted to the jet, or by fracture of the diaphragm. Leakage at the outer edge may be cured by tightening the float-chamber securing screws (6) but fracture or leaking at the inner edge will probably call for a new jet assembly.

The jet may also stick, either up or down, due to dirt between it and its bearing (2), or due to corrosion, The cure is to remove the parts by undoing the jet screw (3), clean, and refit.

Section DDD.2

CARBURETTER REMOVAL

To remove the carburetters from the inlet manifold proceed as follows :—

(1) **Turn the battery master switch to the 'off' position.**

(2) Disconnect the petrol feed pipe from the union on the forward carburetter.

(3) Remove the air cleaners from the carburetter flanges.

(4) The float chamber overflow pipes must be removed.

(5) Disconnect the accelerator link rod from the lever on the carburetter throttle shaft.

(6) Release the vacuum advance pipe from its union on the rear carburetter.

(7) Remove the top of the solenoid and disconnect the cables therefrom. Replace the solenoid top securely.

(8) Slacken the unions securing the external feed pipe, removing the union completely from the manifold.

(9) Remove the four nuts from each carburetter flange and pull the units off the studs together, after taking the bracket, which locates the rear extension of the throttle shaft, off the car bulkhead.

If only one carburetter is to be removed the interconnecting petrol pipe must be released. In this case it is also necessary to split the throttle shaft at the centre connecting clip.

Section DDD.3

MODIFIED STARTING JET NEEDLE SPRING

To improve starting characteristics, the green spring (24) Fig. DDD.3, fitted to the starting jet needle has been changed to a blue spring (Part No. AUC1041) commencing at engine No. 3664. It is recommended that all engines between 2864 and 3664 be checked and that the blue spring be fitted if it is not already incorporated.

Section DDD.4

REVERSION TO HAND CHOKE CONTROL

Commencing at chassis Nos. BN5234 and BT5310, standard type H.D. carburetters and hand choke controls were fitted.

The fitting of the H.D. type carburetter with the cold start device has been discontinued.

Section DDD.5

LOCATION OF FUEL PUMP

The fuel pump on the BT.7 is situated under the left hand seat pan, and is accessible when the seat pan has been removed.

The BN.7 has the fuel pump fitted on the left hand side of the car and access to it is obtained through the hinged portion of the spare wheel floor, as for the batteries (See Section NN.1).

Section DDD.6

CARBURETTERS (Healey 3000 Mk.II)

The carburetters fitted to the Healey 3000 Mk. II are triple S.U. type H.S.4. Each carburetter is mounted on an individual manifold secured to the cylinder head by three studs and nuts, and interconnected by an external balance pipe running above the manifolds.

Description

The H.S.4 carburetter is of the automatically expanding choke type in which the size of the main air passage (or choke) over the jet, and the effective area of the jet, are variable according to the degree of throttle opening used on the engine against the prevailing road conditions (which may differ widely from light cruising to heavy pulling).

Therefore, to serve the complete throttle range a single jet only is used, being a simple metal tube sliding in a single bearing bush, fed by fuel along a small-diameter nylon tube leading direct from the base of the float-chamber. The jet is varied in effective area by a tapered fuel metering needle.

Adjustments

Slow-running is governed by the setting of the jet adjusting nuts and the throttle adjusting screws, all of which must be correctly set and synchronized if satisfactory results are to be obtained.

Before blaming the carburetter setting for incorrect slow-running make certain that the trouble is not caused by badly adjusted distributor contact points, faulty plugs, incorrect valve clearance, or faulty valves and springs.

Slow running adjustment and synchronization

After the first long run or so when the engine is fully run in, the slow running may require adjustment. This must only be carried out when the engine has reached its normal running temperature.

As the needle size is determined during engine development, tuning of the carburetters is confined to correct idling setting. Slacken the actuating arms on the throttle spindle inter-connection. Close all throttles fully by unscrewing the throttle adjusting screws, then open each throttle by screwing down the idling adjustment screws one turn.

Remove pistons and suction chambers, and disconnect the jet control cables. Screw the jet adjusting nuts until each jet is flush with the bridge of its carburetter, or as near to this as possible (all jets being in the same relative position to the bridge of their respective carburetters). Replace the pistons and suction chamber assemblies, and check that the pistons fall freely on to the bridge of the carburetters (by use of the piston lifting pins). Turn down the jet adjusting nut two complete turns (12 flats).

Re-start the engine, and adjust the throttle adjusting screws to give the desired idling speed, by moving each throttle adjusting screw an equal amount. By listening to the hiss in the intakes, adjust the throttle adjusting screws until the intensity of the hiss is similar on all intakes. This will synchronize the throttle setting.

When this is satisfactory, the mixture should be adjusted by screwing each jet adjusting nut up or down by the same amount, until the fastest idling speed is obtained consistent with even firing. During this adjusting, it is necessary that the jets are pressed upwards to ensure that they are in contact with the adjusting nuts.

As the mixture is adjusted the engine will probably run faster, and it may therefore be necessary to unscrew the throttle adjusting screws a little, each by the same amount, to reduce the speed.

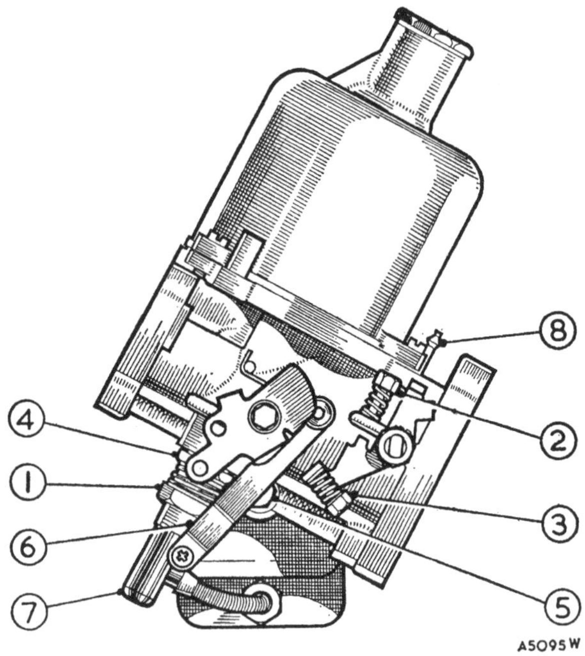

Fig. DDD.4. The H.S.4 type carburetter

1. Jet adjusting nut
2. Throttle adjusting screw.
3. Fast idle adjusting screw.
4. Jet locking nut.
5. Float chamber securing nut.
6. Jet link.
7. Jet head.
8. Vacuum ignition take-off.

Now check the mixture strength by lifting the piston of the front carburetter by approximately $\frac{1}{32}$ in. (·75 mm.) when if:

(*a*) the engine speed increases, this indicates that the mixture strength of the front carburetter is too rich.

(*b*) the engine speed immediately decreases, this indicates that the mixture strength of the front carburetter is too weak.

(*c*) the engine speed momentarily increases very slightly, then the mixture strength of the front carburetter is correct.

Repeat the operation at the centre and rear carburetters, and after adjustment re-check the front carburetter, since all carburetters are inter-dependent.

When the mixture is correct the exhaust note should be regular and even. If it is irregular with a splashy type of misfire and colourless exhaust, the mixture is too weak. If there is a regular or rythmical type of misfire in the exhaust beat, together with a blackish exhaust, then the mixture is too rich.

The carburetter throttle on each carburetter is operated by a lever and pin, with the pin working in a forked lever attached to the throttle spindle. A clearance exists between the pin and the fork, which must be maintained when the throttle is closed and the engine idling, to prevent any load from the accelerator linkage being transferred to the throttle butterfly and spindle.

To set this clearance: with the throttle shaft levers free on the throttle shaft, put a ·012 in. (·305 mm.) feeler between each throttle shaft stop at the top and the carburetter heat shield (see Fig. DDD.5). Move each throttle shaft lever downwards in turn until the lever pin rests lightly on the lower arm of the fork in the carburetter throttle lever. Tighten the clamp bolt of the throttle shaft lever at this position. When all three carburetters have been dealt with, remove the feelers. The pins on the throttle shafts should then have clearance in the forks.

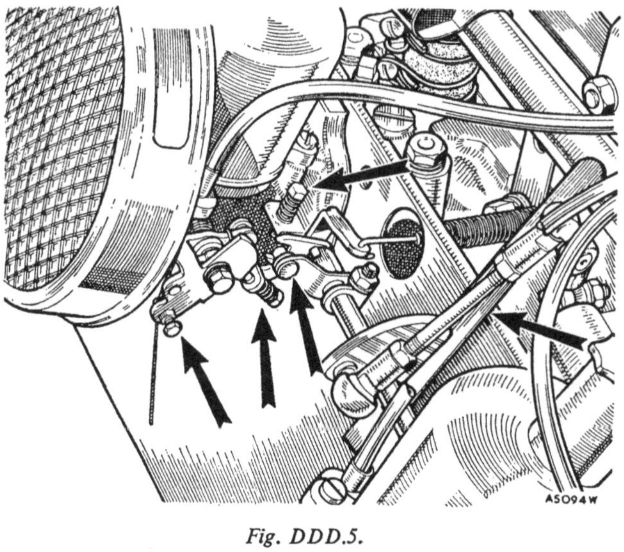

Fig. DDD.5.
The carburetter linkage with a feeler behind the throttle shaft stop and the pin at the bottom of the clearance in the forked lever. The throttle, fast idling and choke cable securing screws are also indicated.

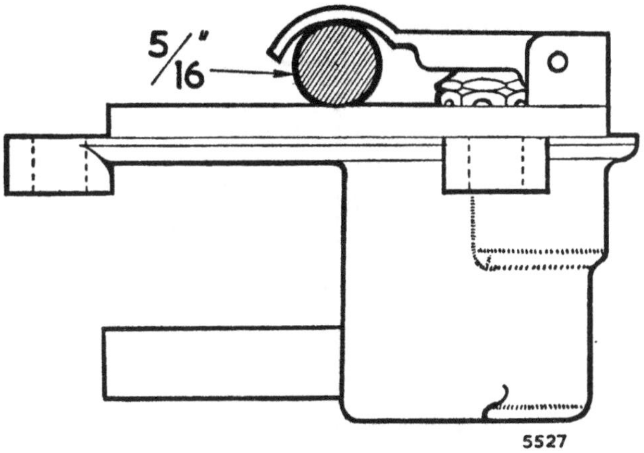

Fig. DDD. 6.
The method of checking the correct adjustment of the float lever.

Re-connect the choke cables, ensuring that the jet heads return against the lower face of the jet adjusting nuts when the choke control is pushed fully in.

Pull out the mixture control knob on the dash panel until the linkage is about to move the carburetter jets (a minimum of $\frac{1}{4}$ in. (6 mm.)) and adjust the fast-idle cam screws to give an engine speed of about 1,000 r.p.m. when hot.

The Float-chamber

The position of the forked lever in the float-chamber must be such that the level of the float (and therefore the height of the fuel at the jet) is correct.

This is checked by inserting a $\frac{5}{16}$ in. (7.94 mm.) round bar between the forked lever and the machined lip of the float-chamber lid. The prongs of the lever should rest on the bar (see Fig. DDD.6) when the needle is on its seating. If this is not so, the lever should be reset at the point where the prongs meet the shank. Care must be taken not to bend the shank, which must be perfectly flat and at right angles to the needle when it is on its seating.

Jet centring

To check the jet for concentricity with the jet needle, set the jet head and the jet adjusting nut in the uppermost position, lift the suction piston with the piston lifting pin and allow the piston to fall. It should fall freely and a definite soft metallic click will be heard as the base of the piston strikes the jet bridge.

If this does not happen with the jet raised, but does occur when the jet is lowered, the jet bearing and jet must be recentred as follows:—

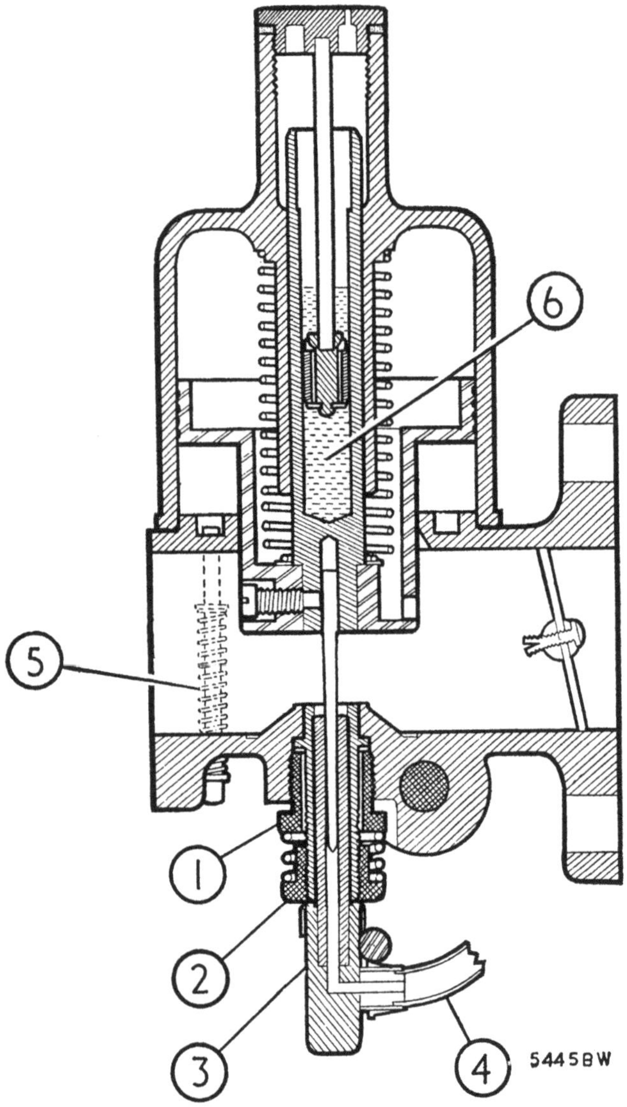

Fig. DDD. 7.
A sectional view of the H.S.4 carburetter

1. Jet locking nut.	*4. Feed tube from float chamber*
2. Jet adjusting nut.	*5. Piston lifting pin.*
3. Jet head.	*6. Damper reservoir.*

Disconnect the link between the jet head and carburetter lever by removing the small Phillips retaining screw from the jet head.

Unscrew the union securing the jet feed tube into the base of the float chamber and withdraw the jet from the jet bearing, complete with feed tube.

Unscrew the jet adjusting nut and remove the lock spring; screw up the nut to its fullest extent and refit the jet head and feed tube. Slacken the jet locking nut until the jet bearing is just free to rotate with finger pressure. Remove the piston damper from the top of the suction chamber body and gently press down the piston on to its stop.

Tighten the jet locking nut, at the same time ensuring that the jet head is held firmly in its uppermost position and at its correct angular relation to the float chamber.

Repeat the check for concentricity both with the jet raised and lowered. If the result is not satisfactory the recentring operation must be repeated until the correct result is obtained.

When the operation is completed, replace the adjusting nut lock spring and the jet operating link.

This adjustment is best effected with the carburetters removed from the engine.

Float needle sticking

If the engine runs unevenly, apparently through lack of fuel, when there is plenty in the tank and the pump is working properly, the probable cause is a sticking float needle. An easy test for this is to disconnect the pipe from the electric pump to the carburetter and switch the ignition on and off quickly while the end of the pipe is directed onto a pad of cloth or into a container.

If fuel is delivered, starvation is almost certainly being caused by the float needle sticking to its seating, and the float chamber lid(s) should therefore be removed and the needle and seating cleaned and refitted.

At the same time it will be advisable to clean out the entire fuel feed system as this trouble is caused by foreign matter in the fuel, and unless this is removed it is likely to recur. It is of no use whatever renewing any of the component parts of the carburetter(s), and the only cure is to make sure that the fuel tank and pipe lines are entirely free from any kind of foreign matter or sticky substance capable of causing this trouble.

Piston sticking

The piston assembly comprises the suction disc and the piston forming the choke, into which is inserted the hardened and ground piston rod which engages in a bearing in the centre of the suction chamber and in which is, in turn, inserted the jet needle. The piston rod running in the bearing is the only part which is in actual contact with any other part, the suction disc, piston, and needle all having suitable clearances to prevent sticking. If sticking does occur the whole assembly should be cleaned carefully and the piston rod lubricated with a spot of thin oil. No oil must be applied to any other part except the piston rod. A sticking piston can be ascertained by removing the piston damper and lifting the piston by pressing the piston lifting pin; the piston should come up quite freely and fall back smartly onto its seating when released. On no account should the piston return spring be stretched or its tension altered in an attempt to improve its rate of return.

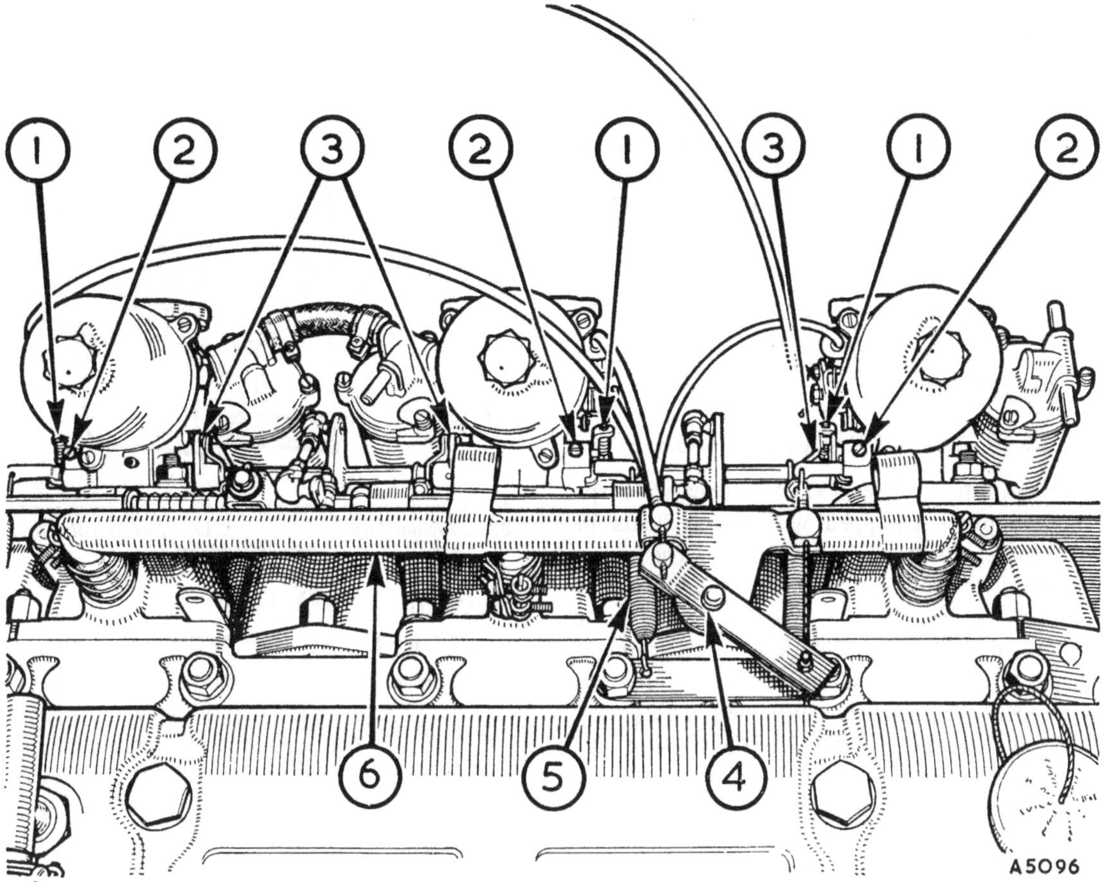

Fig. DDD. 8.
The layout of the triple H.S.4 carburetters.

1. Fast idling adjusting screws. *4. Choke cable relay lever.*
2. Throttle adjusting screws. *5. Throttle return spring.*
3. Throttle operating levers. *6. Balance tube.*

The fuel pipe connecting the front and centre carburetters, and the air cleaners, are removed for clarity.

Water and dirt in the carburetter

Should this be suspected, lift the piston with a pencil; the jet can then be seen. Flood the carburetter and watch the jet; if fuel does not flow freely there is a blockage. To remedy this start the engine, open the throttle, and block up the air inlet momentarily, keeping the throttle open until the engine starts to race.

If the jet is completely blocked it must be removed and thoroughly cleaned.

Float-chamber flooding

This is indicated by fuel flowing from the drain hole in the top of the float chamber lid below the main fuel feed pipe, and is generally caused by grit between the float chamber needle and its guide. The float-chamber lid should be removed and the needle and its guide thoroughly cleaned.

Section DDD.7

CARBURETTER REMOVAL

Turn the battery master switch to the 'off' position.

Disconnect the fuel feed pipe from the front carburetter, the two snap-lock ball joints from the accelerator relay shaft, and the three throttle return springs.

Release the three mixture control cables from the carburetter levers. Slacken the retaining clip and remove the engine breather hose from the rear air cleaner. Pull off the rubber connector for the vacuum ignition control pipe from the top of the rear carburetter body.

Remove the two nuts, spring washers and plain washers securing each carburetter flange and withdraw the three carburetters as one unit.

Detach the throttle interconnecting shafts, remove the fuel pipes and separate the carburetters.

Refitting is a reversal of the removal procedure.

The throttle linkage must be checked and re-adjusted if necessary after refitting.

Section DDD.8

FLOAT CHAMBER OVERFLOW PIPES

Flexible plastic overflow pipes were fitted to each carburetter float chamber from Power Unit No. 29E-H-1092. The float chamber lids were modified to incorporate short overflow nozzles on to which the flexible pipes are a push fit. The overflow pipes may be fitted with the modified lids to earlier 3000 Mk. II cars.

Section DDD.9

CARBURETTERS WITH NYLON FLOATS

Carburetters fitted to later 3000 Mk. II cars incorporated float chambers equipped with nylon floats in place of the metal floats used previously. The nylon floats are integral with the float levers which are attached to the float chamber lids. The nylon float and lever assembly may be interchanged with the earlier metal float and separate lever. Red aluminium tags were used for a time to identify carburetters modified in this way.

To check the float level, hold the float chamber lid and float assembly upside down and place a ⅛ in. (3·18 mm.) diameter bar across the diameter of the machined lip of the float chamber lid, parallel with the float lever hinge pin, and under the float lever (see Fig. DDD.9). The face of the float lever should just rest on the bar when the needle valve is fully on its seating. If it does not do this, carefully reset the angle made between the straight portion of the float lever and its hinge until the correct position is obtained.

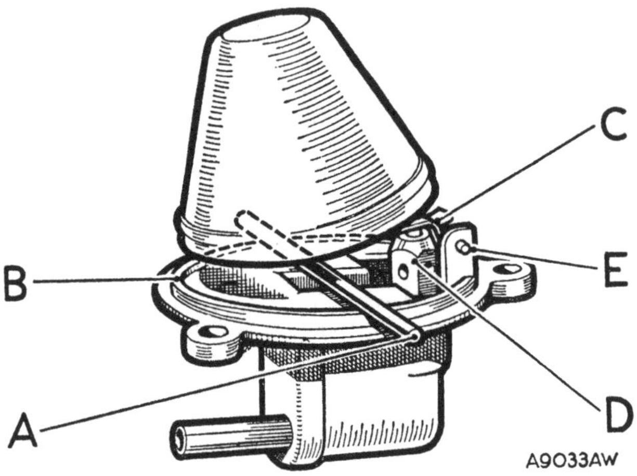

Fig. DDD.9 Checking the nylon float level.
A. *⅛ in. diameter bar.* C. *Float lever resetting point.*
B. *Machined lip.* D. *Needle valve assembly.*
E. *Hinge pin.*

Section DDD.10

REVISED LOCATION OF FUEL PUMP

From Car No. 17547 (BN7) and 17352 (BT7) the fuel pump and fuel lines were transferred from the left-hand side to the right-hand side of the car. The repositioning of these components isolates them from the exhaust system and diminishes any possibility of fuel vaporization. This change involved the introduction of new fuel pipes between the tank and the pump, and between the pump and the flexible pipe leading to the carburetters, new petrol pipe fittings, and associated body modifications.

On the BT7, BJ7 and BJ8 the fuel pump is now accessible when the right-hand rear seat pan has been removed.

Access to the fuel pump on the BN7 is obtained in the same way as before (see Section DDD.5) although it is now located on the right-hand side.

Section DDD.11

CARBURETTERS (CONVERTIBLE MODEL Mk II)

The Austin-Healey 3000 Mk. II Sports Convertible (Series BJ7) is equipped with twin S.U. carburetters, type HS6. Each carburetter is attached by four studs and nuts to a detachable one-piece six port induction manifold (see Section A.36). The carburetter float chambers incorporate nylon floats (see Section DDD.9) and are fitted with flexible overflow pipes (see Section DDD.8).

The construction and servicing of the HS6 carburetter are basically similar to that of the HS4 type described in Section DDD.6 For the description and servicing procedure applicable to the carburetters fitted to Convertible models, refer to Section DDD.6, but note the following differences.

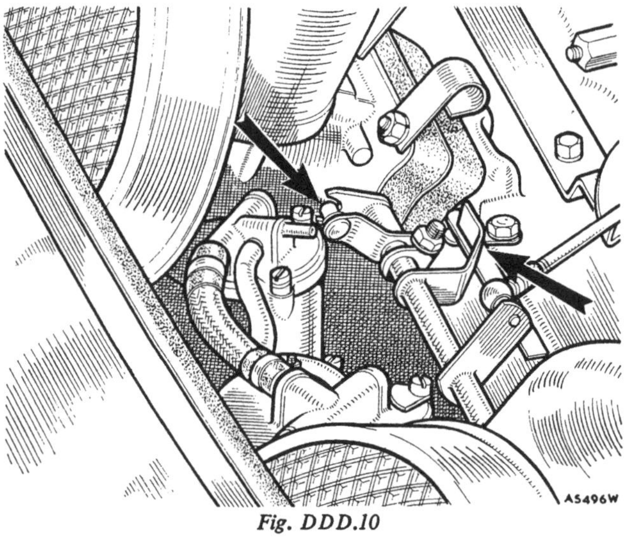

Fig. DDD.10
The HS6 carburetter linkage with a feeler below the throttle shaft stop, and the pin at the bottom of the clearance in the forked lever.

Throttle linkage adjustment

The instructions given under '**Slow running adjustment and synchronisation**' in Section DDD.6 for the slow running and mixture adjustment of the triple carburetter layout apply also to the twin HS6 carburetters. When setting the throttle linkage, however, use the method given below.

The throttle on each carburetter is operated by a lever and pin, with the pin working in a forked lever attached to the throttle spindle. A clearance exists between the pin and fork which must be maintained when the throttle is closed and the engine idling, to prevent any load from the accelerator linkage being transferred to the throttle butterfly and spindle.

To set this clearance, with the throttle shaft levers free on the throttle shaft, put a ·012 in. (·305 mm.) feeler between the throttle shaft stop and its abutment on the inlet manifold (see Fig. DDD.10). Move each throttle shaft lever downwards in turn until the lever pin rests lightly on the lower arm of the fork in the carburetter throttle lever. Tighten the clamp bolt of the throttle shaft lever at this position. When both carburetters have been dealt with, remove the feeler. The pins on the throttle shaft levers should then have clearance in the forks.

Re-connect the choke cables, ensuring that the jet heads return against the lower face of the jet adjusting nuts when the choke control is pushed right in.

Pull out the mixture control knob on the dash panel until the linkage is about to move the carburetter jets (a minimum of ¼ in. (6 mm.)) and adjust the fast-idle cam screws to give an engine speed of about 1,000 r.p.m. when hot.

Float chamber fuel level

Refer to Section DDD.9 for the method of checking and adjusting the level of the nylon floats.

Carburetter removal

Turn the battery master switch to the 'off' position.

Disconnect the fuel feed pipe from the front carburetter, the snap-lock ball joint from the accelerator relay shaft, and the two throttle return springs.

Release the two mixture control cables from the carburetter levers. Remove the engine breather hose from the rear air cleaner. Pull off the rubber connector for the ignition vacuum control pipe from the top of the rear carburetter body.

Remove the four nuts, spring washers and plain washers securing each carburetter flange and withdraw the two carburetters as a unit complete with the float chamber overflow pipes.

Detach the throttle interconnecting shaft, remove the fuel bridge pipe, and separate the carburetters.

Reverse the above procedure when replacing the carburetters and, in addition, check the throttle linkage and re-adjust if necessary.

Section DDD.12

CARBURETTERS (CONVERTIBLE MODEL Mk. III)

The Austin-Healey 3000 Mk. II Sports Convertible (Series BJ8) is equipped with twin S.U. carburetters, type HD8. Each carburetter is attached by four studs and nuts to a detachable one-piece six port induction manifold (see Section A.36).

The construction and servicing of the HD8 carburetter are basically similar to that of the HD6 type described in Section D.12. For the description and servicing procedure applicable to the carburetters fitted to Mk. III Convertible models, refer to Sections D.12 and 13.

Section DDD. 13

FUEL PUMP Type AUF 301 (LATER CONVERTIBLE MODEL MK. III)

Removing

Remove the right-hand seat pan, disconnect the inlet and outlet pipes, earth and feed connections. Remove the nuts securing the pump to the mounting bracket and lift away the assembly.

Dismantling

Contact breaker

(1) Remove the insulated sleeve, terminal nut, and connector together with its shakeproof washer. Remove the tape seal (if fitted) and take off the end-cover.
(2) Unscrew the 5 B.A. screw which holds the contact blade to the pedestal and remove the condenser from its clip. Remove the washer, the long coil lead, and the contact blade.

Coil housing and diaphragm

(3) Unscrew the coil housing securing screws, using a thick-bladed screwdriver to avoid damaging the screw heads.
(4) Remove the earthing screw.
(5) Remove the coil housing from the body.
(6) Remove the diaphragm and spindle assembly by unscrewing the diaphragm anti-clockwise until the armature spring pushes the diaphragm away from the coil housing. It is advisable to hold the housing over the bench so that the 11 brass rollers will not fall on the floor. The diaphragm and its spindle are serviced as a unit and should not be separated.

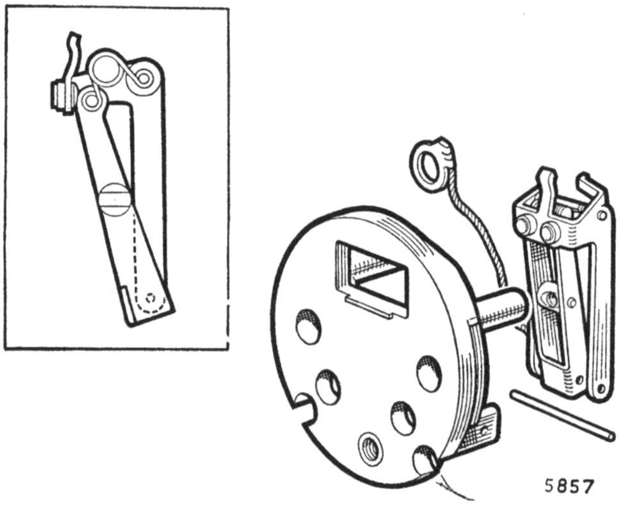

Fig. DDD.11.
*Fitting the rocker assembly to the pedestal. (**Inset**), the correct position of the centre toggle spring.*

Pedestal and rocker

(7) Remove the end-cover seal washer, unscrew the terminal nut, and remove the lead washer. This will have flattened on the terminal tag and is best cut away with cutting pliers or a knife. Unscrew the two 2 B.A. screws holding the pedestal to the coil housing and remove the earth terminal tag. Tip the pedestal and withdraw the terminal stud from the terminal tag. Remove the pedestal and rocker mechanism.

(8) Push out the hardened steel pin which holds the rocker mechanism to the pedestal and separate the two.

Body and valves

(9) Unscrew the two Phillips screws securing the valve clamp plate, remove the valve caps, valves, sealing washers, and filter.

NOTE.—Dismantling of the delivery flow-smoothing device should only be undertaken if its operation is faulty, and if the necessary equipment for pressure-testing after assembly is available. On this understanding proceed as follows:

(10) Remove the four B.A. screws to release the air bottle cover. Remove the 'O' ring, plastic diaphragm, and joint washers.

(11) Remove the single 2 B.A. screw, securing the inlet air bottle cover. Remove the cover and gasket.

(12) Unscrew the inlet and outlet connections.

Inspecting

If gum has formed, the parts in contact with the fuel may have become coated with a substance similar to varnish. This has a strong stale smell and may attack the neoprene diaphragm. Brass and steel parts so affected can be cleaned by being boiled in a 20 per cent solution of caustic soda, dipped in a strong nitric acid solution, and finally washed in boiling water. Light alloy parts must be well soaked in methylated spirits and then cleaned.

(13) Clean the pump and inspect for cracks, damaged joint faces, and threads.

(14) Examine the plastic valve assemblies for kinks or damage to the valve plates. They can best be checked by blowing and sucking with the mouth.

(15) Check that the narrow tongue on the valve cage, which is bent over to retain the valve and to prevent it being forced out of position, has not been distorted but allows a valve lift of approximately $\frac{1}{16}$ in. (1·6 mm.).

(16) Examine the delivery air bottle diaphragm and end cover for damage. If in doubt renew the diaphragm.

(17) Examine the inlet air bottle cover for damage.

(18) Examine the valve recesses in the body for damage and corrosion; if it is impossible to remove the corrosion, or if the recess is pitted, the body must be discarded.

(19) Clean the filter with a brush and examine for fractures, renew if necessary.

(20) Examine the coil lead tag for security and the lead insulation for damage.

(21) Examine the contact breaker points for signs of burning and pitting; if this is evident, the rocker assembly and spring blade must be renewed.

(22) Examine the pedestal for cracks or other damage, in particular to the narrow ridge on the edge of the rectangular hole on which the contact blade rests.

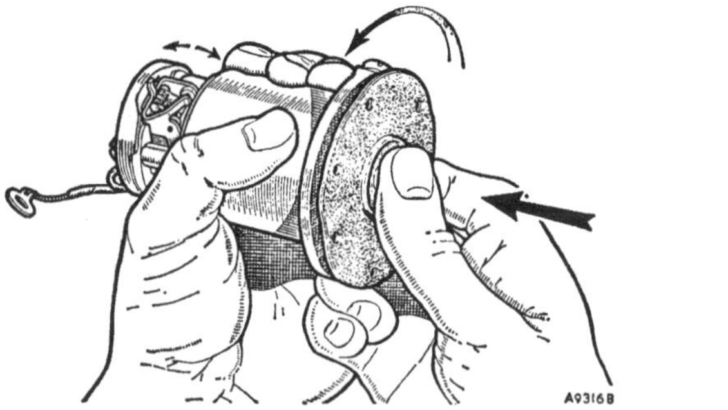

Fig. DDD.12.
Setting the diaphragm.

THE AUF 301 FUEL PUMP COMPONENTS

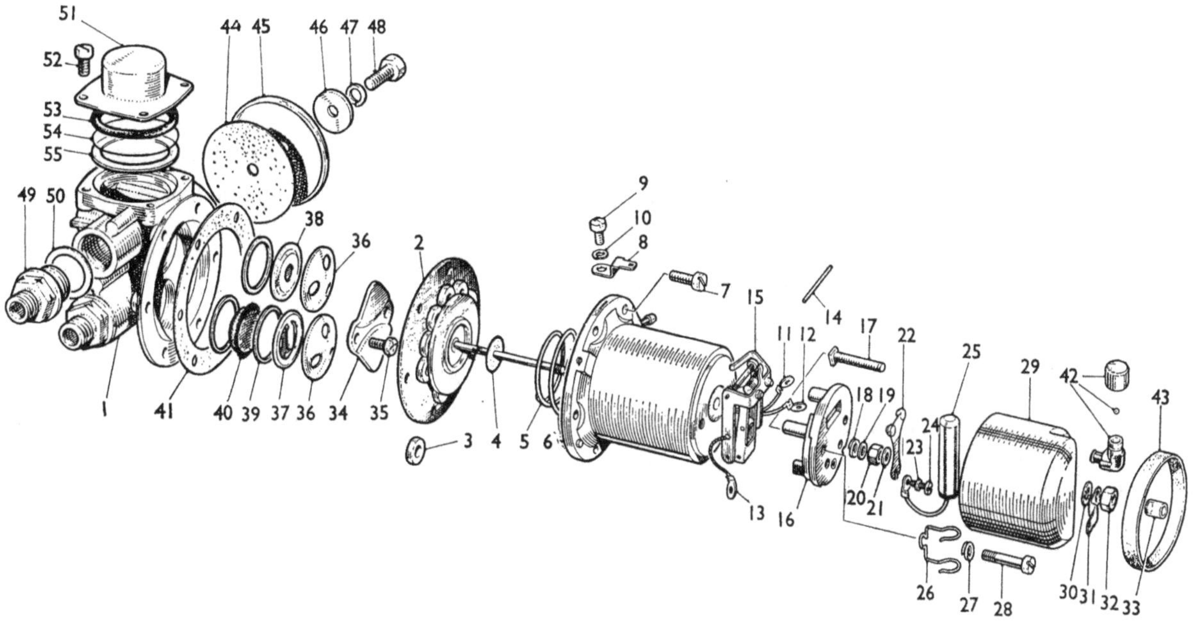

No.	Description	No.	Description	No.	Description
1.	Pump body.	19.	Lead washer.	38.	Outlet valve.
2.	Diaphragm and spindle assembly.	20.	Terminal nut.	39.	Sealing washer.
3.	Armature centralizing roller.	21.	End-cover seal washer.	40.	Filter.
4.	Impact washer.	22.	Contact blade.	41.	Diaphragm gasket.
5.	Armature spring.	23.	Washer.	42.	Vent valve.
6.	Coil housing.	24.	Contact blade screw.	43.	Sealing band.
7.	Set screw.	25.	Condenser.	44.	Inlet air bottle cover joint.
8.	Earth connector.	26.	Condenser clip.	45.	Inlet air bottle cover.
9.	Set screw.	27.	Spring washer.	46.	Dished washer.
10.	Spring washer.	28.	Screw.	47.	Spring washer.
11.	Terminal tag.	29.	End-cover.	48.	Cover securing screw.
12.	Terminal tag.	30.	Shakeproof washer.	49.	Outlet connection.
13.	Earth tag.	31.	Connector.	50.	Sealing washer.
14.	Rocker pivot pin.	32.	Nut.	51.	Delivery air bottle.
15.	Rocker mechanism.	33.	Insulating sleeve.	52.	Cover securing screw.
16.	Pedestal.	34.	Clamp plate.	53.	Rubber 'O' ring.
17.	Terminal stud.	35.	Screw.	54.	Plastic diaphragm.
18.	Spring washer.	36.	Valve cap.	55.	Sealing washer.
		37.	Inlet valve.		

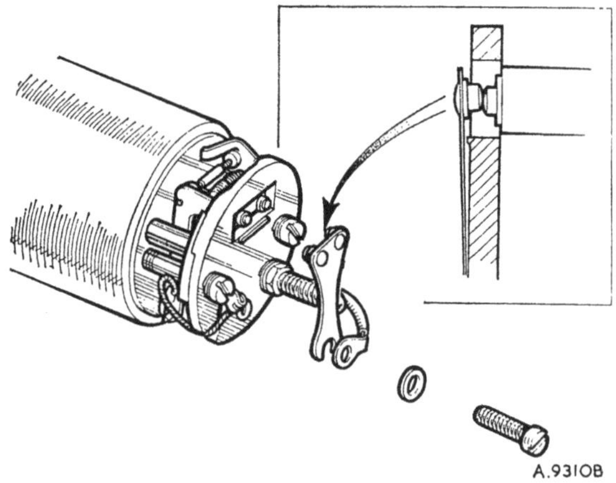

Fig. DDD.13
Setting the correct relative positions of blade and rocker contact points.

(23) If fitted, examine the non-return vent valve in the end cover for damage, ensure that the small ball valve is free to move.

(24) Examine the diaphragm for signs of deterioration.

(25) Renew the following parts: all fibre and cork washers, gaskets, and 'O' section sealing rings, rollers showing signs of wear on periphery, damaged bolts, and unions.

Reassembling

Pedestal and rocker

NOTE.—The steel pin which secures the rocker mechanism to the pedestal is specially hardened and must only be replaced by a genuine S.U. part.

(26) Invert the pedestal and fit the rocker assembly to it by pushing the steel pin through the small holes in the rockers and pedestal struts. Then position the centre toggle so that, with the inner rocker spindle in tension against the rear of the contact points, the centre spring is above the spindle on which the white rollers run. This positioning is important to obtain the correct 'throw-over' action. It is also essential that the rockers are perfectly free to swing on the pivot pin and that the arms are not binding on the legs of the pedestal.

If necessary the rockers can be squared up with a pair of thin-nosed pliers.

(27) Assemble the square-headed 2 B.A. terminal stud to the pedestal, the back of which is recessed to take the square head.

(28) Assemble the 2 B.A. spring washers, and put the terminal stud through the 2 B.A. terminal tag, then fit the lead washer and the coned nut with its coned face to the lead washer. (This makes better contact than an ordinary flat washer and nut). Tighten the 2 B.A. nut and finally add the end-cover seal washer.

(29) Assemble the pedestal to the coil housing by fitting the two 2 B.A. pedestal screws, ensuring that the spring washer on the left-hand screw (9 o'clock position) is between the pedestal and the earthing tag. Fit the condenser wire clip base under the earthing tag.

(30) Tighten the screws, taking care to prevent the earthing tag from turning as this would strain or break the earthing flex. Do not overtighten the screws or the pedestal will crack.

Do not fit the contact blade at this stage.

Diaphragm assembly

(31) Place the armature spring into the coil housing with its large diameter towards the coil.

(32) Before fitting the diaphragm make sure that the impact washer is fitted to the armature. (This is a small neoprene washer that fits in the armature recess). Do not use jointing compound on the diaphragm.

(33) Fit the diaphragm by inserting the spindle in the hole in the coil and screwing it into the threaded trunnion in the centre of the rocker assembly.

(34) Screw the diaphragm until the rocker will not 'throw-over'; this must not be confused with jamming the armature on the coil housing internal steps.

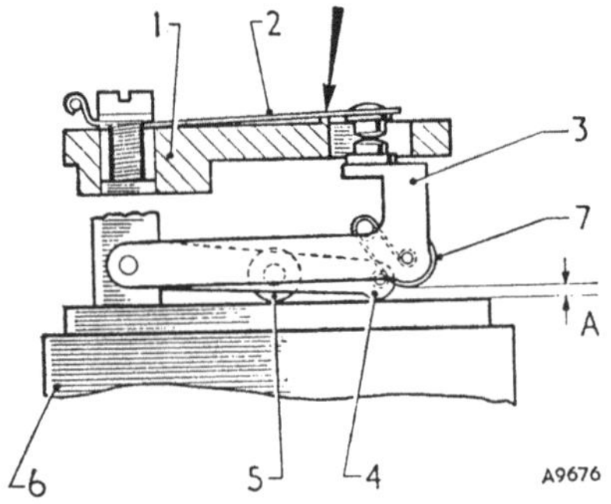

Fig. DDD.14
The contact gap setting on earlier-type rocker assemblies.

1. Pedestal.
2. Contact blade.
3. Outer rocker.
4. Inner rocker.
5. Trunnion.
6. Coil housing.

A. *·030 in (·8 mm.)*

(35) Fit the 11 brass centralizing rollers by turning back the diaphragm edge and dropping the rollers into the coil recess. The pump should be held rocker end downwards to prevent the rollers from falling out.

On later-type rocker mechanisms with adjustable fingers fit the contact blade and adjust the finger settings as described under those headings, then carefully remove the contact blade.

(36) Holding the coil housing assembly in an approximately horizontal position, push the diaphragm spindle in, firmly but steadily. Unscrew the diaphragm, pressing and releasing with the thumb until the rocker just 'throws over'. Now turn the diaphragm back (unscrew) to the nearest hole and again 4 holes (two-thirds of a complete turn). The diaphragm is now correctly set.

(37) Press the centre of the armature and fit the retaining fork at the back of the rocker assembly. This is done to prevent the rollers from falling out when the coil housing is placed on the bench prior to fitting the body, and is not intended to stretch the diaphragm before tightening the body screws.

Body components

(38) The valve assemblies are retained in the body by a clamp plate secured with self-tapping screws. The inlet valve recess in the body is deeper than the outlet recess to allow for the filter and extra washer. Another feature of these pumps is the incorporation of an air bottle on the inlet and a flow-smoothing device on the delivery side.

The inlet air bottle is a chamber in the body casting blanked off by a single cover and joint washer held by a single screw. The delivery flow-smoothing device is formed by a perforated metal plate which is in contact with a plastic barrier backed by a rubber diaphragm, all held in position by a spring and end-cap retained by a vented cover. This assembly seals the delivery chamber in the body.

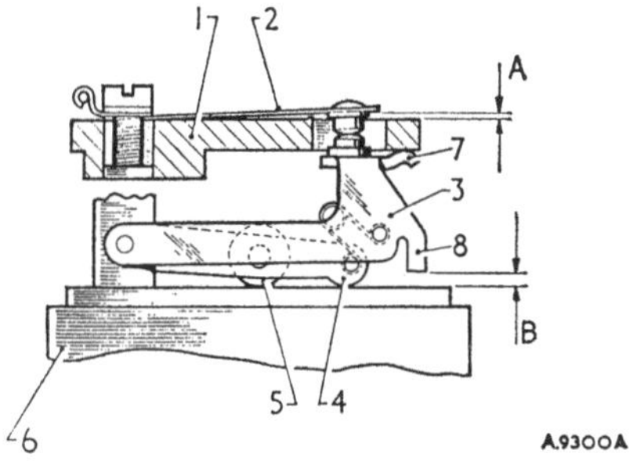

Fig. DDD.15
The contact gap setting on modified rocker assemblies.

1. Pedestal.
2. Contact blade.
3. Outer rocker.
4. Inner rocker.
5. Trunnion.
6. Coil housing.

A ·035 in. (·9 mm.).
B. ·070 in. (1·8 mm.).

(39) Screw in the inlet and outlet connections with their sealing rings. Assemble the outlet valve components into the outlet recess in the following order: first a joint washer, then the valve, tongue side downwards, then the valve cap.

(40) Assemble the inlet valve into the inlet recess as follows: first a joint washer, then the filter dome side downwards, then another joint washer, followed by the valve assembly tongue side upwards, then the valve cap.

(41) Take care that both valve assemblies nest down into their respective recesses, place the clamp plate on top, and tighten down firmly to the body with the two screws.

(42) Replace the inlet air bottle cover with its joint washer and tighten down the central screw.

(43) Place the sealing washer in the bottom of the delivery air bottle, place the plastic diaphragm dome side downwards, then add the 'O' section sealing ring and tighten down the cap with its four screws.

Body attachment

(44) Fit the joint washer to the body, aligning the screw holes.

(45) Offer up the coil housing to the body, ensuring correct seating between them.

(46) Line up the six securing screw holes, making sure that the cast lugs on the coil housing are at the bottom, insert the six 2 B.A. screws finger-tight. Fit the earthing screw with its Lucar connector.

(47) **Remove the roller-retaining fork before tightening the body securing screws,** making sure that the rollers retain their position; a displaced roller will cut the diaphragm. It is not necessary to stretch the diaphragm before tightening the securing screws. Tighten the screws in sequence as they appear diametrically opposite each other.

Contact blade

(48) Fit the contact blade and coil lead to the pedestal with the 5 B.A. washer and screw. When a condenser is fitted the tag on it is placed under the coil lead tag.

(49) Adjust the contact blade so that the contact points on it are a little above the contact points on the rocker when the points are closed, also that when the contact points make or break, one pair of point wipes over the centre line of the other in a symmetrical manner. As the contact blade is

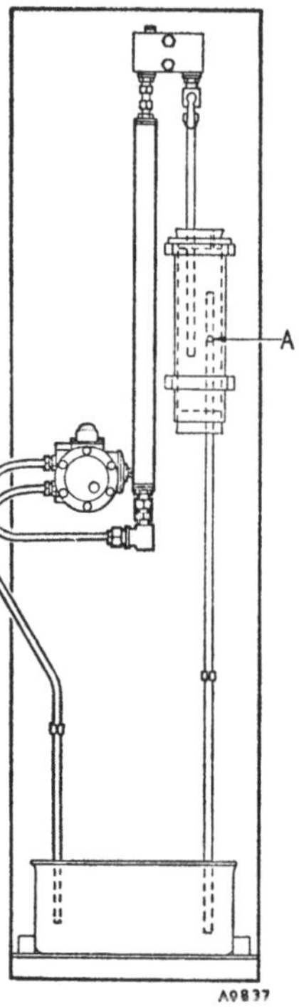

Fig. DDD.16
A checking rig for the fuel pump is obtainable from the S.U. Carburetter Co. Ltd.
1. The $\frac{5}{32}$ in. (4 mm.) dia. hole is 2 in. (51 mm.) below the top of the pipe.

provided with a slot for the attachment screw, some degree of adjustment is possible.

(50) Tighten the contact blade attachment screw when the correct setting is obtained.

Contact gap setting

(51) Check that when the outer rocker is pressed onto the coil housing, the contact blade rests on the narrow rib or ridge which projects slightly above the main face of the pedestal. If it does not, slacken the contact blade attachment screw, swing the blade clear of the pedestal, and bend it downwards a sufficient amount so that when repositioned it rests against the rib.

Earlier-type rocker assemblies

(52) Check the gap between the points indirectly by carefully holding the contact blade against the rib on the pedestal without pressing against the tip. Then check if a ·030 in. (·76 mm.) feeler will pass between the fibre rollers and the face of the coil housing. If necessary, the tip of the blade can be set to correct the gap.

Modified rocker assemblies

(53) Check the lift of the contact blade above the top of the pedestal with a feeler gauge, bending the stop-finger beneath the pedestal, if necessary to obtain a lift of ·035±·005 in. (·9±·13 mm.).

(54) Check the gap between rocker finger and coil housing with a feeler gauge, bending the stop-finger, if necessary, to obtain a gap of ·070±·005 in. (1·8±·13 mm.).

End cover

(55) Tuck all spare cable into position so that it cannot foul the rocker mechanism. Ensure that the end-cover seal washer is in position on the terminal stud, fit the bakelite end-cover and lock washer, secure with the brass nut, fit the terminal tag or connector, and the insulated sleeve.

(56) The pump is now ready for test. After test replace the rubber sealing band over the end cover gap and seal with adhesive tape. This must be retained when the pump is not mounted internally in a moisture-free region.

Testing

(57) Check that the pump points are correctly gapped.

(58) In order to observe the action of the contact breaker assemblies fit a cut-away cover to the pump.

S.U. test rig

(59) Mount the pump on the test rig using the appropriate adaptor set, and connect to a 12-volt battery.

Churchill test rig

(60) Secure the pump in the clamping ring with the outlet connection uppermost. Connect to a 12-volt battery and with the switch in the 'OFF' position clip the connector to the pump. Connect the delivery and suction pipes to the pump.

(61) Use clean paraffin (kerosene) in the sump pan to a depth of at least 2 in. (51 mm.).

Priming

(62) Unscrew the regulator valve (Churchill only), switch on, and the pump should prime from dry in 10 to 15 seconds. Allow the pump to run for a minute to stabilize the flow.

Air leak check

(63) When the pump is first started air bubbles will be mixed with the fluid discharged into the flow meter (S.U.) or the flow glass (Churchill). These bubbles should cease after the pump has been running for a minute or two. If bubbles continue to appear there is an air leak in the suction side of the pump or the connecting unions, and must be rectified.

Valve seat check

(64) Operate the pump for about 10 minutes and then turn the tap (S.U.), or regulator valve (Churchill) right off. The pump should not heat for at least 20 seconds. If pumping action takes place within 20 seconds the inlet valve is not seating correctly and must be renewed.

Delivery check

(65) *S.U.*

Partially open the tap and check that fuel is delivered to the glass jar. Gradually depress the spring blade to reduce the stroke; the pump should continue working with increasing frequency until it eventually stops due to there being no gap left between the points.

Churchill

Adjust the regulator valve to give the feet of paraffin (kerosene) reading in the pressure gauge (see **'GENERAL DATA'**). When the correct reading has been obtained the pump flow may be read directly from the appropriate colour scale on the flow glass (see **'GENERAL DATA'**).

Reduced voltage check

(66) Operate the pump as described in '*Delivery check*' and reduce the voltage to 9·5 by incorporating a variable resistor and voltmeter on the feed side of the testing circuit and check that the pump is functioning satisfactorily.

Sparking check

(67) Switch on the pump and check for excessive sparking between the points. A small degree of sparking is permissible, but a special leak wire in the solenoid winding is designed to reduce sparking to a minimum. If excessive sparking is evident the solenoid assembly must be renewed.

Refitting

Reverse the removal procedure.

Fault diagnosis

1. *Suspected fuel feed failure*

Disconnect the fuel line at the carburetter and check for flow.

(*a*) If normal, examine for obstructed float-chamber needle seating or gummed needle.

(*b*) If normal initially, but diminishing rapidly and accompanied by slow pump operation, check for correct tank venting by removing the filler cap. Inadequate venting causes a slow power stroke, and excessive burning of the contact points.

(*c*) If reduced flow is accompanied by slow operation of the pump, check for any restriction on the inlet side of the pump, such as a clogged filter, which should be removed and cleaned. In the case of reduced flow with rapid operation of the pump, check for an air leak on the suction side, dirt under the valves, or faulty valve sealing washers.

(*d*) If no flow, check for:

(i) *Electrical supply*

Disconnect the lead from the terminal and test for an electrical supply.

(ii) *Faulty contact points*

If electrical supply is satisfactory check that the tungsten points are in contact. The lead should then be replaced on the terminal and a short piece of bared wire put across the contacts. If the pump then makes one stroke the fault is due to dirt, corrosion, or maladjustment of the tungsten points.

(iii) *Obstructed pipeline between fuel tank and pump*

The inlet pipe should be disconnected; if the pump then operates, trouble is due to a restriction in the pipeline between the pump and the tank. This may be cleared by the use of compressed air after removing the fuel tank

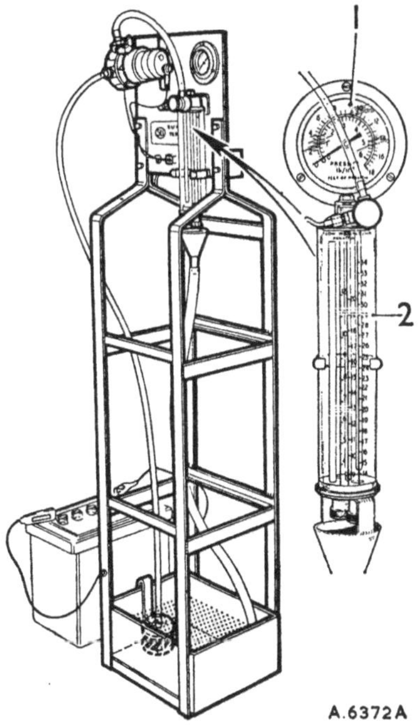

Fig. DDD.17

A checking rig for S.U. fuel pump available from V. L. Churchill and Co. Ltd. The rig measures output in gallons of paraffin (kerosene) per hour, against required suction and delivery heads.

1. Pressure gauge. 2. Flow glass.

filler cap. Do not pass compressed air through the pump, as this would cause serious damage to the valves.

(iv) *Faulty diaphragm action*

If the previous operations fail to locate the trouble, stiffening of the diaphragm fabric or abnormal friction in the rocker 'throw-over' mechanism is to be suspected. To remedy these faults, the coil housing should be removed and the diaphragm flexed a few times taking care not to lose any of the 11 rollers (when fitted) under it. Prior to reassembly, it is advisable to apply a little thin oil to the 'throw-over' spring spindles at a point where they pivot in the brass rockers. The diaphragm armature assembly should then be assembled and set in accordance with the instructions given under that heading.

2. *Noisy pump*

Air leaks. If the pump is noisy in operation, an air leak at one or other of the suction lines may be the cause. Such a leak may be checked by disconnecting the fuel pipe from the carburetter and allowing the pump to discharge into a suitable container with the end of the pipe submerged. Continuous bubbles at this point will indicate an air leak. Rectify the fault as follows:

(*a*) Check that all connections from the fuel tank to the pump are in good order.

(*b*) Check that the inlet union is tight.

(*c*) Check that the coil housing securing screws are well and evenly tightened. Air leaks on the suction side cause rapid operation of the pump and are the most frequent cause of failure.

3. *Pump operates without delivering fuel*

If the pump operates without delivering fuel the most likely causes are:

(*a*) A serious air leak on the suction side, or,

(*b*) Foreign matter lodged under one of the valves, particularly under the inlet valve; remove and clean.

To remedy (*a*) see para 2 above.

SECTION E

CLUTCH

SERIES BN4

Section E.1

DESCRIPTION

The clutch is a Borg & Beck single dry-plate-type operated hydraulically. A steel cover bolted to the flywheel encloses the driven plate, the pressure plate, the pressure springs, and the release levers. The driven plate, to which the friction linings are riveted, incorporates springs assembled around the hub to absorb power shocks and torsional vibration. The pressure springs force the pressure plate against the friction linings, gripping the driven plate between the pressure plate and the engine flywheel. When the clutch pedal is depressed, the release bearing is moved forward against the release plate which bears against the three levers. Each release lever is pivoted on a floating pin, which remains stationary in the lever and rolls across a short, flat portion of the enlarged hole in the eyebolt. The outer ends of the eyebolts extend through holes in the clutch cover and are fitted with adjusting nuts, by means of which each lever is located and locked in position. The outer or shorter ends of the release levers engage the pressure plate lugs by means of struts which provide knife-edge contact between the outer ends of the levers and the pressure plate lugs, so eliminating friction at this point. Pressure applied by the release bearing causes the pressure plate to be pulled away from the driven plate, compressing the pressure springs which are assembled between the pressure plate and the clutch cover. As the friction linings wear, the pressure plate moves closer to the flywheel face and the outer or shorter ends of the release levers follow. This causes the inner or longer ends of the levers to travel farther towards the gearbox and decreases the clearance between the release lever plate and the release bearing. This is automatically compensated unless the master cylinder has been disturbed.

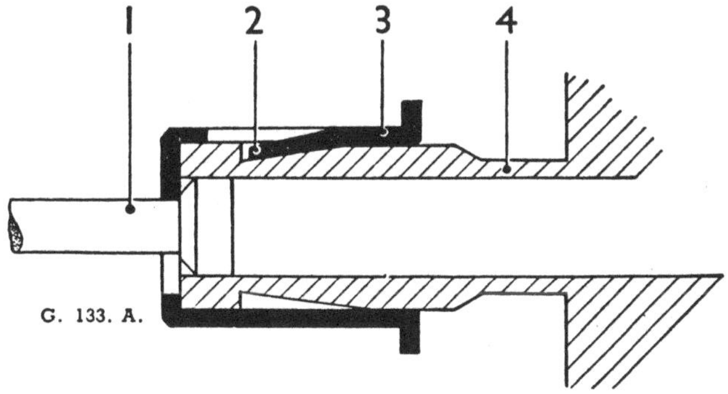

Fig. E.1 Diagrammatic section of master cylinder.

1. Valve stem. *3. Thimble.*
2. Thimble leaf. *4. Plunger.*

When the clutch pedal is depressed, fluid pressure is transmitted through the master cylinder to the slave cylinder, which is mounted on the clutch housing, moving the slave cylinder piston, and push rod. As the push-rod is connected to the lower arm of the clutch withdrawal lever, thereby the clutch is released. The push rod is non-adjustable.

Section E.2

ADJUSTMENT IN VEHICLE

Owing to the hydraulic design of the clutch controls no adjustment is necessary to the clutch pedal.

Section E.3

CLUTCH ASSEMBLY

To Remove

(1) Remove the gearbox as described in Section F.
(2) Slacken the clutch cover screws a turn at a time by diagonal selection until the spring pressure is relieved, when the screws can be taken out and the clutch removed.

To Dismantle

When dismantling the clutch cover assembly the following parts should be suitably marked so that they can be reassembled in exactly the same relative positions to each other to preserve the balance and adjustment—the cover, the lugs on the pressure plate, and the release levers.

The clutch Tool No. 18G 99A provides an efficient and speedy means of dismantling, reassembling and adjusting the clutch with a high degree of accuracy. The tool is universal, and a chart detailing the sizes of spacing washers and distance-pieces for particular types of clutch is provided on the inside of the metal container lid. Proceed as follows :—

(1) Detach the retaining springs from the release lever plate and remove the springs and plate.
(2) Rest the tool base plate on a flat surface, ensure that it is clean, and place upon it spacing washers as directed by the chart. For this 9 in. (22·9 cm.) clutch select three washers and place them in position "D" on the base plate.
(3) Position the clutch on the three spacing washers so that the hole in the clutch cover align with the tapped holes in the base plate, with the release levers as close to the spring washers as possible.
(4) Insert the tool setscrews, tightening them a little at a time in a diagonal pattern, until the cover is firmly and evenly secured to the base plate. This is most important if the best results are to be achieved

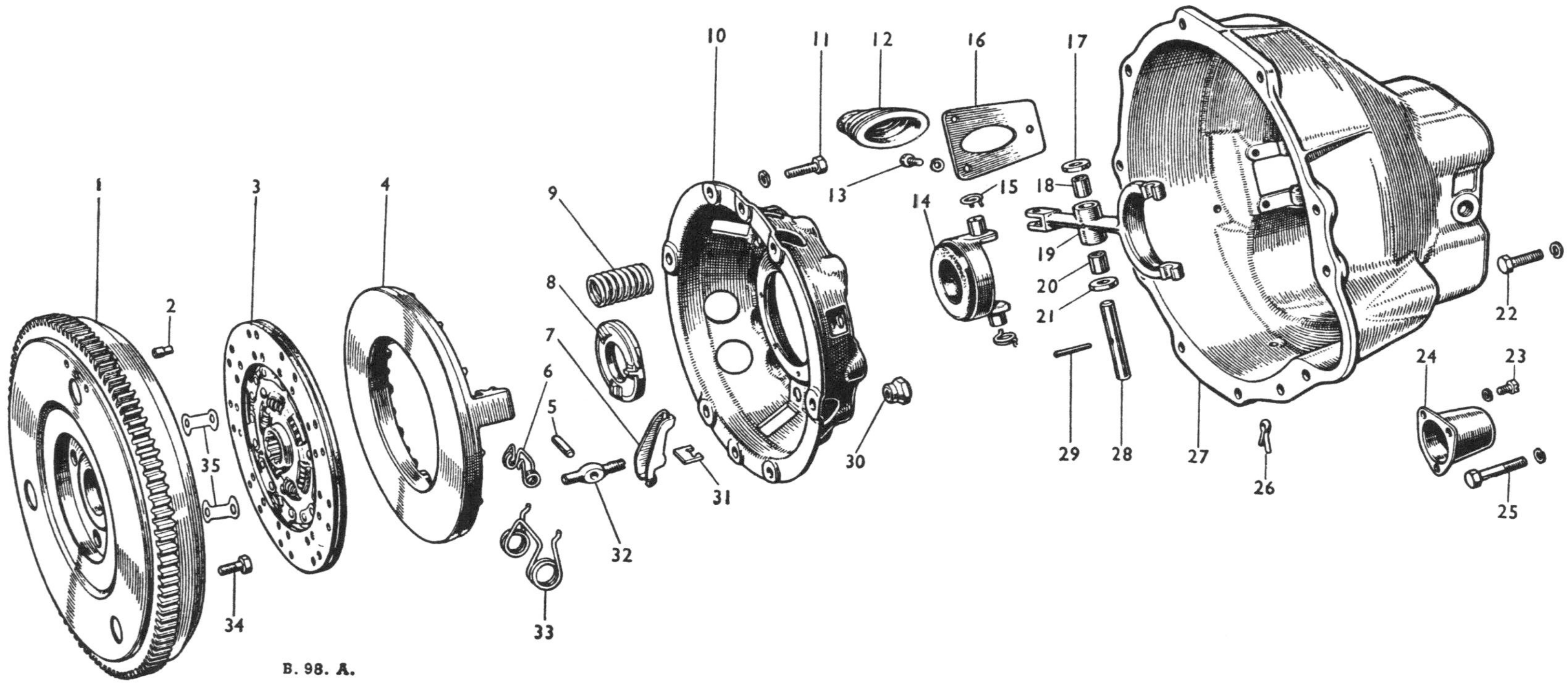

Fig. E.2. The clutch exploded.

1. *Flywheel.*
2. *Locating peg.*
3. *Clutch plate with lining.*
4. *Pressure plate.*
5. *Release lever pin.*
6. *Release lever retainer.*
7. *Release lever.*
8. *Release lever plate.*
9. *Pressure plate spring.*
10. *Clutch cover.*
11. *Cover setpin.*
12. *Fork and lever seal.*
13. *Retaining plate screw.*
14. *Release bearing.*
15. *Release bearing retainer spring.*
16. *Seal retaining plate.*
17. *Fork and lever thrust washer.*
18. *Fork and lever shaft bush.*
19. *Clutch fork and lever.*
20. *Fork and lever shaft bush.*
21. *Fork and lever thrust washer.*
22. *Clutch to gearbox setpin.*
23. *Starter cover screw.*
24. *Cover.*
25. *Clutch to gearbox setpin.*
26. *Split pin for drain hole.*
27. *Clutch housing.*
28. *Fork and lever shaft.*
29. *Taper pin.*
30. *Eye bolt nut.*
31. *Release lever strut.*
32. *Eye bolt.*
33. *Anti-rattle spring.*
34. *Flywheel to crankshaft bolt.*
35. *Lockwashers.*

(5) Remove the three eyebolt adjusting nuts, sheering away the peening by initial pressure.

(6) Unscrew the setscrews securing the clutch cover to the base plate in a diagonal pattern, releasing the pressure on the clutch springs gradually and evenly. Lift off the cover and remove the pressure springs.

(7) To remove the release levers, remove the anti-rattle springs, grasp the lever and eyebolt between the thumb and fingers, so that the inner end of the lever and the threaded end of the eyebolt are as near together as possible, keeping the eyebolt pin seated in its socket in the lever. The strut can then be lifted over the ridge onto the end of the lever, making it possible to lift the eyebolt off the pressure plate.

(8) Clean the clutch parts carefully. If the linings are to be used again they should not be allowed to come in contact with cleaning fluids.

(9) Examine the friction linings for wear or loose rivets and check the driven plate for uneven or worn splines, distortion or signs of fatigue cracks. Generally, it is not desirable to fit new friction linings on the original driven plate because refaced driven plates often are distorted or otherwise impaired and produce unsatisfactory clutch action. If renewing old worn linings, the rivets should be drilled out, not punched out.

(10) After refacing, mount the driven plate on a mandrel between centres and check for "run-out" by means of a dial gauge, set as near to the edge as possible. Where the "run-out" exceeds ·015 in. (·38 mm.), true the plate by prising it in the requisite direction after finding the high spots.

(11) Examine the machined face of the pressure plate. If this is badly grooved and rough, the surface may be reground until the grooves disappear.

(12) Examine the machined surface of the release lever plate. If this is badly grooved, renew the plate. A new plate will also be necesssary if the surfaces on the reverse side of the plate, which are in contact with the tips of the release levers, are worn down.

(13) Examine the tips of the release levers which bear on the back of the release lever plate. A small amount of worn flat surface is permissible, but if this is excessive the lever should be renewed. Check for excessive wear in the groove in which the strut bears. Examine carefully the "U"-shaped depression in the lever into which fits the eyebolt floating pin. If the metal here has worn at all thin, the lever must be renewed as there is a danger of it breaking under load with disastrous results to the whole clutch mechanism.

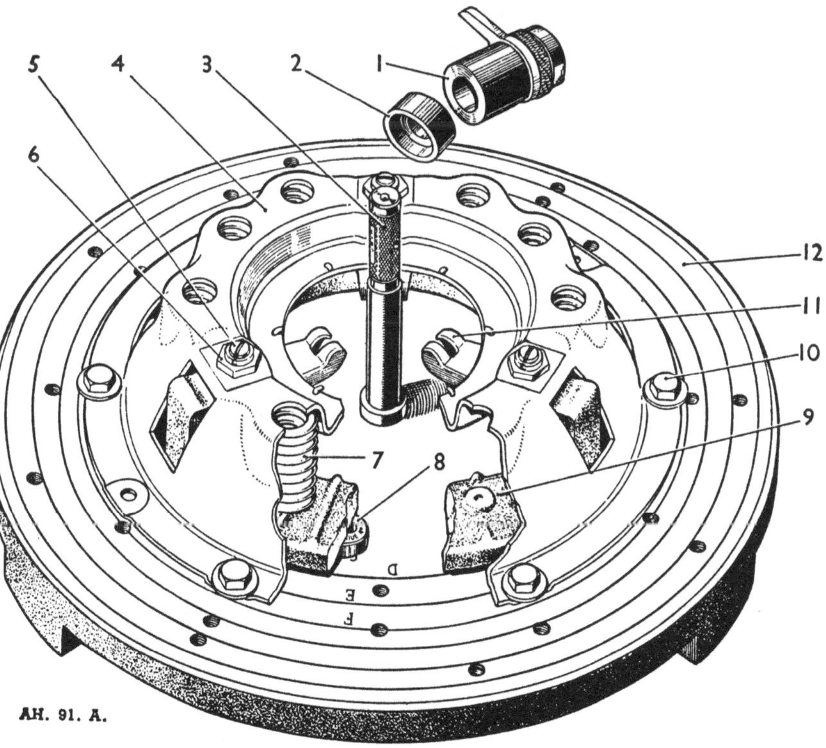

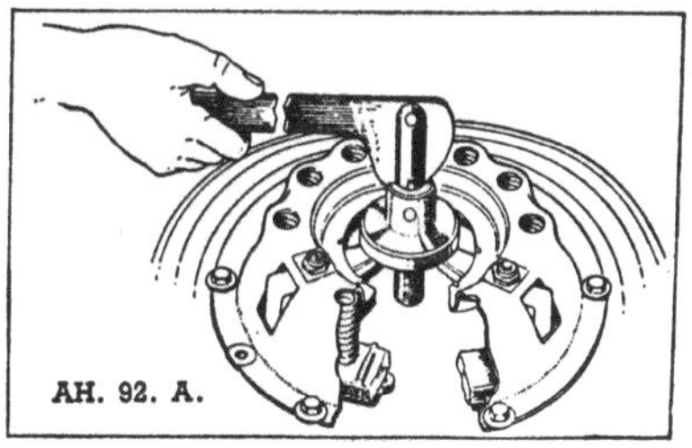

Fig. E.3. Clutch assembly tool.

1. *Height finger.*
2. *Distance piece.*
3. *Centre pillar.*
4. *Clutch cover.*
5. *Eyebolt.*
6. *Eyebolt locknut.*
7. *Thrust spring.*
8. *Spacing washer.*
9. *Pressure plate.*
10. *Setscrew.*
11. *Release lever.*
12. *Base plate.*

Inset shows the clutch tool actuating mechanism in use.

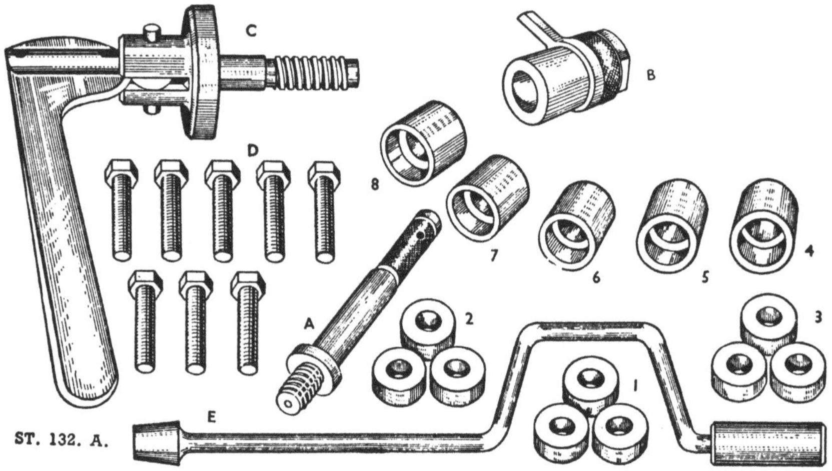

Fig. E.4. Parts of clutch assembly tool.
A. Centre pillar.
B. Height finger.
C. Actuating mechanism.
D. Setpins.
E. Speedbrace.
1., 2., 3. Spacing washer
4., 5., 6., 7., 8. Distance pieces.

(14) Examine each eyebolt for flats on the surface which fits into the pressure plate. If it is a loose fit, it must be renewed. The same applies to the eyebolt floating pin where it passes through the eyebolt. It should be a free fit, but not too loose.

(15) Examine the release bearing for cracks or bad pitting, also measure the amount of bearing standing proud of the metal cup. If the bearing is cracked or badly pitted, or there is $\frac{1}{16}$ in. (1·6 mm.) or less of bearing standing proud of the cup, the cup and bearing must be renewed.

16) Examine the pressure springs for weakness or distortion and renew if necessary. Renew in sets only.

(17) Examine the clutch withdrawal shaft for slackness in the bushes. Renew the bushes if necessary.

To Reassemble

Before reassembly note the positions of the marked parts and make sure to replace them in their original locations to preserve balance, unless the parts have been renewed. Using Tool No. 18G 99A, proceed as follows:—

(1) Position the pressure plate on the three spring washers on the base plate as described under **"To Dismantle".**
Install the release lever, eyebolt and eyebolt pin, holding the threaded end of the eyebolt and the inner end of the lever as close together as possible. With the other hand insert the strut in the slots in the pressure plate sufficiently to allow the plain end of the eyebolt to be inserted in the hole provided in the pressure plate. Move the strut upwards into the slots in the pressure plate lug, over the ridge on the short end of the lever, and drop it into the groove formed in the lever. Fit the remaining release levers in a similar manner. A very slight smear of grease should be applied to the release lever pins, contact faces of the struts, eyebolt seats in the clutch cover, drive lug sides on the pressure plate and the plain end of the eyebolts.

(3) Place the pressure springs on the bosses on the pressure plate.

(4) Lower the cover over the assembled parts, ensuring that the anti-rattle springs are in position and that the tops of the pressure springs are directly under their seats in the cover. In addition the machined portions of the pressure plate lugs must be directly under the slots in the cover through which they will pass.

(5) Insert the tool setscrews through the cover holes and screw them into the base plate in a diagonal pattern, a little at a time, to prevent distortion. Guide the eyebolts and pressure plate lugs through the holes in the clutch cover during this gradual tightening down.

(6) Screw the adjusting nuts onto the eyebolts.

The clutch must now be adjusted still using the clutch assembly tool. With the clutch bolted to the tool base plate as on completion of assembly, proceed as follows:—

(7) Screw the actuator into the base plate and pump the handle a dozen times to settle the clutch mechanism. Remove the actuator.

(8) Screw the tool centre pillar into the base plate and select a distance-piece, Code No. 7, as shown on the chart. Place the distance piece over the centre pillar with its recessed face-downwards.

(9) Place the gauge height finger over the centre pillar.

(10) Adjust the height of the release levers by tightening or loosening the eyebolts until the height finger when rotated, just contacts the highest point on the tips of the release levers. Press downwards on the height finger to ensure that it bears squarely on the adaptor while rotating.

(11) Remove the height finger and pillar, and screw in the actuator to the base plate. Operate the clutch several times to enable the components to settle on their knife edges. Remove the actuator and replace the centre pillar, distance-piece and height finger. Readjust the release levers if necessary. Repeat the procedure to ensure that the release levers are finally seated, and gauge once more.

(12) Remove the centre pillar, distance-piece and height finger and peen over the release lever adjusting nuts.

(13) Fit the release lever plate on the tips of the release levers and secure it by the three retaining springs.

(14) Release the tool setscrews in diagonal sequence a little at a time, relieving pressure slowly and evenly. Remove the clutch assembly from the base plate.

To Replace

Before installing the clutch assembly the engine flywheel should be checked for misalignment (see Section A). To install the clutch proceed as follows :—

(1) Hold the clutch cover assembly and driven plate on the flywheel and screw in the cover securing bolts finger-tight. Note that the splines in the hub of the driven plate are chamfered at one end to permit ready entry of the first motion shaft splines. The longer side of the driven plate hub, with the chamfered splines, should be toward the rear.

(2) Insert a pilot shaft or an aligning arbor No. 18G 79, through the clutch cover and driven plate hub so that the pilot enters the spigot bearing in the rear end of the engine crankshaft. This will centralise the driven plate.

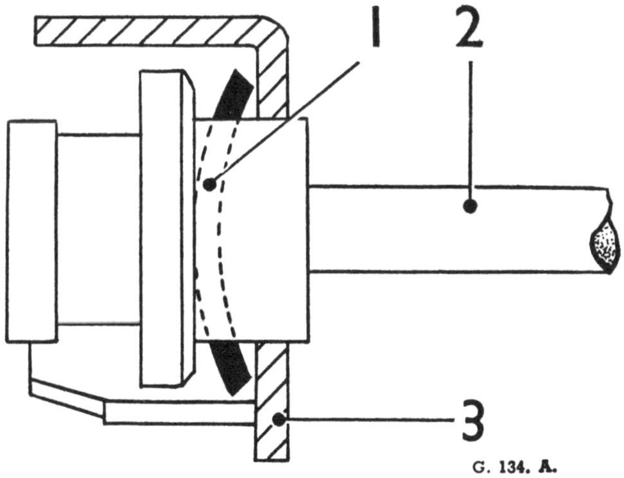

Fig. E.5. Further section of master cylinder.
1. Washer. 2. Valve stem. 3. Valve spacer.

(3) Tighten the clutch cover securing bolts a turn at a time in diagonal sequence to avoid distorting the cover.

(4) Remove the pilot shaft or aligning arbor.

(5) Install the gearbox (see Section F.).

Section E.4

CLUTCH PEDAL

To Remove

(1) The clutch and brake pedal linkages are mounted in a common bracket and thus have to be released as a unit.

(2) Inside the car disconnect the clutch and brake cylinder levers from their master cylinder push rods by removing the clevis pins.

(3) Working under the bonnet unscrew the six securing setpins sufficiently to allow the clutch and brake pedal linkage bracket to be withdrawn from inside the car.

(4) Release the clutch and brake pedal return springs.

(5) Unscrew the nut securing the clutch and brake pedal shaft and withdraw the shaft to release the clutch and brake pedal levers together with their distance piece.

(6) Inspect the lever bushes for wear and renew if necessary.

To Replace

Replacement is the reverse of the procedure "To Remove".

Section E.5

MASTER CYLINDER

Description

The master cylinder consists of an alloy body with a polished finish bore, and reservoir with cap. The inner assembly is made up of the push rod, dished washer, circlip, plunger, plunger seal, spring thimble, plunger return spring, valve spacer, spring washer, valve stem and valve seal. The open end of the cylinder is protected by a rubber dust cover.

Dismantling the Clutch Master Cylinder

(1) Release the master cylinder push rod from the clutch pedal as described in Section E.4.

(2) Disconnect the pressure pipe union from the cylinder and remove the securing bolts, then the master cylinder and fluid reservoir may be withdrawn complete from the car.

(3) Remove the filler cap and drain out the fluid. Pull back the rubber dust cover and remove the circlip with a pair of long nosed pliers. The push rod and dished washer can then be removed.

(4) When the push rod has been removed the plunger with seal attached will be exposed; remove the plunger assembly complete. The assembly can be separated by lifting the thimble leaf over the shouldered end of the plunger.

(5) Depress the plunger return spring allowing the valve stem to slide through the elongated hole of the thimble thus releasing the tension on the spring.

(6) Remove the thimble, spring and valve complete.

(7) Detach the valve spacer, taking care not to lose the spacer spring washer which is located under the valve head. Remove the seal.

(8) Examine all parts, especially the seals, for wear or distortion and replace with new parts where necessary.

Assembling the Clutch Master Cylinder

(1) Replace the valve seal so that the flat side is correctly seated on the valve head.

(2) The spring washer should then be located with the dome side against the underside of the valve head, and held in position by the valve spacer, the legs of which face towards the valve seal.

(3) Replace the plunger return spring centrally on the spacer, insert the thimble into the spring and depress until the valve stem engages through the elongated hole of the thimble, making sure the stem is correctly located in the centre of the thimble. Check that the spring is still central on the spacer.

(4) Refit a new plunger seal with the flat of the seal seated against the face of the plunger. Insert the reduced end of the plunger into the thimble until the thimble leaf engages under the shoulder of the plunger. Press home the thimble leaf.

(5) Smear the assembly with the recommended brake fluid, and insert the assembly into the bore of the cylinder valve, end first, easing the plunger seal lips in the bore.

(6) Replace the push rod with the dished side of the washer under the spherical head, into the cylinder followed by the circlip which engages into the groove machined in the cylinder body.

(7) Replace the rubber dust cover and refit the whole unit into its aperture in the scuttle, not forgetting to fit the packing washer first. Secure the unit by means of the two bolts on the flange and refit the pressure pipe union into the cylinder.

(8) Reconnect the push rod fork with its corresponding hole in the clutch pedal lever, securing it with the circlip.

(9) Bleed the hydraulic system.

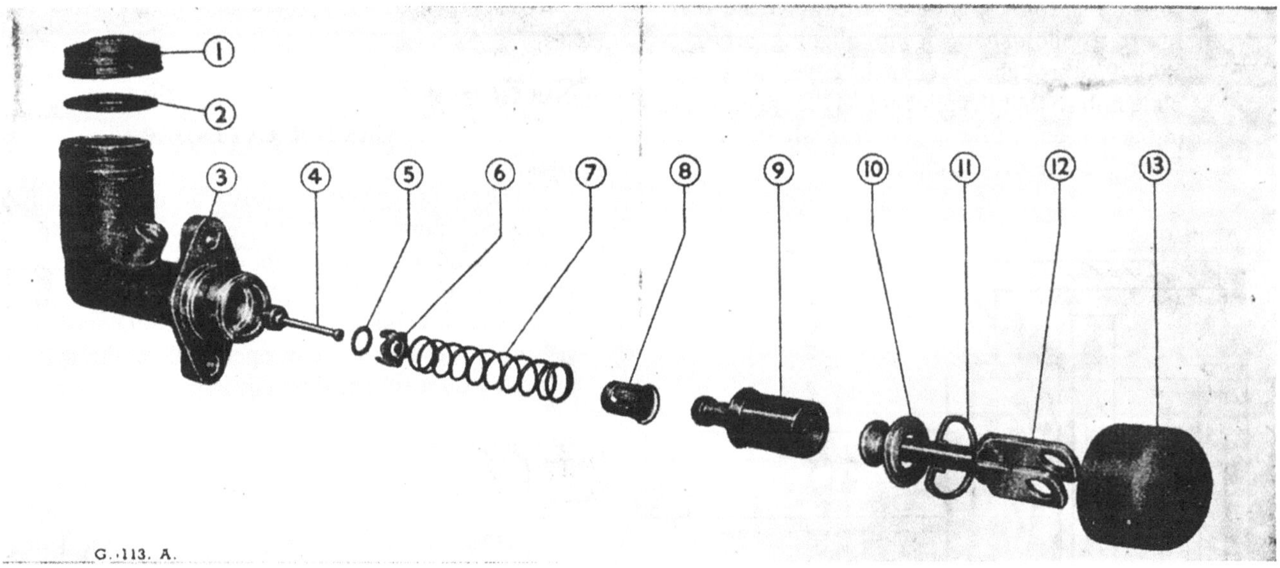

Fig. E.6. Components of master cylinder and reservoir.

1. *Filler cap.*
2. *Washer.*
3. *Master cylinder.*
4. *Valve stem.*
5. *Spring washer.*
6. *Valve spacer.*
7. *Return spring.*
8. *Thimble.*
9. *Plunger.*
10. *Dished washer.*
11. *Circlip.*
12. *Fork.*
13. *Dust cover.*

Section E.6

SLAVE CYLINDER

Description

The cylinder is bolted to the clutch housing and comprises a piston, rubber cup, cup filler, spring, push-rod and bleeder screw. Fluid from the master cylinder is delivered through a flexible hose leading from a union in a bracket on the longitudinal member.

To Remove

(1) Place a receptacle to catch the fluid and remove the flexible hose from the slave cylinder. Note that the thicker washer on the hose connection is nearest the cylinder.
(2) Remove the split pin and clevis pin from the clutch withdrawal lever jaw end, thus freeing the slave cylinder push rod.
(3) Remove the two bolts and spring washers securing the cylinder to the clutch housing.

To Dismantle

(1) Remove all dirt from the outside of the cylinder.
(2) Remove the rubber dust cap from the bleed nipple, attach a bleed tube, open the bleed screw three-quarters of a turn and pump the clutch pedal until all the fluid has been drained into a clean container.
(3) Unscrew the pressure pipe union at the cylinder and remove the setpins from the flange. The slave cylinder can now be removed.
(4) Remove the rubber cover and if an air line is available, blow out the piston and seal. The spring can also be removed.
(5) Clean the slave cylinder components, **using only hydraulic fluid or alcohol.** The main casting may be cleaned with any of the normal cleaning fluids, but all traces of the cleaning fluid must be dried out.
(6) Dry off and examine all rubber components and renew them if they are swollen, distorted or split. If there is any doubt at all as to their condition they must be renewed.
(7) Inspect the piston and cylinder bore for wear and scores, and renew them as necessary.

Assembling the Slave Cylinder

(1) Place the seal into the stem of the piston, with the back of the seal against the piston.
(2) Replace the springs with the small end on the stem, smear well with the recommended fluid and insert into the cylinder.
(3) Replace the rubber dust cover and mount the cylinder in position, making sure the push rod enters the hole in the rubber boot.

To Replace

(1) Secure the cylinder to the clutch housing, and screw in the pipe union.
(2) Bleed the clutch hydraulic system as described in Section E.7.

Section E.7

BLEEDING THE CLUTCH SYSTEM

(1) Remove the bleed screw dust cap at the slave cylinder, open the bleed screw approximately three-quarters of a turn and attach a tube immersing the open end into a clean receptacle containing a small amount of brake fluid.
(2) Fill the master cylinder with Castrol Girling Brake Fluid Amber, and by using slow, full strokes, pump the clutch pedal until the fluid entering the container is free from air bubbles.
(3) On a down stroke of the pedal, screw up the bleed screw, remove the bleed tube and replace the dust cap.

Section E.8 FAULT DIAGNOSIS

Symptom	No.	Possible Fault
(a) Drag or Spin	1	Oil or grease on driven plate linings
	2	Bent engine backplate
	3	Misalignment between engine and first motion shaft
	4	Leaking operating cylinder, pipe line or air in system
	5	Driven plate hub binding on first motion shaft splines
	6	First motion shaft binding on its spigot bush
	7	Distorted clutch plate
	8	Warped or damaged pressure plate or clutch cover
	9	Broken clutch plate linings
	10	Dirt or foreign matter in clutch
(b) Fierceness or Snatch		Check 1, 2 and 3 in (a)
	1	Check 4 in (a)
	2	Worn clutch linings
(c) Slip		Check 1, 2 and 3 in (a)
		Check 1 in (b)
	1	Weak thrust springs
	2	Weak anti-rattle springs
(d) Judder		Check 1, 2 and 3 in (a)
	1	Pressure plate out of parallel with flywheel face
	2	Friction facing contact area not evenly distributed
	3	Bent first motion shaft
	4	Buckled driven plate
	5	Faulty engine or gearbox rubber mountings
	6	Worn shackles
	7	Weak rear springs
	8	Propeller shaft bolts loose
	9	Loose rear spring clips
(e) Rattle		Check 3 in (d)
	1	Damaged driven plate, *i.e.* broken springs, etc.
	2	Worn parts of release mechanism
	3	Excessive transmission backlash
	4	Wear in transmission bearings
	5	Release bearing loose on fork
(f) Tick or Knock	1	Worn first motion shaft bush
	2	Badly worn centre plate hub splines
	3	Out of line thrust plate
	4	Faulty bendix drive on starter
	5	Loose flywheel
(g) Driven Plate Fracture		Check 2 and 3 in (a)
	1	Drag and metal fatigue due to hanging gearbox in driven plate

SECTION EE

CLUTCH

SERIES BN6

NOTE

For details of the clutch fitted to BN6 cars refer to Section E.

SECTION EEE

CLUTCH

Mk. I and II (SERIES BN7 and BT7)
AND Mk. II and Mk. III (SERIES BJ7 and BJ8)

NOTE

For all other service and repair procedures see Section E.

Section EEE.1

MASTER CYLINDER

Description

The master cylinder consists of an alloy body with a polished bore to which are connected the fluid inlet and outlet ports. The fluid reservoir is connected to the inlet port at the end of the cylinder by a length of steel pipe. Connected to the outlet port in the top of the cylinder body is the pressure pipe to the clutch slave cylinder. The inner assembly is made up of the push rod, dished washer, circlip, plunger, plunger seal, end seal, spring thimble, plunger return spring, valve spacer, spring washer, valve stem and valve seal. The open end of the cylinder is protected by a rubber dust cover.

Dismantling the Clutch Master Cylinder

(1) Release the master cylinder push rod from the clutch pedal as described in Section E.4.

(2) Disconnect the inlet and pressure pipe unions from the cylinder and remove the securing bolts, then withdraw the master cylinder.

(3) Drain the fluid from the cylinder. Pull back the rubber dust cover and remove the circlip with a pair of long nosed pliers. The push rod and dished washer can then be removed.

(4) When the push rod has been removed the plunger with seal attached will be exposed; remove the plunger assembly complete. The assembly can be separated by lifting the thimble leaf over the shouldered end of the plunger (Fig. E.1).

(5) Depress the plunger return spring and allow the valve stem to slide through the elongated hole of the thimble thus releasing the tension on the spring.

(6) Remove the thimble, spring and valve complete.

(7) Detach the valve spacer taking care not to lose the spacer spring washer located under the valve head. Remove the seal.

(8) Examine all parts, especially the seals, for wear or distortion and replace with new parts where necessary.

Assembling the Clutch Master Cylinder

(1) Replace the valve seal so that the flat side is correctly seated on the valve head.

(2) Locate the spring washer with the domed side against the underside of the valve head, and hold it in position with the valve spacer, the legs of

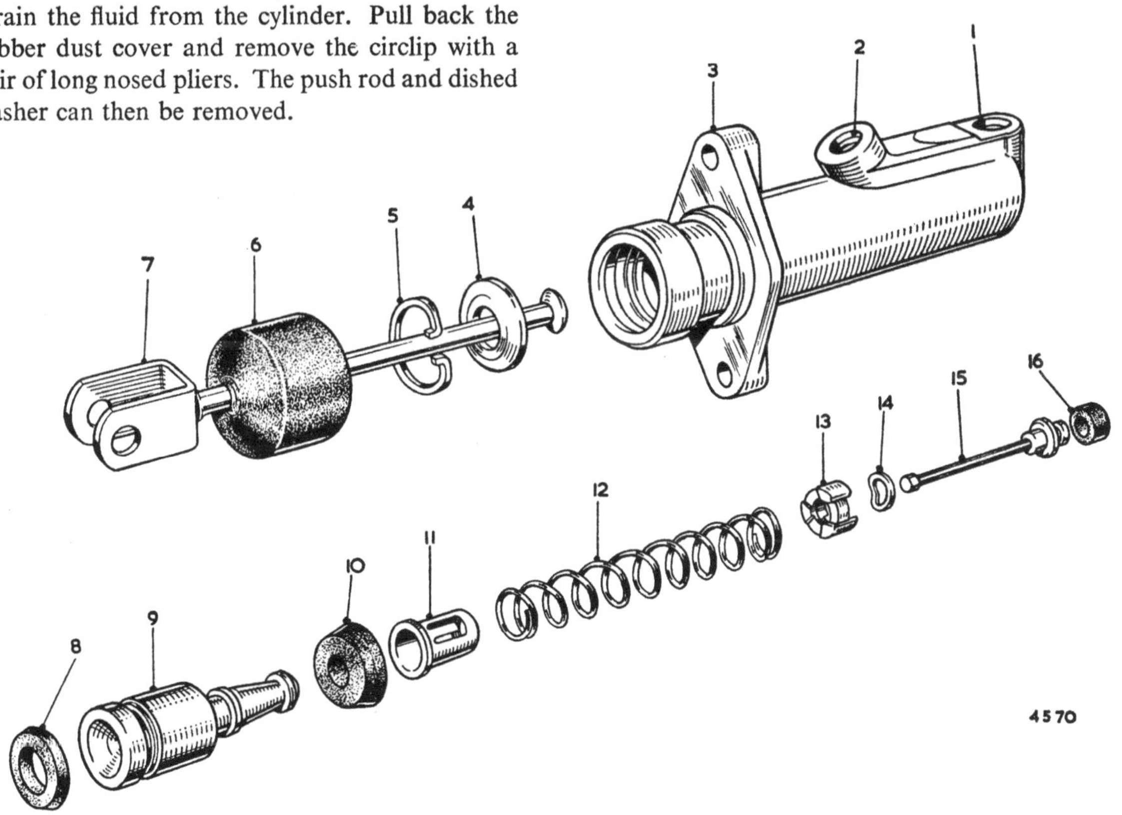

Fig. EEE.1. Components of the master cylinder.

1. *Fluid inlet.*
2. *Fluid outlet*
3. *Master cylinder.*
4. *Dished washer.*
5. *Circlip.*
6. *Dust cover.*
7. *Push rod.*
8. *End seal.*
9. *Plunger.*
10. *Plunger seal.*
11. *Thimble.*
12. *Return spring.*
13. *Valve spacer.*
14. *Spring washer.*
15. *Valve stem.*
16. *Valve seal.*

which should face towards the valve seal. (**Fig. E.5**).

(3) Replace the plunger return spring centrally in the spacer, insert the thimble into the spring and depress until the valve stem engages through the elongated hole of the thimble, making sure the stem is correctly located in the centre of the thimble. Check that the spring is still central on the spacer.

(4) Refit a new plunger seal with the flat of the seal seated against the face of the plunger. Insert the reduced end of the plunger into the thimble until the thimble leaf engages under the shoulder of the plunger. Press home the thimble leaf.

(5) Smear the assembly with Girling red rubber grease or with the recommended hydraulic fluid. Insert the assembly into the bore of the cylinder, valve end first, easing the plunger seal lips into the bore.

(6) Replace the push rod in the cylinder with the dished side of the washer under the spherical head, followed by the circlip which engages in the groove machined in the cylinder body.

(7) Replace the rubber dust cover and refit the whole unit into its aperture in the scuttle, not forgetting to fit the packing washer first. Secure the unit by means of the two bolts on the flange and refit the pressure pipe and inlet pipe unions to the cylinder.

(8) Reconnect the push rod fork with its corresponding hole in the brake pedal lever, securing it with the clevis pin.

(9) Bleed the hydraulic system.

Section EEE. 2

DIAPHRAGM CLUTCH

Description

On later cars from Engine No. 29F/4898 a diaphragm clutch is fitted in which the clutch mechanism is hydraulically operated and consists of a driven plate, a pressure plate, and a diaphragm spring and cover assembly. The cover is bolted to the flywheel and encloses the driven plate, pressure plate, and diaphragm spring.

The hydraulic system comprises a master cylinder coupled to a slave cylinder which operates the clutch release mechanism.

Clutch assembly

The driven plate comprises a splined hub connected to a flexible steel plate by a spring mounting. The annular friction facings are riveted to the plate and damper springs are assembled around the hub to absorb power shocks and torsional vibration.

The diaphragm spring is interposed between two annular rings which provide fulcrum points for the diaphragm when it is flexed. The rings and the diaphragm are located and secured to the cover by nine equally spaced rivets. Three clips that engage the outer edge of the diaphragm are bolted to the pressure plate. The bolts pass through three straps which are riveted to the inside of the cover; the straps prevent the diaphragm and the pressure plate from rotating in relation to the cover.

A release plate having an annular thrust ring is fitted to the outer face of the diaphragm and is retained by a circlip. The release bearing is graphite and is mounted in a cup which fits into the fork of the clutch withdrawal lever. The cup is held in position by two spring retainers.

Removing

Remove the gearbox as described in Section F.

Loosen each of the bolts securing the clutch assembly to the flywheel by slackening them a turn at a time until spring pressure is released. The clutch cover can now be disengaged from the dowels on the flywheel and the assembly removed.

Dismantling

Remove the circlip securing the release plate to the diaphragm and lift the plate from the diaphragm.

Unscrew the three screws securing the clips to the pressure plate, a turn at a time, until the diaphragm contacts the cover. Remove the screws, clips, and washers and the pressure plate.

Rotate the release bearing spring retainers through 90° and withdraw the bearing from the withdrawal lever fork.

Assembling

Assembly is a reversal of the dismantling sequence, but ensure that the release bearing retainers are correctly located and that the spring clip bolts are tightened to the correct torque figure as given under **'GENERAL DATA'**.

Replacing

Position the driven plate assembly on the flywheel with the large end of the hub away from the flywheel.

Centralize the plate by using Service tool 18G79, which fits the splined hub of the driven plate and the pivot bearing in the flywheel. As an alternative a spare first motion shaft can be used.

Locate the cover assembly on the flywheel dowels and secure it with the bolts; tighten the bolts down a turn at a time by diametrical selection. Do not remove the centralizer until all bolts are securely tightened.

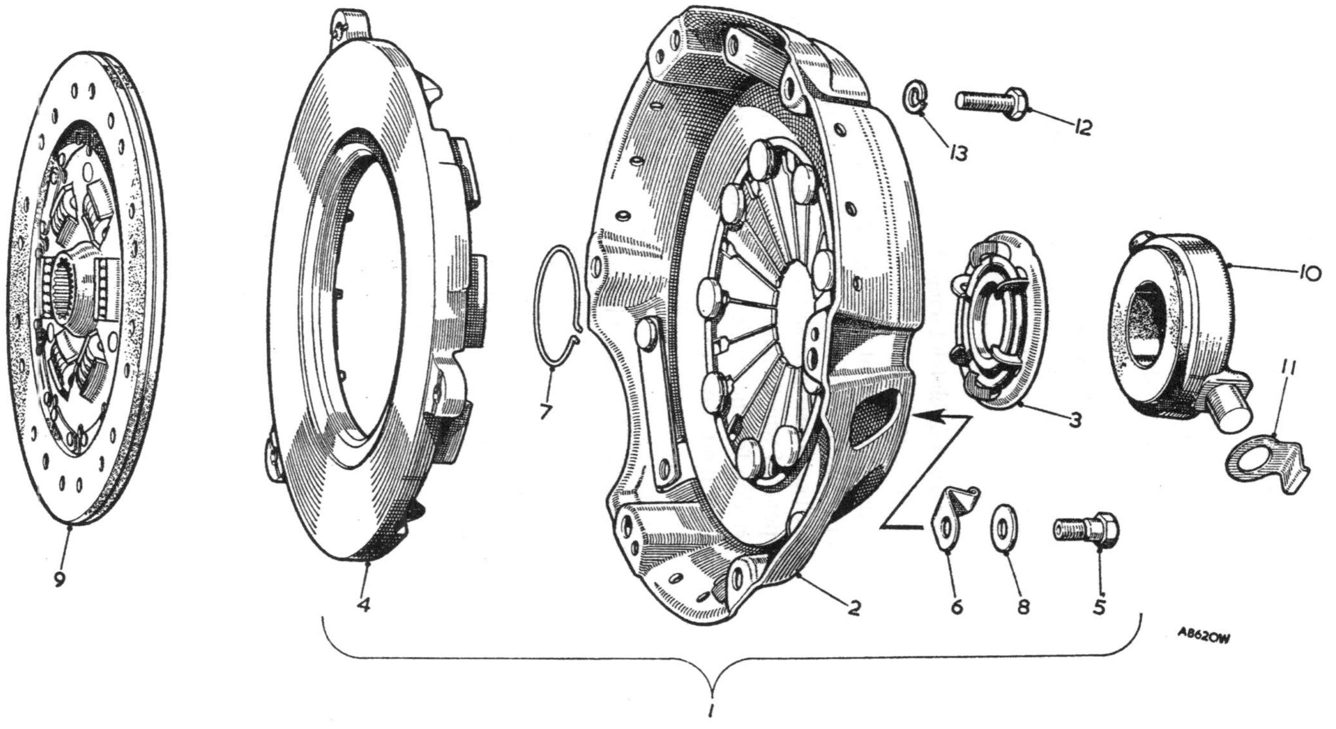

Fig. EEE.1. The diaphragm clutch components.

No.	Description	No.	Description
1.	*Cover assembly.*	8.	*Washer—strap.*
2.	*Cover with straps and diaphragm spring.*	9.	*Plate assembly—driven.*
3.	*Plate—release.*	10.	*Bearing assembly—release.*
4.	*Plate—pressure.*	11.	*Retainer—bearing.*
5.	*Bolt—strap.*	12.	*Screw—clutch to flywheel.*
6.	*Clip—pressure plate.*	13.	*Washer for screw—spring*
7.	*Circlip—release plate.*		

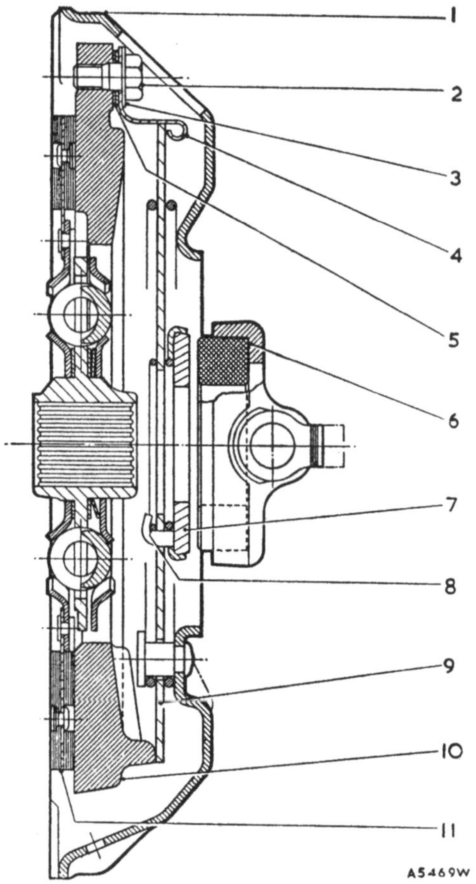

Fig. EEE.2. A section through the clutch.

1. *Cover.*
2. *Strap bolt.*
3. *Washer.*
4. *Clip.*
5. *Strap.*
6. *Release bearing.*
7. *Release plate.*
8. *Circlip—release plate.*
9. *Diaphragm spring.*
10. *Pressure plate.*
11. *Driven plate.*

Remove the clutch centralizer and refit the gearbox. The weight of the engine must be supported during refitting in order to avoid strain on the first motion shaft and distortion or displacement of the release plate or driven plate assembly.

Section EEE. 3

SERVICING THE DIAPHRAGM CLUTCH

Driven plates

It is important that neither oil nor grease should contact the clutch facings.

It is essential to install a complete driven plate assembly when the renewal of the friction surfaces is required. If the facings have worn to such an extent as to warrant renewal, then slight wear will have taken place on the splines, and also on the torque reaction springs and their seatings. The question of balance and concentricity is also involved. Under no circumstances is it satisfactory to repair or rectify faults in clutch driven plate centres.

Tolerances

Wear on the working faces of the driven plate is about ·001 in. (·02 mm.) per 1,000 miles (1600 km.) under normal running conditions. The accuracy of the alignment of the face of the driven plate must be within ·015 in. (·38 mm.).

Condition of clutch facings in service

It is natural to assume that a rough surface will give a higher frictional value against slipping than a polished one, but this is not necessarily correct. A roughened surface consists of small hills and dales, only the 'high-spots' of which make contact. As the amount of useful friction for the purpose of taking up the drive is dependent upon the area in actual contact, it is obvious that a perfectly smooth face is required to transmit the maximum amount of power for a given surface area.

Since non-metallic facings of the moulded asbestos type have been introduced in service the polished surface is common, but it must not be confused with the glazed surface which is sometimes encountered due to conditions to be detailed subsequently. The ideally smooth or polished condition will therefore provide proper surface contact, but a glazed surface entirely alters the frictional value of the facing and will result in excessive clutch slip. These two conditions might be simply illustrated by comparison between a piece of smoothly finished wood and one with a varnished surface; in the former the contact is made directly by the original material, whereas in the latter instance a film of dry varnish is interposed between the contact surfaces and actual contact is made by the varnish.

If the clutch has been in use for some time under satisfactory conditions, the surface of the facings assumes a high polish through which the grain of the material can be seen clearly. This polished facing is of light colour when in perfect condition.

Should oil in small quantities gain access to the clutch and find its way onto the facings, it will be burnt off as a result of the heat generated by the slipping occurring under normal starting conditions. The burning of this small quantity of lubricant has the effect of gradually darkening the facings, but provided the polish of the facing remains such that the grain of the material can be distinguished clearly, it has little effect on clutch performance.

Should increased quantities of oil obtain access to the facing, then one of two conditions, or a combination of these, may arise, depending upon the nature of the oil.

(1) The oil may burn off and leave a carbon deposit on the surface of the facings, which assume a high glaze, producing further slip. This is a very definite, though very thin, deposit, and in general it hides the grain of the material.

(2) The oil may partially burn and leave a resinous deposit on the facings. This has a tendency to produce a fierce clutch, and may also cause excessive 'spinning' due to the tendency of the face of the linings to adhere to the surface of the flywheel or pressure plate.

(3) There may be a combination of conditions (1) and (2) which produces a tendency to 'judder' on such engagement.

Still greater quantities of oil produce a dark and soaked appearance of the facings, and the result will be further slip, accompanied by fierceness or 'juddering'.

If the conditions enumerated above are experienced, the clutch driven plate should be replaced by a new one. **The cause of the presence of the oil must be traced and removed.** It is, of course, necessary for the clutch and flywheel to be cleaned out thoroughly before assembly.

Where the graphite release bearing ring is badly worn in service a complete replacement assembly should be fitted, returning the old assembly for salvage of the metal cup. These graphite rings are inserted into their metal cup by heating the metal cup to a cherry red, then forcing the graphite ring into position. Immediately the ring is forced into position the whole should be quenched in oil. Alignment of the thrust pad in relation to its face and the trunnions should be within ·005 in. (·12 mm.).

In almost every case of rapid wear on the splines of the clutch driven plate misalignment is responsible.

Looseness of the driven plate on the splined shaft results in noticeable backlash in the clutch. Misalignment also puts undue stress on the driven member, and may result in the hub breaking loose from the plate, with consequent total failure of the clutch.

It may also be responsible for a fierce chattering or dragging of the clutch, which makes gear-changing difficult. In cases of persistent difficulty it is advisable to check the flywheel for truth with a dial indicator. The dial reading should not vary more than ·003 in. (·07 mm.) anywhere on the flywheel face.

SECTION F

GEARBOX

SERIES BN4

Section F.1

DESCRIPTION

The gearbox has four forward speeds and one reverse, and synchromesh is incorporated on second, third and top gears.

Top gear is a direct drive; third and second are in constant mesh; first and reverse are obtained by sliding spur pinions.

Section F.2

LUBRICATION

The gearbox oil level should be checked by the dipstick at the recommended mileage and topped up if necessary.

The filler plug, which incorporates the dipstick, is located beneath a rubber cover, and is accessible when the floor mat and rubber cover have been raised.

Drain plugs are provided in the base of the gearbox and overdrive. Ensure that the drain plugs are kept clean and do not forget to replace them after draining.

The capacity of the gearbox is given in **"General Data"**.

Section F.3

REMOVAL AND REPLACEMENT

(1) Turn the battery master switch, which is situated inside the luggage compartment, to the "off" position.

(2) Inside the car remove the seat cushions and release the clips securing the padded arm rest to the central tunnel.

(3) Unclip and roll back the carpet over the short gearbox tunnel to expose the twelve screws securing the tunnel to the body of the car. Unscrew the setscrews and remove the tunnel and its carpeting.

(4) Unscrew the six setscrews, three on either side, which secure the carpet covered bulkhead and remove the bulkhead.

(5) Using a suitable tool tap back the locking washer on the propeller shaft flange bolts and remove the bolts.

(6) Unscrew the four setpins from the gearbox mounting brackets (see Fig. A.3, Section A), also unscrew the speedometer cable at its connection to the gearbox.

Note.—When an overdrive gearbox is fitted it will also be necessary to unclip the cable to the gearbox switch and release it at its terminal on the switch.

(7) Working beneath the vehicle remove setpins (1) Fig. A.4, Section A, and unscrew the nuts (2) and (3) to release the stabiliser bar.

(8) Detach the clutch slave cylinder from the gearbox bell housing by removing the two securing setpins. The slave cylinder push rod is released from the clutch operating lever by the removal of the securing clevis pin.

(9) Remove the starter motor as described in Section N.

(10) Place suitable supports underneath the gearbox bell housing and engine sump, and unscrew the nuts, bolts and setpins securing the bell housing to the engine backplate.

(11) Withdraw the gearbox first motion shaft from the flywheel bearing and clutch by gently easing the gearbox rearwards.

If the unit does not detach itself readily it will be necessary to raise the rear of the engine.

(12) The replacement of the gearbox is a reversal of the removal procedure.

Section F.4

DISMANTLING

(1) Remove the dipstick. Unscrew the breather from the overdrive unit, if fitted. Drain the oil from the gearbox and overdrive by removing the drain plug beneath each unit.

(2) Unscrew the speedometer drive from the right-hand side of the rear extension.

(3) Unscrew the seven short and one long bolt and remove the clutch housing.

(4) Remove the three nuts threaded on studs mounted on the gear lever cup. With the removal of these nuts the cup may be withdrawn together with the three washers and three distance pieces located on the studs.

(5) Withdraw the gear lever from the gearbox.

(6) Unscrew the thirteen bolts securing the side cover to the gearbox housing and remove the cover; there are two dowels locating the cover. Take care not to lose the three selector balls and springs which will be released as the cover is withdrawn.

(7) Unscrew the eight bolts and remove the rear extension.

Note.—For models fitted with overdrive.

Once the overdrive unit has been separated from the gearbox (see Section G.10), the removal of the

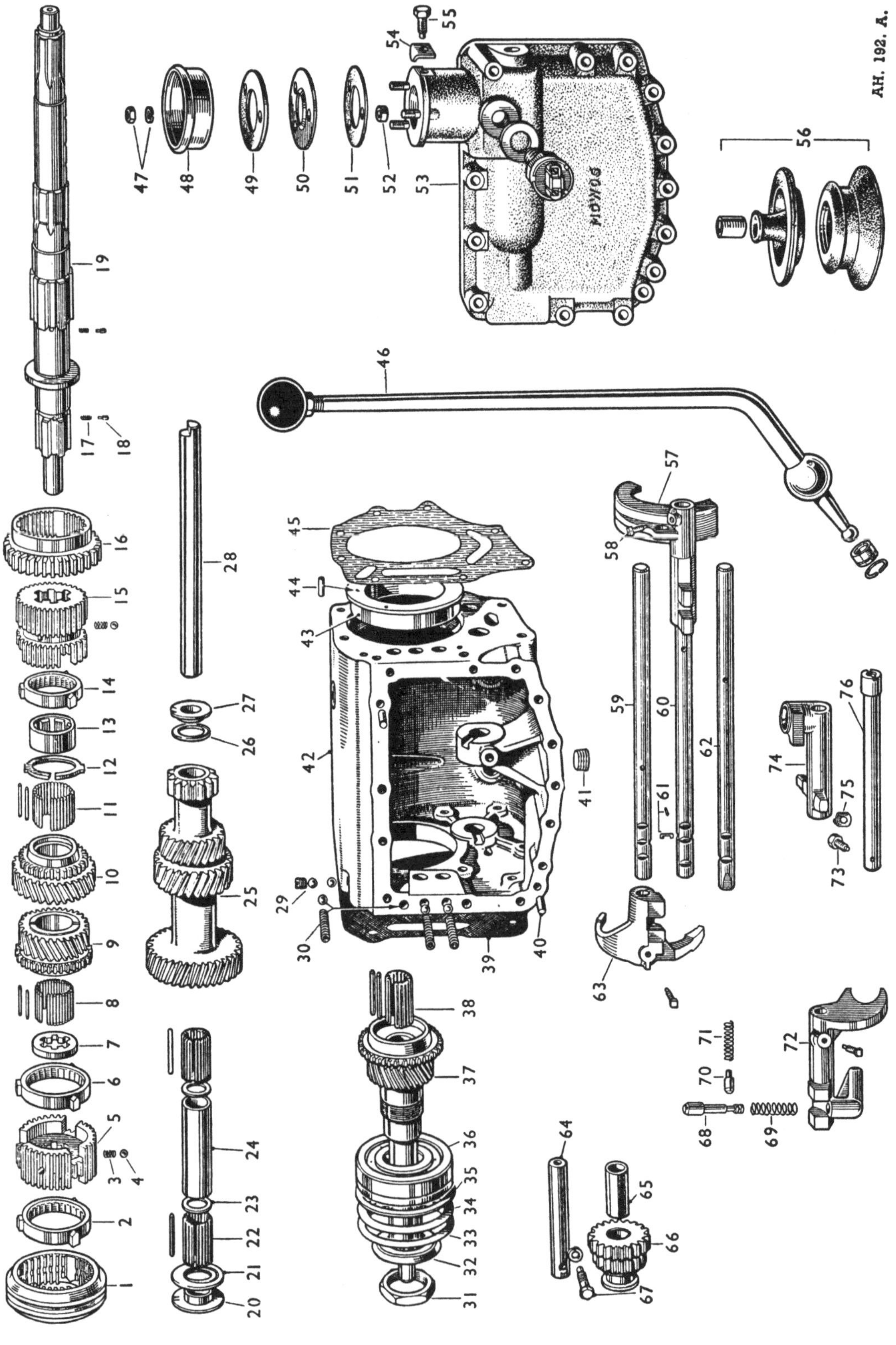

Fig. F.1. Components of the Gearbox.

Fig. F.1. Components of the Gearbox.

1. *Synchromesh sleeve.*
2. *Baulking ring.*
3. *Synchronizer spring.*
4. *Synchronizer ball.*
5. *3rd and 4th speed synchronizer.*
6. *Baulking ring.*
7. *Locking plate.*
8. *Needle rollers.*
9. *Third speed gear.*
10. *Second speed gear.*
11. *Needle rollers.*
12. *Gear washer.*
13. *Locking plate.*
14. *Baulking ring.*
15. *2nd speed synchronizer.*
16. *First speed gear.*
17. *Plunger spring.*
18. *Gear plunger.*
19. *Main shaft.*
20. *Thrust plate.*
21. *Thrust washer.*
22. *Needle rollers.*
23. *Washer, roller.*
24. *Spacer, roller.*
25. *Laygear.*
26. *Washer.*
27. *Thrust plate.*
28. *Layshaft*
29. *Interlocking balls.*
30. *Selector ball and spring.*
31. *Bearing nut.*
32. *Bearing nut lockwasher.*
33. *Bearing spring plate.*
34. *Bearing plate.*
35. *Bearing circlip.*
36. *First motion shaft bearing.*
37. *First motion shaft.*
38. *Needle rollers.*
39. *Joint washer.*
40. *Side cover dowel.*
41. *Drain plug.*
42. *Gearbox casing.*
43. *Bearing housing.*
44. *Locating peg.*
45. *Joint washer.*
46. *Gear lever.*
47. *Nut and washer.*
48. *Cup.*
49. *Rubber washer (thick).*
50. *Steel washer.*
51. *Rubber washer (thin).*
52. *Distance piece.*
53. *Side cover.*
54. *Washer.*
55. *Gear lever locating screw.*
56. *Rubber dust covers.*
57. *1st and 2nd speed fork.*
58. *Screw for fork.*
59. *3rd and 4th speed fork rod.*
60. *1st and 2nd speed fork rod.*
61. *Interlocking pin and rivet.*
62. *Reverse fork rod.*
63. *3rd and 4th speed fork.*
64. *Reverse shaft.*
65. *Bush.*
66. *Reverse gear.*
67. *Locking screw.*
68. *Selector plunger.*
69. *Selector plunger spring.*
70. *Detent plunger.*
71. *Detent plunger spring.*
72. *Reverse fork.*
73. *Control shaft locating screw.*
74. *Locking washer.*
75. *Control shaft.*
76. *Control lever.*

adapter plate is accomplished by unscrewing the eight nuts in the recess in the adapter plate.

The overdrive pump cam should slide freely along the third motion shaft thus giving access to the circlip holding the distance piece to the rear adapter plate. Remove the circlip and slide the distance piece off the shaft. The adapter plate should now pull away from the gearbox, together with the rear main bearing. It may be necessary for one operator to hold the gearbox vertically by the adapter plate whilst a second operator taps the third motion shaft until the ball race in the adapter plate is free of the shaft.

(8) Cut the locking wires and unscrew the fork retaining screws. Remove the shifter shafts and forks in the following order:—
   (*a*) The reverse shaft and fork together with its selector and detent plungers and springs.
   (*b*) Top gear shifter shaft only.
   (*c*) First and second shaft and fork.
   (*d*) Top gear fork.

   Take care not to lose the two interlock balls, normally located one at each side of the centre shifter shaft, which will be released when the shaft is removed.

(9) Unscrew the reverse shaft locating screw and push out the shaft; lift the gear from the box.

(10) Tap out the layshaft and allow the gear to rest in the bottom of the box.

(11) Withdraw the first motion shaft assembly; note that there are 16 spigot rollers.

(12) Withdraw the mainshaft rearwards.

(13) Lift out the layshaft gear and thrust washers.

Section F.5

DISMANTLING THE MAINSHAFT

(1) Slide the top and third gear hub and interceptors from the forward end.

(2) Depress the plunger locating the third gear locking plate, rotate the plate to line up the splines and slide it from the shaft. Extract the plunger and spring, and slide off the third speed gear and its 32 rollers.

(3) Unscrew the main shaft nut; remove the nut, locking washer, speedometer drive gear, bearing with housing and distance collar.

(4) Slide the first and second speed hub, second speed interceptor and first speed gear rearwards from the shaft; if the first speed gear is withdrawn from the hub, take care to hold the balls and springs located in holes in the hub.

(5) Depress the second gear locking collar plunger and rotate the collar to line up the splines; slide the collar from the shaft and extract the two halves of the second gear washer, retaining the spring and plunger.

(6) Withdraw the second speed gear and its 33 rollers from the shaft.

(7) To dismantle the first motion shaft assembly, tap up the locking tab, unscrew the nut and remove the bearing.

NOTE:—The method of dismantling and reassembling the overdrive gearbox is the same as that described for the standard gearbox, with the exception that no speedometer drive gear or locking washer and nut is fitted.

Section F.6

REASSEMBLY

Mainshaft

(1) Smear the shaft with grease and assemble the 33 second speed gear rollers; slide the second gear into position.

(2) Replace the plunger and spring. Fit the two halves of the second gear washer and slide the collar on to the splines. Depress the plunger and push the collar into position, locating the lugs of the washer in the cut-outs of the collar; rotate the collar to bring the splines out of line.

(3) Replace the balls and springs in the second and first speed hub; depress the balls and slide the first speed gear on to the hub; refit the assembly to the shaft.

(4) Refit the bearing distance collar, the bearing and housing, the speedometer drive gear key and gear, locking washer and nut. Tighten the nut and tap over the locking washer.

(5) Fit the third gear and its 32 rollers to the shaft; replace the plunger and spring and the third speed locking plate; rotate the plate to bring the splines out of line.

(6) Fit the balls and springs to the top and third speed hub and slide the striking dog into position on the hub.

(7) Replace the hub, striking dog and interceptors on the shaft.

Fig. F.2. Assembling gear and synchronizer.
1. Gear. 2. Baulking ring. 3. Synchronizer.

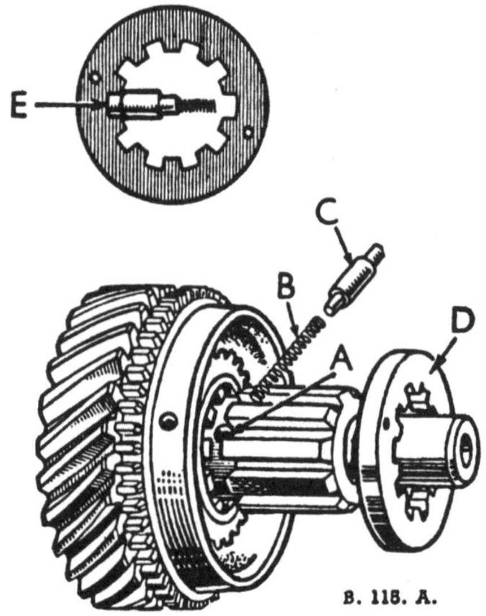

Fig. F.3. Securing the third motion shaft gears.
A. Hole for spring C. Location peg.
B. Spring. D. Locking washer.
E. Peg located in washer.

Layshaft

(1) Fit the distance tube to the layshaft gear with a washer at each end of the tube.

(2) Smear the rollers with grease and position them in the gear. Place the thrust washers and plates in position at each end of the gear.

(3) To retain the rollers in position, a length of round bar of layshaft diameter and just long enough to hold the thrust washers and plates, should be inserted in the gear assembly.

(4) Place the gear in the box and allow it to rest at the bottom.

Gearbox

(1) Insert the mainshaft assembly from the rear of the box.

(2) Position the first motion shaft rollers and the first motion shaft assembly in the box.

(3) Lift the layshaft gear into position, locating the thrust washer tags in the grooves provided. Push the layshaft through the housing and gear, and withdraw the retaining bar as the shaft pushes it out of the gear. The cut-away portion of the shaft must be aligned to fit the groove in the bell housing provided to prevent the layshaft from turning.

(4) Refit the reverse gear and shaft and tighten the setscrew. Place the top gear shifter fork in the box. Replace the first and second gear shifter fork and shaft.

(5) Replace one interlock ball above the first and second shifter shaft and insert the top gear shifter shaft.

(6) Position the remaining interlock ball, holding it with grease and refit the reverse fork and shaft together with its selector and detent plungers and springs.

(7) Screw in the fork setscrews, tighten up and wire.

(8) Bolt the rear extension into position, using a new gasket if necessary. Note that the plain bearing plate is fitted against the bearing.

Note.—For models fitted with overdrive :—

Slide the adapter plate, together with its bearing and paper joint washer, along the third motion shaft. Fit and tighten down the eight setpins securing the adapter plate to the gearbox.

Fit the distance piece which covers the space between the rear main bearing and the groove allocated for the circlip, and fix on the latter. Refit overdrive (see Section G.10).

(9) Refit the selector balls to the holes in the gearbox housing and the springs in the holes in the side cover.

(10) The gear lever together with its cup, washers and distance pieces may now be attached to the side cover. Ensure that the ball of the lever makes a good fit with its mating socket.

(11) Refit the cover, fitting a new gasket as required. Observe that the top right-hand setpin is longer than the other twelve.

(12) Refit the clutch housing with plain bearing plate against the bearing.

Refit the speedometer drive, breather and dipstick.

Section F.7

MODIFIED GEARS

Commencing at engine number 11342, the gears were modified to increase their rigidity. In earlier versions of this modification the laygear was fitted with plain bushes, but this was later replaced by a layshaft assembly with needle roller bearings.

The modified gears are only interchangeable with earlier types in complete sets.

SECTION FF

GEARBOX

SERIES BN6

For details of the gearbox fitted to BN6 cars refer to Section F.

SECTION FFF

GEARBOX

Mk. I and II (SERIES BN7 and BT7)
AND Mk. II and Mk. III (SERIES BJ7 and BJ8)

NOTE

For all other repair and service procedures see Section F.

Section FFF.1

CENTRAL CONTROL GEARBOX

The power unit for later Mark II BN7 and BT7 cars incorporated a gearbox with a centrally mounted remote control change speed lever in place of the side mounted lever fitted previously. This type of gearbox, with accompanying body and gearbox cover modifications, was used from Car Numbers BN7 16039 and BT7 15881.

The following two sections give the necessary information for dismantling and assembling the central control box.

Section FFF.2

DISMANTLING THE CENTRAL CONTROL GEARBOX

Unscrew the eight short bolts and one long bolt and remove the clutch housing complete with the clutch operating mechanism. Withdraw the drive gear bearing plain and spring plates.

Remove the oil level indicator from the gearbox top cover, unscrew the 12 bolts from the top of the gearbox and lift off the cover. Note that the two cover securing bolts nearest the change speed lever turret are longer. Ensure that the three detent springs positioned in the gearbox casing under the front edge of the cover are not lost.

If it is necessary to remove the change speed lever from the gearbox top cover, release the circlip, washer, and conical spring from the change speed lever turret. Using a small diameter punch, drive the two rollpins in turn into the $\frac{3}{16}$ in. (·187 mm.) holes on each side of the change speed turret. This will cause them to move into the bore of the lever ball. Lift out the lever and retrieve the roll pins from the ball end.

Remove the three detent springs. Cut the locking wires and unscrew the striking fork retaining screws. Hold the gear shifter shafts in the neutral position—this will prevent the interlock balls from operating—and withdraw the third and fourth speed shifter shaft retrieving the detent ball that will drop down into the shaft bore at the front of the gearbox casing. Withdraw the remaining shifter shafts retrieving their detent balls and the two interlocking balls located between the shafts at the front of the gearbox casing. Lift out the three striking forks.

If it is necessary to remove the reverse selector plunger from the reverse striking fork, extract the split pin to release the plunger and spring which in turn will release a detent plunger and spring.

Undo the propeller shaft flange nut, using Service Tool 18G34A to prevent the flange from turning, and pull the flange from the gearbox mainshaft splines. Unscrew the speedometer pinion housing from the gearbox rear extension casing and remove it complete with pinion. Unscrew the eight bolts and remove the gearbox rear extension casing from the main casing. Withdraw the mainshaft bearing plain and spring plates.

In the case of gearboxes fitted with overdrive, unscrew the eight nuts holding the overdrive adaptor plate to the back of the gearbox. Do not disturb the joint between the overdrive and the adaptor plate. Pull the overdrive and adaptor plate away from the gearbox and over the mainshaft. Slide the overdrive oil pump cam off the mainshaft. The mainshaft bearing distance collar and circlip will now be exposed, but may be left in position when dismantling the gearbox. If, when dismantling the mainshaft assembly, it is found to be necessary to remove the mainshaft bearing from the shaft, the circlip and distance collar will have to be withdrawn first. Withdraw the mainshaft bearing plain and spring plates.

Unscrew the reverse idler gear shaft locating screw. Withdraw the shaft and lift out the reverse idler gear.

Push the layshaft forward and pull it out from the front of the gearbox casing. Lower the layshaft gear unit to the bottom of the casing.

Mark the position of the locating peg on the mainshaft bearing housing in relation to the gearbox casing, so that on reassembly the peg may be aligned correctly to locate in the hole provided for it in the rear extension casing or the overdrive adaptor plate. Withdraw the complete mainshaft assembly from the rear of the gearbox casing.

Take out the 18 needle rollers from the rear of the drive gear. Use a suitable brass drift to drive the bearing forwards from its housing and withdraw the drive gear assembly from the front of the gearbox casing.

Lift out the layshaft gear unit and thrust washers.

To dismantle the main shaft assembly follow the instructions given in Section F.5.

Section FFF.3

ASSEMBLING THE CENTRAL CONTROL GEARBOX

To assemble the mainshaft follow the instructions given under "**Mainshaft**" in Section F.6.

Fit the spacer to the layshaft gear unit with a roller washer positioned at each end. Smear the needle rollers

THE GEARBOX COMPONENTS
(Gearbox without overdrive)

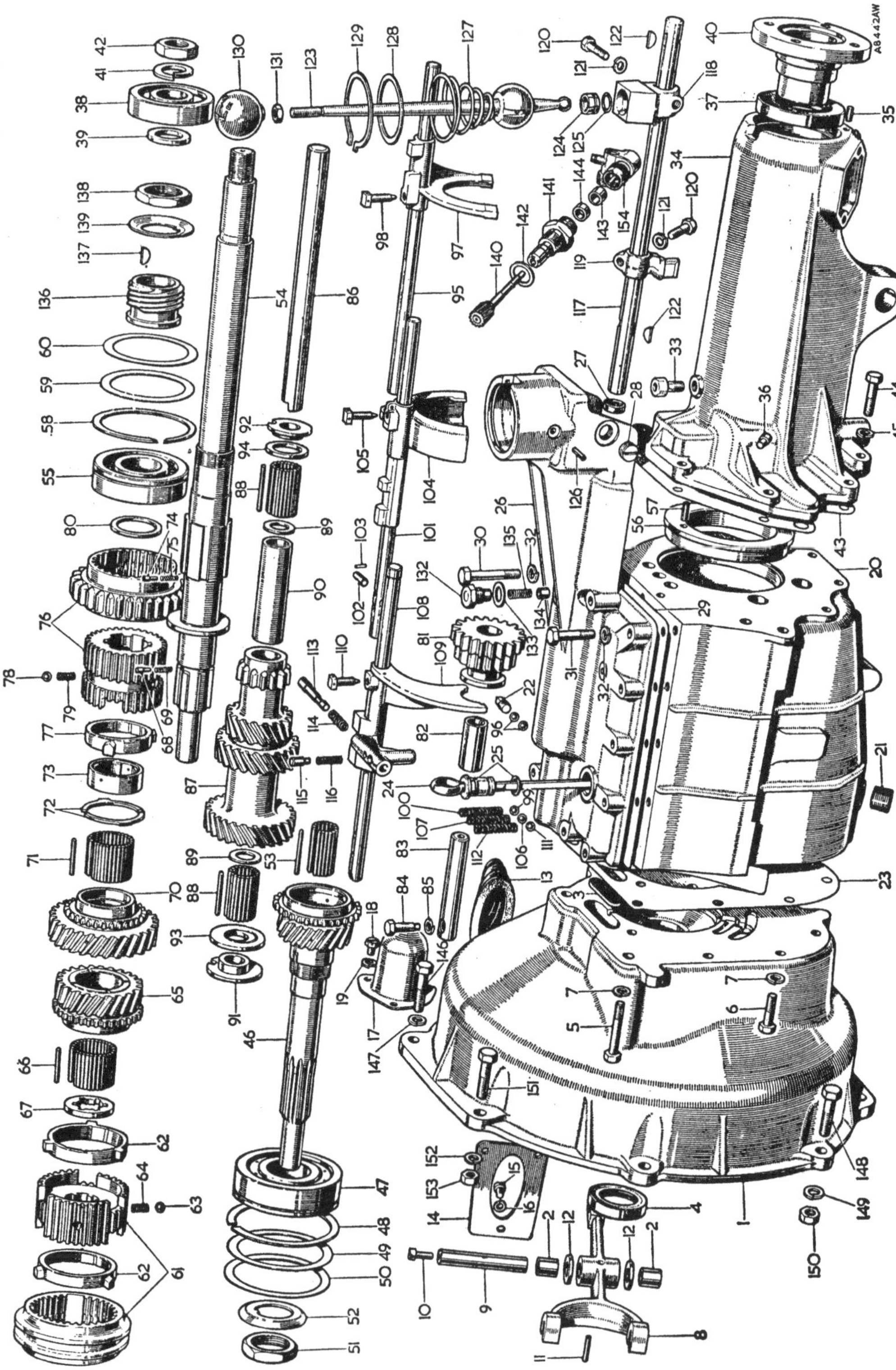

KEY TO THE GEARBOX COMPONENTS
(Gearbox without overdrive)

No.	Description
1.	*Clutch housing.*
2.	*Fork and lever shaft bush.*
3.	*Buffer pad.*
4.	*Oil seal.*
5.	*Bolt—long.*
6.	*Bolt—short.*
7.	*Spring washer.*
8.	*Clutch fork and lever.*
9.	*Fork and lever shaft.*
10.	*Clutch withdrawal fork screw.*
11.	*Taper pin.*
12.	*Thrust washer for fork and lever.*
13.	*Fork and lever seal.*
14.	*Seal retaining plate.*
15.	*Retaining plate screw.*
16.	*Spring washer.*
17.	*Starter end cover.*
18.	*End cover screw.*
19.	*Spring washer.*
20.	*Gearbox case.*
21.	*Oil drain plug.*
22.	*Interlock ball hole plug.*
23.	*Case to clutch housing joint.*
24.	*Oil level indicator.*
25.	*Rubber grommet.*
26.	*Gearbox top cover.*
27.	*Cover oil seal.*
28.	*Cover plug.*
29.	*Cover to gearbox joint.*
30.	*Bolt—long.*
31.	*Bolt—short.*
32.	*Spring washer.*
33.	*Gearbox breather.*
34.	*Gearbox extension casing.*
35.	*Casing taper plug.*
36.	*Speedometer pinion thrust button.*
37.	*Oil seal.*
38.	*Bearing.*
39.	*Bearing washer.*
40.	*Coupling flange.*

No.	Description
41.	*Spring washer.*
42.	*Flange nut.*
43.	*Casing to gearbox joint.*
44.	*Bolt—casing to gearbox.*
45.	*Spring washer.*
46.	*Drive gear.*
47.	*Bearing for drive gear.*
48.	*Bearing circlip.*
49.	*Bearing plate.*
50.	*Bearing plate (spring).*
51.	*Bearing nut.*
52.	*Lock washer.*
53.	*Roller for drive gear.*
54.	*Mainshaft.*
55.	*Mainshaft bearing.*
56.	*Bearing housing.*
57.	*Locating peg.*
58.	*Bearing circlip.*
59.	*Bearing plate.*
60.	*Bearing plate (spring).*
61.	*Top and third sliding hub with striking dog.*
62.	*Sliding hub interceptor*
63.	*Sliding hub ball.*
64.	*Ball spring.*
65.	*Third speed gear.*
66.	*Gear roller.*
67.	*Locking plate.*
68.	*Gear plunger.*
69.	*Plunger spring.*
70.	*Second speed gear.*
71.	*Gear roller.*
72.	*Gear washer.*
73.	*Locking plate.*
74.	*Gear plunger.*
75.	*Plunger spring.*
76.	*First speed gear with first and second sliding hub.*
77.	*Sliding hub interceptor.*
78.	*Sliding hub ball.*

No.	Description
79.	*Ball spring.*
80.	*Mainshaft distance collar.*
81.	*Reverse gear.*
82.	*Gear bush.*
83.	*Gear shaft.*
84.	*Shaft retaining screw.*
85.	*Spring washer.*
86.	*Layshaft.*
87.	*Layshaft gear unit.*
88.	*Gear unit roller.*
89.	*Roller washer.*
90.	*Roller spacer.*
91.	*Gear unit thrust plate—front.*
92.	*Gear unit thrust plate—rear.*
93.	*Gear unit thrust washer—front.*
94.	*Gear unit thrust washer—rear.*
95.	*Top and third shifter shaft.*
96.	*Shaft interlocking ball.*
97.	*Top and third striking fork.*
98.	*Screw for striking fork.*
99.	*Shifter shaft ball.*
100.	*Ball spring.*
101.	*First and second shifter shaft.*
102.	*Shaft interlocking pin.*
103.	*Interlocking pin rivet.*
104.	*First and second striking fork.*
105.	*Screw for striking fork.*
106.	*Shifter shaft ball.*
107.	*Ball spring.*
108.	*Reverse shifter shaft.*
109.	*Reverse striking fork.*
110.	*Screw for striking fork.*
111.	*Shifter shaft ball.*
112.	*Ball spring.*
113.	*Reverse selector plunger.*
114.	*Plunger spring.*
115.	*Detent plunger.*
116.	*Detent plunger spring.*
117.	*Remote control shaft.*
118.	*Change speed lever shaft.*

No.	Description
119.	*Selector lever.*
120.	*Selector lever and change speed lever socket screw.*
121.	*Spring washer.*
122.	*Selector lever and change speed lever socket key.*
123.	*Change speed lever.*
124.	*Lever bush.*
125.	*Circlip for bush.*
126.	*Rollpin.*
127.	*Ball end retaining spring.*
128.	*Spring washer.*
129.	*Circlip.*
130.	*Change speed lever knob.*
131.	*Locknut for knob.*
132.	*Plunger retaining plug.*
133.	*Plug washer.*
134.	*Plunger.*
135.	*Plunger spring.*
136.	*Speedometer gear.*
137.	*Key for gear.*
138.	*Locknut for gear.*
139.	*Lockwasher for gear.*
140.	*Speedometer pinion.*
141.	*Pinion bearing.*
142.	*Washer for bearing.*
143.	*Pinion distance collar.*
144.	*Pinion oil seal.*
145.	*Reverse switch hole plug.*
146.	*Clutch housing bolt—long.*
147.	*Spring washer.*
148.	*Clutch housing bolt—short.*
149.	*Spring washer.*
150.	*Nut.*
151.	*Clutch housing dowel bolt.*
152.	*Spring washer for dowel bolt.*
153.	*Nut for dowel bolt.*
154.	*Speedometer drive adaptor box.*

THE GEARBOX COMPONENTS

(Gearbox with overdrive)

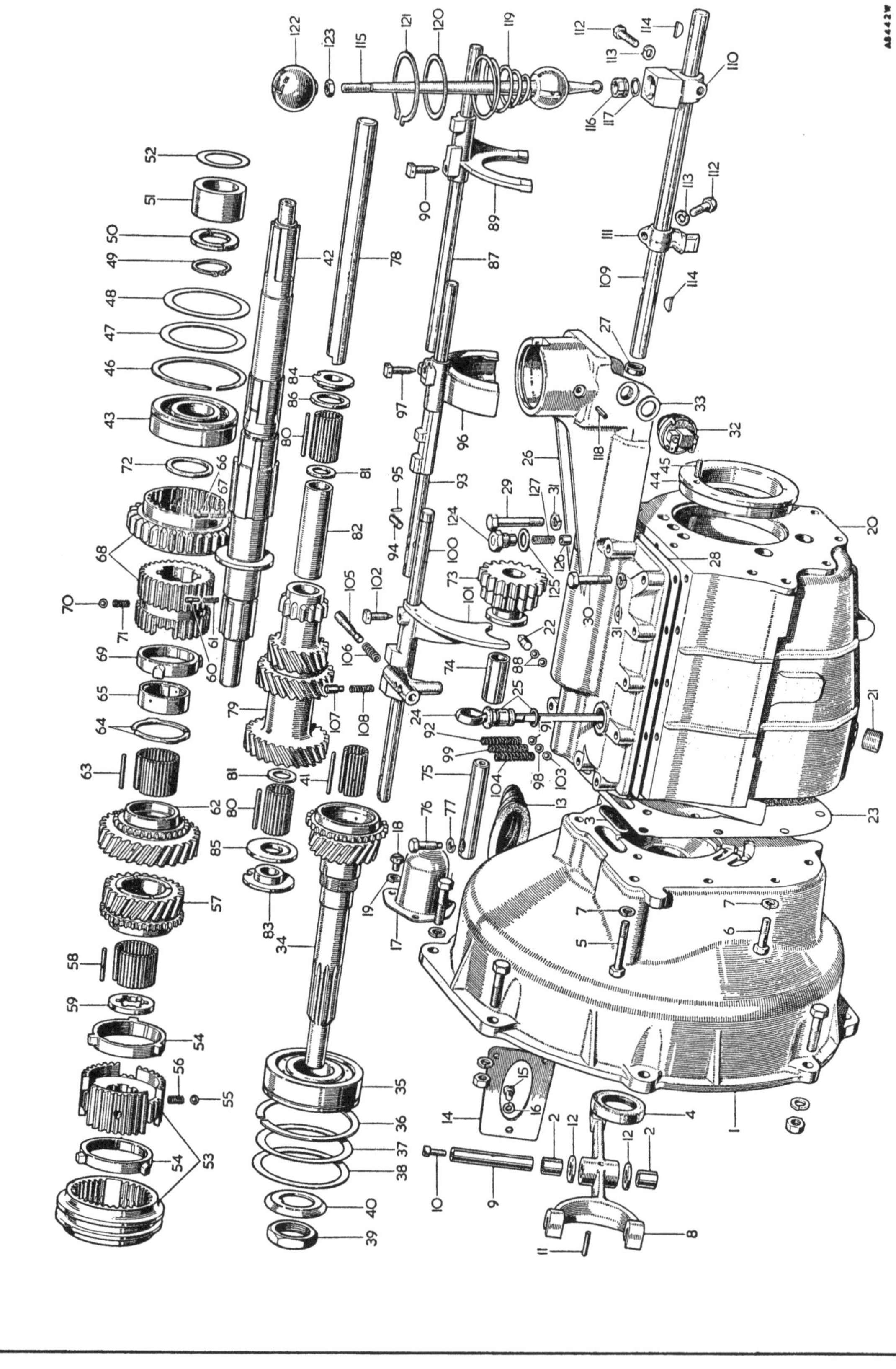

KEY TO THE GEARBOX COMPONENTS

(Gearbox with overdrive)

No.	Description
1.	*Clutch housing.*
2.	*Fork and lever shaft bush.*
3.	*Buffer pad.*
4.	*Oil seal.*
5.	*Bolt—long.*
6.	*Bolt—short.*
7.	*Spring washer.*
8.	*Clutch fork and lever.*
9.	*Fork and lever shaft.*
10.	*Clutch withdrawal fork, screw.*
11.	*Taper pin.*
12.	*Thrust washer for fork and lever.*
13.	*Fork and lever seal.*
14.	*Seal retaining plate.*
15.	*Retaining plate screw.*
16.	*Spring washer.*
17.	*Starter end cover.*
18.	*End cover screw.*
19.	*Spring washer.*
20.	*Gearbox case.*
21.	*Oil drain plug.*
22.	*Interlock ball hole plug.*
23.	*Case to clutch housing joint.*
24.	*Oil level indicator.*
25.	*Rubber grommet.*
26.	*Gearbox top cover.*
27.	*Cover oil seal.*
28.	*Cover to gearbox joint.*
29.	*Bolt—long.*
30.	*Bolt—short.*
31.	*Spring washer.*
32.	*Overdrive switch.*
33.	*Joint for switch.*

No.	Description
34.	*Drive gear.*
35.	*Bearing for drive gear.*
36.	*Circlip for bearing.*
37.	*Plate for bearing.*
38.	*Plate for bearing (spring).*
39.	*Nut for bearing.*
40.	*Lock washer.*
41.	*Drive gear roller.*
42.	*Mainshaft.*
43.	*Mainshaft bearing.*
44.	*Bearing housing.*
45.	*Locating peg.*
46.	*Bearing circlip.*
47.	*Plate for bearing.*
48.	*Plate for bearing (spring).*
49.	*Mainshaft circlip.*
50.	*Bearing abutment collar.*
51.	*Abutment collar retaining ring.*
52.	*Shim.*
53.	*Top and third sliding hub with striking dog.*
54.	*Sliding hub interceptor.*
55.	*Sliding hub ball.*
56.	*Ball spring.*
57.	*Third speed gear.*
58.	*Roller for gear.*
59.	*Locking plate.*
60.	*Gear plunger.*
61.	*Plunger spring.*
62.	*Second speed gear.*
63.	*Roller for gear.*
64.	*Gear washer.*
65.	*Locking plate.*

No.	Description
66.	*Gear plunger.*
67.	*Plunger spring.*
68.	*First speed gear with first and second sliding hub.*
69.	*Sliding hub interceptor.*
70.	*Sliding hub ball.*
71.	*Ball spring.*
72.	*Mainshaft distance collar.*
73.	*Reverse gear.*
74.	*Gear bush.*
75.	*Gear shaft.*
76.	*Shaft retaining screw.*
77.	*Spring washer.*
78.	*Layshaft.*
79.	*Layshaft gear unit.*
80.	*Gear unit roller.*
81.	*Roller washer.*
82.	*Roller spacer.*
83.	*Gear unit thrust plate—front.*
84.	*Gear unit thrust plate—rear.*
85.	*Gear unit thrust washer—front.*
86.	*Gear unit thrust washer—rear.*
87.	*Top and third shifter shaft.*
88.	*Shaft interlocking ball.*
89.	*Top and third striking fork.*
90.	*Screw for striking fork.*
91.	*Shifter shaft ball.*
92.	*Ball spring.*
93.	*First and second shifter shaft.*
94.	*Shaft interlocking pin.*
95.	*Interlocking pin rivet.*
96.	*First and second striking fork.*
97.	*Screw for striking fork.*

No.	Description
98.	*Shifter shaft ball.*
99.	*Ball spring.*
100.	*Reverse shifter shaft.*
101.	*Reverse striking fork.*
102.	*Screw for striking fork.*
103.	*Shifter shaft ball.*
104.	*Ball spring.*
105.	*Reverse selector plunger.*
106.	*Plunger spring.*
107.	*Detent plunger.*
108.	*Detent plunger spring.*
109.	*Remote control shaft.*
110.	*Change speed lever rocket.*
111.	*Selector lever.*
112.	*Selector lever and change speed lever socket screw.*
113.	*Spring washer.*
114.	*Selector lever and change speed lever socket key.*
115.	*Change speed lever.*
116.	*Lever bush.*
117.	*Circlip for bush.*
118.	*Rollpin.*
119.	*Ball end retaining spring.*
120.	*Washer for spring.*
121.	*Circlip.*
122.	*Change speed lever knob.*
123.	*Locknut for knob.*
124.	*Plunger retaining plug.*
125.	*Plug washer.*
126.	*Plunger.*
127.	*Plunger spring.*

with grease and place them in the ends of the gear unit (23 at each end).

Assemble the gear unit front and rear thrust washers and plates, and position them in the gearbox with grease ensuring that their tags engage the grooves in the gearbox casing.

Place the gear unit in the gearbox and allow it to rest in the bottom.

Smear the 18 mainshaft spigot rollers with grease and place them in the drive gear. Insert the drive gear through the front of the casing and press the bearing into position.

Fit the mainshaft assembly from the rear of the gearbox casing ensuring that the sliding dog and interceptors are in position on the third and fourth speed synchronizing hub. Align the mainshaft bearing housing locating peg with the mark made on the gearbox casing when dismantling, and press the bearing housing into position.

Lift the layshaft gear unit into position and push the layshaft through the front of the gearbox casing making sure that the thrust washers and needle rollers remain in their correct locations.

Place the reverse idler gear in position, push the gear shaft into its housing and secure it in position with its locating screw and washer.

With a non-overdrive gearbox fit the mainshaft bearing plain and spring plates with the plain plate against the bearing. Bolt the gearbox rear extension into position, first ensuring that the rear bearing washer is in position on the mainshaft. Screw in the speedometer pinion and housing. Push the propeller shaft flange onto the mainshaft splines and secure it with the washer and nut, using Service Tool 18G34A to prevent the flange from turning while tightening the nut.

If the reverse selector plunger assembly has been removed from the reverse gear striking fork, refit the detent plunger and spring to the fork. Press down the detent plunger and push in the selector plunger and spring. Fit a new split pin in the end of the plunger. Make sure that the selector plunger, when pressed down into the fork, returns freely.

Place the three gear striking forks in position. Install the reverse gear shifter shaft and secure it to the reverse gear striking fork ensuring that the dowelled end of the locating screw engages in the hole in the shaft. Place the shifter shaft interlock ball into its location between the reverse gear and the first and second speed shifter shaft bores at the front of the gearbox casing.

Hold the reverse striking fork in the neutral position and install the first and second speed gear shifter shaft, noting that this shaft is fitted with an interlocking pin. Fit the fork locating screw and place the second interlock ball into its location between the first and second speed gear and the third and fourth speed gear and the third and fourth speed gear shifter shaft bores.

Holding the first and second speed gear striking fork in the neutral position, install the third and fourth speed gear shifter shaft. Fit the fork locating screw. Tighten the three striking fork locating screws and secure them in position with new locking wires.

To fit the change speed lever to the gearbox top cover, fit the two roll pins into the diametrically opposed $\frac{3}{16}$ in. (·187 mm.) holes on each side of the change speed lever turret. Before driving them in too far, place the change speed lever into its seating with the ball notches opposite the holes. Drive in the rollpins until their ends are level with the bottom of the counterbores on each side of the turret. Place the conical spring, washer, and circlip over the change speed lever. Press the spring down and engage the circlip in the groove in the turret.

Place the shifter shaft detent balls and springs into their bores and refit the gearbox top cover, locating the remote control rod selector arm in the striking forks. Replace the oil level indicator.

In the case of gearboxes fitted with overdrive, the components of the overdrive unit must be checked for alignment, using Service Tool 18G185, before the unit is offered up to the gearbox. Use a long thin screwdriver to align by eye the splines of the uni-directional clutch (see Fig. G.7) with those of the planet carrier. If the splines are not in line it will be found that the splined hub of the uni-directional clutch may be turned with the screwdriver blade in an anti-clockwise direction to the required position. Insert the dummy mainshaft (Service Tool 18G185). Gently turn the coupling flange to and fro while holding the tool to assist in engaging the shaft of the tool with the splines of the planet carrier and uni-directional clutch. Make quite sure that the tool shaft has gone right home into the spigot bush, checking this by using the screwdriver blade as a depth gauge. Place the oil pump operating cam in position in the overdrive unit on the central bushing of the body with the lowest part of the cam in contact with the oil pump plunger. A smear of grease will help to retain the cam in the correct position. Place the mainshaft bearing plain and spring plates in the recess in the adaptor plate (plain plate towards the bearing). Engage top gear in the gearbox and carefully fit the overdrive and adaptor plate over the mainshaft. Thread the mainshaft through the oil pump cam and into the centre bushing of the overdrive body, making sure that the pump cam is not disturbed from its correct position in relation to the pump plunger. Gently turn the first motion shaft to and fro to feel the mainshaft splines into the planet carrier, uni-directional clutch, and oil pump cam. When these parts are correctly aligned the studs on the rear face of the gearbox will have entered the holes in the adaptor plate. Press the overdrive unit right home, if necessary lightly tapping the propeller shaft driving flange with a copper hammer while turning the first motion shaft of the gearbox.

NOTE.—**If any difficulty is experienced in fitting the overdrive it is probable that one of the components has become misaligned. The overdrive must be removed and the components re-aligned with Service Tool 18G185.**

Turn the layshaft so that its stepped end is aligned to engage the groove in the clutch housing. Position the drive gear bearing plain and spring plates in the recess in the rear face of the clutch housing (plain plate towards the bearing). Refit the clutch housing to the main casing taking care not to damage the oil seal on the first motion shaft splines.

SECTION G

OVERDRIVE

(Series BN4)

Section G.1

LUBRICATION

The lubricating oil in the overdrive unit is common with that in the gearbox and the level should be checked with the gearbox dipstick.

It is essential that an approved lubricant be used when refilling. Trouble may be experienced if some types of extreme pressure lubricants are used because the planet gears act as a centrifuge to separate the additives from the oil.

Recommended lubricants are given in Section Q. It should be emphasised that any hydraulically controlled transmission must have clean oil at all times and great care must be taken to avoid the entry of dirt whenever any part of the casing is opened.

At the recommended mileage check the oil level of the gearbox and overdrive and top up if necessary through the gearbox dipstick hole.

In addition to the normal drain plug fitted to the gearbox the overdrive unit incorporates a plug at its base which gives access to a filter.

Occasionally remove the overdrive oil pump filter and clean the filter gauze by washing in petrol. The filter is accessible through the drain plug hole and is secured by a central set bolt.

Refilling of the complete system (gearbox and overdrive) is accomplished through the gearbox filler plug. The capacity of the combined gearbox and overdrive unit is 6¼ pints (7·5 U.S. pints; 3·5 litres).

After draining, ¼ pint of oil will remain in the overdrive hydraulic system, so that only 6 pints will be needed for refilling. If the overdrive has been dismantled the total of 6¼ pints will be required.

After refilling the gearbox and overdrive with oil, recheck the level after the car has been run, as a certain amount of oil will be retained in the hydraulic system of the overdrive unit.

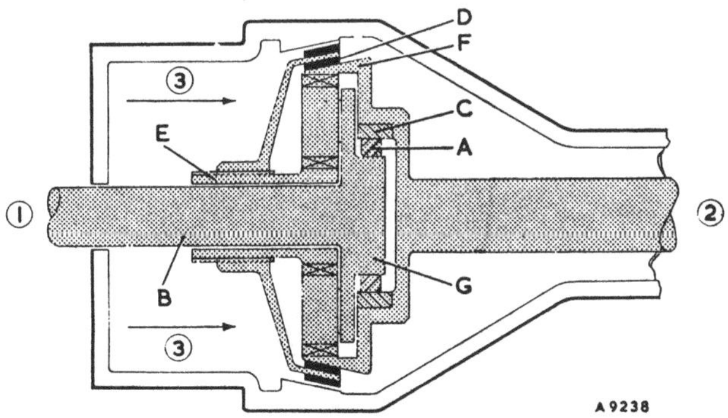

1. From gearbox. 2. To propeller shaft. 3. Hydraulic pressure.

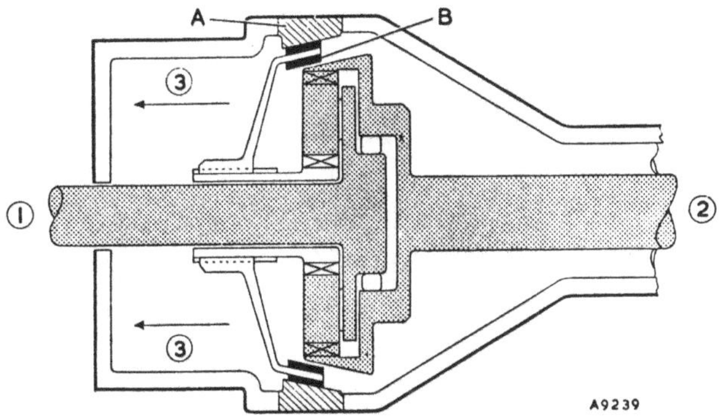

1. From gearbox. 2. To propeller shaft. 3. Spring pressure.

Section G.2

WORKING DESCRIPTION

The overdrive unit comprises a hydraulically controlled epicyclic gear housed in a casing which is directly attached to an extension at the rear of the gearbox.

The synchromesh gearbox third motion shaft is extended and carries at its end the inner member of an uni-directional clutch (see Fig. G.1). The outer member of this clutch is carried in the combined annulus and output shaft.

Also mounted on the third motion shaft are the planet carrier G and a freely rotatable sun wheel. Splined to a forward extension E of the sun wheel and sliding thereon is a cone clutch member D, the inner lining of which engages the outside of the annulus F while the outer lining engages a cast-iron brake ring sandwiched between the front and rear parts of the unit housing.

A number of compression springs is used to hold the cone clutch in contact with the annulus, locking the sun wheel to the latter so that the entire gear train rotates as a solid unit, giving direct drive. In this condition the drive is taken through the uni-directional clutch, the cone clutch taking over-run and reverse torque, as without it there would be a free-wheel condition.

The spring pressure can be overcome through the medium of two pistons, working in cylinders formed in the unit housing, supplied with oil under pressure from a hydraulic accumulator. This hydraulic pressure causes the cone clutch to engage the stationary brake ring (A Fig. G.2) and bring the sun wheel to rest, allowing the annulus to over-run the uni-directional clutch and give an increased speed to the output shaft, i.e. "overdrive".

When changing from overdrive to direct gear, if the accelerator pedal is released (as in a change down for engine braking) the cone clutch, being oil immersed, takes up smoothly. If the accelerator pedal is not

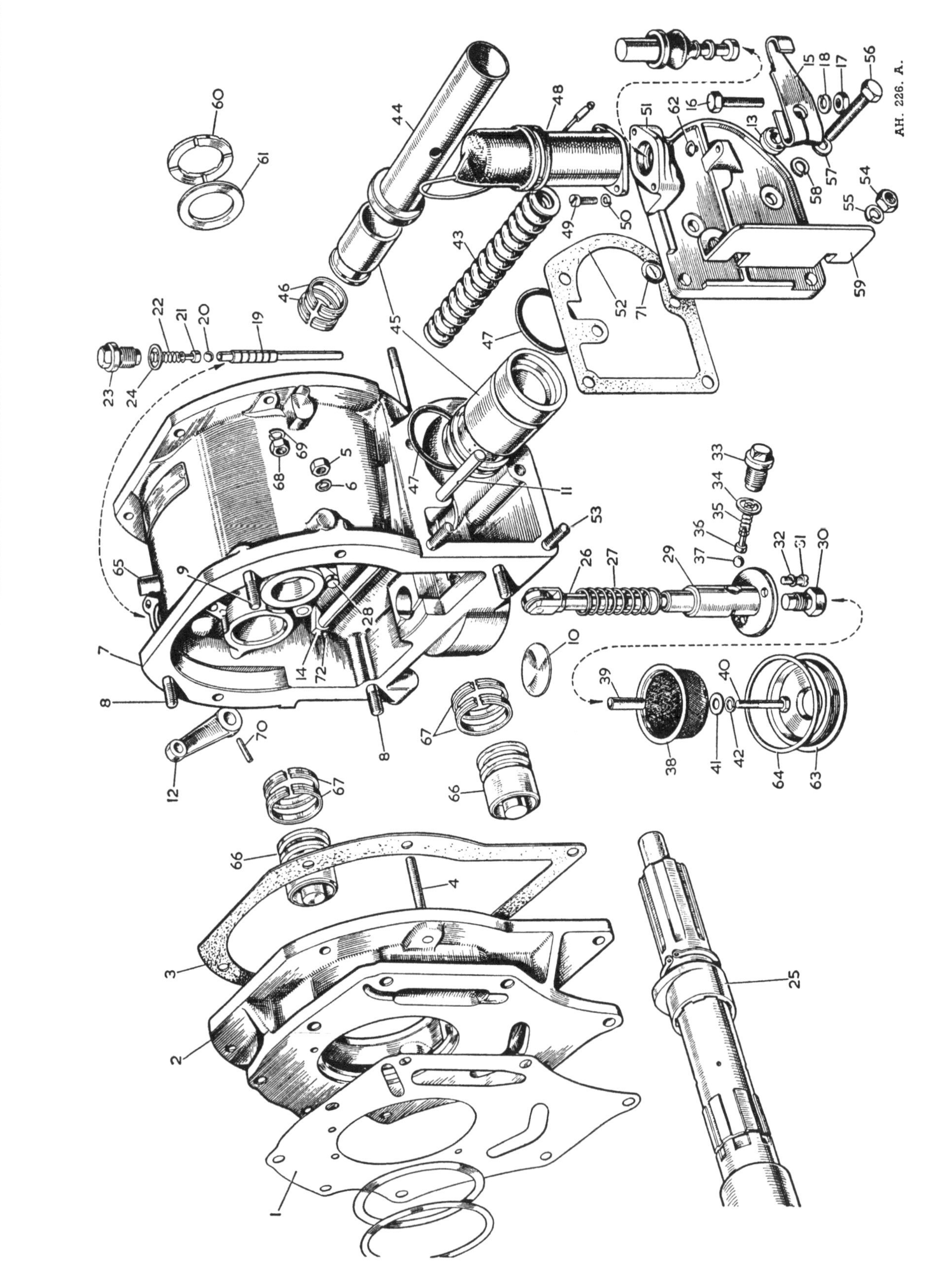

Fig. G.3. Components of overdrive front casing.

Fig. G.3. Components of overdrive front casing.

1. *Joint washer.*
2. *Adapter plate.*
3. *Joint washer.*
4. *Locating stud.*
5. *Nut.*
6. *Spring washer.*
7. *Main casing.*
8. *Stud.*
9. *Stud.*
10. *Welch plug.*
11. *Valve operating shaft.*
12. *Setting lever.*
13. *Collar.*
14. *Shaft cam.*
15. *Solenoid lever.*
16. *Adjusting screw.*
17. *Nut.*
18. *Washer.*
19. *Valve push rod.*
20. *Ball valve.*
21. *Ball valve plunger.*
22. *Valve spring.*
23. *Valve plug.*
24. *Copper washer.*
25. *Third motion shaft.*
26. *Pump plunger.*
27. *Plunger spring.*
28. *Guide peg.*
29. *Pump body.*
30. *Pump body plug.*
31. *Body screw.*
32. *Spring washer.*
33. *Valve plug.*
34. *Plug washer.*
35. *Valve spring.*
36. *Ball valve plunger.*
37. *Ball valve.*
38. *Pump filter.*
39. *Distance piece.*
40. *Filter bolt.*
41. *Plain washer.*
42. *Spring washer.*
43. *Accumulator spring.*
44. *Distance tube.*
45. *Piston assembly.*
46. *Piston rings.*
47. *Rubber rings.*
48. *Solenoid unit.*
49. *Unit screw.*
50. *Spring washer.*
51. *Solenoid lever housing.*
52. *Joint washer.*
53. *Stud.*
54. *Nut.*
55. *Spring washer.*
56. *Setpin.*
57. *Plain washer.*
58. *Spring washer.*
59. *Solenoid shield.*
60. *Thrust washer.*
61. *Spacing washer.*
62. *Rubber stop.*
63. *Drain plug.*
64. *Drain plug washer.*
65. *Breather.*
66. *Piston.*
67. *Piston rings.*
68. *Nut.*
69. *Spring washer.*
70. *Cotter pin.*
71. *Oil seal.*
72. *Peg.*

released, when contact between the cone clutch and brake ring is broken, the unit still operates momentarily in its overdrive ratio as engine speed and road speed remain unchanged. But the load on the engine is released and it begins to accelerate, speeding up the sun wheel from rest until, just at the instant when its speed synchronises with the speed of the annulus, the whole unit revolves solidly and the uni-directional clutch takes up the drive once more. The movement of the cone clutch is deliberately slowed down so that the uni-directional clutch is driving before the cone clutch contacts, ensuring a perfectly self-synchronised change.

Section G.3

CONSTRUCTION

The third motion shaft of the synchromesh gearbox is extended to carry first a cam operating the oil pump and then a steady bearing with opposed plain bushes carried in the front housing. Next is the sun wheel of the epicyclic gear carried on a Clevite bush, and beyond this the shaft is splined to take the planet carrier and uni-directional clutch. The end of the shaft is reduced and carried in a plain bush in the output shaft. The latter is supported in the rear housing by two ball bearings. The clutch member slides on the splines of the sun wheel extension to contact either the annulus or a cast iron brake ring forming part of the unit housing.

To the hub of the cone clutch member is secured a ball bearing housed in a flanged ring. This ring carries on its forward face a number of pegs acting as guides to compression springs by which the ring, and with it the clutch member, is held against the annulus. The springs prevent free-wheeling on over-run and are of sufficient strength to handle reverse torque. Also secured to the ring are four studs picking up two bridge pieces against which bear two pistons operating in cylinders formed in the unit housing. The cylinders are connected through a valve to an accumulator in which pressure is maintained by the oil pump. The operating pistons are fitted with synthetic rubber sealing rings, and the accumulator piston with three piece cast iron rings.

When the valve is open, oil under pressure is admitted to the cylinders and pushes the pistons forward to engage the overdrive clutch. Closing the valve cuts off the supply of oil to the cylinders and allows it to escape. Under the influence of the springs the clutch member moves back to engage direct drive position. The escape of oil from the cylinders is deliberately restricted so that the clutch takes about half a second to move over.

The sun wheel and pinions are cyanide case-hardened and the annulus heat-treated. Gear teeth are helical. The pinions run on needle roller bearings on a case-hardened pin.

The outer ring of the uni-directional clutch is pressed into the annulus member. The clutch itself is of the caged roller type, loaded by a lock-type spring made of round wire.

The hydraulic system is supplied with oil by a plunger type pump operated by a cam on the gearbox third motion shaft. The pump body is pressed into the front housing and delivers oil through a non-return valve to the accumulator cylinder, in which a piston moves back against a compression spring until the required pressure is reached when relief holes are uncovered. From the relief holes the oil is led through drilled passages to an annular groove between the two steady bushes on the gearbox third motion shaft.

Radial holes in the shaft collect the oil and deliver it along an axial drilling to other radial holes in the shaft from which it is fed to the sun wheel bush, thrust washers, planet carrier and planet pins.

From the accumulator, oil under pressure is supplied to the operating valve chamber. This forms an enlargement at the top of a vertical bore and contains a ball valve, the ball seating downwards thus preventing oil from circulating to the operating cylinders. The valve is a hollow spindle sliding in the bore, its top end reduced and carrying a seating for the ball, which is then lifted, admitting oil to the operating cylinders and moving the pistons forward to engage the overdrive clutch.

When the valve is lowered the ball is allowed to come on to its seating in the housing, cutting off pressure to the cylinders.

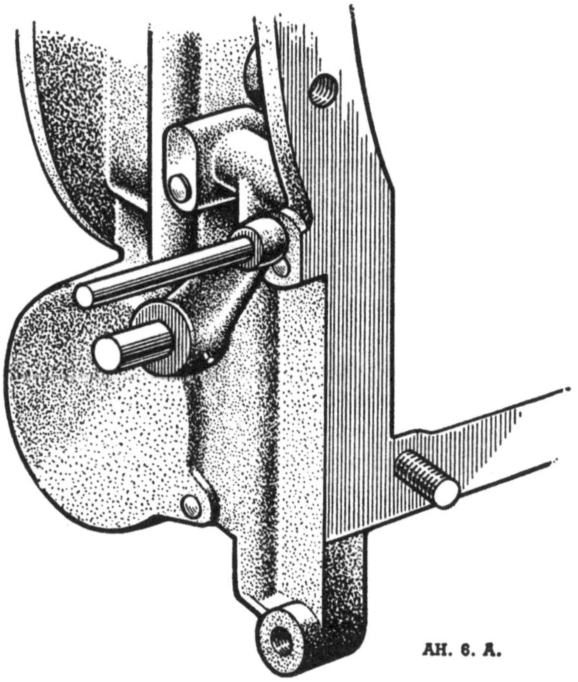

Fig. G.4. Valve setting lever.

Further movement of the valve brings it out of contact with the ball, allowing the oil from the cylinders to escape down the inside of the valve to discharge into the sump. The cone member then moves back under the influence of the clutch springs.

Section G.4

SERVICING IN POSITION

When the overdrive does not operate properly it is advisable first to check the level of oil and, if below the requisite level, top up with fresh oil and test the unit again before making any further investigations.

Before commencing any dismantling operations it is important that the hydraulic pressure is released from the system. Do this by operating the overdrive 10 to 12 times.

As the unit is fitted with a speed responsive control it will be found more convenient to carry out this operation by moving the valve setting lever manually.

Section G.5

GUIDE TO SERVICE DIAGNOSIS

Overdrive does not engage

(1) Insufficient oil in unit.
(2) Failure of switches or wiring (check solenoid operation visually).
(3) Solenoid setting out of adjustment.
(4) Insufficient hydraulic pressure due to pump non-return valve setting incorrectly (probably dirt on the seat).
(5) Insufficient hydraulic pressure due to worn accumulator.
(6) Damaged parts within the unit requiring removal and inspection of the assembly.

Overdrive does not release

(1) Control mechanism out of adjustment.
(2) Solenoid sticking.
(3) Blocked restrictor jet in operating valve.
(4) Solenoid setting out of adjustment.
(5) Clutch sticking.

NOTE:—Do not attempt to reverse the car: severe damage to the overdrive may result.

Clutch slip in overdrive

(1) Insufficient oil in unit.
(2) Solenoid setting out of adjustment.
(3) Insufficient hydraulic pressure due to pump non-return valve setting incorrectly.
(4) Insufficient hydraulic pressure due to worn accumulator piston, worn accumulator housing, or leaking 'O' ring.
(5) Operating valve seating incorrectly.
(6) Worn or glazed clutch lining.

Clutch slip in reverse or free-wheel condition on overrun

(1) Solenoid setting out of adjustment.
(2) Blocked restrictor jet in operating valve.

Section G.6

OPERATING VALVE

Having gained access to the unit through the floor, unscrew the valve plug and remove the spring and plunger. The ball valve will then be seen inside the valve chamber. The ball should be lifted $\frac{1}{32}$ in. (·8 mm.) off its seat when the overdrive control is operated.

As the unit is fitted with a speed responsive control the appropriate parts of the electrical circuit must be shorted out in order to operate the control.

If the ball does not lift by this amount the fault lies in the control mechanism. Located on the right-hand side of the unit and pivoting on the valve operating cross shaft, which passes right through the housing, is a valve setting lever. In its outer end is a $\frac{3}{16}$ in. (4·7 mm.) diameter hole which corresponds with a similar

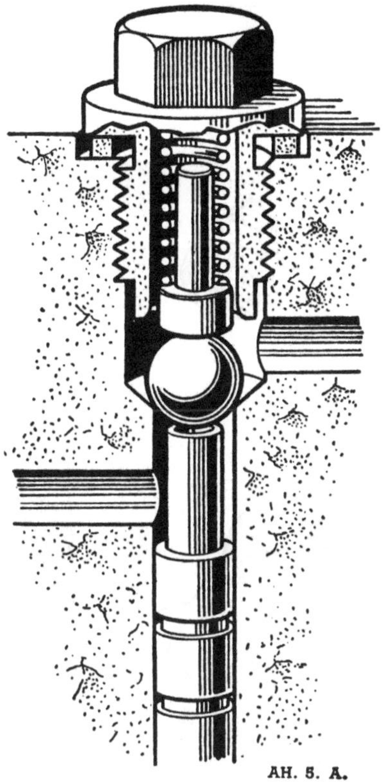

Fig. G.5. The operating valve.

hole in the housing when the unit is in "overdrive" (i.e. when the ball is lifted $\frac{1}{32}$ in. (·8 mm.) off the valve seat).

If the two holes do not line up, adjust the control mechanism until a $\frac{3}{16}$ in. (4·7 mm.) diameter rod can be inserted through the setting lever into the hole in the housing.

A small magnet will be found useful for removing the ball from the valve chamber. The valve can be withdrawn by inserting a length of $\frac{1}{8}$ in. (3·2 mm.) welding rod, but care must be taken not to damage the ball seating at the end of the valve. Near the bottom of the valve will be seen a small hole breaking through to the centre drilling. This is the jet for restricting the exhaust of oil from the operating cylinders. Ensure that this jet is not choked.

Section G.7

HYDRAULIC SYSTEM

If the unit fails to operate and the ball valve is found to be seating and lifting correctly check that the pump is functioning.

Jack up the rear wheels of the car, then with the engine ticking over and the valve plug removed, engage top gear. Watch for oil being pumped into the valve chamber. If none appears then the pump is not functioning.

The pump (Fig. G.6) described above, is of the plunger type and delivers oil via a non-return valve to the accumulator. Possible sources of trouble are (1) failure of the non-return valve due to foreign matter on the seat or to a broken valve spring and (2) breakage of the spring holding the pump plunger in contact with the cam.

The pump is self priming, but failure to deliver oil after the system has been drained and refilled indicates that the air bleed is choked causing air to be trapped inside the pump.

In the unlikely event of this happening it will be necessary to remove the pump and clean the flat on the pump body and the bore of the casting into which it fits.

Section G.8

PUMP VALVE

Access to the pump valve is gained through a cover on the left-hand side of the unit. Proceed as follows:—

(1) Remove drain plug and drain off oil.
(2) Remove solenoid.
(3) Slacken off clamping bolt in operating lever and remove lever, complete with solenoid plunger.
(4) Remove distance collar from valve operating shaft.
(5) The solenoid bracket is secured by two $\frac{5}{16}$ in. (7·9 mm.) studs and two $\frac{5}{16}$ in. diameter bolts,

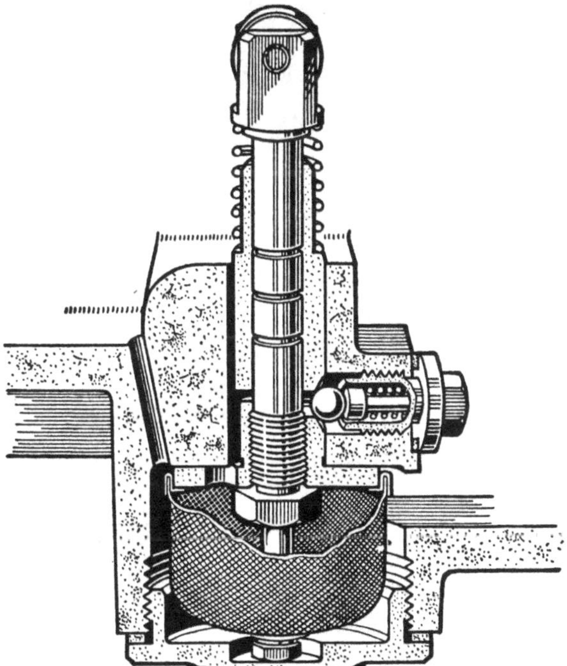

Fig. G.6. The pump in cut-away form.

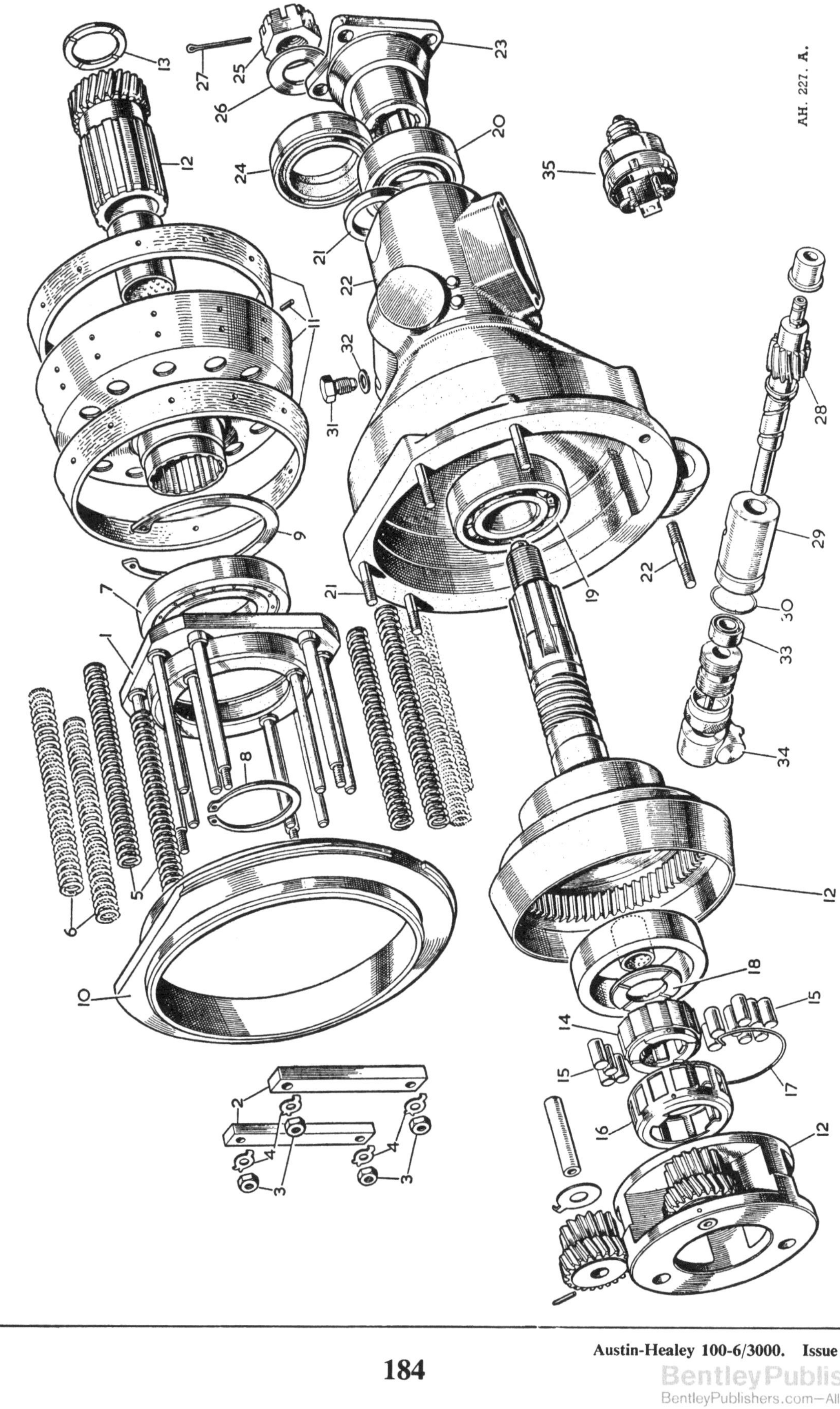

Fig. G.7. Components of overdrive rear casing

Fig. G.7. Components of overdrive rear casing.

1. Clutch thrust ring.	10. Brake ring.	19. Inner bearing.	28. Speedometer spindle.
2. Bridge pieces.	11. Clutch assembly.	20. Outer bearing.	29. Spindle sleeve.
3. Nuts.	12. Sun wheel assembly.	22. Rear housing.	30. Washer.
4. Locking washers.	13. Thrust washer.	21. Spacing washer.	31. Locking peg.
5. Clutch spring (long).	14. Uni-directional clutch.	23. Driving flange.	32. Washer.
6. Clutch spring (short).	15. Rollers.	24. Oil seal.	33. Oil seal.
7. Front bearing.	16. Outer casing.	25. Flange nut.	34. Spindle adaptor.
8. Circlip (small).	17. Securing clip.	26. Washer.	35. Overdrive switch.
9. Circlip (large).	18. Thrust washer.	27. Split pin.	

the heads of which are painted red, **remove the nuts from the studs before touching the bolts. This is important.** The two bolts should now be slackened off together, releasing the tension on the accumulator spring.

(6) Remove the solenoid bracket.

(7) Unscrew the valve cap and take out the spring, plunger and ball.

Clean the valve ball and seat with non-lint rag. Reseat the ball by tapping it on its seat with a light hammer and drift. Reassembly is the reverse of the above operations. Ensure that the soft copper washer between the valve cap and pump housing is nipped up tightly to prevent oil leakage.

It will now be necessary to reset the valve operating lever. Proceed as follows :—

Before clamping up the valve operating shaft rotate the shaft until a $\frac{3}{16}$ in. (4·7 mm.) diameter pin can be inserted through the valve setting lever on the offside of the unit into the corresponding hole in the casing. Leave the pin in position. Energise the solenoid by pulling off the snap connector at the solenoid terminal and connecting up with a short lead direct to the negative terminal of the battery. Check that the plunger travels the full extent of its stroke. Hold the solenoid lever lightly against the bottom shoulder on the solenoid plunger. Retighten the lever clamping bolt taking care that there is ·008 to ·010 in. (·20 to ·25 mm.) end float in the cross shaft. Remove the locating pin from the setting lever. Operate the solenoid several times to check for correct working.

To Dismantle the Pump

Proceed as follows

(1) Remove the drain plug and drain off oil.

(2) Remove pump valve as previously described.

(3) Remove the filter after unscrewing the securing bolt.

(4) Take out the two cheese head screws securing the pump body flange and extract the pump body. A special extractor tool (18G 183) is available for this purpose. This screws into the bottom of the pump body in the place of the screwed plug.

Assembly of the Pump

Line up the pump body so that the inlet port and holes for securing screws register with the corresponding holes in the housing, and tap the pump body home. Use service tool 18G 184.

The pump plunger is prevented from rotation when in position by a guide peg carried in the front casing. When assembling the pump the plunger should be inserted with the flat on its head facing the rear of the unit. It is possible to guide it past the guide peg by means of a screwdriver inserted through the side of the casing.

Replace the plug in the bottom of the pump body, ensuring that it is screwed home tightly.

Section G.9

HYDRAULIC PRESSURE

A working oil pressure of 470 to 490 lbs. per sq. in. (33·04 to 34·45 kg./cm.^{2}) is required.

A hydraulic pressure gauge complete with fittings (Service Tool 18G251) is obtainable from B.M.C. Service Ltd.

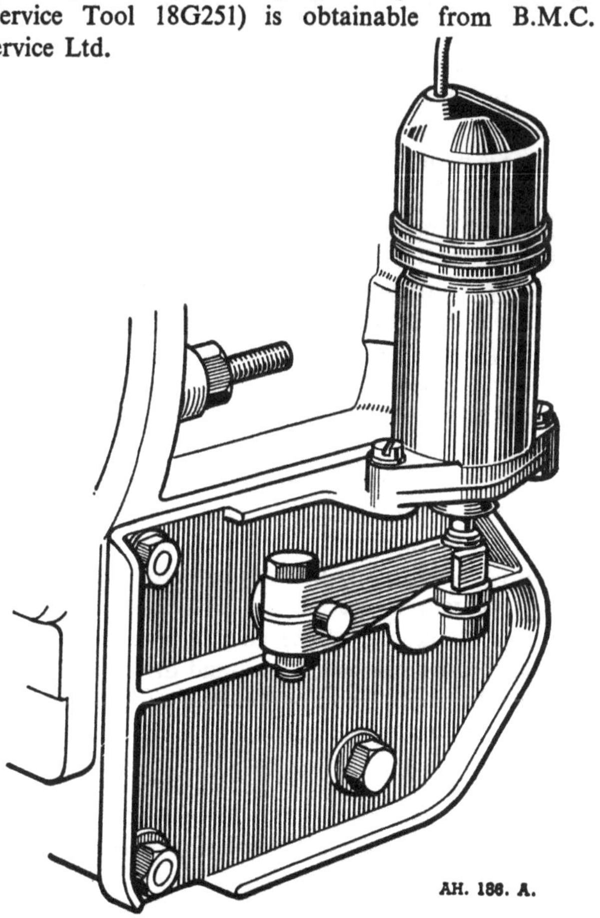

Fig. G.8. Showing the solenoid plunger bolt resting on the rubber stop

Section G.10

DISMANTLING AND REASSEMBLING UNIT

Dismantling

Should trouble arise necessitating dismantling of the unit to a degree further than has already been described, it will be necessary to remove the unit from the car.

Whilst it is possible to lift out the overdrive alone from the car, it is advised that the gearbox and overdrive be removed as a single unit. It is far easier to refit the overdrive to the gearbox when the assembly is on a bench as the extended third motion shaft must be lined up with the splines of the uni-directional clutch.

The unit is split at the adaptor plate which is attached to the front casing by six $\frac{5}{16}$ in. (7·9 mm.) studs, two of which are extra long. The four nuts on the shorter studs should be removed before those on the longer ones are touched. The latter should be unscrewed together releasing the compression of the clutch springs. The unit can then be drawn off the mainshaft, leaving the adapter plate attached to the gearbox.

Remove the clutch springs from their pins. The two bridge pieces against which the operating pistons bear can now be removed. Each is secured by two $\frac{1}{4}$ in. nuts locked by tab washers. Withdraw the two operating pistons.

As the adaptor plate is now separated from the unit the pump valve can be dismantled without removing the side cover (solenoid bracket) from the casing and there is no need to disturb the latter unless it is necessary to remove the accumulator piston and spring.

Remove the six $\frac{5}{16}$ in. (7·9 mm.) nuts securing the two halves of the casing and separate them, removing the brake ring which is spigoted into the two pieces. Lift out the planet carrier assembly. Remove the clutch sliding member complete with the thrust ring and bearing, the sun wheel and thrust washers. Take out the inner member of the uni-directional clutch, the rollers, cage, etc.

If it is necessary to remove the planet gears from the carrier the three split pins securing the planet bearing shafts must be extracted before the latter can be knocked out to release the gears and needle roller bearings.

To remove the annulus, first take off the coupling flange at the rear of the unit, remove the speedometer gear, centrifugal switch, etc., and drive out the annulus from the back. The front bearing will come away on the shaft leaving the rear bearing in the casing.

Inspection

Each part should be thoroughly inspected after the unit is dismantled and cleaned to ensure which parts should be replaced. It is important to appreciate the difference between parts which are worn sufficiently to affect the operation of the unit and those which are merely "worn in".

(1) Inspect the front casing for cracks, damage, etc. Examine the bores of the operating cylinders and accumulator for scores and wear. Check for leaks from plugged ends of the oil passages. Ensure that the welch washer beneath the accumulator bore is tight and not leaking. Inspect the support bushes in the centre bore for wear and damage.

(2) Examine the clutch sliding member assembly. Ensure that the clutch linings are not burned or worn. Inspect the pins for the clutch springs and bridge pieces and see that they are tight in the thrust ring and not distorted. Ensure that the ball bearing is in good condition and rotates freely. See that the sliding member slides easily on the splines of the sun wheel.

AH. 168. A.

Fig. G.9. Using tool 18G 178 for assembling the roller clutch.

(3) Check the clutch springs for distortion or collapse.

(4) Inspect the teeth of the gear train for damage. If the sun wheel bush is worn, the gear will have to be replaced since it is not possible to fit a new bush in service because it has to be bored true to the pitch line of the teeth.

(5) Examine the steel and bronze thrust washers.

(6) See that the rollers of the uni-directional clutch are not chipped and that the inner and outer members of the clutch are free from damage. Make sure that the member is tight in the annulus. Ensure that the spring is free from distortion.

(7) Inspect the ball bearings on the output shaft and see that there is no roughness when they are rotated slowly.

(8) Ensure that there are no nicks or burrs on the mainshaft splines and that the oil holes are open and clean.

(9) Inspect the oil pump for wear on the pump plunger and roller pin. Ensure that the plunger

spring is not distorted. Its free length is 2 in. (5·08 cm.). Inspect the valve seat and ball and make sure that they are free from nicks and scratches.

(10) Check the operating valve for distortion and damage and see that it slides easily in its bore in the front casing.

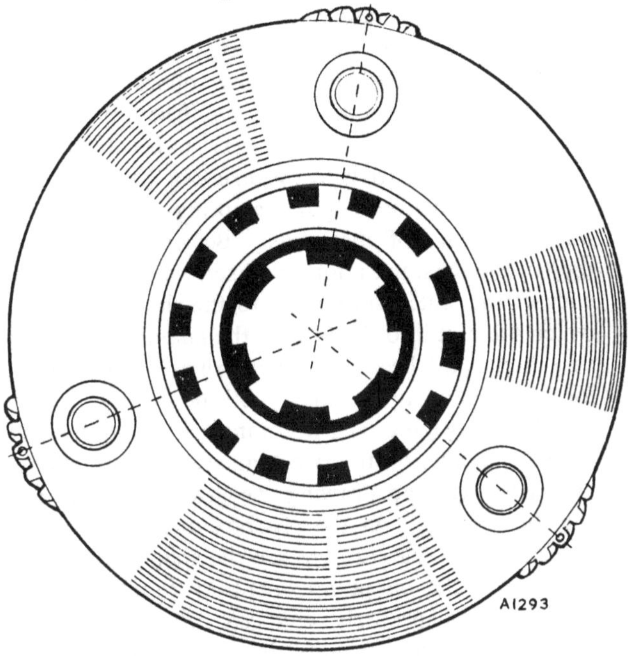

Fig. G.10. Before assembling the planet carrier to the sun wheel rotate the planet wheels until the punch marks are in the position shown.

Reassembling the Unit

The unit can be reassembled after all the parts have been thoroughly cleaned and checked to ensure that none are damaged or worn.

Assemble the annulus into the rear casing, not forgetting the spacing washer which fits between a shoulder on the shaft and the rear ball bearing. This washer is available in different thicknesses for selective assembly and should allow no end float of the annulus (output shaft) and no pre-loading of the bearings.

Selective washers are available in the following sizes :—

·146 in. ± ·0005 in. (3·7 mm. ± ·01)
·151 in. ± ·0005 in. (4·3 mm. ± ·01)
·156 in. ± ·0005 in. (3·9 mm. ± ·01)
·160 in. ± ·0005 in. (4·1 mm. ± ·01)
·161 in. ± ·0005 in. (4·09 mm. ± ·01)

Replace the thrust washer and uni-directional clutch inner member with its rollers and cage. The fixture (Fig. G.9) is for retaining the rollers in position when assembling the clutch. Ensure that the spring is fitted correctly so that the cage urges the rollers up the ramps on the inner member.

Fit the pump cam on to the gearbox mainshaft, offer up the front housing to the cover plate and secure temporarily with two nuts. In order to determine the amount of end float of the sun wheel, which should be ·008 in. to ·014 in. (·20 mm. to ·35 mm.) an extra thrust washer of known thickness should be assembled with the two normally used in front of the sun wheel.

The gear train must be assembled in the following way so that the planet wheels will mesh with the sun wheel and at the same time allow their compound teeth to mesh correctly with the annulus. One tooth on each planet wheel is punch-marked, and the planet wheels must be turned in the carrier so that the punch marks are radially outwards as shown in Fig. G.10. With the planet wheels aligned in this way, fit the planet carrier over the sun wheel. The position of the planet wheels in relation to each other ensures that the second set of teeth on the planet wheels will mesh with the annulus teeth. Offer up the assembly to the annulus turning the planet carrier until the locating peg on the inner member of the uni-directional clutch enters the corresponding hole in the planet carrier. This lines up the splines in the two members.

Assemble the brake ring to the front casing then offer up the front and rear assemblies, leaving out the clutch sliding member with its springs, etc. The gap between the flanges of the brake ring and rear casing should be measured. This gap will be less than the thickness of the extra thrust washer by the amount of end float of the sun wheel. If this is between the limits specified the unit may be stripped down again and re-assembled without the extra thrust washer. The clutch sliding member bridge pieces, etc., must now be replaced. The compression of the springs is taken up on the two long studs between the front casing and adapter plate.

If the indicated end float is more, or less, than that required it must be adjusted by replacing the steel thrust washer at the front of the sun wheel by one of less or greater thickness, as required. Washers of varying thickness are stocked for this purpose.

Seven sizes are available, as follows :—

·113 in. to ·114 in. (2·8 mm. to 2·9 mm.)
·107 in. to ·108 in. (2·72 mm. to 2·74 mm.)
·101 in. to ·102 in. (2·56 mm. to 2·59 mm.)
·095 in. to ·096 in. (2·4 mm. to 2·44 mm.)
·089 in. to ·090 in. (2·26 mm. to 2·28 mm.)
·083 in. to ·084 in. (2·1 mm. to 2·13 mm.)
·077 in. to ·078 in. (1·9 mm. to 1·98 mm.)

Care must be taken to ensure that the thrust washers at the front and rear of the sun wheel are replaced in their correct positions. At the front of the sun wheel the steel washer fits next to the head of the support bush in the housing and the bronze washer between the steel one and the sun wheel. At the rear, the steel washer is sandwiched between the two bronze washers.

Grip the mounting flange of the overdrive unit in a vice, so that the unit is upright, and insert a dummy shaft 18G 185 or a spare mainshaft if the dummy shaft is not available, so that the sun wheel and thrust washers, planet carrier and roller clutch line up with each other; a long thin screwdriver should be used to line by eye the

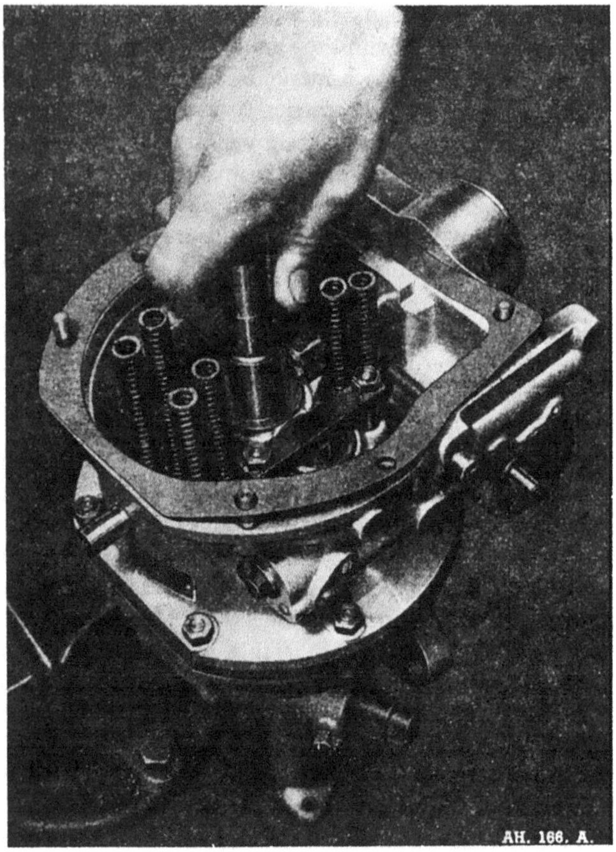

Fig. G.11. Centralising the gears with dummy mainshaft.

splines in the planet carrier and the roller clutch before inserting the dummy shaft. Gently turn the coupling flange to and fro while holding the dummy shaft, to assist in feeling the shaft into the splines of the planet carrier and roller clutch. Make sure that the dummy shaft has gone right into the spigot bush, checking this by using the screwdriver blade as a depth gauge.

Make quite sure that the clutch springs are in their correct positions—the 4¼ in. (108 mm.) long springs are the inner ones, and the 4½ in. (115 mm.) ones are the outer. This is most important because if any of the springs are in the wrong position they will become "coil bound" when the adapter plate is in place and restrict the movement of the sliding clutch so that overdrive will not engage.

Place the oil pump operating cam in position on top of the centre bushing with the lowest part of the cam in contact with the oil pump plunger and also place the paper joint washer in position.

The gearbox, with top gear engaged, should now be lifted by hand on to the overdrive unit, carefully threading the mainshaft through the oil pump cam and into the centre bushing in the body of the overdrive unit. Gently turn the first motion shaft to and fro to assist in "feeling" the mainshaft into the splines of the planet carrier. When the mainshaft is sufficiently entered for the gearbox to come to rest against the clutch springs with the two long studs just protruding through the holes in the overdrive body, put the spring washers and nuts on to the end of the studs. Before commencing to tighten the nuts, use a long thin screwdriver to guide the ends of the clutch springs on to the short locating pegs which are cast into the face of the adapter plate—this is very important because if the springs are not properly located they may become "coil bound" and prevent overdrive engaging. Now commence simultaneously to tighten the nuts on the two long studs, compressing the clutch springs and drawing the gearbox and overdrive together evenly. As the gearbox and overdrive come together watch carefully to see the splines on the mainshaft enter the oil pump operating cam and that the cam remains properly engaged with the oil pump plunger. If the two units do not pull together easily with only the resistance of the clutch springs being felt as the two nuts are tightened, stop tightening immediately. Gently rotate the gearbox first motion shaft in a clockwise direction whilst holding the overdrive coupling flange stationary until the mainshaft is felt to enter the roller clutch. The tightening of the nuts on the two long studs can then be completed, and the nuts fitted and tightened on to the four short studs.

NOTE: The gearbox mainshaft should enter the overdrive easily, provided that the lining-up procedure previously described is carried out and the unit is not disturbed. If any difficulty is experienced it is probable that one of the components has become misaligned, and the gearbox should be removed and the overdrive re-aligned with the dummy shaft.

Section G.11

OVERDRIVE RELAY SYSTEM

Description

Engagement of overdrive is controlled electrically through a manually operated toggle switch. The circuit shown in Fig. G.11 includes the following components :

(i) Relay, model SB40. An electro-magnetic switch used with item (ii) to enable an interlocking

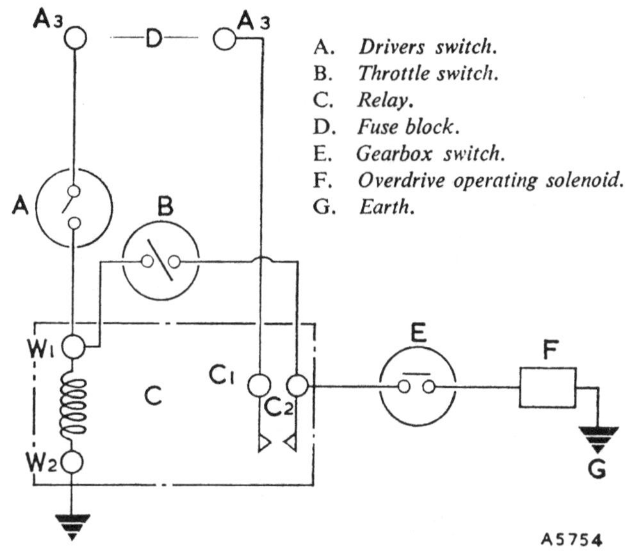

A. *Drivers switch.*
B. *Throttle switch.*
C. *Relay.*
D. *Fuse block.*
E. *Gearbox switch.*
F. *Overdrive operating solenoid.*
G. *Earth.*

Fig. G.12. Wiring diagram for the overdrive electrical circuit

safeguard to be incorporated against changing out of overdrive with throttle closed.

(ii) Throttle Switch, model RTS1. A lever-operated semi-rotary normally closed switch used in conjunction with item (i) to override the toggle switch under closed throttle conditions.

(iii) Gear Switch, model SS10. A small plunger-operated switch allowing overdrive to be engaged only in the two highest forward-gear positions.

(iv) Solenoid Unit, model TGS1. An electro-magnetic actuator to engage overdrive mechanism by opening hydraulic control valve.

Operation

When the toggle switch contacts are closed, current flows by way of the ignition switch and fuse unit supply terminal A3 to energize the relay operating coil. Closure of the relay contacts connects terminal A3 to the gear switch and, providing one of the two higher ratio gears is engaged, will energize the solenoid unit and effect a change from direct drive to overdrive.

Overdrive will be maintained until the solenoid unit is de-energized.

Change from overdrive to direct drive is effected either by selecting a low gear (when the gear switch contacts will open) or by turning the toggle switch to off with open throttle (when the relay contacts will open).

If effected with closed throttle, a change from overdrive to direct drive could result in a shock to the transmission. An interlocking circuit is therefore incorporated to override the toggle switch under closed throttle conditions. Under these conditions, the throttle switch contacts provide an alternative supply circuit to the relay operating coil.

Maintenance

Regular attention should be paid to wiring and connections. Damaged cabling must be replaced and loose terminals tightened, including the relay and solenoid unit earthing connections.

Section G.12

FAULT TRACING

The Solenoid Unit

With the engine stopped, neutral gear engaged, and the ignition switched on, disconnect the solenoid connection. Using a jumper lead, momentarily connect the solenoid to fuse unit supply terminal A3. The solenoid should be heard to operate. If no sound is heard, the solenoid is defective or incorrectly adjusted to the operating linkage. Remake the connection.

The Gear Switch

Engage top gear, depress the throttle pedal and momentarily connect relay terminal C2 to terminal A3. The solenoid should be heard to operate. If no sound is heard, the gear switch is defective. Re-engage neutral gear.

The Relay Coil

Momentarily connect relay terminal W1 to terminal A3. The relay should be heard to operate. If no sound is heard, the relay is defective.

The Toggle Switch

Operate the toggle switch. The relay should be heard to operate. If no sound is heard, the toggle switch is defective.

The Relay Contacts

With top gear engaged, toggle switch closed and throttle switch open, the solenoid should be heard to operate. If no sound is heard, the relay is defective.

The Throttle Switch

Engage top gear and close the toggle switch. Open the toggle switch and slowly depress the accelerator. The solenoid should be energized from zero to one-fifth throttle. If the solenoid is heard to release under one-fifth throttle, the throttle switch must be checked.

Throttle switch adjustment

The setting of the throttle switch is critical and incorrect adjustment will result in the overdrive disengaging when the car slows down with the throttle closed, accompanied by a noticeable braking effect.

The switch will normally only require adjustment after the carburetter or accelerator controls have been adjusted. The method of switch checking and adjustment. is as follows.

Connect a low consumption test lamp (a 12 volt 2·2 watt fascia panel light bulb is suitable) between the top terminal "A" (Fig. G.13) and a convenient earthing point.

The bulb should light when the overdrive and the ignition are both switched on, and the gear lever is set in the third or top gear position.

When the overdrive is switched off, the bulb should remain alight with the throttle still closed.

Progressively open the throttle by means of the accelerator pedal until the light goes out. Check the position of the throttle opening when this occurs: it should be one-fifth open. This position of the throttle has been reached when a $\frac{3}{16}$ in. (5 mm.) diameter rod can be just passed between the throttle stop screw and the stop lever on the HD type of carburetter, or when a feeler gauge of ·048 in. (1·22 mm.) thickness can be inserted between the throttle stop screw and the stop on the H4 type of carburetter used on earlier engines.

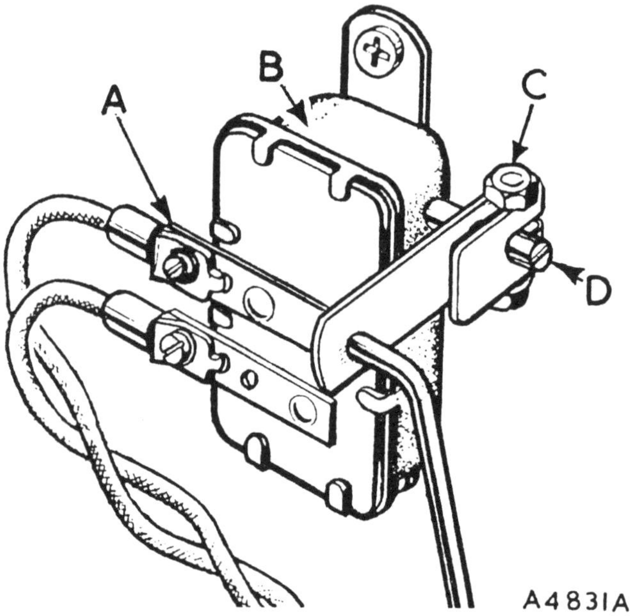

Fig. G.13. The overdrive throttle switch showing:

A. *Switch terminal.*
B. *Switch body.*
C. *Lever clamping bolt.*
D. *Operating shaft.*

If the switch requires adjustment, slacken the lever clamping bolt "C" (Fig. G.13) and turn the switch operating shaft "D" (Fig. G.13) with a screwdriver until the setting is correct: the end of the shaft is slotted for this purpose.

SECTION GG

OVERDRIVE

(Series BN6)

For details of the overdrive unit fitted to BN6 cars refer to Section G.

SECTION GGG

OVERDRIVE

Mk. I and II (SERIES BN7 and BT7)
AND Mk. II and Mk. III (SERIES BJ7 and BJ8)

Section GGG.1

OVERDRIVE UNIT

The overdrive unit is basically similar to that described in Section G. The following changes have, however, been made:

(1) A re-designed filter has been fitted and is retained in the body by a boss on the inside of the drain plug. The filter is accessible through the drain plug hole.

(2) New operating pistons have been fitted with synthetic rubber sealing rings and the accumulator piston with three piece cast iron rings.

(3) The pinions are fitted with needle roller bearings in lieu of "Clevite" bushes.

(4) The outer ring of the uni-directional clutch is not riveted.

(5) An additional selective washer 0·160±·0005 in. thick is available.

(6) A re-designed solenoid bracket and adapter plate has been fitted.

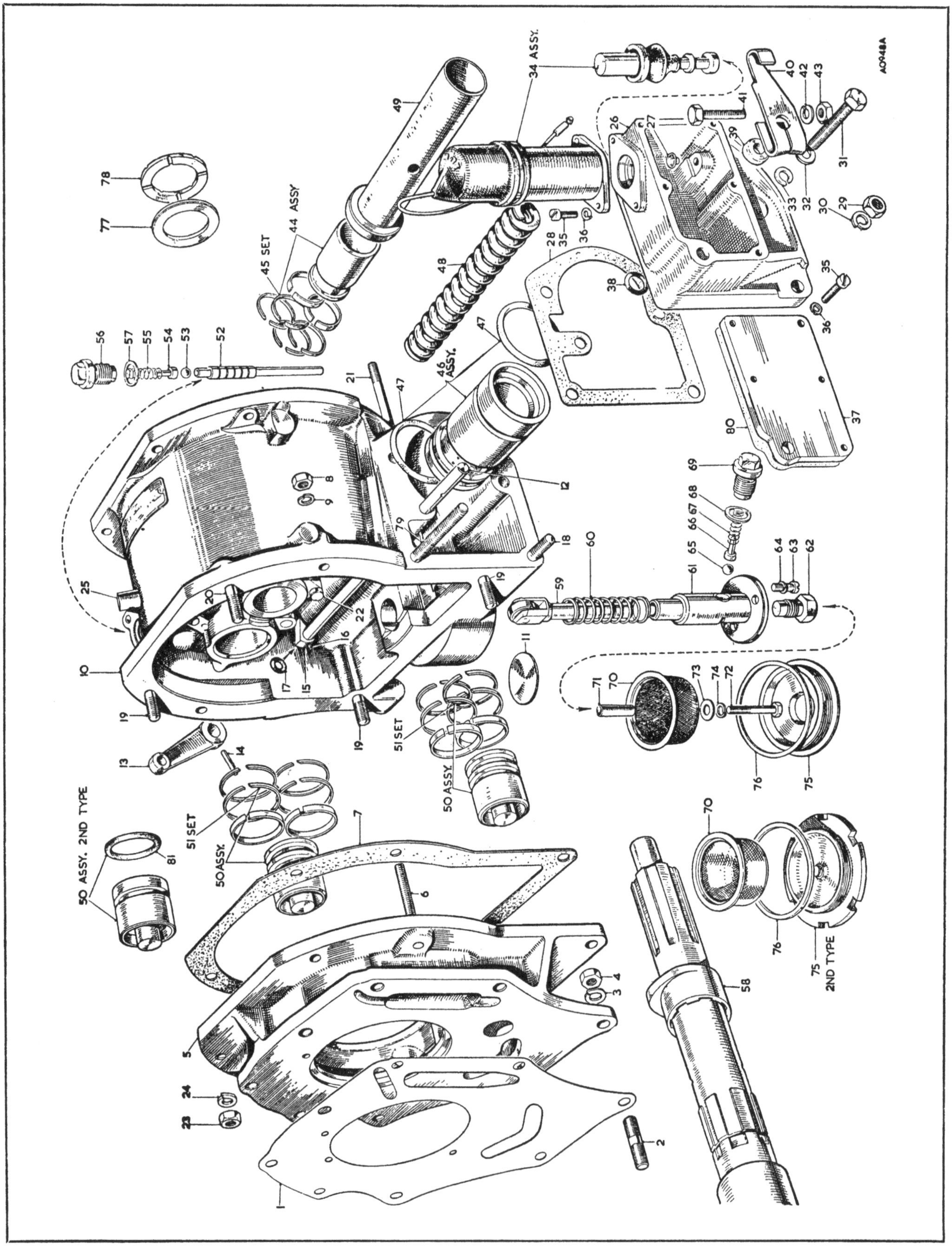
A0948A
34 ASSY.
44 ASSY
45 SET
46 ASSY.
51 SET
50 ASSY.
50 ASSY. 2ND TYPE
75 2ND TYPE

No.	Description	No.	Description	No.	Description
1.	*Joint—overdrive unit.*	28.	*Joint—bracket to case.*	55.	*Spring for plunger.*
2.	*Stud—overdrive unit.*	29.	*Nut—bracket to casing.*	56.	*Plug for valve—screwed.*
3.	*Washer for stud (spring).*	30.	*Washer for nut (spring).*	57.	*Washer for plug.*
4.	*Nut for stud.*	31.	*Screw—bracket to casing.*	58.	*Cam—oil pump.*
5.	*Plate—adaptor.*	32.	*Washer for screw.*	59.	*Plunger assembly—oil pump.*
6.	*Stud—adaptor to casing.*	33.	*Washer for screw (spring).*	60.	*Spring for plunger.*
7.	*Joint—adaptor to casing.*	34.	*Solenoid.*	61.	*Body—oil pump.*
8.	*Nut—plate stud to casing.*	35.	*Screw—solenoid to bracket.*	62.	*Plug for body—screwed.*
9.	*Washer for stud (spring).*	36.	*Washer for screw (spring).*	63.	*Screw—body to front casing.*
10.	*Casing—front.*	37.	*Plate—solenoid bracket cover.*	64.	*Washer for screw (spring)*
11.	*Plug.*	38.	*Seal, oil—valve operating shaft.*	65.	*Ball valve.*
12.	*Shaft—valve operating.*	39.	*Distance collar.*	66.	*Plunger, ball.*
13.	*Lever for shaft.*	40.	*Lever—solenoid.*	67.	*Spring for plunger.*
14.	*Pin—lever to shaft.*	41.	*Screw—lever to spindle.*	68.	*Washer for valve plug (copper).*
15.	*Cam—valve operating.*	42.	*Washer for screw (spring).*	69.	*Plug—valve.*
16.	*Pin for cam.*	43.	*Nut for screw.*	70.	*Strainer—oil pump.*
17.	*Seal for shaft.*	44.	*Piston assembly—accumulator.*	71.	*Distance tube for strainer.*
18.	*Stud for solenoid bracket.*	45.	*Ring—piston.*	72.	*Bolt for strainer.*
19.	*Stud for plate—long.*	46.	*Housing assembly—accumulator.*	73.	*Washer for bolt.*
20.	*Stud for plate—short.*	47.	*Ring, rubber.*	74.	*Washer for bolt (spring).*
21.	*Stud—front to rear casing.*	48.	*Spring—accumulator pressure.*	75.	*Plug—oil drain.*
22.	*Peg—oil pump plunger guide.*	49.	*Tube for spring.*	76.	*Washer for plug.*
23.	*Nut—front casing to adaptor plate.*	50.	*Piston—assembly—operating.*	77.	*Washer, steel.*
24.	*Washer for nut (spring).*	51.	*Ring, piston.*	78.	*Washer—phosphor bronze.*
25.	*Breather.*	52.	*Valve, operating.*	79.	*Stud for solenoid bracket.*
26.	*Bracket solenoid.*	53.	*Ball for valve.*	80.	*Joint for cover plate.*
27.	*Stop (rubber).*	54.	*Plunger, ball.*	81.	*Ring—piston.*

Key to the overdrive front casing components.

SECTION H

PROPELLER SHAFT

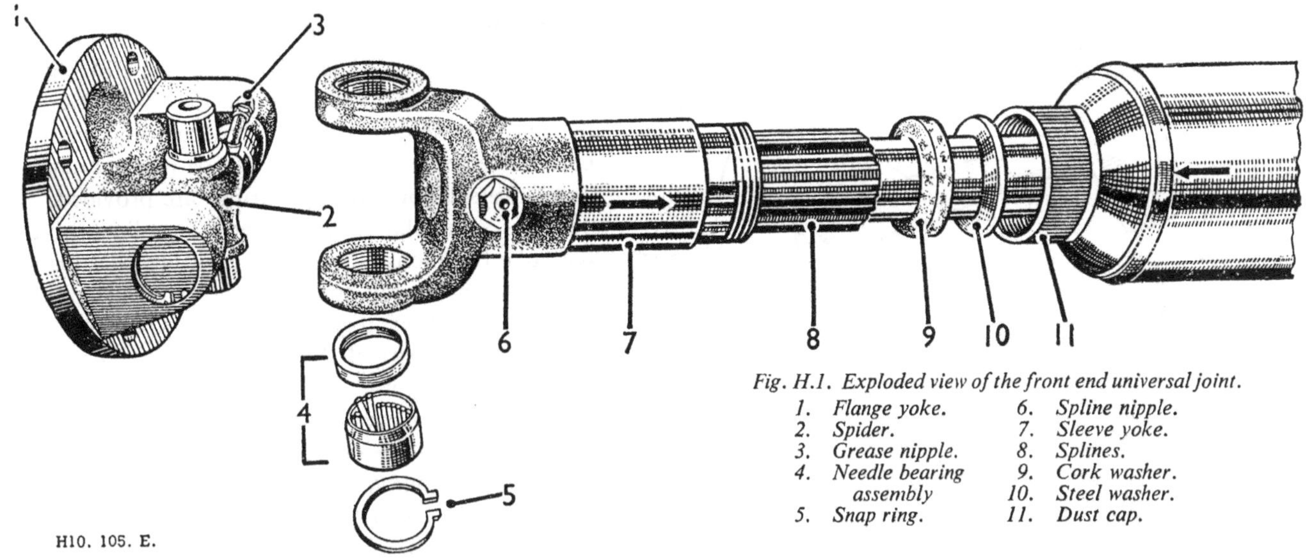

Fig. H.1. Exploded view of the front end universal joint.

1. Flange yoke.	*6. Spline nipple.*
2. Spider.	*7. Sleeve yoke.*
3. Grease nipple.	*8. Splines.*
4. Needle bearing assembly	*9. Cork washer.*
5. Snap ring.	*10. Steel washer.*
	11. Dust cap.

Section H.1

DESCRIPTION

The propeller shaft and universal joints (Fig. H.1) are of Hardy Spicer manufacture.

The fore and aft movement of the rear axle and other components is allowed for by a sliding spline between the propeller shaft and gearbox unit. Each universal joint consists of a centre spider, four needle roller bearings and two yokes. Reference to the Lubrication Chart (end of Manual) shows the location of the joints.

Section H.2

LUBRICATION

A nipple is fitted to each centre spider for the lubrication of the bearings. Reference to fig. H.2 shows that the central lubricant chamber is connected to four reservoirs and to the needle roller bearing assemblies.

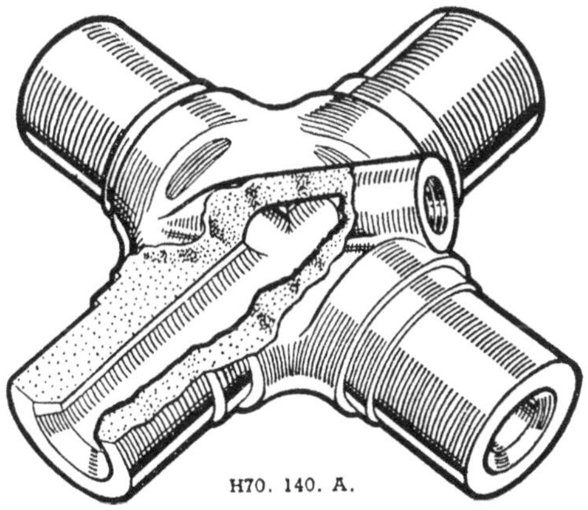

Fig. H.2. Showing oil channels in a joint spider.

The needle roller bearings are filled with lubricant on assembly. A nipple is provided on the sleeve yoke of the sliding spline joint for lubrication of the splines.

If a large amount of grease exudes from the cork seals the joint should be dismantled and new seals fitted.

After dismantling, and before reassembly, the inside splines of the sleeve yoke should be smeared liberally with grease.

Section H.3

PROPELLER SHAFT ASSEMBLY

Tests for Wear

(1) Wear on the thrust faces is located by testing the lift in the joint, either by hand, or by using a length of wood suitably supported.

(2) Any circumferential movement of the shaft relative to the flange yokes, indicates wear in the needle roller bearings, or the sliding spline.

Removal of Complete Assembly

Before removal of the propeller shaft can be effected, the gearbox tunnel must be removed.

The removal procedure for the propeller shaft is as follows:—

(1) Mark the propeller shaft and companion flanges on the gearbox and axle to facilitate replacement in the same position.

(2) Support the shaft near the sliding joint, then withdraw the bolts from the gearbox companion flange.

(3) Unscrew, by hand, the dust cap at the rear of the sliding joint. Slide the splined sleeve yoke about

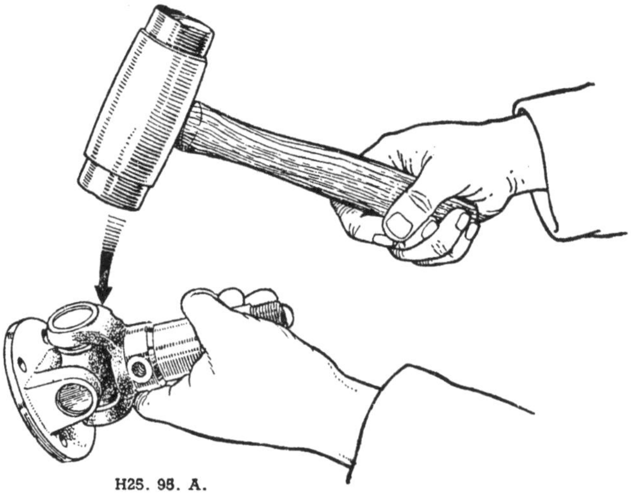

Fig. H.3. Tapping the joint to extract bearing.

half an inch rearwards, to disengage the pilot flanges.

(4) Remove the four nuts and bolts securing the rear flange yoke from the axle companion flange and lower the propeller shaft to the ground.

Dismantling

The following directions apply to both universal joints of the propeller shaft except for the fact that the front joint can be separated from the shaft, whereas the rear joint has one yoke permanently fixed to the tube. Unscrew the dust cap at the rear of the sliding spline and pull the splined sleeve from the main shaft. Retain the dust cover and washers.

(1) Clean away the enamel from all the snap rings and bearing faces, to ensure easy extraction of the bearings.

(2) Remove the snap rings by pressing together the ends of the rings and extract with a screwdriver. if the ring does not come out easily, tap the bearing face lightly to relieve the pressure against the ring.

(3) Hold the splined end of the shaft in one hand and tap the radius of the yoke with a lead or copper hammer (see fig. H.3) when it will be found that the bearing will begin to emerge. If difficulty is experienced, use a small bar to tap the bearing from the inside, taking care not to damage the race itself. Turn the yoke over and extract the bearing with the fingers (see fig. H.4), being careful not to lose any of the needles.

(4) Repeat this operation for the other bearing, and the splined yoke can be removed from the spider (see fig. H.5).

(5) Using a support and directions as above remove the spider from the other yoke.

Examination and Checking for Wear

After long usage the parts most likely to show signs of wear are the bearing races and the spider journals of the universal joints. Should looseness or stress marks be observed, the assembly should be renewed complete, as no oversize journals or bearings are provided.

It is essential that bearing races are a light drive fit in the yoke trunnions. Should any ovality be apparent in the trunnion bearing holes, new yokes must be fitted.

With reference to wear of the cross holes in a fixed yoke, which is part of the tubular shaft assembly, only in cases of emergency should this be replaced. It should normally be renewed with a complete tubular shaft assembly. The other parts likely to show signs of wear are the splined sleeve yoke, or splined stub shaft. A total of ·004 in. (·1mm.) circumferential movement, measured on the outside diameter of the spline, should not be exceeded. Should the splined stub shaft require renewing, this must be dealt with in the same way as the fixed yoke, *i.e.* a replacement tubular shaft assembly fitted.

Reassembly

(1) See that all drilled holes in the journals of the universal joints are cleaned out and filled with lubricant.

(2) Assemble the needle rollers in the bearing races and fill with grease. Should difficulty be experienced in assembly, smear the walls of the races with grease to retain the needle rollers in place.

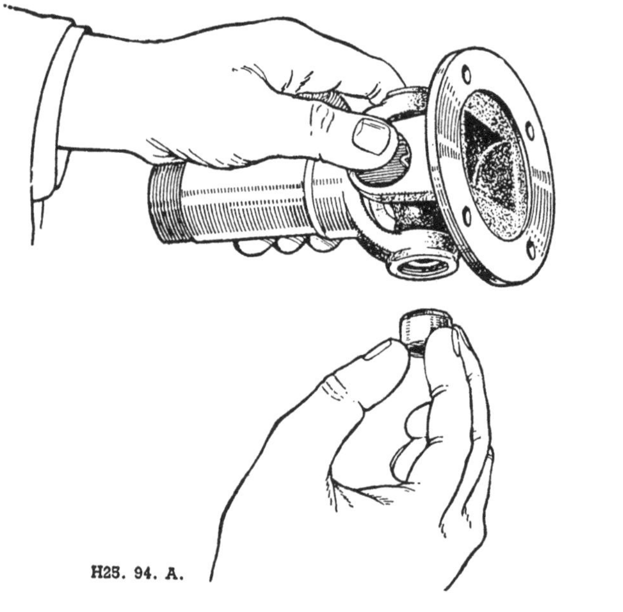

Fig. H.4. Withdrawing a bearing cup

(3) Insert the spider in the flange yoke.

(4) Using a soft-nosed drift about $\frac{1}{32}$ in. smaller in diameter than the hole in the yoke, tap the bearing in position. It is essential that bearing races are a light drive in the yoke trunnion.

(5) Repeat this operation for the other three bearings. The spider journal shoulders should be coated with shellac prior to fitting the retainers to ensure a good seal.

(6) If the joint appears to bind, tap lightly with a wooden mallet which will relieve any pressure of the bearings on the end of the journals. When replacing the sliding joint on the shaft, be sure that the trunnions in the sliding and fixed yoke are in line. This can be checked by observing that arrows marked on the splined sleeve yoke and the splined stub shaft are in line. It is advisable to renew cork washers and washer retainers on spider journals, using a tubular drift.

(7) Place the dust cover, steel washer, and cork washer over the splines of the shaft. Slide the splined sleeve onto the shaft making sure that the front and rear universal joint spiders are lying in the same plane. To achieve this condition the arrows on the sleeve and shaft must be in line.

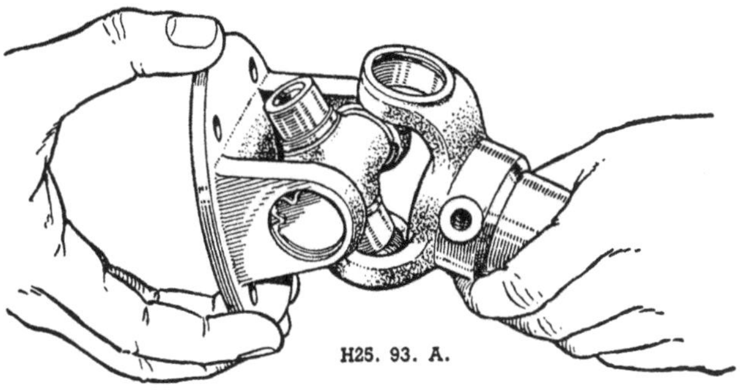

Fig. H.5. Separating the joint.

Replacing the Shaft Assembly

(1) Wipe the companion flange and flange yoke faces clean, to ensure that the pilot flange registers properly and the joint faces bed evenly all round.

(2) The sliding joint must be at the gearbox end.

(3) Align the marks made on all the flanges when removing the propeller shaft.

(4) Insert the bolts, and see that the nuts are tightened evenly all round and are securely locked.

SECTION J

REAR AXLE AND SUSPENSION

Section J.1

LUBRICATION

For the lubrication of the hypoid axle use lubricants only from approved sources, as tabulated in Section Q. Do not, in any circumstances mix brands of hypoid lubricant. If there is any doubt as to the oil previously used, drain and flush the axle before finally filling up with new hypoid oil. Do not use paraffin as a flushing medium.

The filler plug is situated on the rear side of the axle, and the drain plug in the bottom of the banjo casing.

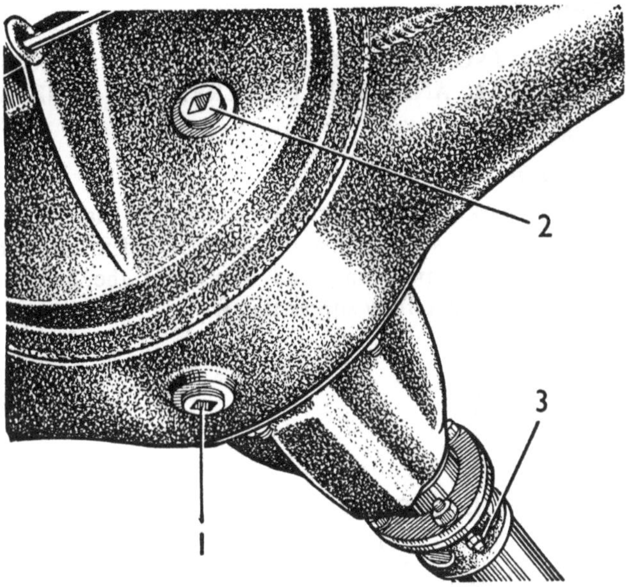

Fig. J.1. Rear axle.
1. Drain plug. 2. Filler plug. 3. Propeller shaft universal nipple

Section J.2

AXLE UNIT

To Remove and Replace

(1) Loosen the wheel nuts or hub caps, then jack-up the car and place supports under frame members just forward of the rear springs front anchorage. Take off both wheels after removing the wheel nuts or hub caps.

(2) Working from under the car, unscrew the four self-locking nuts and remove the bolts (U.N.F.) securing the propeller shaft flange to axle pinion flange.

(3) Disconnect the handbrake cable from the axle. This is accomplished by unscrewing it from its link to the brake balance lever, and unscrewing the nut holding its outer casing to the axle.

(4) The hydraulic brake pipe at the rear axle is detached from the flexible pipe at the union just forward of the right hand shock absorber.

(5) Unscrew the nuts securing the shock absorber links to the axle mounting brackets. Do not attempt to remove the links as this operation will prove much easier when freeing the axle.

(6) Remove the self-locking nuts from the spring clips ("U" bolts) which secure the axle to the springs. Observe that a fibre pad is situated between the axle and spring.

(7) Disconnect the tiebar at its axle anchorage by unscrewing its securing nuts.

(8) With the axle free, the connecting links from the shock absorbers should be detached.

(9) Remove the rubber block fixed between the axle and the left hand chassis frame. It is not necessary to detach the corresponding block on the right hand chassis frame.

(10) The complete axle should be removed from the right-hand side of the car. Take care not to damage other components, particularly the petrol pump.

(11) Installing the axle is the reverse of the above operations.

On re-assembling, it is advisable to jack-up the springs to meet the axle thus locating the spring centre bolt properly. Remember to fit the fibre pad.

When assembly is complete adjust the handbrake if required and bleed the hydraulic brake system all round.

Section J.3

AXLE SHAFTS

To Remove and Replace

(1) Loosen the wheel nuts or hub cap of the wheel concerned before jacking-up the car.

(2) Remove the wheels after further unscrewing the wheel nuts or hub caps.

(3) Take out the two drum locating screws, using a screwdriver.

Note.—If wire wheels are fitted it will be necessary to remove the five self-locking nuts, which secure the rear hub extension, to gain access to the two drum locating screws.

(4) The drum can be tapped off the hub and brake linings, provided the handbrake is released and the brake shoes are not adjusted so closely as to bind on the drum.

Should the brake linings hold the drum when the handbrake is released, it will be found necessary to slacken the brake shoe adjuster a few notches.

(5) Remove the axle shaft retaining screw and withdraw the axle shaft. Should the paper washer be damaged it must be renewed when re-assembling.

(6) Replacement is the reversal of the above operations, but ensure that the bearing spacer is in position. Adjust the rear brakes, and if hand brake adjustment is necessary refer to Section M.1.

Section J.4

HUBS

To Remove and Replace

(1) Remove the drum, axle shaft and bearing spacer.

(2) Knock back the tab of the locking washer and unscrew the nut with Service Tool 18G 258.

(3) Tilt the lock washer to disengage the key from the slot in the threaded portion of the axle casing; remove the washer.

(4) The hub can then be withdrawn with a suitable extractor such as Service Tool 18G 220 with adaptors 'A', 'D' and 'E'. The bearing and oil seal will be withdrawn with the hub.

(5) Fit a new oil seal using Service Tool 18G 134 and adaptor 18G 134AQ.

(6) The bearing is not adjustable and is replaced in one straightforward operation.

When re-assembling it is essential that the outer face of the bearing spacer should protrude from ·001 in. (·025 mm.) to ·004 in. (·1 mm.) beyond the outer face of the hub and the paper washer, when the bearing is pressed into position. This ensures that the bearing is gripped between the abutment shoulder in the hub and the driving flange of the axle shaft.

Section J.5

REMOVING AND REPLACING THE PINION OIL SEAL AND THE DIFFERENTIAL PINIONS

Pinion Oil Seal

Removal

(1) Mark the propeller shaft and pinion shaft driving flanges so that they can be replaced in the same relative positions, and disconnect the propeller shaft, carefully supporting it.

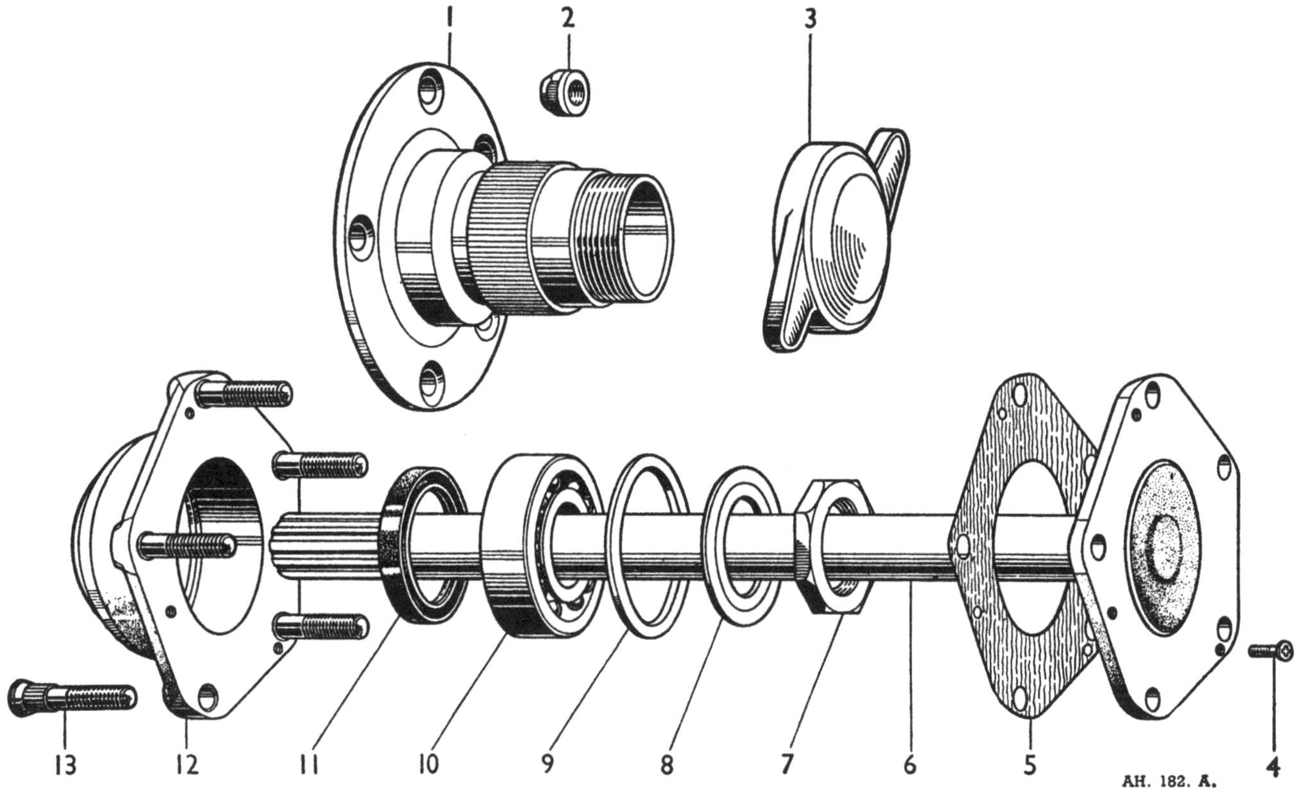

Fig. J.2. Rear axle hub assembly (with wire wheel hub extension).

1. Hub extension. 2. Securing nut. 3. Hub cap. 4. Securing screw. 5. Joint washer. 6. Half shaft. 7. Hub locknut. 8. Hub lockwasher. 9. Bearing spacer. 10. Hub bearing. 11. Oil seal. 12. Hub casing. 13. Hub extension stud.

(2) Using Service Tool 18G 34A to prevent the pinion flange from turning, remove the nut and spring washer. Withdraw the flange and pressed end-cover from the pinion shaft using Service Tool 18G2.

(3) Extract the oil seal from the casing.

Replacement

(1) Press a new seal into the casing with the sealing edge facing inwards.

(2) Replace the driving flange and end-cover, taking care not to damage the edge of the oil seal, and tighten the nut with a torque wrench to a reading of 140 lb. ft. (19·4 kg.m.)

(3) Reconnect the propeller shaft, taking care to fit the two flanges with the locating marks in alignment.

Differential Pinions

Removal

(1) Drain the oil from the axle.

(2) Remove the axle shafts as detailed in Section J.3.

(3) Mark the propeller shaft and pinion driving flanges so that they can be replaced in the same relative positions, and disconnect the propeller shaft.

(4) Remove the twelve nuts and spring washers securing the gear carrier to the axle case, and withdraw the complete carrier assembly.

(5) Having marked the differential bearing housing caps so that they can be replaced in their original positions, remove the four nuts and spring washers. Withdraw the bearing caps and the differential cage from the carrier. Note the thickness of the spacer collars fitted between the **outer** ring of each bearing and the differential carrier.

(6) Tap out the dowel pin locating the differential pinion shaft. The diameter of the pin is $\frac{3}{16}$ in. (4·8 mm.) and it must be tapped out from the crown wheel side of the differential cage as the hole into which it fits has a smaller diameter at the crown wheel end to prevent the pin passing right through. It may be necessary to clean out the metal peened over the entry hole with a $\frac{3}{16}$ in. drill in order to facilitate removal of the dowel pin.

(7) Drive out the differential pinion shaft and remove the pinions and thrust washers from the differential cage.

Replacement

Examine the pinions and thrust washers for wear and renew as required. Reassembly is a reversal of the above proceedure. Care must be taken to peen over the dowel pin entry hole in order to retain the dowel pin securely in position. Refill the axle with oil to Ref. B (page Q.1.).

NOTE.—If it proves necessary to fit any new parts other than those detailed in this section the axle assembly must be set up as in Section J.7.

Section J.6

DISMANTLING THE AXLE

Remove the differential carrier from the axle case as detailed in **'Differential Pinions'** of Section J.5.

It is strongly recommended that operators should make use of the Service Tools that are available. These tools have been specially designed to render easy operation and to prevent damage to the parts concerned.

Crown Wheel and Bearing Removal

(1) Remove the bearing caps, having marked them so that they can be replaced in their original positions. Lever out the differential cage and bearings. Note the spacer collars fitted between the outer ring of each bearing and the differential carrier.

(2) Remove the lockplates and bolts, and take off the crown wheel.

(3) The ball races should be a tight fit on each end of the differential cage and if found to be loose, a new cage will be needed. Service Tool 18G 47C and adaptor 18G 47R should be used to remove the bearings.

Bevel Pinion and Bearing Extraction

(1) Using wrench 18G 34A to keep the bevel pinion flange from turning, remove the nut and spring washer. Withdraw the flange using extractor 18G 2.

(2) Drive out the bevel pinion rearwards through the carrier using a soft metal drift. The pinion will take the inner race and rollers of the rear bearing, distance piece and shims leaving the front bearing and oil seal in the carrier. Extractor 18G12A with adaptor 12F or Service Tool 18G 285 can be used to draw the rear inner race and rollers off the bevel pinion. Retain the pinion head washer.

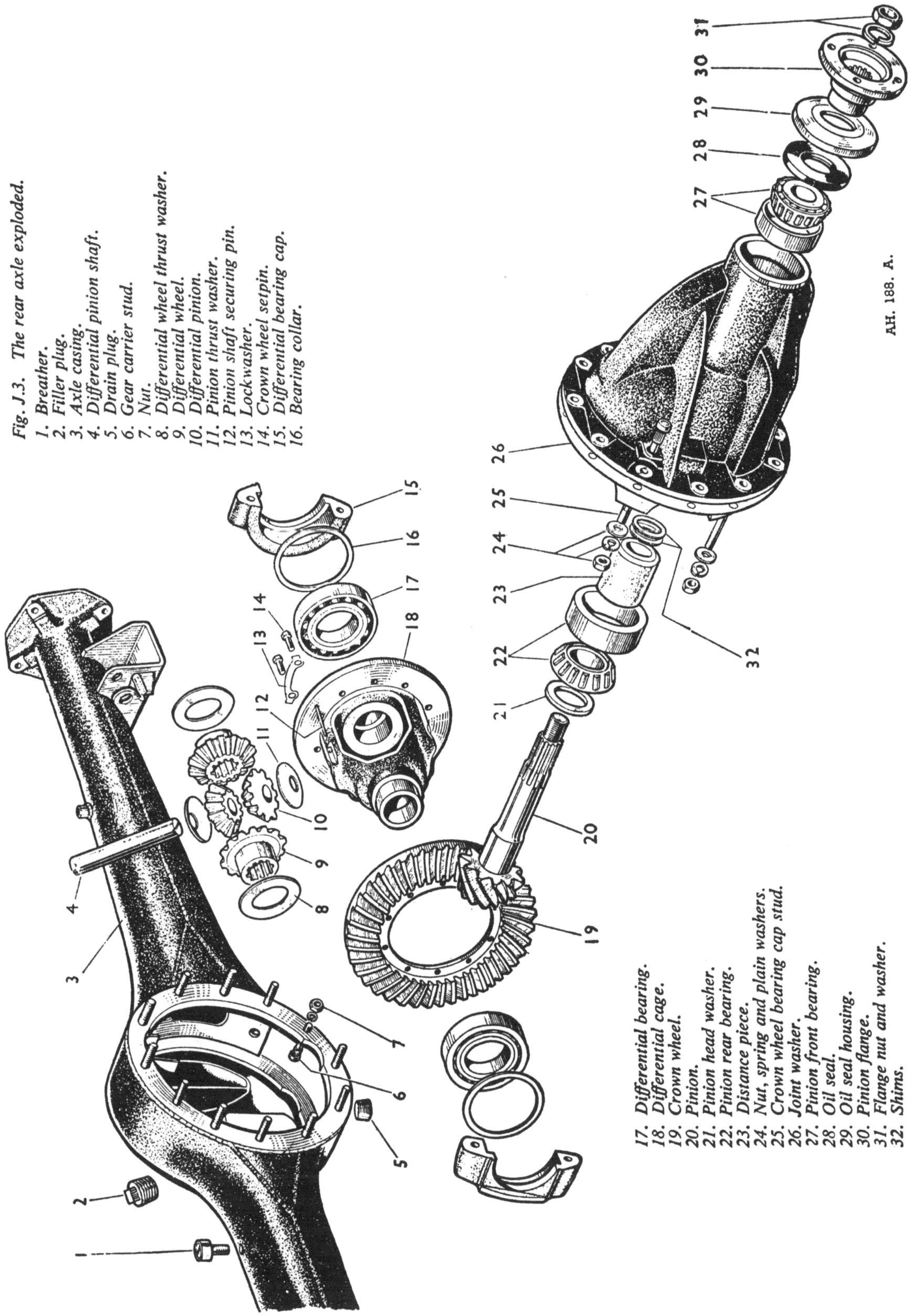

Fig. J.3. The rear axle exploded.

1. *Breather.*
2. *Filler plug.*
3. *Axle casing.*
4. *Differential pinion shaft.*
5. *Drain plug.*
6. *Gear carrier stud.*
7. *Nut.*
8. *Differential wheel thrust washer.*
9. *Differential wheel.*
10. *Differential pinion.*
11. *Pinion thrust washer.*
12. *Pinion shaft securing pin.*
13. *Lockwasher.*
14. *Crown wheel setpin.*
15. *Differential bearing cap.*
16. *Bearing collar.*
17. *Differential bearing.*
18. *Differential cage.*
19. *Crown wheel.*
20. *Pinion.*
21. *Pinion head washer.*
22. *Pinion rear bearing.*
23. *Distance piece.*
24. *Nut, spring and plain washers.*
25. *Crown wheel bearing cap stud.*
26. *Joint washer.*
27. *Pinion front bearing.*
28. *Oil seal.*
29. *Oil seal housing.*
30. *Pinion flange.*
31. *Flange nut and washer.*
32. *Shims.*

(3) Remove the oil seal from the gear carrier and extract the front pinion bearing inner race and rollers with the fingers. The outer races of both bearings can be withdrawn with Service Tool 18G 264 using adaptors 18G 264D and 18G 264H.

Section J.7

ASSEMBLING AND SETTING THE CROWN WHEEL AND PINION

Apart from fitting parts as detailed in Section J.3, J.4 and J.5 it is not permissible to fit any new parts (e.g. crown wheel and pinion, pinion bearings, differential bearings etc.) to the axle assembly without working through the procedure given in this section. Furthermore, if a new crown wheel or a new pinion is needed, **a mated pair—crown wheel and pinion—must be fitted.**

Replacing and Adjustment

Fitting a new crown wheel and pinion involves four distinct operations:

Fig. J.4. The differential carrier.

(1) Setting the position of the pinion.
(2) Adjusting the pinion bearing pre-load.
(3) Adjusting the differential bearing pre-load.
(4) Adjusting the backlash between the gears.

To carry out these operations correctly, the following special tools are required; the bevel pinion setting and differential bearing gauge, Service Tool 18G 191B, the pinion bearing outer race remover and replacer, Service Tool 18G 264 with adaptors 18G 264D and 18G 264H, Service Tool 18G 285 and the pre-load checking tool, Service Tool 18G 207.

Setting the Pinion Position

(1) Fit the bearing outer races to the gear carrier, using Service Tool 18G 264 with adaptors 18G 264D and 18G 264H

(2) Smooth off the pinion head with an oil stone, but do not erase any figures that are etched on the pinion head.

(3) Refit the pinion head washer; if the original washer is damaged or not available, select a washer from the middle of the range of thicknesses.

(4) Fit the inner race of the rear bearings to the pinion shaft using Service Tool 18G 285 and position the pinion in the gear carrier without the shims, bearing spacer and oil seal. Fit the inner race of the front bearing.

Fig. J.5. Gauging the depth of the differential bearing housings.

(5) Refit the universal joint driving flange and tighten the nut gradually until a pre-load figure of 13 to 15 lb. in. (·15 to ·173 kg. m.) is obtained.

(6) Adjust the dial indicator to zero on the machined step 'C' of the setting block (Service Tool 18G 191B).

(7) Remove the keep disc from the base of the magnet; clean the pinion head and place the magnet and dial indicator in position. Move the indicator arm until the foot of the gauge rests on the centre of the differential bearing bore at

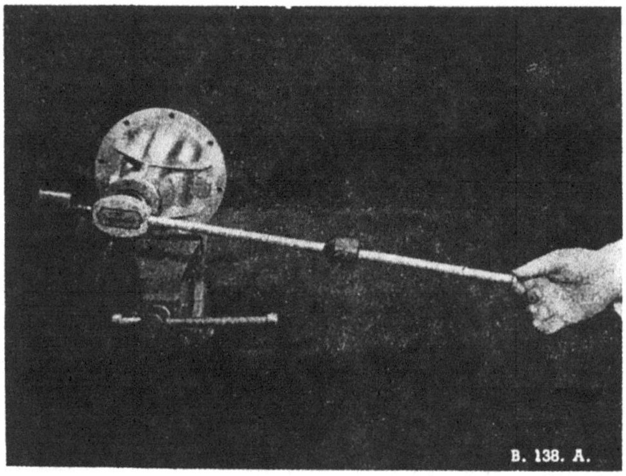

Fig. J.6. *Checking the bevel pinion bearing pre-load (Service Tool No. 18G 207).*

one side and tighten the knurled locking screw. Obtain the maximum depth reading and note any variation from the zero setting.

Repeat the check in the opposite bearing bore. Add the two variations together and divide by two to obtain a mean reading.

(8) Take into account any variation in pinion head thickness. This will be shown as an unbracketed figure etched on the pinion head and will always be minus(—). If no unbracketed figure is shown, the pinion head is of nominal thickness.

Using the mean clock gauge reading obtained and the unbracketed pinion head figure (if any), the following calculation can be made:

(*a*) **If the clock reading is minus** add the clock reading to the pinion head marking, the resulting sum being minus. **Reduce** the washer thickness by this amount.

Example:

Clock reading	—·002 in.
Pinion marking	—·005 in.
Variation from nominal	—·007 in.

Reduce the washer thickness by this amount.

(*b*) **If the clock reading is plus** and numerically less than the pinion head marking, **reduce** the washer thickness by the difference.

Example:

Pinion marking	—·005 in.
Clock reading	+·003 in.
Variation from nominal	—·002 in.

Reduce the washer thickness by this amount.

(*c*) **If the clock reading is plus** and numerically greater than the pinion head marking, **increase** the washer thickness by the difference.

Example:

Clock reading	+·008 in.
Pinion marking	—·003 in.
Variation from nominal	+·005 in.

Increase the washer thickness by this amount.

The only cases where no alterations are required to the washer thickness are when the clock reading is **plus** and **numerically equal** to the unbracketed pinion head marking, or when the clock reading is zero and there is no unbracketed marking on the pinion head.

(9) Allowance must finally be made for the mounting distance marked on the pinion head in a rectangular bracket. Proceeds as follows:

If the marking is a **plus** figure, **reduce** the washer thickness by an equal amount.

If the marking is a **minus** figure, **increase** the washer thickness by an equal amount.

A tolerance of ·001 in. is allowed in the thickness of the washer finally fitted.

Fit a washer of this thickness to the pinion with the chamfer towards the pinion head.

Adjusting Pinion Bearing Pre-load

(1) Assemble the pinion, bearings, bearing spacer and shims to the gear carrier ; fit the oil seal and driving flange.

Fig. J.7. *Illustrating the machining tolerances for the differential bearing housings as marked by the factory inspector.*

(2) Tighten the flange nut gradually to a torque wrench reading of 140 lb. ft. (19·4 kg.m.), checking the pre-load at intervals to ensure that it does not exceed 18 lb. in. (2·49 kg.m.), *i.e.* 3 lb in. greater than the previous figure as the oil seal is now fitted.

(3) If the pre-load is too great more shims must be added, and if too small the thickness of the shimming must be decreased.

Adjusting Differential Bearing Pre-load

Units marked with variations. The differential bearings must be pre-loaded and this is done by "pinching" them to the extent of ·002 in. on each bearing, the "pinch" being obtained by varying the thickness of the bearing distance collar fitted between each bearing outer ring and the register in the differential carrier. The collar thickness is calculated as shown below.

In making the necessary calculations, machining variations, and variations in bearing width must be taken into account. Machining variations are stamped on the component: bearing width variations must be measured.

The dimensions involved in pre-loading the differential bearings are illustrated in fig. J.8, and it is emphasised that it is the variation from nominal on each dimension which is important and referred to in the formula used.

The dimensions are :—

(1) From the centre line of the differential to the bearing register on the left-hand side of the gear carrier.
Variation: stamped on the carrier.

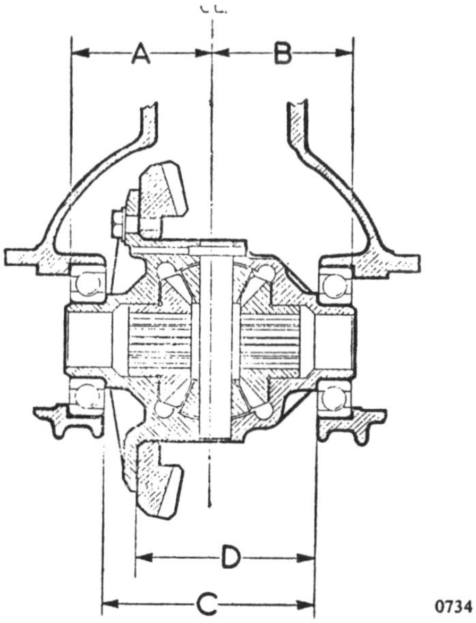

Fig. J.8. Illustrates the points from which the calculations must be made to determine the collar thickness for the bearings on each side of the carrier.

Fig. J.9. Checking differential bearing width with Service Tool No. 18G 191B.

(2) From the centre line of the differential to the bearing register on the right-hand side of the carrier.
Variation: stamped on the carrier.

(3) From the bearing register on one side of the differential cage to the register on the opposite side.
Variation: stamped on the cage.

(4) From the rear face of the crown wheel to the bearing register on the opposite side.
Variation stamped on the cage.

To calculate the collar thickness :—

Left-hand side:

Formula: A+D—C+·1815 in. (4·610 mm.).

Substitute the dimensional variations for the letters in the formula. The result is the thickness of the collar required at the left-hand side to compensate for machining variations and to give the necessary pinch, **with bearings of standard width.** The width of the bearing must now be checked and any variation from standard added to or subtracted from the collar thickness. If the bearing width is under standard, that amount must be added to the collar thickness, and *vice versa.*

Fig. J.10. *Checking crown wheel to pinion backlash (Service Tool No. 18G 191B).*

Table of Washer and Shim Thickness

Pinion head washer thicknesses	·208 in. to ·222 in. in steps of ·002 in.
Pinion bearing pre-load shims	·004 in. to ·012 in. in steps of ·002 in., plus ·020 in. and ·030 in.
Crown wheel bearing collars	·175 in. to ·193 in. in steps of ·002 in.
Pinion bearing pre-load ...	13 to 15 lb. in. without oil seal ; 16 to 18 lb. in. with oil seal.
Crown wheel bearing pinch...	·002 in. each side.

To Check Bearing Width

(1) Rest the bearing on the small surface plate of Tool No. 18G 191B with the inner race over the recess and the thrust face downwards.

(2) Place the magnet on the surface plate and set the dial indicator to zero on the step marked "C" of the small gauge block ; this is the width of a standard bearing. Transfer the indicator to the plain surface of the bearing inner race and, holding the race down against the balls, note the reading on the dial. A **negative** reading shows the additional thickness to be **added** to the collar at this side ; a **positive** reading, the thickness to be **subtracted.**

Right-hand side:

Formula: B—D+·1825 in. (4·634 mm.).

The procedure is the same as that for the left-hand side.

When a framed number is marked on the crown wheel, e.g. +2, it must be taken into account before assembling the collars to the differential carrier. This mark assists in relating the crown wheel with the pinion.

If, for example, the mark is +2 then the spacer thickness must be increased on the right-hand side by ·002 in. (·05 mm.) and decreased on the left-hand side (crown wheel side) by the same amount. If the marking is —2, the spacer thickness must be decreased on the right-hand side and increased on the left-hand side by the same amount.

Units not marked with variations: Some early models are fitted with differentials bearing no markings except the correct backlash for that particular pair of gears. The differential in such a case can be set as follows:—

(1) Fit the differential to the carrier with a distance collar at each side.
By trial and error select collars of thicknesses such that the differential with bearings and collars just fits into the carrier without slack and without pinching the bearings.

(2) Remove the unit and add ·002 in. to the thickness of the collar at each side to give the required pre-load.

(3) Fit the unit to the carrier and bolt up.

(4) Check and adjust the backlash as detailed below.

Adjusting Backlash

(1) Assemble the bearings 'Thrust' side outwards to the differential cage.

(2) Bolt the crown wheel to the differential cage, but do not knock over the locking tabs. Tighten the nuts to a torque wrench reading of 60 lb. ft. (8·3 kg. m.).

(3) Mount the assembly on two "V" blocks and check the amount of run out of the crown wheel as it is rotated. by means of a suitably mounted dial indicator.

(4) The maximum permissible run out is ·002 in. (·05 mm.) and any greater irregularity must be corrected. Detach the crown wheel and examine the joint faces on the flange of the differential case and crown wheel for any particles of dirt.

(5) When the parts are thoroughly cleaned it is unlikely that the crown wheel will not run true.

(6) Tighten the bolts to the correct torque wrench reading and knock over the locking tabs.

(7) Refit the differential to the gear carrier with the collars of calculated thickness. Tighten the bearing caps to a torque wrench reading of 65 lb. ft. (8·99 kg. m.).

(8) Mount the dial indicator on the magnet bracket so that an accurate measurement of the backlash can be taken. The correct figure for the backlash to be used with any particular crown wheel and pinion is etched on the rear face of the crown wheel concerned and must be adhered to strictly.

(9) Vary the backlash by decreasing the thickness of the collar at one side and increasing the thickness of the collar at one side and increasing the thickness of the collar at the other side by the same amount, thus moving the crown wheel into or out of mesh as required. The total thickness of the two collars must not be changed.

Section J.8

REAR SPRINGS

Description

The road springs are of the semi-elliptical type. The rear ends pivot in shackles to allow for variation in the effective lengths of the springs as they are flexed on load or rebound. The front ends of the springs are mounted in rigid brackets on the chassis longitudinal members. Driving and braking forces are transmitted from the axles to the chassis by this end of the springs.

Two rubber buffers attached to the axle limit any excessive upward or bump movement of the axle.

The rear spring dampers are of the lever, hydraulic type and are mounted to brackets on the chassis longitudinal members. The levers are attached to brackets on the axle. A filler plug is located in the top plate of each rear damper.

Maintenance

(1) Examine and tighten, if necessary the spring "U" bolts.

(2) Examine the oil level in the rear spring dampers and top up if necessary.

(3) Clean the springs and wipe them with an oily rag.

(4) Examine the springs for fractures and the bushes for wear.

To Remove

(1) Jack up the car on that side from which the spring is to be removed.

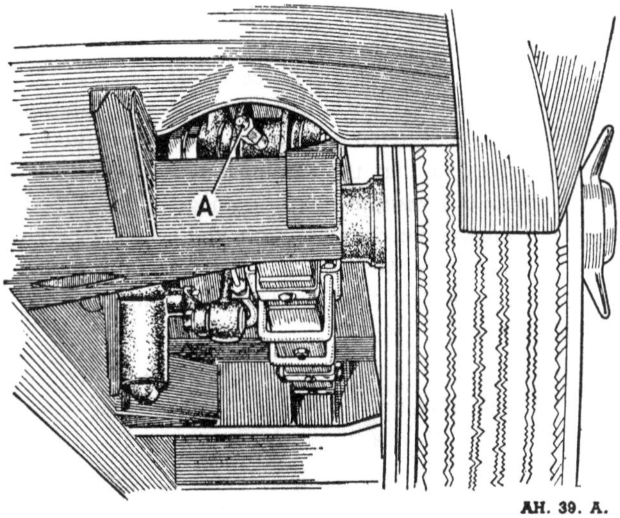

Fig. J.11.
This illustration shows the position of the rear spring lubricator beneath the luggage compartment.

(2) Pack up the chassis rear cross member with suitable supports, placing the supports as near to the spring rear anchorage as possible.

(3) Place a screw jack under the centre of the spring to relieve the tension.

(4) Remove the respective wheel.

(5) Using a box spanner release the four self-locking nuts from the "U" bolts which secure the spring to the axle tube.

(6) Detach the nut and spring washer on the inside of the upper rear shackle, and the locknut, spring washer and nut on the inside of the lower rear shackle.

(7) Remove the shackle inside connecting link and extract the top and bottom shackle pins, together with the outside link.

(8) At the forward end of the spring detach the anchor pin by removing the nut and spring washer on the inside of the pin and drive the pin clear.

(9) Remove the supporting jack from under the spring to withdraw the latter from the car.

To Dismantle

(1) Grip the spring in a vice jaws against the top and bottom leaves, adjacent to the centre bolt.
(2) Unscrew the clips.
(3) Unscrew the nut from the centre pin and withdraw the pin.
(4) Open the vice when the spring leaves, together with the zinc interleaving will separate.
(5) Examine the zinc interleaves for signs of failure or cracks.

To Reassemble

(1) Replace the spring in a vice.
(2) Utilising a rod of similiar diameter to the clamping bolt and having a tapered end, position the leaves so that the clamping bolt can be readily replaced without risk of damage to the thread.
(3) Replace the clamping bolt and secure by its nut and spring washer.
(4) Refit the leaf clips renewing their pins if necessary.

Fig. J.12.
Spring rear shackle assembly in exploded form.

Section J.9

SHOCK ABSORBERS

General Description

The shock absorbers are of the hydraulic double acting piston type. All the working parts are submerged in oil. They are carefully set before dispatch and cannot be adjusted without special equipment. Any attempt to dismantle them will seriously affect their operation and performance. Should adjustment or repair be necessary they must be returned to their makers.

Maintenance

The maintenance of the hydraulic shock absorbers should include a periodical examination of their anchorages to the body frame and axle brackets. The fixing bolts and nuts must be tightened as necessary.

The cheese-headed screws securing the cover-plates must be kept fully tightened to prevent leakage of the fluid.

When checking the fluid level all road dirt must be carefully cleared away from the vicinity of the filler plugs before the plugs are removed. This is most important as it is absolutely vital that no dirt or foreign matter should enter the operating champer.

The correct fluid level is just below the filler plug threads.

The use of Armstrong Super (thin) Shock Absorber Oil is recommended. When this is not available any good quality mineral oil to Specification S.A.E. 20/20 W is acceptable. This alternative is not suitable for low temperature operation.

To Remove

(1) Remove the nut and spring washer that secure the shock absorber lever to the link arm.
(2) Withdraw the two fixing setpins from the shock absorber body and chassis bracket.
(3) Remove the shock absorber, threading the lever over the link arm bolt.

To Replace

The replacement of a rear shock absorber is a reversal of the removal procedure. Ensure that the link is above the arm when refitting the unit to the chassis and axle. When handling shock absorbers that have been removed from the chassis, it is important to keep the assemblies upright otherwise air may enter the working chamber and so cause erratic resistance.

Connecting Link Bushes

The rubber bushes integral with both ends of the connecting link which joins the shock absorber to the rear axle cannot be renewed. When these bushes are worn the arm must be renewed complete.

Section J.10

TIE-ROD ASSEMBLY

A tie-rod that is mounted in rubber bushes between a bracket welded to the axle casing and a bracket welded to the chassis frame prevents lateral motion between the axle and the frame.

To renew the rubber bushes, remove the self-locking nuts, washers and outer bushes from the ends of the tie-rod, free the rear axle as described in Section J.2 paras. (1) to (7) and remove the rod. Then remove the inner rubber bushes.

Replacement of the tie-rod is a reversal of this procedure.

SECTION K

STEERING

SERIES BN4

Section K.1

DESCRIPTION

The steering gear is a unit of extreme simplicity. The steering tube revolves a cam, which, in turn, engages with a taper peg fitted to a rocker shaft. This assembly is enclosed in an oil tight casing which carries two ball bearings at either end of the cam. These bearings are designed to carry radial and thrust loads.

When the steering wheel is turned the tube revolves the cam, which, in turn, causes the taper peg to move over a predetermined arc, thus giving the rocker shaft its desired motion. Attached to the rocker shaft is a steering side and cross tube lever, which links up with the steering linkage.

The steering is of the "three cross tube" type, having a centre cross tube connecting the steering side and cross tube lever to the arm on the idler shaft. Two shorter side tubes, one on either side, connect the steering arms to the steering gear and idler levers respectively.

Section K.3

ADJUSTMENTS IN THE VEHICLE

The following adjustments maintain the performance of the steering at its maximum and consist of aligning the front wheels and taking up backlash in the steering gear. Proceed as detailed below.

(1) Front wheel alignment is governed by four factors—camber, castor, swivel pin inclination and wheel toe-in. The correct camber and swivel pin angles are built into the front suspension and will change only if the suspension is distorted by accidental damage. It is most important that the front wheels should toe-in $\frac{1}{16}$ in. (1·6mm) to $\frac{1}{8}$ in. (3 mm.), and this is governed by the angle of the track-rod arms and the length of the track-rod. An adjustment is provided so that the track-rod may be lengthened or shortened to maintain the correct alignment. The track-rod should not be adjusted to correct a bent track-rod arm.

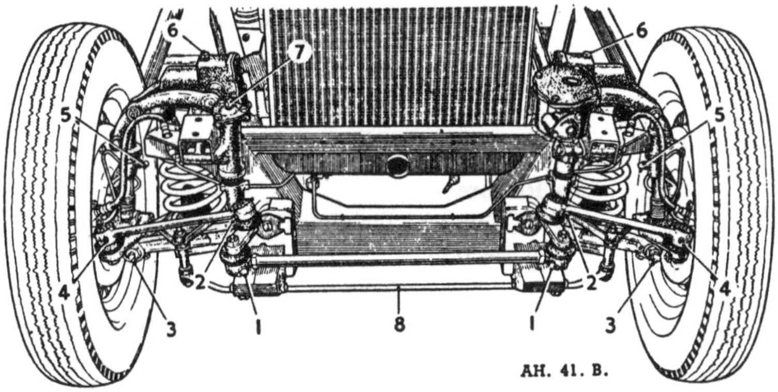

Fig. K.1. Showing the front suspension, steering layout and lubrication points.

1. *Cross tube connections.*
2. *Side rod inner connections.*
3. *Lower link.*
4. *Side rod outer connections.*
5. *Swivel pin.*
6. *Shock absorber.*
7. *Steering idler.*
8. *Anti-roll bar (no lubrication required).*

Section K.2

MAINTENANCE

Lubrication of the grease nipples on the steering connections and swivel bearings is most important to maintain accurate steering.

At the recommended mileage use the grease gun filled to Ref. C (page Q.1) to charge the following points with lubricant:—

(*a*) Steering rods and cross tube—6 nipples.

(*b*) Lower wishbone arm outer bearing—2 nipples.

(*c*) Swivel pin bushes—4 nipples.

The steering box and steering idler should be topped up with recommended oil to the top of the filler plug opening.

The track is best adjusted by means of a Dunlop Optical Alignment Gauge, particulars of which can be obtained from the Dunlop Rubber Co. Ltd., Fort Dunlop, Erdington, Birmingham, England.

The cross tube is threaded right-hand at one end and left-hand at the other, so that the track adjustment can be made by simply rotating the tube in the required direction after releasing the locknuts. Always re-tighten the locknuts at each end of the cross tube after an adjustment has been made.

The side-rods are non-adjustable.

When adjusting the track the following precautions should be observed :—

(*a*) The car should have come to rest from a forward movement. This ensures as far as possible that the wheels are in their natural running position.

(*b*) It is preferable for alignment to be checked with car laden.

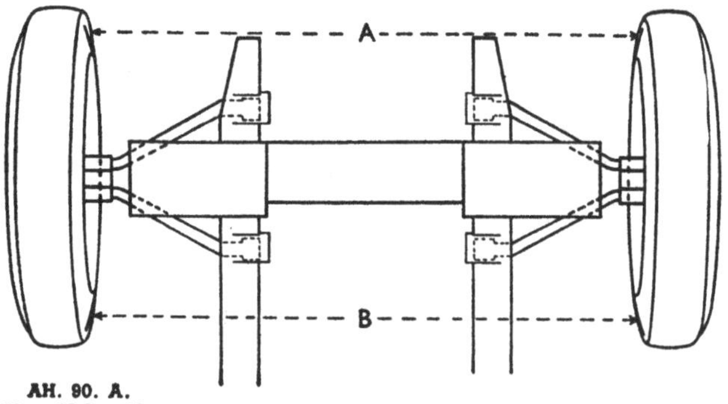

AH. 90. A.

Fig. K.2. The toe-in must be adjusted so that A is $\frac{1}{16}$ in. to $\frac{1}{8}$ in. less than B.

(*c*) With conventional base-bar tyre alignment gauges measurements in front of and behind the wheel centres should be taken at the same points on the tyres or rim flanges. This is achieved by marking the tyres where the first reading is taken and moving the car forwards approximately half a road wheel revolution before taking the second reading at the same points. With the Dunlop Optical Gauge two or three readings should be taken with the car moved forwards to different positions—180° road wheel turn for two readings and 120° for three readings. An average figure should then be calculated.

Wheels and tyres vary laterally within their manufacturing tolerances, or as the result of service, and alignment figures obtained without moving the car are unreliable.

(2) Steering cam bearing adjustment should be carried out to eliminate all perceptible end play. To adjust the cam bearings, proceed as follows :—

(*a*) From underneath the vehicle disconnect the side rod from the steering lever to free the gear of all loads.

(*b*) Disconnect the flashing indicator switch and horn push cables at the snap connectors behind the radiator grille and, from inside the vehicle, gently draw out the indicator switch and horn push, see Section N.31, until the cables have been drawn into the stator tube, being thus protected from oil.

(*c*) Place an oil tray under the steering box.

(*d*) Remove the nut and olive from the end cover and remove the end cover by unscrewing the four retaining bolts.

(*e*) Add or remove shims as necessary to obtain the correct adjustment. The steering wheel should turn freely when held lightly at the rim with the thumb and forefinger, but should have no end play.

(3) Rocker-shaft adjustment should be carried out after adjusting the cam bearings (described above).

(*a*) With the side rod still disconnected from the steering lever, slacken the adjusting screw locknut and screw in the adjusting screw.

(*b*) Check for backlash by exerting a light pressure on the lower end of the steering lever alternatively in both directions, while an assistant turns the steering wheel slowly from lock to lock. It will be noticed that the amount of slackness is not constant, there being less slackness in the centre than in the full lock position. If slackness appears at all positions of the drop arm, the adjusting screw should be screwed in farther. After further adjustment, test again in the same manner. The correct adjustment is such that a "tight spot" will barely be apparent as the steering wheel is removed past the centre position, with no backlash at the steering drop arm. At this position tighten the adjusting screw locknut.

(*c*) Refill the steering box with the correct grade of oil.

(*d*) Reconnect the side rod.

Section K.4

STEERING GEAR ASSEMBLY

To Remove

(1) Remove the horn quadrant as described in Section N.31.

(2) Remove the nut from the centre of the steering wheel hub, and pull off the steering wheel.

(3) Prize off the circlip, exposed to view, and then release the locking ring behind the steering wheel hub. (Adjustable type column only.)

Fig. K.3. Showing the steering column being manoeuvred out through the radiator grille aperture.

(4) Pull the steering wheel clear of the column, followed by the telescopic spring and locating collar. (Adjustable type columns only.)

(5) From behind the fascia release the two-piece clamping bracket supporting the top end of the column.

(6) Remove the radiator as described in Section C.

(7) Remove the radiator grille as described in Section P.

(8) There are two sealing plates, one on each side of the scuttle, through which the steering column passes, release each plate by undoing the four metal thread screws.

(9) Jack up the front of the car and remove the front wheels.

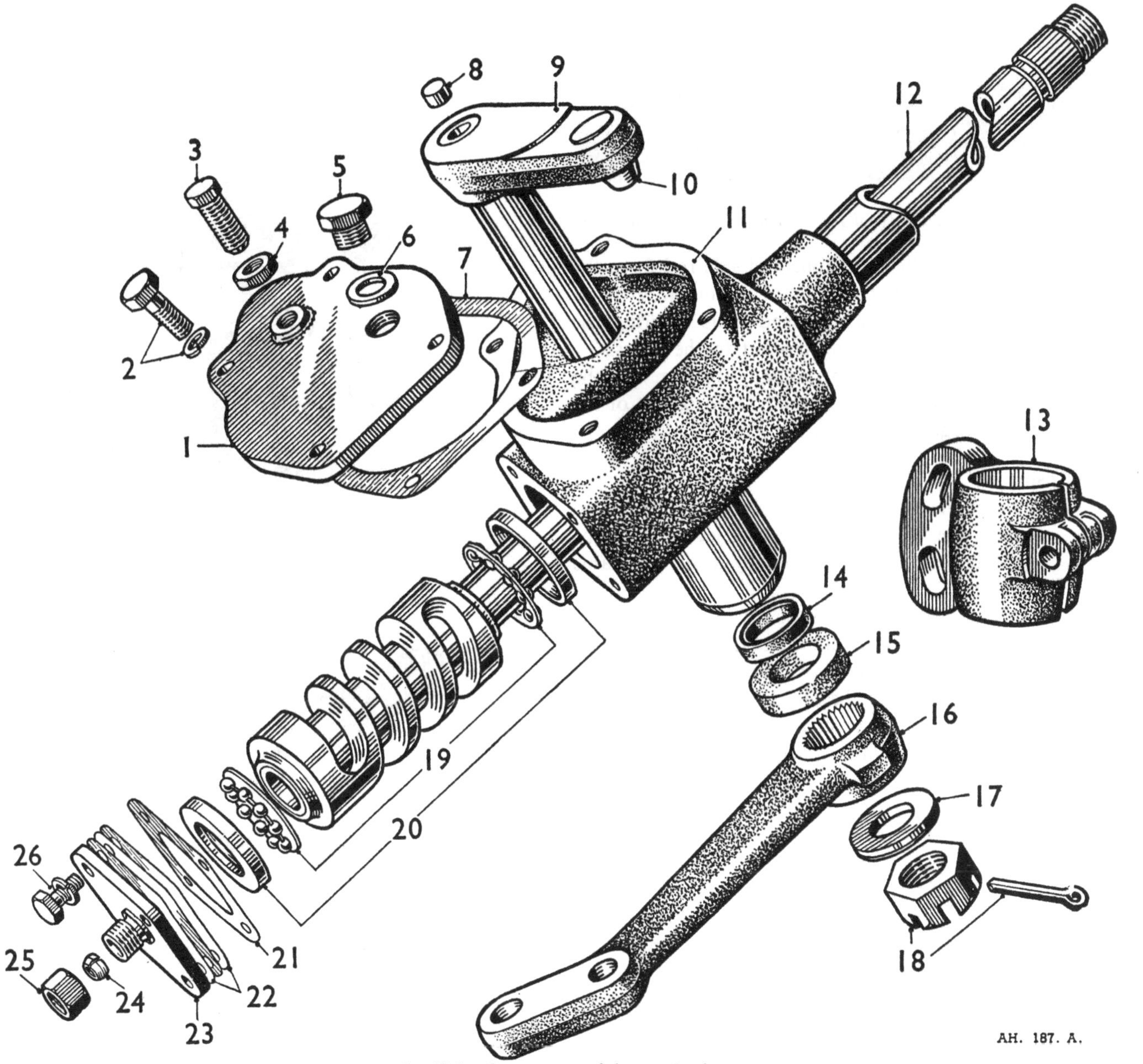

Fig. K.4. Components of the steering box

1. *Top cover.*
2. *Setpin and washer.*
3. *Adjusting screw.*
4. *Locknut.*
5. *Filler plug.*
6. *Washer.*
7. *Joint washer.*
8. *Adjusting screw stop.*
9. *Follower peg screw.*
10. *Follower peg.*
11. *Steering box.*
12. *Inner column.*
13. *Steering box bracket.*
14. *Oil seal.*
15. *Dust excluder.*
16. *Steering lever.*
17. *Washer.*
18. *Castellated nut and washer.*
19. *Inner races.*
20. *Outer races.*
21. *Joint washer.*
22. *Adjusting shims.*
23. *End cover.*
24. *Olive.*
25. *Stator tube nut.*
26. *Setpin and washer.*

Fig. K.5. Using Service Tool 18G 75A to remove steering lever.

(10) Disconnect the cross tube and side rod from the steering lever as described in Section K.6.
(11) Unscrew the three nuts and bolts securing the steering box mounting bracket to the chassis.
(12) Manœuvre the steering column together with the steering box downwards and forwards out of the radiator grille opening, see Fig. K3.

To Dismantle

(1) Extract the split pin and unscrew the castellated nut at the base of the steering lever. Pull the steering lever off the splines by using a suitable extractor.
(2) Unscrew the four setscrews securing the steering box top cover plate, and remove the plate.
(3) Turn the steering gear over and suitably support the top face when the rocker shaft can be lightly tapped out using a soft metal drift.

NOTE.—**The follower peg, situated in the rocker, is a pressed fit. The peg is peened over at the top to ensure complete security and should only be removed if showing an appreciable amount of wear.**

(4) Release the nut and olive at the steering box end of the column allowing the oil to drain into a suitable receptacle and withdraw the long stator tube.
(5) Remove the four setpins holding the end cover plate in position and release the end cover.
(6) Up end the complete unit so that the steering box is uppermost.
(7) Displace the worm and its two ball bearings by bumping the end of the inner shaft on a piece of wood placed on the floor.
(8) Withdraw the complete inner column from the casing via the open end of the steering box.
(9) Extract the ball race at the top of the outer casing of the column by pulling it upwards by hand, or if it is tight, ease it from the column with a screwdriver behind the protruding lip. Replacing the ball race merely entails pushing it into place.
(10) Clean all components in paraffin and blow them dry with compressed air.
(11) Examine the rocker shaft, rocker shaft bush and splines for wear.
(12) Examine the steering column shaft cam for excessive wear in the grooves.
(13) Carefully examine the steering lever for cracks and accidental damage.

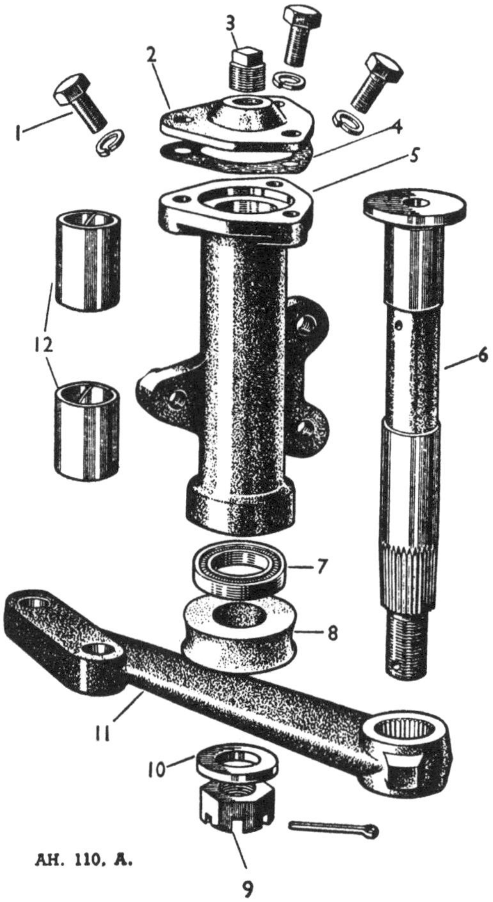

Fig. K.6. Steering idler exploded.

1. Cap setpin.	*7. Oil seal.*
2. Idler cap.	*8. Dust excluder.*
3. Oil plug.	*9. Castellated nut.*
4. Joint washer.	*10. Plain washer.*
5. Idler body.	*11. Steering lever.*
6. Idler shaft.	*12. Bush bearings.*

To Reassemble

Reassembly is a reversal of the removal procedure giving particular attention to adjustments as described in Section K.3. Before refitting the top cover plate, screw back the adjuster.

To Replace

The replacement of the steering gear is a reversal of the procedure "To Remove", but observe the following precautions :—

(1) Carefully align the steering column so that no bending stress is imposed upon it before tightening the support brackets.
(2) Make sure that the steering wheel is in the centre of travel and the front wheels are in the straight ahead position when installing the side rod.
(3) Tighten the steering wheel securing nut to the recommended torque tightness (see General Data).

Section K.5

STEERING IDLER

To Remove

(2) Disconnect the side and cross tubes from their connections at the idler lever.

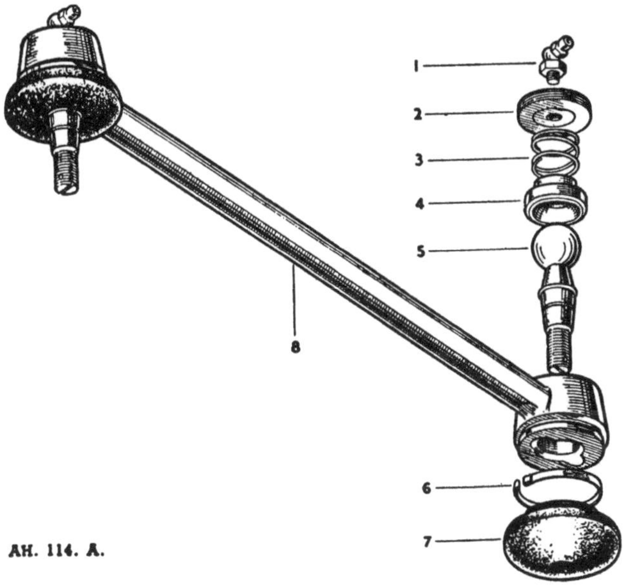

Fig. K.7.
An exploded view of a non-adjustable type ball and socket connection.
1. Oil nipple. 2. Screwed fixing plate. 3. Spring. 4. Socket. 5. Ball. 6. Spring clip. 7. Rubber boot. 8. Side rod.

(2) Unscrew the three setpins which secure the idler to its mounting bracket.
(3) Lift the idler and its lever clear of the body.

To Dismantle

(1) Unscrew the three setscrews securing the idler cap to the idler body.
(2) Remove the split pin, castellated nut, idler lever and dust excluder from the base of the idler body.
(3) Pull the idler shaft upwards through the idler body taking care not to damage the oil seal.

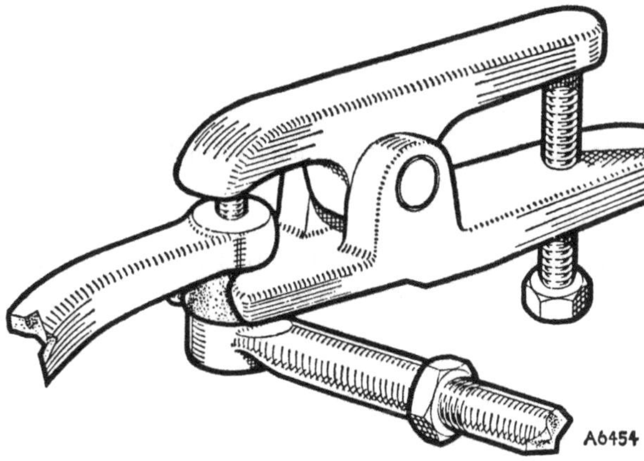

Fig. K.8
Separating the ball joint from a steering arm using Service Tool 18G 1063.

(4) Tap out the oil seal.
(5) Check the two bush bearings for wear and renew if necessary.

To Assemble and Replace

The assembling and replacement is a reversal of the procedure to dismantle and remove. Ensure that an oil tight seal is maintained at the base of the idler body.

Idler shaft end float is adjustable by means of joint washers fitted beneath the top cap. Fit sufficient joint washers to enable the idler shaft to turn freely without end-float when the top cap set screws are fully tightened.

Section K.6

SIDE RODS

To Remove

(1) Withdraw the split pins and remove the nuts from the ball pins at the steering lever end and the swivel arm end of the side rod.
(2) Loosen the ball pins from the steering lever and swivel arm by using Service Tool 18G 1063; see Fig. K.8.

To Dismantle

(1) Remove the dust covers from the ball pins and sockets by releasing the clips and levering the covers off. Further dismantling of the socket assemblies is not permissible.
(2) Check the ball pins for wear. They must be tight enough to prevent end play, yet loose enough to allow free movement. Renew as complete assemblies if necessary.
(3) Renew the dust covers if damaged.
(4) Examine the side rod for damage. Renew if it is bent or damaged.

Section K.7

CROSS TUBE

To Remove

The removal of the cross tube from the steering lever and the idler lever is similar to that of the side rods.

To Dismantle

(1) Slacken the socket locknuts at each end of the cross tube and unscrew the socket assemblies which are screwed left-hand and right-hand respectively.

(2) Follow the procedure described in "To Dismantle" Section K.6.

NOTE.—The procedures to assemble and replace the side rods and cross tube is the reverse of the procedures for removing and dismantling them, with the precaution that after replacing the cross tube the "toe-in" should be checked as described in Section K.3.

SECTION KK

STEERING

SERIES BN6

NOTE

For details of the steering mechanism fitted to BN6 cars refer to Section K.

SECTION KKK

STEERING

Mk. I and II (SERIES BN7 and BT7)
AND Mk. II and Mk. III (SERIES BJ7 and BJ8)

NOTE

This section should be used in conjunction with Section K.

Section KKK.1

NYLON-SEATED BALL JOINTS

Nylon-seated ball joints, which are sealed in manufacture and therefore require no further lubrication, were introduced on 3000 Mk. II cars from Car No. BT7 19191.

It is essential that no dirt or abrasive matter enters the nylon ball joint; in the event of a rubber boot being torn or damaged in service it is probable that the ball joint will have been left exposed, and it is therefore imperative that both the ball joint and the boot are renewed.

If damage to the boot occurs whilst the steering side- or cross-rod is being removed in the workshop, only a new rubber boot need be fitted, provided the ball joint is clean. Smear the area adjacent to the joint with a little Dextragrease Super G.P. before assembling the boot.

Section KKK.2

STEERING-COLUMN LOCK AND IGNITION/STARTER SWITCH

Cars exported to Germany and Sweden are fitted with a combined ignition/starter switch and steering-column lock mounted on the steering-column.

On cars fitted with the lock a sleeve integral with the inner column is slotted to permit engagement of the lock tongue: the outer column is also slotted to allow the lock tongue to pass through. A hole drilled in the upper surface of the outer column locates the steering lock bracket. The bracket is secured by two bolts each waisted below the head to permit removal of the heads by shear action during assembly.

To remove the lock, disconnect the battery and the ignition/starter switch connections and turn the key to the 'GARAGE' position to unlock the steering. Free the steering-column assembly as described in Section K.4 and remove the lock securing bolts with an 'easy-out' extractor.

SECTION L

FRONT SUSPENSION AND FRONT HUBS

SERIES BN4

Section L.1

INDEPENDENT FRONT SUSPENSION

Description

The independent front suspension is known as the "wishbone" type, since the top and bottom linkages roughly conform to the shape of a wishbone. Between these two wishbones is the coil spring, held under compression between the top spring plate and the lower spring plate which is secured to the lower wishbone by four bolts.

The top wishbone is formed by the lever arms of a double-acting hydraulic shock absorber which is anchored to the top spring plate bracket by four bolts. At the swivel end, the top wishbone is secured to the swivel pin trunnion by means of a fulcrum pin and tapered rubber bushes. The bottom wishbone is secured by a single lower link spindle and tapered rubber bushes to two mounting plates, bolted to the front suspension member, and by two screwed bushes and a screwed fulcrum pin to the lower end of the swivel pin.

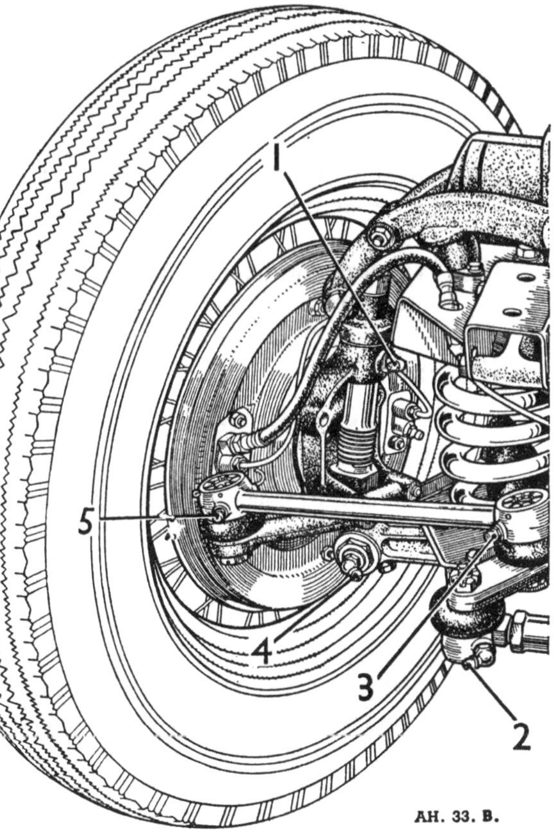

Fig. L.1. Front suspension lubrication.

1. Swivel pin. 2 and 5. Side rod connections. 3. Cross tube connection. 4. Lower link bushes.

Checking for Wear

The following tests should be made to check for wear in various components of the front suspension unit.

(1) Wear of the swivel pin, or bushes, or both, may be checked by jacking up the front of the car and endeavouring to rock the wheel by grasping opposite points of the tyre in a vertical position. If any sideways movement can be detected between the swivel axle assembly, the swivel pin or the swivel pin bushes are worn and must be stripped for examination.

(2) Up and down, or sideways movement of the shock absorber cross shaft, relative to the shock absorber casting, denotes wear of the shock absorber shaft bearings which can only be remedied by refitting a new shock absorber. These bearings are best checked when the suspension is dismantled and when with some freedom of movement, it is possible to move the top wishbone arms, which are attached at their inner ends of the shock absorber cross shaft.

(3) The rubber bearing bushes used for the upper wishbone arm outer bearings and for the lower wishbone arm inner bearings may in time deteriorate and need renewing. Excessive sideways movement in either of these bearings would denote softening of the rubber bushes.

(4) The screwed bushes or the screwed trunnion fulcrum pin of the lower wishbone arm outer bearing assembly may develop excess free play due to wear of either of these parts. This assembly can best be checked when the suspension has been dismantled.

Section L.2

COIL SPRINGS

To Remove

(1) Place a jack under the chassis front cross-member and raise the car until the front wheels are clear of the ground.

(2) Remove the appropriate wheel by unscrewing the five securing nuts, or the "knock-on" hub cap if wire wheels are fitted.

(3) Release the compression of the coil spring using Service Tool 18G 37. If this tool is not available use two $\frac{3}{8}$ in. B.S.F. slave bolts of high-tensile steel, 4 in. long and threaded their entire length.

(4) The bottom spring plate is secured to the suspension lower links by four self-locking nuts. Unscrew the nuts from two diagonally opposite bolts, remove the bolts and insert the two slave bolts in the vacated holes.

Fig. L.2. Illustrating the coil spring compressor (Service Tool 18G. 37) in position.

(5) Screw two nuts down securely onto the slave bolts and remove the remaining two short nuts and bolts.
(6) Unscrew the nuts from the slave bolts, each, a little at a time.
(7) When the spring is fully extended, release the bolts and remove the spring plate and coil spring.

To Replace

The replacement of the coil springs is a reversal of the procedure to remove, making sure that all securing nuts and bolts are fully tightened.

Section L.3

FRONT SUSPENSION

To Remove

(1) Remove the coil spring as described in Section L.2.
(2) Disconnect the steering side tube from the steering arm by withdrawing the split pin and unscrewing the nut, see Section K.
(3) Disconnect the flexible brake pipe from the brake backplate, tying it to some higher point to prevent unnecessary loss of fluid.
(4) With the suspension unit supported, remove the fulcrum pins securing the lower wishbone arms to their brackets on the frame, taking care to retrieve the two rubber bushes and special washers from each bearing.
(5) Unscrew the four setpins securing the shock absorber to the top spring bracket.
(6) Lift the suspension unit clear of the car.

To Dismantle

(1) Unscrew the nut from the clamping bolt connecting the top wishbone arms together.
(2) Remove the split pin and nut from the upper trunnion fulcrum pin on the outer end of the top wishbone arms.
(3) The forward arm (left-hand suspension unit) of the top wishbone is secured to the shock absorber spindle by a clamping bolt. Slacken the clamping bolt and partially withdraw the arm. The trunnion fulcrum pin can now be withdrawn and the shock absorber removed complete with the top wishbone arms.
(4) Withdraw the rubber bearing from each end of the upper trunnion. The bearings fit into a groove in the swivel pin and must be taken out before the swivel pin is removed.
(5) Take out the split pin and unscrew the nut from the top of the swivel pin.
(6) Remove the upper trunnion and the three thrust washers and lift off the swivel axle and hub assembly.
(7) Detach the cork washer from the lower end of the swivel pin.
(8) Slacken the nut on each of the half moon cotters located in the ends of the lower wishbone arms, screw out the two threaded bushes and detach the arms.
(9) Unscrew the nut from the cotter, located in the centre of the lower trunnion, and tap out the cotter.
(10) Withdraw the fulcrum pin and remove the cork washer from each end of the trunnion.

Section L.4

EXAMINATION FOR WEAR

Swivel Pin

(1) Carefully examine the swivel pin for wear by checking for ovality with a micrometer.
(2) If the pin does not show any appreciable wear renewal of the swivel bushes may effect a satisfactory cure. The bushes can easily be driven out or replaced with a suitable drift.

NOTE.—When refitting the top bush the greasing hole must locate with the grease hole in the swivel housing. The second bush must be flush with the recessed housing and protrude about $\frac{1}{8}$ in. above the lower housing upper face.

(3) Ream the bushes from the bottom as necessary using Service Tools 18G 64 and 18G 65.
(4) Check the efficiency of the dust covers and renew if necessary.

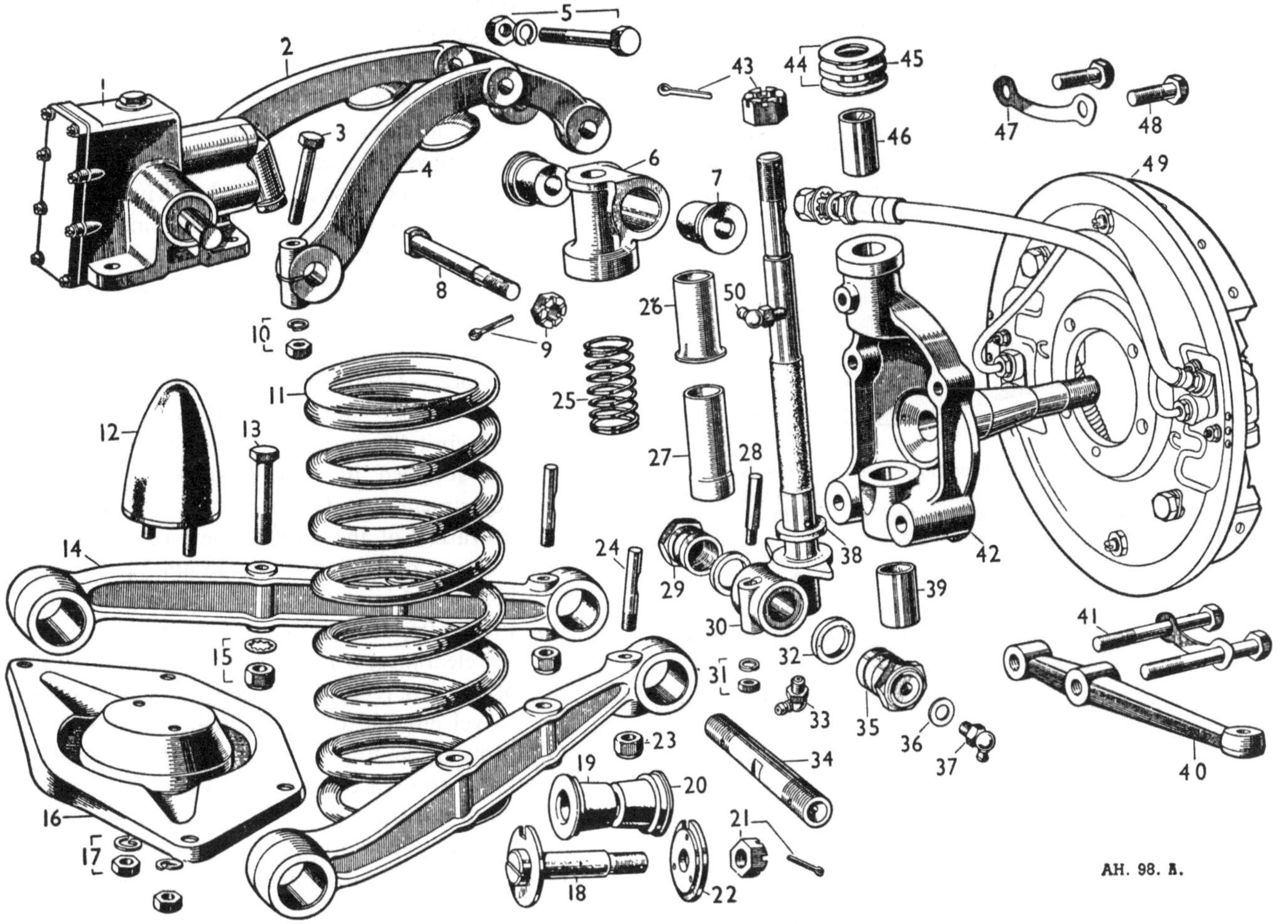

Fig. L.3. Components of the independent front suspension.

1. Shock absorber.
2. Rear top wishbone arm.
3. Clamping bolt for front wishbone arm.
4. Front top wishbone arm.
5. Joining bolt for top wishbone arms.
6. Upper trunnion link.
7. Trunnion rubber bearing.
8. Upper trunnion fulcrum pin.
9. Fulcrum locking nut and split pin.
10. Nut and washer for clamping bolt.
11. Coil spring.
12. Rebound rubber bumper.
13. Spring plate bolt.
14. Rear lower wishbone arm.
15. Simmonds nut and lockwasher.
16. Spring plate.
17. Rebound bumper nut and washer.
18. Fulcrum pin for inner lower bearing.
19. An inner lower rubber bearing.
20. An outer lower rubber bearing.
21. Fulcrum pin nut and split pin.
22. Fulcrum pin special washer.
23. Nut for bush cotter.
24. Bush cotter.
25. Swivel pin dust cover spring.
26. Upper dust cover.
27. Lower dust cover.
28. Cotter for fulcrum pin.
29. Rear screwed bush.
30. Swivel pin and lower trunnion.
31. Nut and washer.
32. Cork ring.
33. Trunnion oil nipple.
34. Screwed fulcrum pin.
35. Front screw bush.
36. Flat washer.
37. Oil nipple.
38. Cork ring.
39. Swivel axle lower bush.
40. Steering arm.
41. Steering arm setpin.
42. Swivel axle.
43. Swivel pin nut and split pin.
44. Staybrite washers.
45. Oilite washer.
46. Swivel axle upper bush.
47. Back plate setpin lockwasher.
48. Back plate setpin.
49. Back plate assembly.
50. Swivel pin oil nipple.

Wishbone Arm Screwed Bush Bearing

(1) Test to see if the screwed bushes can be moved backwards or forwards on the fulcrum pin thread.

(2) If such movement is detected replace the bushes and if movement is still detected replace the fulcrum pin.

Shock Absorbers

(1) If any up and down or sideways movement of the cross shaft is detected the shock absorbers must be completely renewed.

(2) Carefully examine the shock absorbers for any leaks and test for effective damping. Secure the

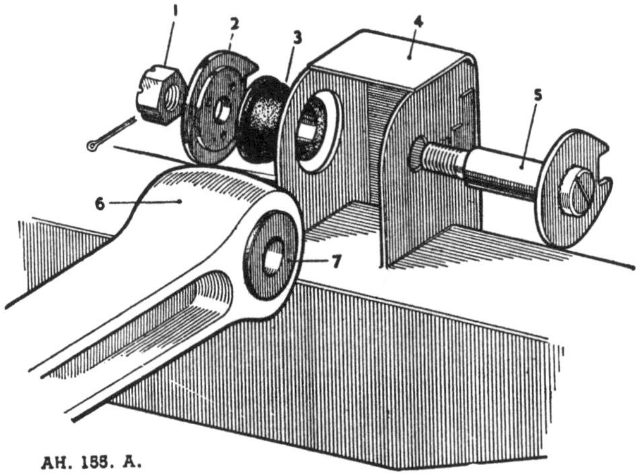

Fig. L.4. Lower wishbone inner bearing assembly.

1. *Castellated nut.*
2. *Special washer.*
3. *Bush.*
4. *Mounting bracket.*
5. *Fulcrum pin.*
6. *Wishbone arm.*
7. *Bush.*

shock absorber mounting plate in a vice and move the wishbone arms up and down through a complete stroke. A moderate resistance should be felt throughout the whole stroke.

If resistance is erratic it may mean that the fluid level is too low and that there are air-locks in the shock absorber. To rectify this remove the shock absorber filler plug and maintain the fluid at the correct level whilst the arms are moved steadily up and down through full strokes. If this treatment does not effect a cure the shock absorber must be renewed as a complete unit.

Section L.5

FRONT SUSPENSION

To Reassemble

(1) Fit the screwed fulcrum pin into the lower trunnion at the bottom end of the swivel pin, ensuring that it is centralised and secured by means of its cotter pin.

(2) Fit a cork ring into the recess provided at each end of the lower trunnion and place the lower wishbone arms in their respective positions. Ensure that the half-moon cotters are correctly positioned to receive the steel bushes which should be greased and partially screwed home.

(3) Service Tool 18G 56 should be used to ensure that the alignment of the lower wishbone arm is correct. If this Service Tool is not available bolt the lower spring plate securely in position.

(4) Screw the threaded bushes home evenly, then slacken them back one flat. Finally secure the bushes by tightening the nuts on each of the half-moon cotters. Do not overtighten the cotter nuts as this may cause distortion of the bushes. If the assembly has been correctly carried out it will be possible to insert a ·002 in. feeler gauge between the inner shoulder of the bush and the outer face of the wishbone arm on each side. The lower trunnion assembly should now operate freely in the screwed bushes.

(5) Place the cork washer on the swivel pin with its chamfered face downwards and smear the swivel pin with a little clean engine oil.

(6) Position the swivel axle and hub assembly on the swivel pin.

(7) Refit the thrust washers.

Note.—The three thrust washers are made up of an "Oilite" washer interposed between two "Staybrite" washers. The "Staybrite" washers are supplied in varying thicknesses to permit adjustment, as it is necessary to provide easy operation of the swivel axle with the minimum amount of lift; the maximum permissible lift being ·002 in.

(8) Fit the upper trunnion and swivel nut, and check the clearance, correcting it if necessary by means of the "Staybrite" washers. Then slacken the swivel pin nut to permit further assembly.

(9) Moisten the upper trunnion rubber bearings with water and place them in position.

(10) Place the trunnion, with its bearings, in position between the two upper wishbone arms.

(11) Refit the fulcrum pin, re-position and tighten the slackened upper wishbone arm to the shock absorber arms.

NOTE.—The swivel pin and upper trunnion fulcrum pin nuts must not be tightened at this stage.

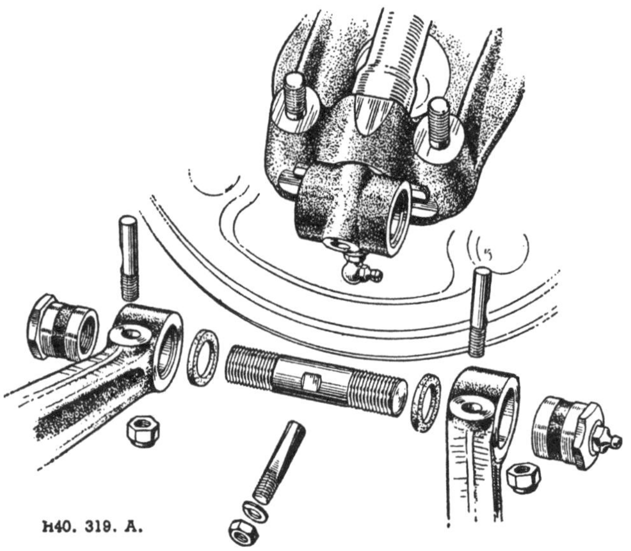

Fig. L.5. This exploded view shows the screwed bush housing assembly at the lower end of the wishbone arms.

To Replace

(1) Fit one rubber bearing to each of the suspension lower links, on the side which corresponds to the small hole in each of the frame brackets.

(2) Raise the links to the frame brackets, insert the fulcrum pins and slide the second bearing and special washer over the protruding end of each pin. Fit the nut but do not screw it home. Position the shock absorber on its top bracket and partially tighten the four setscrews.

(3) The assembly must next be set in the normal loaded position. This can be accomplished by placing a distance piece between the shock absorber wishbone arm and the upper spring plate at a point opposite the rubber buffer. The length of the distance piece must be 2 in.

(4) Tighten the nuts on the fulcrum pins securing the lower wishbone arms to the frame brackets. Do not forget to lock them with the split pins.

(5) Tighten the four setscrews securing the shock absorber to its bracket on the frame.

(6) Tighten the upper trunnion fulcrum pin nut and secure with a split pin.

(7) Tighten the swivel pin nut and lock with a split pin.

(8) Service Tool 18G 56, or the lower spring plate, whichever used, should now be removed from the lower wishbone arms and the coil spring refitted as described in Section L.2.

(9) Connect the brake fluid pipe to the brake back-plate, secure the steering side tube to the steering arm, refit the road wheel, lower the car to the ground and remove the distance piece used to retain the suspension in the normal loaded position.

(10) Finally, bleed the brakes as described in Section M.

Fig. L.7. When building up the suspension, the arms must be correctly set by the distance piece A (2 in.) before the various bearings are tightened.

Section L.6

CASTOR AND CAMBER ANGLES AND SWIVEL PIN INCLINATION

Description

The castor and camber angles and the swivel pin inclination are three design settings of the front suspension assembly. They have a very important bearing on the steering and general riding of the car. Each of these settings is determined by the machining and assembly of the component parts during manufacture. They are not therefore adjustable.

However, should the car suffer damage to the suspension affecting these settings, the various angles must be verified to ascertain whether replacements are necessary.

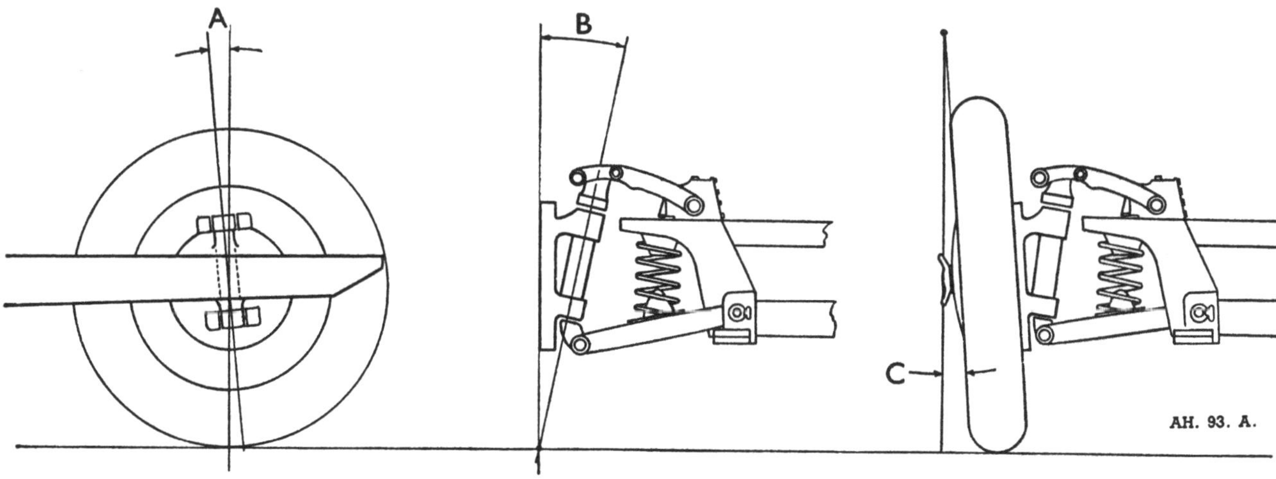

Fig. L.6.

A. Castor angle 2°. B. Swivel pin inclination 6½°. C. Camber angle 1°

Camber Angle

This is the outward tilt of the wheel and a rough check can be made by measuring the distance from the outside wall of the tyre, immediately below the hub, to a plumb line hanging from the outside wall of the tyre above the hub. The distance must be the same on both wheels. Before making this test, it is very important to ensure that the tyres are in a uniform condition and at the same pressure. Also that the car is unladen and on level ground.

Damage to the upper and lower wishbone arms may well affect the camber angle.

Castor Angle

This is the tilt of the swivel pin when viewed from the side of the car. This also is only likely to be affected by damage to the upper and lower wishbone arms.

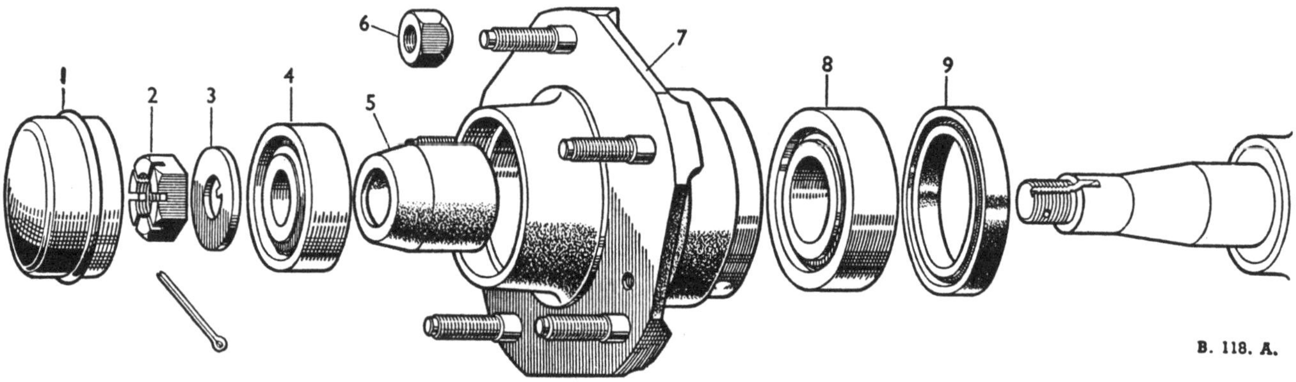

Fig. L.8. Front hub exploded (for disc wheels).

1. *Hub cap.*
2. *Castellated nut.*
3. *Locating washer.*
4. *Outer bearing.*
5. *Distance piece.*
6. *Wheel nut.*
7. *Hub.*
8. *Inner bearing.*
9. *Oil seal.*

Swivel Pin Inclination

This is the tilt of the swivel pin when viewed from the front of the car and is again only likely to be affected by damage to the wishbone arms.

A useful tool which can be used for checking these settings is the Dunlop "wheel camber, castor and swivel gauge". With the car standing on level ground this gauge will give readings enabling the castor, camber and swivel pin angles to be quickly verified.

Section L.7

FRONT HUBS
(Disc Wheels)

To Check for Wear

The inner and outer ball bearings of the front hub are non-adjustable, the amount of thrust being determined by a distance piece. To check for wear of these bearings, the car should be jacked until the wheel of the front hub is clear of the ground. Then grasp the tyre with both hands in the vertical position and rock the wheel. Movement between the wheel and the back plate denotes wear of the hub bearings. Should a very positive movement be apparent, the front hub bearings will need renewing.

To Remove and Dismantle

(1) Jack the car until the wheel is clear of the ground and then place blocks under the independent suspension spring plate. Lower the car on to the blocks.

(2) Remove the wheel and the countersunk screw holding the brake drum. If the drum appears to bind on the brake shoes, the shoe adjusters should be slackened.

(3) Lever off the hub cap, and then extract the split pin from the swivel axle locking nut. Using a box spanner and tommy bar remove the axle nut and ease the flat washer, under the nut, clear of the axle thread.

(4) The front hub can now be withdrawn by using an extractor, 18G 220, which fits over the wheel studs. The hub is withdrawn complete with the inner and outer bearings, the distance piece and the oil seal. Should the inner bearing race remain on the swivel axle it can be removed by carefully inserting a narrow rod into the two small holes, in turn, in each side of the swivel axle and tapping the race lightly.

(5) With the hub removed, the outer bearing and the distance piece can be dismantled by inserting a drift through the inner bearing and gently tapping the outer bearing clear of the hub. The inner bearing and oil seal can then be removed by inserting the drift from the opposite side of the hub.

(6) The removal of the brake backplate is described fully in the section on brakes.

To Assemble and Replace

(1) Repack the inner ball bearing race with a **recommended** grease and insert it into the hub with the side of the race marked "thrust" facing the distance piece.

(2) Insert the distance piece so that the domed end faces the outer bearing.

(3) Pack the outer bearing with a **recommended** grease and replace the bearing in the hub so that the "thrust" side faces the distance piece. Use a soft metal drift to replace both bearings, tapping them gently and alternately on diametrically opposite sides of the bearing to ensure they move evenly into their respective housings on the hub.

Fig. L.9. Front hub exploded (for wire wheels).

1. Grease cup.
2. Axle nut.
3. Split pin.
4. Washer.
5. Outer bearing.
6. Bearing outer race.
7. Hub.
8. Bearing outer race.
9. Inner bearings.
10. Oil seal.
11. Swivel axle.

Inset shows distance piece and shims.

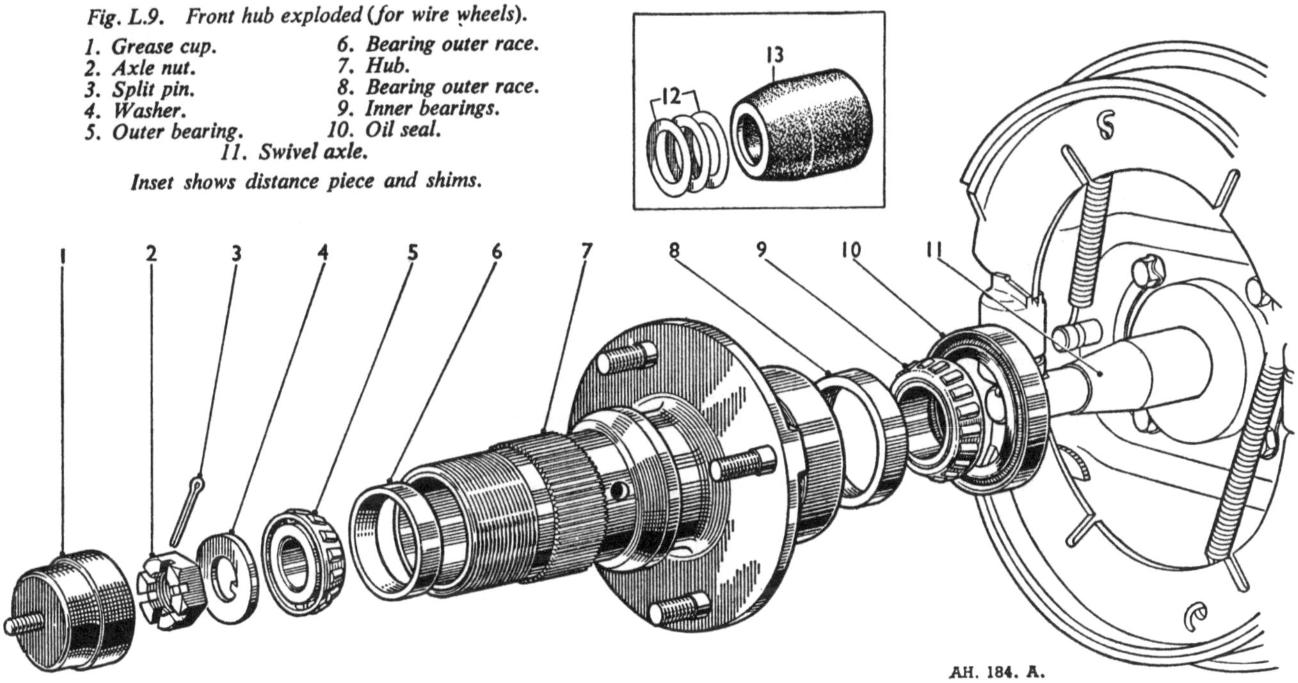

(4) Fill the cavity between the inner bearing and the oil seal with a recommended grease and replace the hub oil seal. Renew the seal if it is damaged.

(5) Replace the hub on the swivel axle, using a hollow drift which will bear evenly on both the inner and outer races of the outer hub bearing. Gently tap the hub into position until the inner race bears against the shoulder on the swivel axle.

(6) Place the swivel axle flat washer into position and tighten the nut. The split pin should be inserted to lock the nut.

(7) Tap the hub cap on to the hub: **do not put grease in the hub or cap.**

(8) Replace the brake drum and secure with the countersunk screw. It is important that the drum is fully home before this screw is tightened and, if necessary, the drum should be pressed in position by tightening two wheel nuts.

(9) Refit the wheel. The wheel nuts are best finally tightened when the car is off the jacking blocks, but re-adjust the brake shoes if necessary before the car is lowered to the ground.

Section L.8

FRONT HUBS
(Wire Wheels)

To Check for Wear

The inner and outer bearings of the front hub are of the taper roller type and are therefore adjustable. To check for wear of these bearings the car should be jacked up until the wheel of the front hub to be checked, is clear of the ground. Movement between the wheel and the back plate denotes wear of the hub bearings. Should a very positive movement be apparent, the front hub bearings will need renewing.

To Remove and Dismantle

(1) Jack up the car until the wheel is clear of the ground and then place blocks under the independent spring plate. Lower the car on to the blocks.

(2) Remove the "knock-on" hub cap (direction of rotation marked on cap) and pull the wheel off the splines.

(3) Release the nuts and washers holding the brake drum, then gently tap the brake drum clear of the

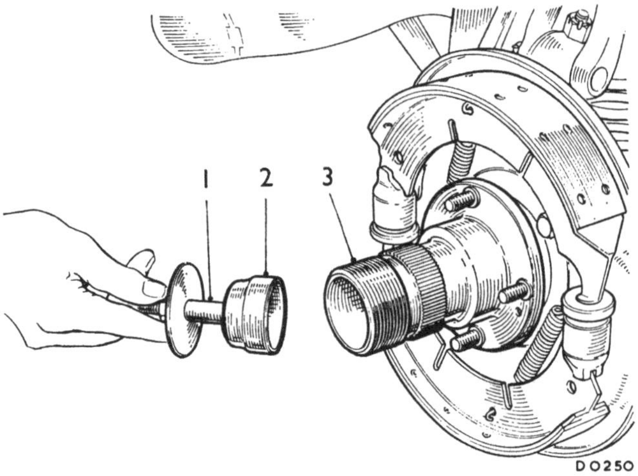

Fig. L.10. Showing the grease cap removed.
1. Extractor. 2. Grease cup. 3. Hub.

front hub assembly. If the drum appears to bind on the brake shoes, the shoe adjusters should be slackened.

(4) Use the extractor provided in the tool kit to extract the grease retaining cup from within the hub.
(5) Straighten the end of the split pin and then prise it out through the hole provided in the hub.
(6) Using a box spanner and tommy bar remove the hub securing nut and flat washer from the swivel axle.
(7) Withdraw the front hub using an extractor. It is preferable to use an extractor which screws into position on the hub cap thread (see Fig. L.12), but an extractor which locates over the hub studs may also be used. The hub is withdrawn complete with the inner and outer bearings and oil seal.
(8) With the hub removed, dismantle the outer bearing by inserting a drift through the inner bearing and gently tapping the outer bearing clear of the hub. The inner bearing and oil seal can then be removed by inserting the drift from the opposite side of the hub.

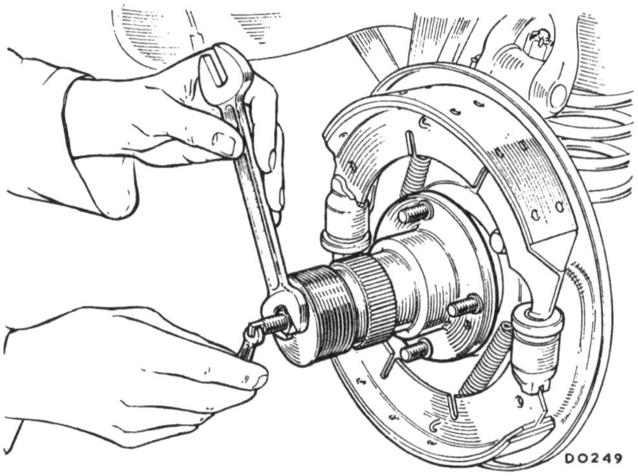

Fig. L.11. Using the special extractor to remove the grease retaining cup.

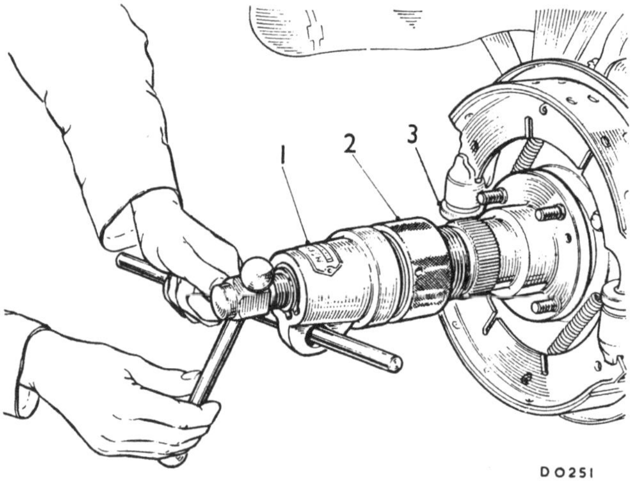

Fig. L.12. Extracting the front hub (3) with extractor 18G 8 (1) and adaptor 18G 8H or 18G 8K (2). Centre screw extension 18G 8J must also be used.

To Reassemble and Replace

The end-float in the hub bearings must be checked and adjusted whenever the hub has been dismantled for attention or when play in the hub bearings has become excessive. The end-float is adjustable by means of shims situated between the outer bearing and the bearing distance piece.

(1) Press the two bearing outer rings into the hub. Insert the inner race and rollers of the inner bearing and the bearing spacer into the hub packing the bearing and the cavity between the inner bearing and the oil seal with a recommended grease.
(2) Fit the oil seal to the hub and mount the hub assembly on the stub axle. Pack the inner race and rollers of the outer bearing with a recommended grease and position them in the hub **without fitting shims at this stage.**
(3) Fit the stub axle nut and washer. Tighten the nut and at the same time rotate the hub back and

forth until there is noticeable drag. This ensures that the bearing cones are properly seated.

(4) Unscrew and remove the stub axle nut. Extract the washer and the centre of the outer bearing. Insert a sufficient thickness of shims **to produce an excessive amount of end-float.** Note the total thickness of shims used. Replace the bearing centre, the washer, and tighten the stub axle nut.

(5) Measure accurately the total amount of end-float in the bearings. Remove the stub axle nut, washer, and outer bearing centre. Reduce the number of shims to eliminate end-float, while still allowing the hub to rotate freely, when the stub axle nut has been refitted and tightened to a torque wrench reading of 40 to 70 lb/ft. (5·53 to 9·68 kg.m.). Latitude for this reading is given so that the nut may be aligned with the split pin hole in the stub axle.

(6) Insert a new split pin through the hole provided in the hub and lock the stub axle nut.

(7) Remove any surplus grease to allow room for expansion and, using a drift, tap the grease retaining cap gently but firmly up against the outer bearing. Do not put grease in the cap.

(8) Replace the brake drum and secure with the four spring washers and self-locking nuts.

(9) Grease the wheel hub splines, refit the road wheel and replace the knock-on hub cap.

Section L.9 FAULT DIAGNOSIS

Symptom	No.	Possible Fault
(a) Wheel Wobble	1	Unbalanced wheels and tyres
	2	Slack steering connections
	3	Incorrect steering geometry
	4	Excessive play in steering gear
	5	Broken or weak front springs
	6	Loose idler mounting or worn idler shaft
	7	Worn hub bearings
	8	Loose or broken shackles
(b) Wander		Check 2, 3, 4 and 8 in (a)
	1	Broken spring clips
	2	Front suspension and rear axle mounting points out of alignment
	3	Uneven tyre pressures
	4	Uneven tyre wear
	5	Weak shock absorbers or springs
(c) Heavy Steering		Check 3 in (a)
	1	Excessively low tyre pressures
	2	Insufficient lubricant in steering box
	3	Insufficient idler lubrication
	4	"Dry" steering connections
	5	Out of track
	6	Incorrectly adjusted steering gear
	7	Misaligned steering column
(d) Tyre Squeal		Check 3 in (a)
		Check 1 in (c)

Section L.10

SHOCK ABSORBERS

For a general description and details of maintenance and topping-up shock absorbers refer to Section J.9. See also items (2) and (3) in **"Checking for Wear"** of Section L.1 and **"Shock Absorbers"** of Section L.4 for information on the testing of shock absorbers.

To Remove

(1) Jack up the car and place stands under the chassis in safe positions.

(2) Remove the road wheel. Place a jack beneath the outer end of the lower wishbone arm and raise it until the shock absorber arms are clear of their rebound rubber.

(3) Remove the clamp bolt connecting the two shock absorber arms together.

(4) Remove the split pin and castellated nut on the upper fulcrum pin and withdraw the pin.

(5) One arm of each shock absorber unit is secured to the shock absorber spindle by a clamp bolt. When the clamp bolt has been removed the arm may be partially withdrawn. This allows the trunnion link and its rubber bushes to be separated easily from the shock absorber arms.

(6) Retrieve the trunnion link rubber bushes.

(7) Once the four shock absorber fixing bolts and their spring washers have been unscrewed the unit may be removed from car.

NOTE.—The jack must be left in position under the suspension wishbone while the top link remains disconnected in order to keep the coil spring securely in position and to avoid straining the steering connections.

To Replace

Replacement is a reversal of the above procedure, but attention must be given to the following points:

(1) Having bolted the shock absorber to the chassis frame and before fitting the upper trunnion fulcrum pin, work the arms of the unit three or four times through their full travel to expel any air which may have found its way into the operating chamber.

(2) The fulcrum pin bushes must be renewed if softening of the rubber or side movement is evident.

(3) Fit the trunnion with its bushes between the shock absorber arms and refit the fulcrum pin before pushing the loosened arm home on the shock absorber spindle and replacing the clamp bolt.

(4) Tighten the fulcrum pin nut and the clamp bolt connecting the two shock absorber arms only when the load is on the suspension, i.e., with the jack in position under the lower suspension arm or a 2 in. (5·08 mm.) distance piece interposed between the shock absorber arm and the chassis frame (see Fig. L.7).

SECTION LL

FRONT SUSPENSION AND FRONT HUBS

SERIES BN6

NOTE

For details of the front suspension and front hubs fitted to BN6 cars refer to Section L.

SECTION LLL

FRONT SUSPENSION AND FRONT HUBS

Mk. I and II (SERIES BN7 and BT7)
AND Mk. II and Mk. III (SERIES BJ7 and BJ8)

NOTE

This section should be used in conjunction with Section L, particular attention being given to Sections L.1, 2, 3, 4, 5, 6, 9 and 10.

Section LLL.1

FRONT HUBS

To Check for Wear

The inner and outer bearings of the front hub are of the taper roller type and are therefore adjustable. To check for wear of these bearings the car should be jacked up until the wheel of the front hub to be checked is clear of the ground. Movement between the brake disc and the steering arm denotes wear of the hub bearings or incorrect adjustment. Should a very positive movement be apparent, the front hub bearings will need renewing. The amount of movement present may be checked by a dial gauge.

(6) Using a box spanner and tommy bar remove the hub securing nut and flat washer from the swivel axle.

(7) Withdraw the front hub using one of the following procedures, but do not attempt to remove the hub by pulling on the brake disc. Withdraw the hub complete with the inner and outer bearings and oil seal.

Disc wheels

Withdraw the front hub using Service Tool 18G 220 as described in Section L.7, item 4.

Wire wheels

Screw Service Tool 18G 363 on to the hub cap thread (12 T.P.I.): the tool is marked for right and left-hand threads.

From chassis No. 26705 (Mk. III) use Service Tool 18G 1032, the right hand thread end of the tool being knurled to distinguish it from 18G 363 (as the hub now has 8 T.P.I.).

Fig. LLL.1. Front hub exploded and brake disc.

1. Grease cup.
2. Axle nut.
3. Split pin.
4. Washer.
5. Outer bearing.
6. Bearing outer race.
7. Hub.
8. Bearing outer race.
9. Inner bearing.
10. Oil seal.
11. Brake disc.

Inset shows distance piece and shims.

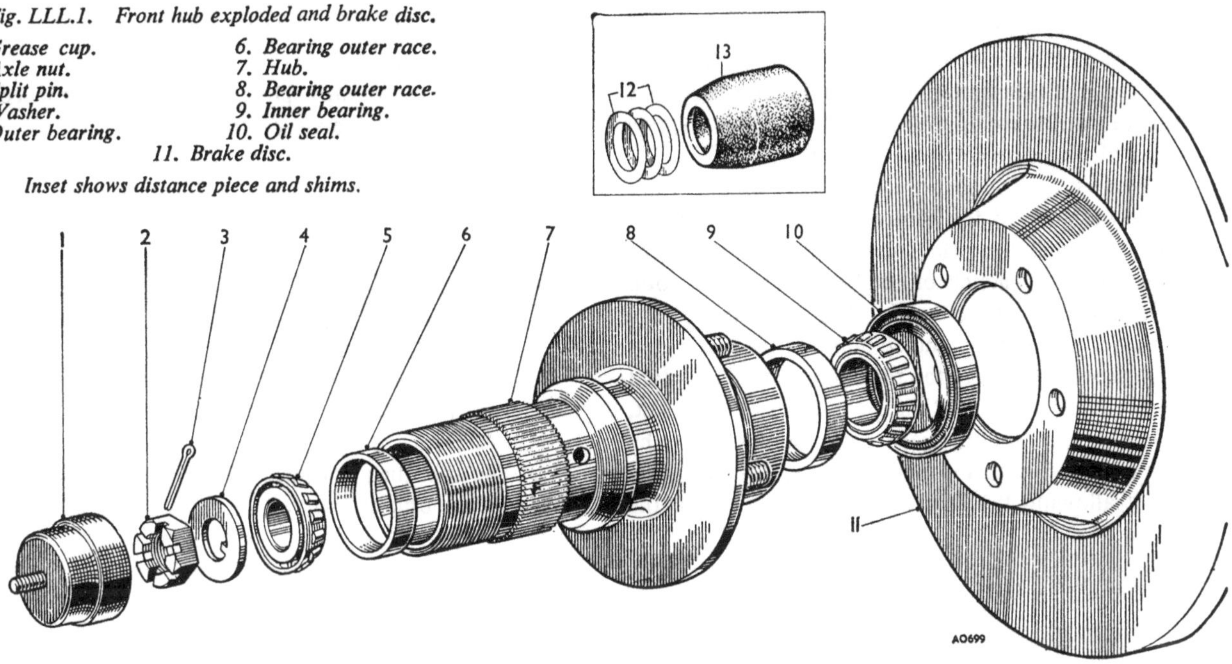

To Remove and Dismantle

(1) Jack up the car until the wheel is clear of the ground and then place blocks under the spring plate. Lower the car on to the blocks.

(2) *Wire wheels*

Remove the "knock-on" hub cap (direction of rotation marked on cap) and pull the wheel off the splines.

Disc wheels

Remove the hub cover, unscrew the securing nuts and pull the wheel off the hub wheel studs.

(3) Remove the brake calliper unit as described in Section MMM.7.

(4) Use the extractor provided in the tool kit to extract the grease retaining cup from within the hub (Fig. L.11).

(5) Straighten the end of the split pin and then prise it out through the hole provided in the hub.

(8) With the hub removed, withdraw the inner race and rollers of the outer bearing together with the shims fitted between the bearing and the distance piece. Remove the oil seal. Withdraw the inner race and rollers of the inner bearing and the bearing distance piece. The bearing outer rings may be removed from the hub using Service Tool 18G260 with adaptors 18G260J (inner bearing) and 18G260K (outer bearing).

To Reassemble and Replace

The end-float in the hub bearings must be checked and adjusted whenever the hub has been dismantled for attention or when play in the hub bearings has become excessive. The end-float is adjustable by means of shims situated between the outer bearing and the bearing distance piece.

(1) Fit the bearing outer rings to the hub using service tool 18G 260 with adaptors 18G 260J and 18G 260K. Insert the inner race and rollers of the inner bearing and the bearing spacer into the hub packing the bearing and the cavity between the inner bearing and the oil seal with a recommended grease.

(2) Fit the oil seal to the hub and mount the hub assembly on the stub axle. Position the inner race and rollers of the outer bearing, packed with a recommended grease, in the hub without fitting shims at this stage.

(3) Fit the stub axle nut and washer. Tighten the nut and at the same time rotate the hub back and forth until there is noticeable drag. This ensures that the bearing cones are properly seated.

(4) Unscrew and remove the stub axle nut. Extract the washer and the centre of the outer bearing. Insert a sufficient thickness of shims **to produce an excessive amount of end-float.** Note the total thickness of shims used. Replace the bearing centre, the washer, and tighten the stub axle nut.

(5) Measure accurately the total amount of end-float in the bearings. Remove the stub axle nut, washer, and outer bearing centre. Reduce the number of shims to eliminate end-float, while still allowing the hub to rotate freely, when the stub axle nut has been refitted and tightened to a torque wrench reading of 40 to 70 lb/ft.(5·53 to 9·68 kg.m.). Latitude for this reading is given so that the nut may be aligned with the split pin hole in the stub axle.

(6) Insert a new split pin through the hole provided in the hub and lock the stub axle nut.

(7) Remove any surplus grease to allow room for expansion and, using a drift, tap the grease retaining cap gently but firmly up against the outer bearing. Do not put grease in the cap.

(8) Replace the calliper assembly and tighten the securing bolts to a torque reading of between 45 to 50 lb./ft. (6·22 to 6·91 kg.m.).

(9) *Wire wheels*
Grease the hub splines, refit the road wheel and tighten the knock-on hub cap.
Disc wheels
Refit the road wheel and tighten the wheel nuts.

Section LLL.2

MODIFIED FRONT SPRINGS

To improve road holding, modified front suspension springs were fitted at car numbers BT7. 10303 ; BN7. 10329.

The modified springs are interchangeable with the earlier type only in pairs.

SECTION M

BRAKES

SERIES BN4

PREVENTIVE MAINTENANCE

To safeguard against the possible effects of wear or deterioration it is recommended that:—

1. Disc brake pads, drum brake linings, hoses and pipes should be examined at intervals no greater than those laid down in the Passport to Service.

2. Brake fluid should be changed completely every 18 months or 24,000 miles (40,000 kms.) whichever is the sooner.

All fluid seals in the hydraulic system and all flexible hoses should be examined and renewed if necessary every 3 years or 40,000 miles (65000 kms.) whichever is the sooner. At the same time the working surface of the pistons and of bores of the master cylinder, wheel cylinders and other slave cylinders should be examined and new parts fitted where necessary.

Care must be taken always to observe the following points:—

(a) At all times use the recommended brake fluid.

(b) Never leave fluid in unsealed containers. It absorbs moisture quickly and this can be dangerous.

(c) Fluid drained from the system or used for bleeding is best discarded.

(d) The necessity for absolute cleanliness throughout cannot be over-emphasised.

Section M.1

DESCRIPTION

The brakes on all four wheels are hydraulically operated by foot pedal application, directly coupled to a master cylinder in which the hydraulic pressure of the brake operating fluid is originated. A supply tank cast integrally with the master cylinder provides a reservoir by which the fluid is replenished, and a pipe line consisting of tube, flexible hose and unions, interconnect the master cylinder and the wheel cylinders.

The pressure generated in the master cylinder by application with the foot pedal is transmitted with equal and undiminished force to all wheel cylinders simultaneously. This moves the pistons outwards, which in turn expand the brake shoes thus producing automatic equalisation, and efficiency in direct proportion to the effort applied at the pedal.

When the pedal is released the brake shoe springs return the shoes which then return the wheel cylinder pistons, and therefore the fluid back into the pipe lines and master cylinder.

An independent mechanical linkage actuated by a handbrake, mounted on the propeller shaft tunnel, operates the rear wheels by mechanical expanders attached to the rear wheel cylinder bodies.

The front brakes are of the two leading shoe types with sliding shoes which ensure automatic centralisation of the brake shoe in operation.

The rear brakes are also fitted with sliding shoes, and incorporate the handbrake mechanism.

Front Brakes

The front brakes are operated by two wheel cylinders situated diametrically opposite each other on the inside of the backplate and interconnected by a bridge pipe on the outside.

Each wheel cylinder consists of a light alloy body containing a spring seal support, seal, steel piston and edges of both shoes making initial contact with the drum. The shoes are allowed to slide and centralise during the actual braking operation which distributes the braking force equally over the lining area ensuring high efficiency and even lining wear.

Adjustment for lining wear is by means of two knurled snail cam adjusters, each operating against a peg at the actuating end of each shoe. Both adjusters turn clockwise to expand the shoes.

The brake shoes rest on supports formed in the backplate and are held in position by two return springs which pass from a hole in the abutment end of each web to a peg fixed to the backplate.

The bleed screw which is incorporated in one cylinder, is provided with a steel ball, this is normally seated firmly on a valve opening in the cylinder. A dust cover is fitted over the screw nipple to exclude dirt and with the removal of this cover and an anti-clockwise turn of the screw the fluid may escape.

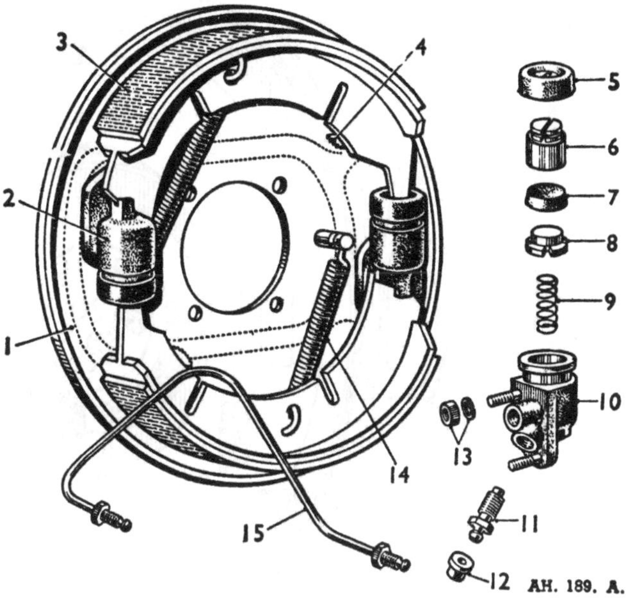

Fig. M.1. Front brake assembly.

1. Backplate.
2. Wheel cylinder.
3. Shoe.
4. Snail cam adjuster.
5. Dust cover.
6. Piston.
7. Seal.
8. Seal support.
9. Spring.
10. Cylinder housing.
11. Bleed screw.
12. Bleed screw cover.
13. Nut and washer.
14. Shoe return spring.
15. Cylinder connecting pipe.

Rear Brakes

The rear brake shoes are not fixed but are allowed to slide and centralise with the same effect as in the front brakes. They are hydraulically operated by a single acting wheel cylinder incorporating the handbrake mechanism. At the cylinder end the leading shoe is located in a slot in the piston while the trailing shoe rests in a slot formed in the cylinder body. At the adjuster end they rest in slots in the adjuster links. Both shoes are supported on the backplate and are held in position by two return springs fitted from shoe to shoe with the shorter spring nearer the adjuster.

The wheel cylinder consists of a light alloy die casting into the end of which moves a piston, with seal in a highly finished bore. In the other end of the housing a slot is machined to carry the trailing shoe. The pivoted handbrake lever projects through the backplate at right angles and operates on the leading shoe. The cylinder is attached to the backplate by a spring clip allowing it to slide laterally.

A bleed screw is incorporated in the cylinder housing with a rubber dust cap over the nipple end.

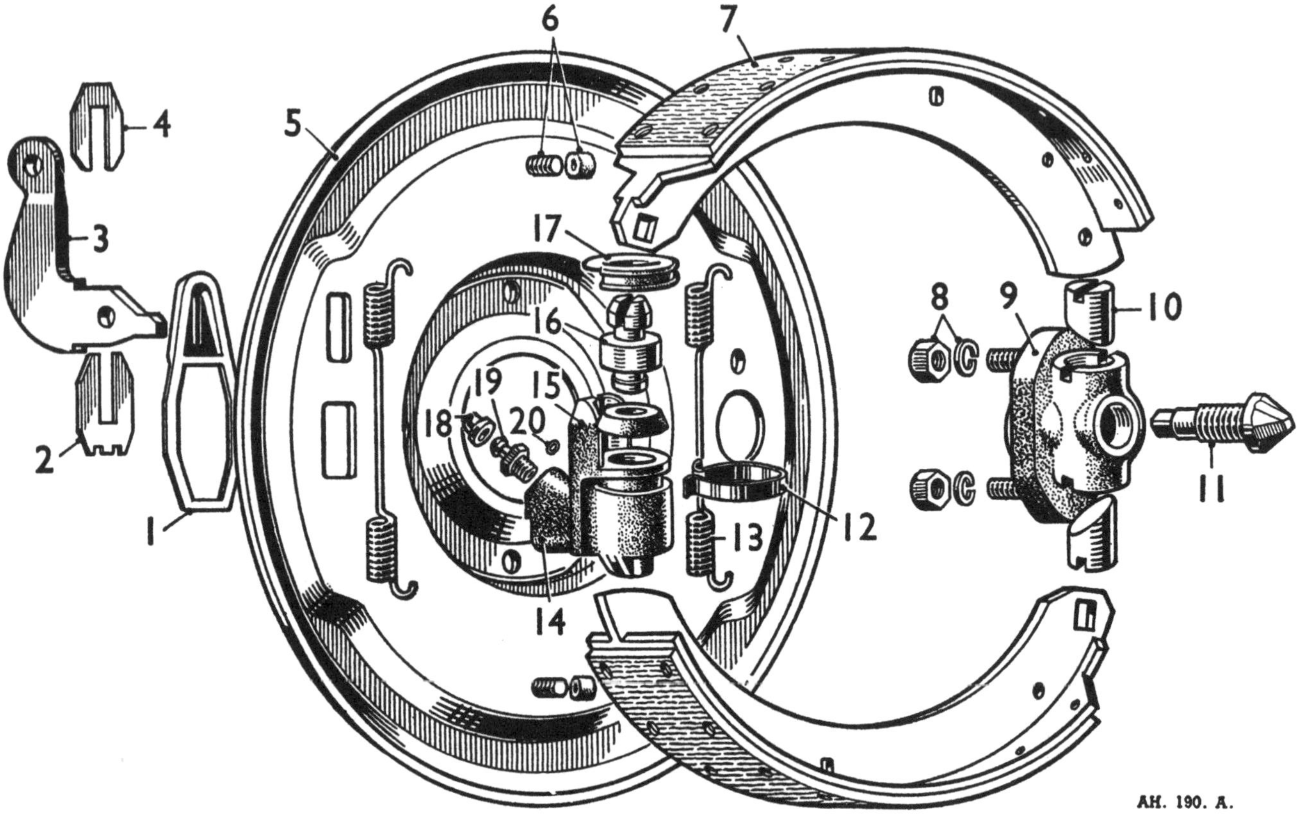

Fig. M.2. Rear brake backplate exploded.

1. *Rubber seal.*
2. *Wheel cylinder locking plate.*
3. *Handbrake lever.*
4. *Wheel cylinder locking plate.*
5. *Backplate.*
6. *Steady post.*
7. *Brake shoe.*
8. *Nut and spring washer.*
9. *Adjuster body.*
10. *Adjuster tappets.*
11. *Adjuster wedge.*
12. *Dust cover clip.*
13. *Shoe return spring.*
14. *Pipe orifice.*
15. *Cylinder body.*
16. *Piston.*
17. *Dust cover.*
18. *Bleed nipple dust cover.*
19. *Bleed nipple.*
20. *Bleed valve ball.*

Adjustment for lining wear is made by the brake shoe adjuster. This has a steel housing which is spigotted and bolted firmly to the inside of the backplate. The housing carries two opposed steel links, the outer end slotted to carry the shoes, and the inclined inner faces bearing on inclined faces of the hardened steel wedge.

The wedge has a threaded spindle with a square end which projects on the outside of the backplate, enabling a spanner to be used for adjustment purposes, by rotating the wedge in a clockwise direction, the links are forced apart and the fulcrum of the brake shoes expanded.

When the brake is applied, the piston under the influence of the hydraulic pressure, moves the leading shoe and the body reacts by sliding on the backplate to operate the trailing shoe.

The handbrake lever is pivoted in the cylinder body, and when operated the lever tip expands the leading shoe, and the pivot moves the cylinder body and with it the trailing shoe.

Handbrake

The handbrake operates on the rear wheels only and is applied by a pull-up type of lever situated on the propeller shaft tunnel. The cable from the control is attached to the compensator mounted on the rear axle. From compensator to the brake levers are transverse rods which are non-adjustable.

Before any adjustment to the handbrake is made the rear brake shoes must be adjusted to the brake drums.

The handbrake linkage is set when leaving the works and should not require any attention. Only when a complete overhaul is necessary should the handbrake linkage require re-setting.

When this is correct the rear shoes should be locked to the drums, the handbrake control just slightly applied, and the cable slackness just removed, by means of adjusting the sleeve nut at the front end of the longitudinal cable.

Section M.2

MAINTENANCE

Replenishment of Hydraulic Fluid

Inspect the supply tank at regular intervals and maintain at the indicated level by the addition of Castrol Girling Brake Fluid Amber.

Exercise great care when adding brake fluid, to prevent dirt or foreign matter entering the system.

IMPORTANT.—Serious consequences may result from the use of incorrect fluids, and on no account should any other than Castrol Girling Brake Fluid Amber be used.

Bleeding the Hydraulic System

Bleeding is necessary when a portion of the hydraulic system has been disconnected, or if the level of the brake fluid has been allowed to fall so low that air has entered the master cylinder.

Pressure bleeding methods are not suitable for Girling systems and are not recommended.

To prepare the brakes for bleeding release the handbrake. Check that each **rear** wheel cylinder is free to slide and turn the adjuster clockwise until the drum is fully locked by the shoes. The wheel cylinder piston will then be pushed right into the bore, with a minimum of air to be expelled. Slacken off the adjusters of the **front** brakes to allow the shoe springs to push the pistons into the wheel cylinder bores, leaving a minimum amount of space for air or fluid. Fill the master cylinder with the recommended fluid and keep it at least half-full throughout the operation, otherwise air will be drawn into the system necessitating a fresh start.

Attach the bleeder tube to the wheel cylinder bleeder screw farthest from the master cylinder and allow the free end of the tube to be submerged in a small quantity of fluid in a clean glass jar.

Open the bleeder screw about three-quarters of a turn.

Begin bleeding with a fairly full stroke of the pedal, allowing it to fly back freely. Lift the floor covering if this prevents free movement of the pedal. One or two slightly faster applications may now be made to advantage. Watch the flow of fluid into the glass jar, and when air bubbles cease to appear close the bleeder screw during the last (slow) pedal application.

If the bleeding of any cylinder continues without success for a considerable time it is possible that air is being drawn in past the bleeder screw threads. In such cases tighten the bleeder screw at the end of each downward stroke of the pedal and allow the pedal to return fully before re-opening it. Close the bleeder screw finally during the last pedal application.

Tighten the bleeder screws to a torque wrench reading of 5 to 7·5 lb. ft. (·69 to 1·03 kg. m.).

Repeat the operation on each wheel, finishing with the wheel nearest the master cylinder.

After bleeding top up the master cylinder to its correct level. Adjust all brakes in the usual manner.

NOTE.—Clean fluid bled from the system must be allowed to stand for at least 24 hours until it is clear of air bubbles before it is used again. Dirty fluid should be discarded.

Adjusting the Brake Shoes

The brakes are adjusted for lining wear, **only** at the brakes themselves, and no alteration should be made to the handbrake cable for this purpose.

Front Brakes. A separate snail cam adjuster is provided for each shoe. Jack up the car until the wheel to be adjusted is clear of the ground, then fully release both the hexagon head adjuster bolts on the outside of the backplate by turning them anti-clockwise.

Turn one of the adjuster bolts clockwise until the brake shoe touches the brake drum, then release the adjuster until the shoe is just free of the drum. Repeat the process for the second adjuster and shoe. Spin the wheel to ensure that the brake shoes are quite free of the drum. Repeat the procedure for the second front wheel.

Rear Brakes. One adjuster is provided for both shoes and the adjustment of both rear wheels is identical. Release the handbrake and jack up the car. Turn the square end of the adjuster on the outside of each rear

Fig. M.3.
Showing the location of the two brake shoe adjusters on a front brake backplate.

brake backplate in a clockwise direction until a resistance is felt. Slacken two clicks and the drum should rotate freely.

Section M.3

FRONT BRAKES

Replacing Brake Shoes

NOTE.—**Always fit Girling "Factory Lined" replacement shoes. These have the correct type of lining and are accurately ground to size. When fitting replacement shoes, fit a new set of shoe return springs.**

(1) Jack up the car and remove road wheels, hub extensions and brake drums.

(2) Lift one shoe out of the abutment slot of one wheel cylinder, then release from the piston slot of the other. (It will be found quite simple to remove the shoe return springs). To prevent the wheel cylinder pistons from expanding it is advisable to place a rubber band round each cylinder. Repeat with the second shoe.

(3) Clean down the backplate, check wheel cylinders for leaks and freedom of motion.

(4) Check adjusters for easy working and turn back (anti-clockwise) to full "off" position. Lubricate where necessary with **Girling (White) Brake Grease.**

(5) Smear the tips of the brake shoe supports on the backplate, and the operating and abutment ends of the new shoes with **Girling (White) Brake Grease.** The (white) brake grease must not be allowed to contact hydraulic cylinders, pistons or rubber parts. Keep all grease off the linings on new replacement shoes and do not handle more than necessary.

(6) Fit new shoe return springs to the new shoes. Place the hooked end of the spring through the hole in the shoe web and the swan neck through the hole in the backplate near the abutment end of the same shoe. Each shoe can be replaced independently. Remove rubber bands from cylinder.

(7) Make sure the drums are clean and free from grease, etc., then re-fit.

(8) Adjust the brakes as described under "Running Maintenance".

(9) Re-fit the road wheels and lower the car to the ground.

Section M.4

REAR BRAKES

Replacing Brake Shoes

Proceed in stages as described for front brakes, paragraphs 1 to 9, substituting the details in the following paragraphs for those bearing the same number.

(2) Lift one of the shoes out of the slots in the adjuster link and wheel cylinder piston. Both shoes can then be removed complete with springs. Place a rubber band round the wheel cylinder to keep piston in place.

(6) Fit the two new shoe return springs to the new shoes (with the shorter spring at adjuster end) from shoe to shoe and between shoe web and backplate. Locate one shoe in the adjuster link and wheel cylinder piston slots, then prise over the opposite shoe into its relative position. Remove rubber band.

NOTE.—**The first shoe has the lining positioned towards the heel of the shoe and on the second shoe towards the toe or operating end in both left-hand and right-hand brake assemblies. Make several hard applications of the brake pedal to ensure all the parts are working satisfactorily and the shoes bedding to the drums, then the brakes should be adjusted as described.**

Immediately after fitting brake shoes the hand brake adjustment should be slackened. It is also advisable to slacken one further click on the rear brake adjuster, to allow for possible lining expansion, reverting to normal adjustment and hand brake adjustment afterwards.

Section M.5

BRAKE PEDAL

To Remove

(1) The brake and clutch pedal linkages are mounted in a common bracket and thus have to be released as a unit.

(2) Inside the car disconnect the brake and clutch cylinder levers from their master cylinder push rods by removing the clevis pins.

(3) Working under the bonnet, unscrew the six securing setpins sufficiently to allow the brake and clutch pedal linkage bracket to be withdrawn from inside the car.

(4) Release the brake and clutch pedal return springs.

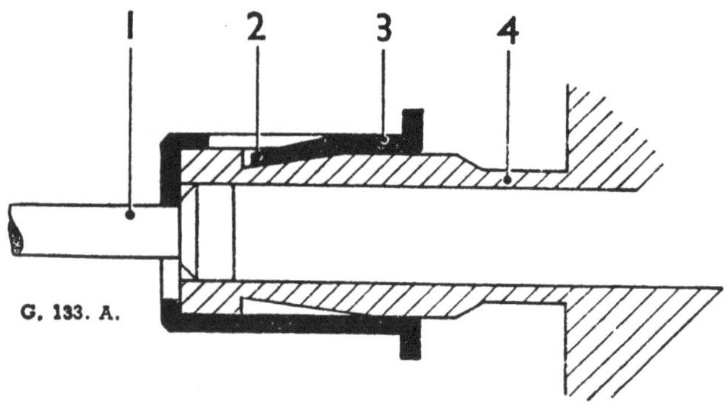

Fig. M.4. Diagrammatic section of master cylinder.

1. *Valve stem.*
2. *Thimble leaf.*
3. *Thimble.*
4. *Plunger.*

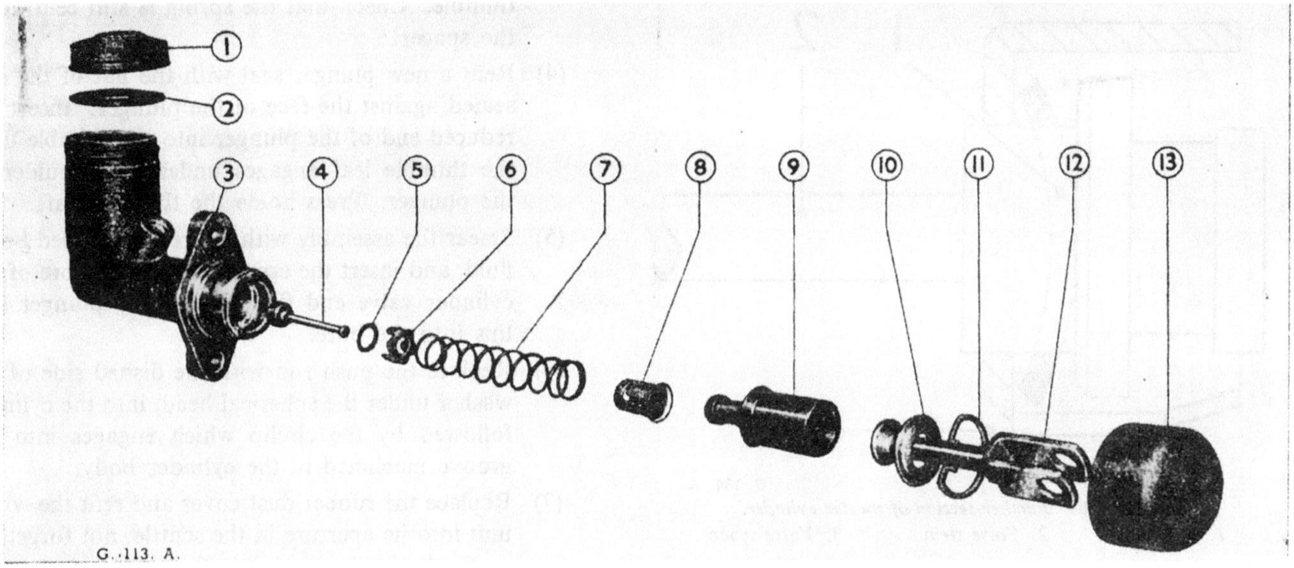

Fig. M.5. Components of the master cylinder and reservoir.

1. Filler cap.
2. Washer.
3. Master cylinder.
4. Valve stem.
5. Spring washer.
6. Valve spacer.
7. Return spring.
8. Thimble.
9. Plunger.
10. Dished washer.
11. Circlip.
12. Fork.
13. Dust cover

(5) Unscrew the nut securing the brake and clutch pedal shaft and withdraw the shaft to release the brake and clutch pedal levers together with their distance piece.

(6) Inspect the lever bushes for wear and renew if necessary.

To Replace

Replacement is the reverse of the procedure **"To Remove"**.

Section M.6

MASTER CYLINDER

Description

The master cylinder consists of an alloy body with a polished finish bore, and reservoir with cap. The inner assembly is made up of the push rod, dished washer circlip, plunger, plunger seal, spring thimble, plunger return spring, valve spacer, spring washer, valve stem and valve seal. The open end of the cylinder is protected by a rubber dust cover.

Dismantling the Brake Master Cylinder

(1) Release the master cylinder push rod from the brake pedal as described in Section M.5.

(2) Disconnect the pressure pipe union from the cylinder and remove the securing bolts, then the master cylinder and fluid reservoir may be withdrawn complete from the car.

(3) Remove the filler cap and drain out the fluid. Pull back the rubber dust cover and remove the circlip with a pair of long nosed pliers. The push rod and dished washer can then be removed.

(4) When the push rod has been removed the plunger with seal attached will be exposed ; remove the plunger assembly complete. The assembly can be separated by lifting the thimble leaf over the shouldered end of the plunger.

(5) Depress the plunger return spring allowing the valve stem to slide through the elongated hole of the thimble thus releasing the tension on the spring.

(6) Remove the thimble, spring and valve complete.

(7) Detach the valve spacer, taking care not to lose the spacer spring washer which is located under the valve head. Remove the seal.

(8) Examine all parts, especially the seals, for wear or distortion and replace with new parts where necessary.

Assembling the Brake Master Cylinder

(1) Replace the valve seal so that the flat side is correctly seated on the valve head.

(2) The spring washer should then be located with the dome side against the underside of the valve

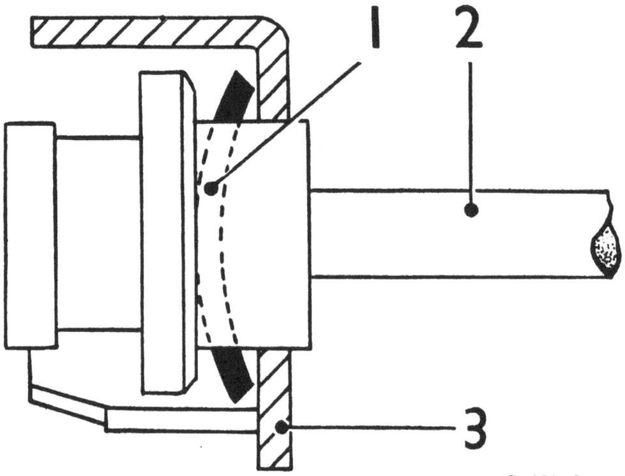

Fig. M.6. Further section of master cylinder.
1. Washer. 2. Valve stem. 3. Valve spacer.

head, and held in position by the valve spacer the legs of which face towards the valve seal.

(3) Replace the plunger return spring centrally on the spacer, insert the thimble into the spring and depress until the valve stem engages through the elongated hole of the thimble, making sure the stem is correctly located in the centre of the thimble. Check that the spring is still central on the spacer.

(4) Refit a new plunger seal with the flat of the seal seated against the face of the plunger. Insert the reduced end of the plunger into the thimble until the thimble leaf engages under the shoulder of the plunger. Press home the thimble leaf.

(5) Smear the assembly with the recommended brake fluid, and insert the assembly into the bore of the cylinder valve end first, easing the plunger seal lips into the bore.

(6) Replace the push rod with the dished side of the washer under the spherical head, into the cylinder followed by the circlip which engages into the groove machined in the cylinder body.

(7) Replace the rubber dust cover and refit the whole unit into its aperture in the scuttle, not forgetting to fit the packing washer first. Secure the unit by means of the two bolts on the flange and refit the pressure pipe union into the cylinder.

(8) Reconnect the push rod fork with its corresponding hole in the brake pedal lever, securing it with the circlip.

(9) Bleed the hydraulic system.

SECTION MM

BRAKES

SERIES BN6

NOTE

For details of the brakes fitted to BN6 cars refer to Section M.

SECTION MMM

BRAKES

Mk. I and II (SERIES BN7 and BT7)
AND Mk. II and Mk. III (SERIES BJ7 and BJ8)

NOTE

This section should be used in conjunction with Section M, particular attention being given to Sections M.2 and M.5

Section MMM.1

GENERAL DESCRIPTION

The brakes on all four wheels are hydraulically operated by foot pedal application, the pedal being directly coupled to a master cylinder in which the hydraulic pressure of the brake operating fluid is originated.

Steel pipe lines, unions and flexible hoses convey the hydraulic pressure from the master cylinder to each wheel cylinder.

The cable actuated hand brake mechanism operates the rear brake shoes only by mechanical expanders attached to the brake cylinder bodies.

Girling calliper type disc brakes are fitted to the front wheel hubs. Each brake consists of two carriers to which friction pads are bonded. The system is self adjusting in operation.

The rear brakes are of the single leading shoe type with sliding shoes which ensure automatic centralisation in operation. Manual adjustment is provided by means of a wedge type adjuster.

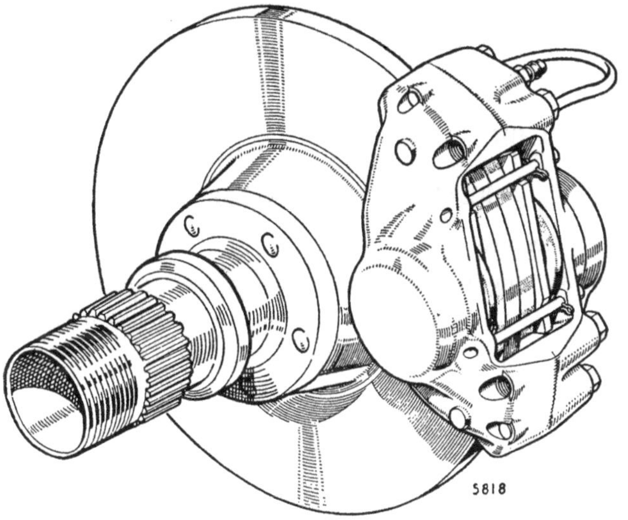

Fig. MMM.1. Front disc brake and hub assembly.

Front Brakes

The front brake unit consists of a hub mounted disc rotating with the wheel, and a braking unit rigidly attached to the swivel axle at the rear. The brake unit is a calliper which straddles the disc and houses two horizontally opposed blind cylinders and the friction pads. Within each cylinder is a rubber sealing ring positioned by a groove in the cylinder body, and a piston protected by a dust cover. A segmental friction pad bonded to a steel backplate is inserted between each piston and the disc. The pads and backplates are secured by retaining pins and spring clips. The pads are self adjusting in operation and should need no attention between replacements.

If any part of the hydraulic system is disconnected the brake lines must be bled, the bleed screw is fitted at the top of the calliper housing on the inside.

NOTE :—**The bridge bolts joining the two halves of the calliper together should not be removed.**

Rear Brakes
(Refer to fig. M.2.)

The brake shoes are allowed to slide and centralise during the actual braking operation which distributes the braking force equally over the lining area ensuring high efficiency and even lining wear. They are hydraulically operated by a single acting wheel cylinder incorporating the handbrake mechanism. At the cylinder end the leading shoe is located in a slot in the piston while the trailing shoe rests in a slot formed in the cylinder body. At the adjuster end the shoes rest in slots in the adjuster links. Both shoes are supported on the backplate and are held in position by the two return springs fitted from shoe to shoe with the shorter spring nearer the adjuster.

The wheel cylinder consists of a light alloy die casting with a highly finished bore in which moves the piston and seal. A slot is machined in the other end of the cylinder body to carry the trailing shoe and at right angles projecting through the backplate is pivoted the handbrake lever. The cylinder is attached to the backplate by a spring clip allowing it to slide laterally.

A bleed screw is incorporated in the cylinder housing with a rubber dust cap cover over the nipple.

Adjustment for lining wear is made by the brake shoe adjuster. This has a steel housing which is spigotted and bolted firmly to the inside of the backplate. The housing carries two opposed steel links, the outer end slotted to carry the shoes, and the inclined inner faces bearing on inclined faces of the hardened steel wedge.

The wedge has a threaded spindle with a square end which projects on the outside of the backplate, enabling a spanner to be used for adjustment purposes, by rotating the wedge in a clockwise direction, the links are forced apart and the fulcrum of the brake shoes expanded.

When the brake is applied, the piston under the influence of the hydraulic pressure, moves the leading shoe and the body reacts by sliding on the backplate to operate the trailing shoe.

The handbrake lever is pivoted in the cylinder body, and when operated the lever tip expands the leading shoe, and the pivot moves the cylinder body which in turn moves the trailing shoe against the brake drum.

Handbrake

The handbrake operates on the rear wheels only

and is applied by the lever alongside the gearbox cover. The cable from the control is attached to the compensator mounted on the rear axle. From the compensator to the brake levers are transverse rods which are non-adjustable.

The handbrake linkage is set when leaving the works and should not require any attention. Only when a complete overhaul is necessary should the handbrake linkage require re-setting.

When this is correct the rear shoes should be locked to the drums, the handbrake control just slightly applied, and the cable slackness just removed, by adjusting the sleeve nut at the front end of the cable.

Section MMM.2

MAINTENANCE

This section should be used in conjunction with Section M.2 particular attention being given to—**"Replenishment of Hydraulic Fluid"** and **"Bleeding the Hydraulic System."**

Front brake adjustment

Wear on the front disc brake friction pads is automatically compensated during braking operations, manual adjustment is therefore not required. In order to maintain peak braking efficiency and at the same time obtain the maximum life from the friction pads, they should be examined at the recommended mileage and if the wear on one pad is greater than the other their operating positions should be changed over.

Rear Brakes are adjusted for wear only at the brakes themselves and on no account should any alteration be made to the handbrake cable for this purpose.

One common adjuster is provided for both shoes and the adjustment for both wheels is identical.

Release the handbrake and jack up the car. Turn the end of the adjuster on the outside of each brake backplate in a clockwise direction until a resistance is felt. Slacken two notches when the drum should rotate freely.

Section MMM.3

FRONT BRAKES

Replacing Friction Pads

When wear has reduced the thickness of the pads to approximately $\frac{1}{8}$in. (3·2mm.) they must be renewed. Under no circumstances should a pad be allowed to wear below $\frac{1}{16}$in. (1·6mm.).

(1) Jack up the car and remove the road wheels.
(2) Remove the spring clips locking the retaining pins in position and draw them back. Pull out the friction pad assemblies.
(3) Clean down the callipers and inspect for fluid leaks.
(4) Push in the pistons to the bottom of the cylinder bores with a suitable lever.
(5) Slip in the new pads and locate them in position with the retaining pins and secure with the spring clips.
(6) Press the brake pedal hard once or twice in order to settle the hydraulic system.

Section MMM.4

REMOVING A CALLIPER UNIT

To Remove

(1) Unscrew the brake pipe union nut in front of its support bracket, disconnect and blank off the pipe.

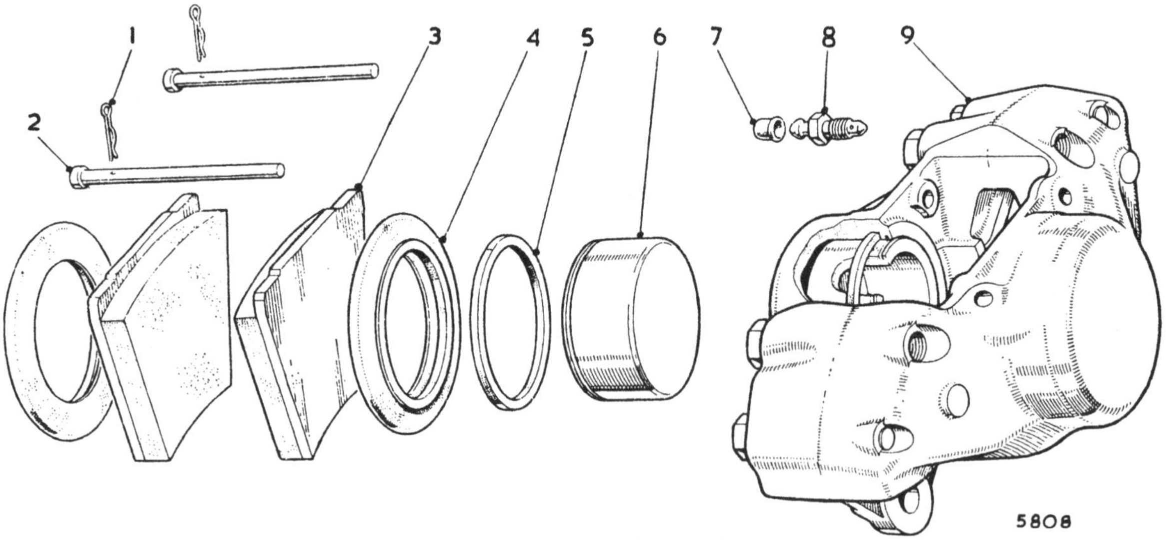

Fig. MMM.2. Front calliper exploded.

1. *Wire clip.*
2. *Retaining pin.*
3. *Lining pad and steel backplate.*
4. *Dust cover.*
5. *Sealing ring.*
6. *Piston.*
7. *Bleed nipple dust cover.*
8. *Bleed nipple.*
9. *Calliper body.*

(2) Remove the two nuts securing the brake hose support bracket and remove the bracket.
(3) Unscrew the two calliper retaining bolts and remove the calliper assembly complete.

To Replace

Replacement is the reverse of the procedure "To Remove".

NOTE.—**Tighten the retaining bolts to a torque reading of between 45 and 50lb. ft. (6·22 and 6·91 kg.m.).**

Section MMM.5

DISMANTLING A CALLIPER UNIT

(1) Remove the calliper from the vehicle as described in section MMM.4
(2) Withdraw the brake pads as described in Section MMM.3 but do not push the pistons to the bottom of their bores.
(3) It is recommended that the unit is thoroughly cleaned before proceeding with dismantling.
(4) Force the pistons out of their bores by connecting the flexible hose to a fluid supply and applying pressure to reject the pistons. Alternatively, push back the dust seal and insert two suitable levers into the seal groove and pull the piston out with an even pressure. Taking care not to damage the groove and piston surfaces.
(5) Disengage the dust cover and remove the internal seal, by inserting a blunt blade along its side and easing it out. Renew the dust cover and internal seal if they show signs of wear.

Clean internally with methylated spirits only and allow to dry. Use brake fluid to clean rubber parts.

NOTE.—**No attempt should be made to remove the bridge bolts joining the two halves of the calliper.**

Section MMM.6

ASSEMBLING A CALLIPER UNIT

(1) Fit the internal seal into the groove in the cylin) der bore with the scraping edge (smaller diameter- innermost.
(2) Locate the lip of the dust cover in the outer groove.
(3) Smear the piston with brake fluid and push it into the bore, closed end first.
(4) Push the piston right home and then engage the outer edge of the cover with the groove in the piston body.
(5) Refit the pad assemblies and lock in position with the retaining pins. (Section MMM.3).
(6) Refit the calliper unit taking care that the disc passes between the two pads. (Section MMM.4).
(7) Connect the brake hose and bleed the system (Section M.2). Check for leaks with the brake fully applied.

Section MMM.7

THE BRAKE DISC

A check should be made to ascertain that the disc is running true. As maximum efficiency can only be attained when the disc run-out is at a minimum.

It must be remembered that run-out at the disc may also be due to the hub bearings being out of adjustment, this item should be checked carefully before condemning a disc (see Section LLL.1).

To Check for Run-out

Clamp the dial indicator to a suitable fixed point on the vehicle with the needle pad bearing on the disc face. Run-out must not exceed ·004 in. (·1mm) total, as excessive run-out will cause knocking back of the pistons which may create judder and increased pedal travel. If there is doubt concerning this condition the disc should be renewed.

To Remove

(1) Remove the calliper unit as described in Section MMM.4 but do not disconnect the hydraulic supply hose. A spacer should be placed between the pads in order to hold the pistons in position.
(2) Dismantle the hub assembly as described in Section LLL.1.
(3) The disc is held to the hub by five nuts and spring washers, after removing these the two components may be separated.

To Replace

(1) Assembling the brake disc to the hub is a reversal of the instructions "To Remove".
(2) Replace the hub on the swivel axle (See Section LLL.1).
(3) Check for run-out as detailed above.
(4) Replace the calliper assembly as described in Section MMM.4
(5) Refit the wheel.

Scoring of brake discs is not detrimental, provided that the scoring is concentric, even, and not excessive. If the discs are heavily or unevenly scored, however, the braking efficiency will be impaired and pad wear increased. In this case the discs may be reground, but this work must be carried out with extreme accuracy and should only be considered if new discs are unobtainable. The ground surface must be quite flat and parallel with the mounting face, and must have a fine finish.

Avoid sharp corners at the inner circumference of the ground area. Either or both sides may be ground, but no more than ·040 in. (1·02 mm.) should be removed from each disc, i.e. after grinding the thickness must not be less than ·335 in. (8·51 mm.).

When fitted to the hub disc must run centrally between the calliper cylinders. To check this remove the friction pads from the calliper and insert feeler gauges between the pad abutments on the calliper body and the faces of the disc. It is permissible for the gap on opposite sides of the disc to differ by ·015 in. (·38 mm.) but there must be no difference between the gaps at the two abutments on the same side of the calliper. This ensures that the calliper is in correct alignment with the disc and that the pads and pistons are square with the disc. Shims must be used at the calliper mounting to correct any discrepancy.

(7) Make sure the drums are clean and free from grease, etc., then re-fit.
(8) Adjust the brakes as described in Section MMM.2.
(9) Re-fit the road wheels and lower the car to the ground.

NOTE.—The first shoe has the lining positioned towards the heel of the shoe and on the second shoe towards the toe or operating end in both left-hand and right-hand brake assemblies. Several hard applications of the brake pedal should be made to ensure all the parts are working satisfactorily and the shoes bedding to the drums, then the brakes should be adjusted as described.

Immediately after fitting replacement shoes it is advisable to slacken one further notch on the brake adjuster to allow for possible lining expansion, reverting to normal adjustment afterwards.

Section MMM.8

REAR BRAKES

Replacing Brake Shoes

NOTE.—**Always fit Girling "Factory Lined" replacement shoes. These have the correct type of lining and are accurately ground to size. When fitting replacement shoes, fit a new set of shoe return springs.**

(1) Jack up the car and remove the road wheels, hub extensions and brake drums.
(2) Lift one of the shoes out of the slots in the adjuster link and wheel cylinder piston. Both shoes can then be removed complete with springs. Place a rubber band round the wheel cylinder to keep the piston in place.
(3) Clean down the backplate, check the wheel cylinders for leaks and freedom of motion.
(4) Check the adjusters for easy working and turn back (anti-clockwise) to full "off" position. Lubricate where necessary with **Girling (White) Brake Grease.**
(5) Smear the tips of the brake shoe supports on the backplate, and the operating and abutment ends of the new shoes with **Girling (White) Brake Grease.** The (white) brake grease must not be allowed to contact hydraulic cylinders, pistons or rubber parts. Keep all grease off the linings of new replacement shoes and do not handle more than necessary.
(6) Fit the two new shoe return springs to the new shoes (with the shorter spring at adjuster end) from shoe to shoe and between shoe web and backplate. Locate one shoe in the adjuster link and wheel cylinder piston slots, then prise over the opposite shoe into its relative position. Remove rubber band.

Section MMM.9

DISMANTLING A WHEEL CYLINDER

(Refer to Fig. M.2)

(1) Jack up the car and remove the road wheel, hub extension and brake drum.
(2) Disconnect the rod from the handbrake lever, remove the brake shoes, disconnect the pressure pipe union from the cylinder, and remove the rubber dust cover from rear of the backplate.
(3) Prise the retaining plate and spring plate apart and tap the retaining plate from beneath the neck of the wheel cylinder.
(4) Withdraw the handbrake lever from between the backplate and wheel cylinder.
(5) Remove the spring plate and distance piece, and finally the wheel cylinder from the backplate.
(6) Examine all parts for wear or distortion and replace with new parts where necessary.

Section MMM.10

ASSEMBLING A WHEEL CYLINDER

(1) Locate the neck of the wheel cylinder in the larger slot and replace the distance piece with cranked lips away from the backplate.
(2) Insert the spring plate between the distance piece and the backplate with cranked lips outwards.
(3) Replace the handbrake lever and tap the retaining plate into position between the distance piece and the spring plate, until located by the spring plate.
(4) Fit the rubber dust cover, connect the pressure hose union and refit the brake rod.
(5) Replace the shoes and brake drum, and bleed the system. Finally refit the road wheel.

Section MMM.11

MASTER CYLINDER

Description

The master cylinder consists of an alloy body with a polished bore, into which is connected the fluid inlet and outlet ports. The fluid reservoir is connected to the inlet port at the end of the cylinder by a length of steel pipe. Connected to the outlet port in the top of the cylinder body is the pressure pipe for the braking system. The inner assembly is made up of the push rod, dished washer, circlip, plunger, plunger seal, end seal, spring thimble, plunger return spring, valve spacer, spring washer, valve stem and valve seal. The open end of the cylinder is protected by a rubber dust cover.

Dismantling the Brake Master Cylinder

(1) Release the master cylinder push rod from the brake pedal as described in Section M.5.
(2) Disconnect the inlet and pressure pipe unions from the cylinder and remove the securing bolts, then withdraw the master cylinder.
(3) Drain off the fluid from inside the cylinder. Pull back the rubber dust cover and remove the circlip with a pair of long nosed pliers. The push rod and dished washer can then be removed.
(4) When the push rod has been removed the plunger with seal attached will be exposed. Remove the plunger assembly complete. The assembly can be separated by lifting the thimble leaf over the shouldered end of the plunger (ref. Fig. M.4.).
(5) Depress the plunger return spring allowing the valve stem to slide through the elongated hole of the thimble thus releasing the tension on the spring.
(6) Remove the thimble, spring and valve complete.
(7) Detach the valve spacer, taking care not to lose the spacer spring washer which is located under the valve head. Remove the seal.
(8) Examine all parts, especially the seals, for wear or distortion and replace with new parts where necessary.

Assembling the Brake Master Cylinder

(1) Replace the valve seal so that the flat side is correctly seated on the valve head.
(2) The spring washer should then be located with the dome side against the underside of the valve head, and held in position by the valve spacer, the legs of which face towards the valve seal. (Ref. Fig. M.7.).
(3) Replace the plunger return spring centrally on the spacer, insert the thimble into the spring and depress

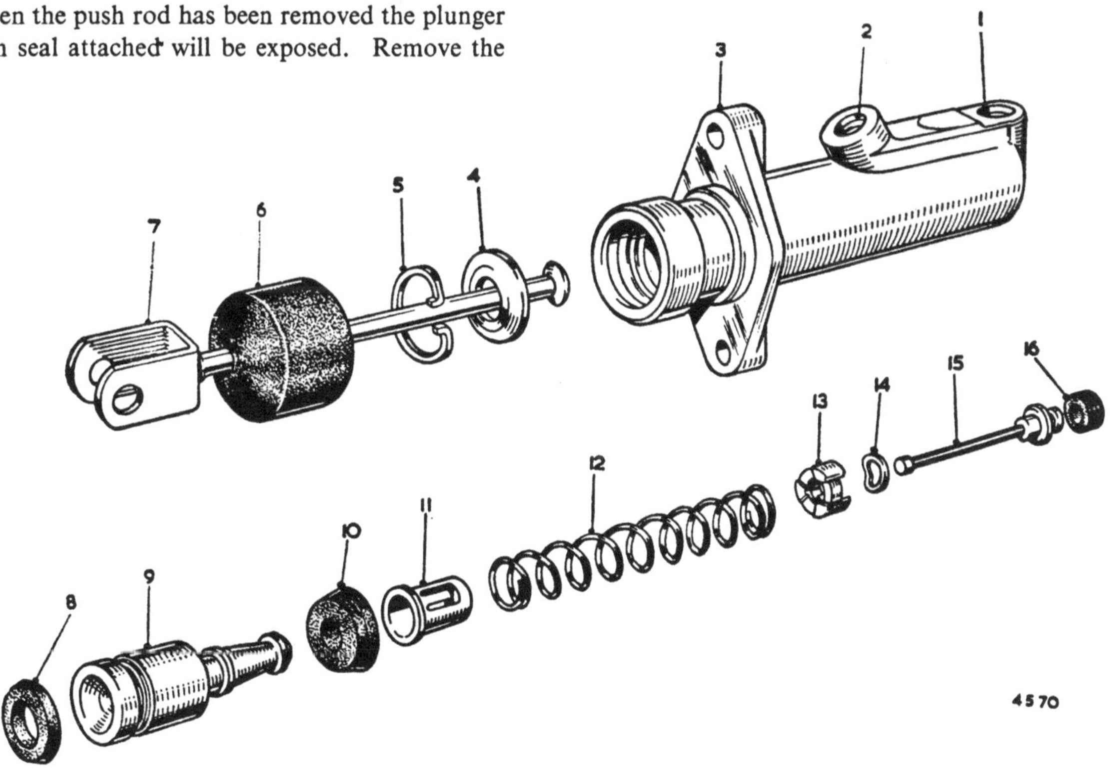

Fig. MMM.3. Components of the master cylinder.

1. Fluid inlet.
2. Fluid outlet.
3. Master cylinder.
4. Dished washer.
5. Circlip.
6. Dust cover.
7. Push rod.
8. End seal.
9. Plunger.
10. Plunger seal.
11. Thimble.
12. Return spring.
13. Valve spacer.
14. Spring washer.
15. Valve stem.
16. Valve seal.

until the valve stem engages through the elongated hole of the thimble, making sure the stem is correctly located in the centre of the thimble. Check that the spring is still central on the spacer.

(4) Refit a new plunger seal with the flat of the seal seated against the face of the plunger. Insert the reduced end of the plunger into the thimble until the thimble leaf engages under the shoulder of the plunger. Press home the thimble leaf.

(5) Smear the assembly with Girling red rubber grease or with the recommended hydraulic fluid. Insert the assembly into the bore of the cylinder, valve end first, easing the plunger seal lips into the bore.

(6) Replace the push rod with the dished side of the washer under the spherical head, into the cylinder followed by the circlip which engages into the groove machined in the cylinder body.

(7) Replace the rubber dust cover and refit the whole unit into its aperture in the scuttle, not forgetting to fit the packing washer first. Secure the unit by means of the two bolts on the flange and refit the pressure pipe and inlet pipe unions to the cylinder.

(8) Reconnect the push rod fork with its corresponding hole in the brake pedal lever, securing it with the circlip

(9) Bleed the system.

Section MMM.12

FAULT DIAGNOSIS

This Section should be used in conjunction with Section M.7.

Symptom	No.	Possible fault
(c) **Brakes Grab or Pull to Side**	1 2 3	Disc out of true Calliper loose Pad loose in calliper
(d) **Dragging Brakes**	1 2	Excessive pad wear Pressure build up in fluid supply.
(f) **Brakes Inefficient**	1 2	Disc out of true Incorrect grade of lining pad.

Section MMM.13

DUST COVERS

To prevent excessive wear of the front brake pads, due to the entry of water and road grit, dust covers have been fitted in production to the front brakes of later cars. This change took place at the following car numbers :

BT7. (Disc wheels) 9088 ; (Wire wheels) 9090.

BN7. (Disc wheels) 9450 ; (Wire wheels) 9453.

This modification can also be carried out on earlier cars.

Section MMM.14

DISC BRAKE PAD SHIMS

Should the front disc brakes develop a high pitched squeal after several applications of the brakes, this can be remedied by fitting four anti-squeal shims (Part No. BHA4195). They must be positioned between each brake pad and piston with the indicating arrow on each shim pointing towards the calliper bleeder screw, i.e. in the direction of the forward rotation of the wheel.

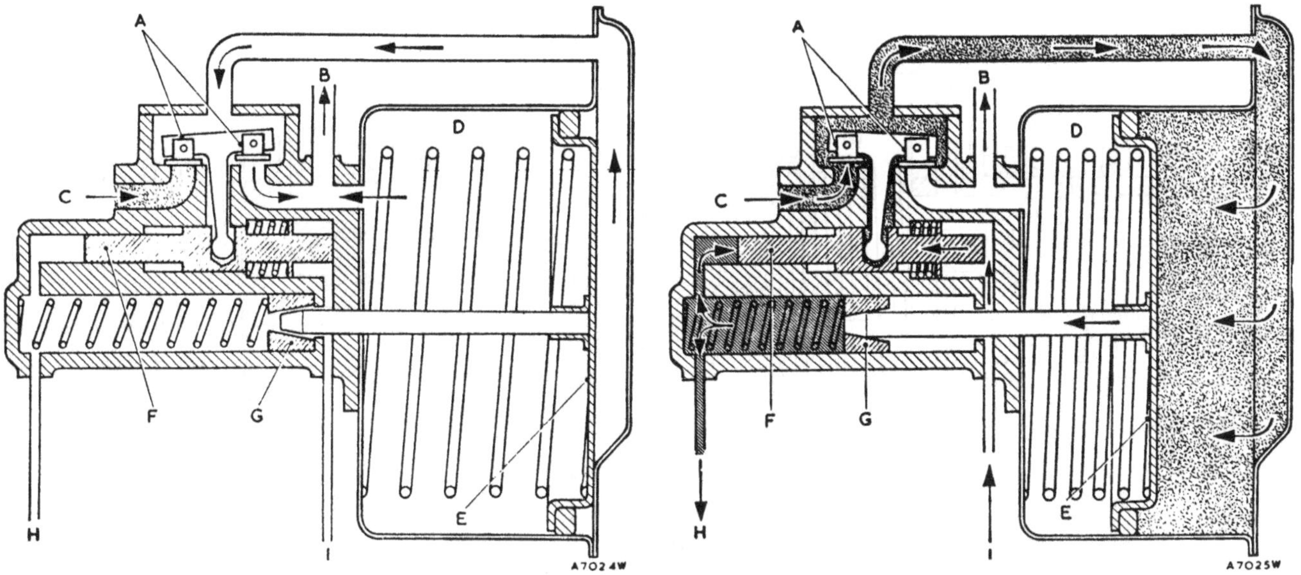

Fig. MMM.4 *Fig. MMM.5*

Vacuum servo unit operation—diagrammatic.

A. *Air valves.*
B. *To inlet manifold.*
C. *Air inlet.*
D. *Vacuum cylinder.*
E. *Vacuum piston.*
F. *Control piston assembly.*
G. *Output piston.*
H. *To wheel cylinders.*
I. *From master cylinder.*

Section MMM.15

BRAKE SERVO
(Optional on 3000 Mk. II cars)

A vacuum servo of Girling manufacture for the braking system was offered as an optional extra from Car Number 15104 on 3000 Mk. II cars, but standard equipment on 3000 Mk. III cars. When this is fitted the following different parts are also embodied in the braking system: a master cylinder with a larger bore (⅞ in.), new front brake calliper units (incorporating new calliper body assemblies, pistons, friction pads, and the anti-squeal shims mentioned in Section MMM.14), and new rear brake shoes.

General Description

The brake servo unit is installed in the hydraulic system between the master cylinder and the wheel cylinders, with the outlet pipe from the master cylinder connected to the servo hydraulic inlet and the servo hydraulic outlet connected to the wheel cylinders. The force required to augment the driver's effort is obtained by admitting atmospheric pressure to a vacuum cylinder containing a piston. The pressure difference thus obtained across the vacuum piston produces a thrust which is used to increase the hydraulic pressure available at the wheel cylinders.

The piston in the vacuum cylinder is normally subjected to the vacuum on both sides. This provides a more rapid response than the direct vacuum type of cylinder, in which the piston is normally subjected to atmospheric pressure on both sides and the vacuum is introduced to one side when a pressure difference is required.

When atmospheric pressure is admitted to the vacuum cylinder by the control valve, the piston drives the piston rod down the hydraulic cylinder, providing a considerable increase in the pressure of fluid at the wheel cylinders. The control valve, operated by the fluid from the master cylinder, exercises control over the pressure increase and the brakes are operated in proportion to the effort applied to the pedal.

Fig. MMM.4 shows the unit at rest with no pressure in the hydraulic system. The valve is open to the engine inlet manifold and the vacuum on both sides of the piston is equal.

When the brake pedal is applied hydraulic pressure is exerted throughout the whole system and equally on both ends of the valve control piston. As one end of the piston is larger than the other, an equal pressure on both ends causes a proportionally greater thrust to be exerted on the large end. The piston therefore moves to the left (see Fig. MMM.5) and the "T" shaped rocking lever opens the valve to the atmosphere. The atmospheric pressure admitted to the right-hand end of the vacuum cylinder drives the piston into the cylinder. The piston rod first seals the central hole in the output pistòn and, continuing the movement into the hydraulic cylinder,

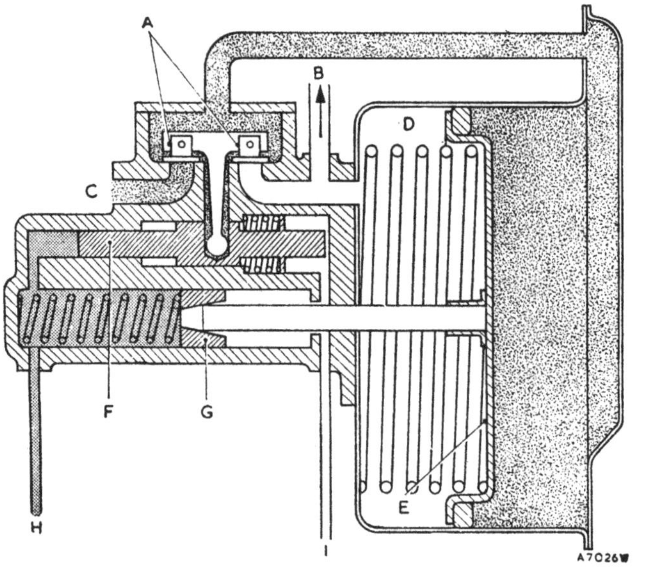

Fig. MMM.6

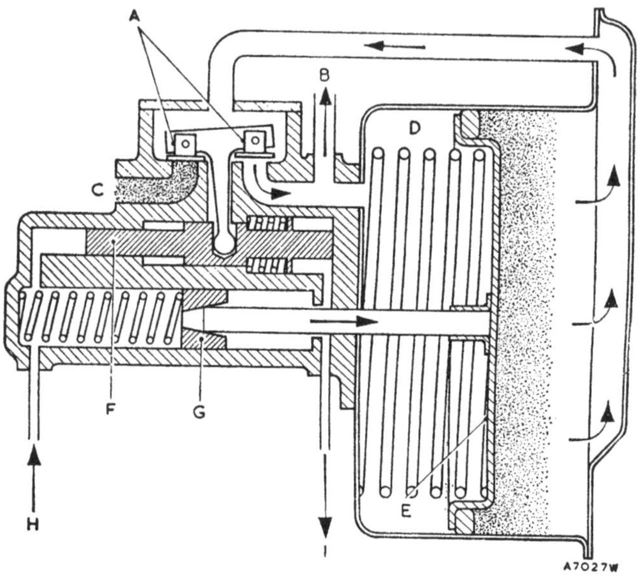

Fig. MMM.7

Vacuum servo unit operation—diagrammatic.

A. *Air valves.*
B. *To inlet manifold.*
C. *Air inlet*
D. *Vacuum cylinder.*
E. *Vacuum piston.*
F. *Control piston assembly.*
G. *Output piston.*
H. *To wheel cylinders.*
I. *From master cylinder.*

applies pressure on the fluid proceeding to the wheel cylinders and to the small end of the valve control piston (see Fig. MMM.5).

Movement of the output piston into its bore continues until the thrust on the small end of the valve control piston by the fluid at high pressure overcomes the thrust of the fluid at low pressure at the large end. The valve control piston is thus moved back, closing the air valve. At this point both valves are closed and the brakes are being held on (see Fig. MMM.6). If the brake pedal is released the fluid pressure is reduced at the large end of the control piston. The control piston moves to the right opening the vacuum valve, and consequently air is drawn out of the vacuum cylinder. As a result the vacuum piston moves towards the end cover and with it the hydraulic output piston, relieving the fluid pressure to the wheel cylinders (see Fig. MMM.7). The piston rod is withdrawn from the output piston seal allowing fluid to flow between the wheel cylinders and supply tank.

If the force on the pedal is increased after arriving at the position shown in Fig. MMM.6, the valve gear operates to give additional assistance from the vacuum piston until the thrust on each end of the control piston balances, or until the limit of available vacuum is reached. Conversely, if the brake pedal force is reduced, the valve gear operates to reduce the pressure at the wheel cylinders until a state of balance at the control piston is reached. In this way the pressure in the wheel cylinders varies in proportion to the effort at the pedal and full control of the braking effort is maintained.

A non-return valve is fitted in the pipeline between the engine inlet manifold and the servo to prevent the entry of air or petrol vapour into the servo in conditions of limited or non-existent manifold depression. The valve, which is serviced as an assembly, is screwed into the body of the servo unit and seated on a gasket.

Dismantling

Absolute cleanliness is essential when dismantling the servo unit.

Hold the unit in a vice by the mounting lugs on the body.

Remove the seven screws and nuts from the vacuum cylinder flange, supporting the cylinder cover against the pressure of the piston return spring. Carefully allow the piston return spring to extend and remove the end cover, gasket, piston and seal assembly, and the return spring. If the piston rod shows signs of scoring, the complete servo unit must be replaced.

Unscrew the three set bolts from inside the closed end of the vacuum cylinder. The body of the servo will now be released from the cylinder. Remove the three set bolts, copper washers, and the clamp plate from inside the cylinder. Pull the body away from the cylinder, easing the vacuum pipe from the grommet in the cylinder flange.

Push the air filter cover spring clip aside and remove the cover. The air filter element is now accessible. Remove the four screws retaining the valve chest cover and take off the cover with the vacuum pipe and gasket. Remove the two screws from inside the valve chest and lift out the valve retaining plate and the valves with their rocking lever. The valve plates are attached to the rocking lever by two separate clips.

Pull the gasket off the face of the body and, by tapping the face on a wooden surface, remove the plug sealing the valve control piston bore (upper bore). The control piston assembly will now be pushed out by its spring. Lift out the complete assembly. To dismantle this assembly, remove the circlip from the large diameter end to release the washer and spring. Lift off the square-section spring abutment washer and remove the two tapered seals.

To remove the components from the output cylinder (lower bore), pull out the end guide bush. Ease up the gland seal with a thin bladed screwdriver and lift out the nylon spacer. Using a pair of long-nosed circlip pliers, remove the circlip from its groove in the bore. The output piston with its stop washer will now be ejected from the bore by the piston return spring. The piston assembly will be found to have two seals: a tapered seal is located in an annular groove in the outside of the piston, and a second washer-like seal is held in place by a metal cup pressed into the end of the piston. The latter, which serves to seal the end of the piston rod when the brakes are applied, cannot be serviced on its own: a new output piston assembly must be fitted.

Assembling

With the servo unit dismantled examine the metal components for signs of corrosion, pitting, or scoring. The piston rod, pistons and bores must be free from scoring or steps. All cleaning must be done with methylated spirit or brake fluid. Never allow oil, grease, paraffin, or trichlorethylene to be used on any hydraulic parts. Lubricate hydraulic parts, pistons, seals and bores with Castrol Girling Brake Fluid Amber, before assembly. Always exercise extreme cleanliness.

Fit a new taper seal to a new output piston assembly with the taper facing the smaller end of the piston. Assemble the return spring, piston and washer to the output cylinder (lower bore). Press these components down into the bore against the tension of the spring and, using suitable circlip pliers, insert the circlip into its groove in the bore. Drop in the seal spacer (large end first), ease in the gland seal with the taper towards the output piston, and push in the guide bush with its flange level with the face of the body.

Fit new seals to the valve control piston, positioning the large tapered seal with the taper facing the spring and the small tapered seal with its taper facing away from the spring. Place the abutment washer, spring and retaining washer in position on the piston. Press the spring down and insert the circlip into its groove. Place the piston assembly into the bore aligning the hole in the piston with the hole in the side of the bore. Fit the valve plates to the valve rocking lever and place the assembly into the valve chest engaging the ball end of the lever in the valve control piston. Position the valve retainer over the valve assembly and secure with the two screws and washers. Check the operation of the valve gear by depressing and releasing the valve operating piston. The valves must move freely, and in the normal position, the valve nearest the body flange should be open, and the other valve closed. Fit a new seal into the groove in the valve control bore plug. Insert the plug into the bore until about $\frac{1}{16}$ in. (1·6 mm.) of the plug stands proud of the body face.

Fit the valve chest cover and vacuum pipe, securing the cover with the four screws. The vacuum cylinder can now be assembled to the body. Place the retaining plate on the vacuum pipe. Insert the gasket between the body and the cylinder and place the cylinder in position, easing the vacuum pipe into the grommet in the cylinder flange. Position the clamp plate inside the cylinder and tighten home the three set bolts with their copper washers to a torque wrench reading of 10 to 12 lb. ft. (1·38 to 1·66 kg. m.).

No lubrication is needed for the piston seal as the seal and cylinder are specially treated during manufacture, but the seal sponge rubber backing ring and the end cover gasket must be renewed. Insert the return spring, piston, and piston seal assembly into the cylinder, taking care not to damage the rod or the central bearing guide bush. Press the piston home, fit the end cover with a new gasket, and secure the cover and vacuum pipe retaining plate with the seven screws and nuts.

Fit the moulded cellular air filter element. This should be renewed each time replacement brake shoes are fitted. Place the cover over the element and press the spring retaining clip into position.

Removing

The servo unit is mounted under the right-hand front wing behind the road wheel. When removing it from the car proceed as follows.

From underneath the bonnet, disconnect the top end of the rubber vacuum hose from the vacuum pipe adjacent to the master cylinders.

Unscrew the servo hydraulic inlet pipe union from the three-way connector situated next to the windtone horn on the wing valance.

COMPONENTS OF THE SERVO ASSISTED BRAKING SYSTEM

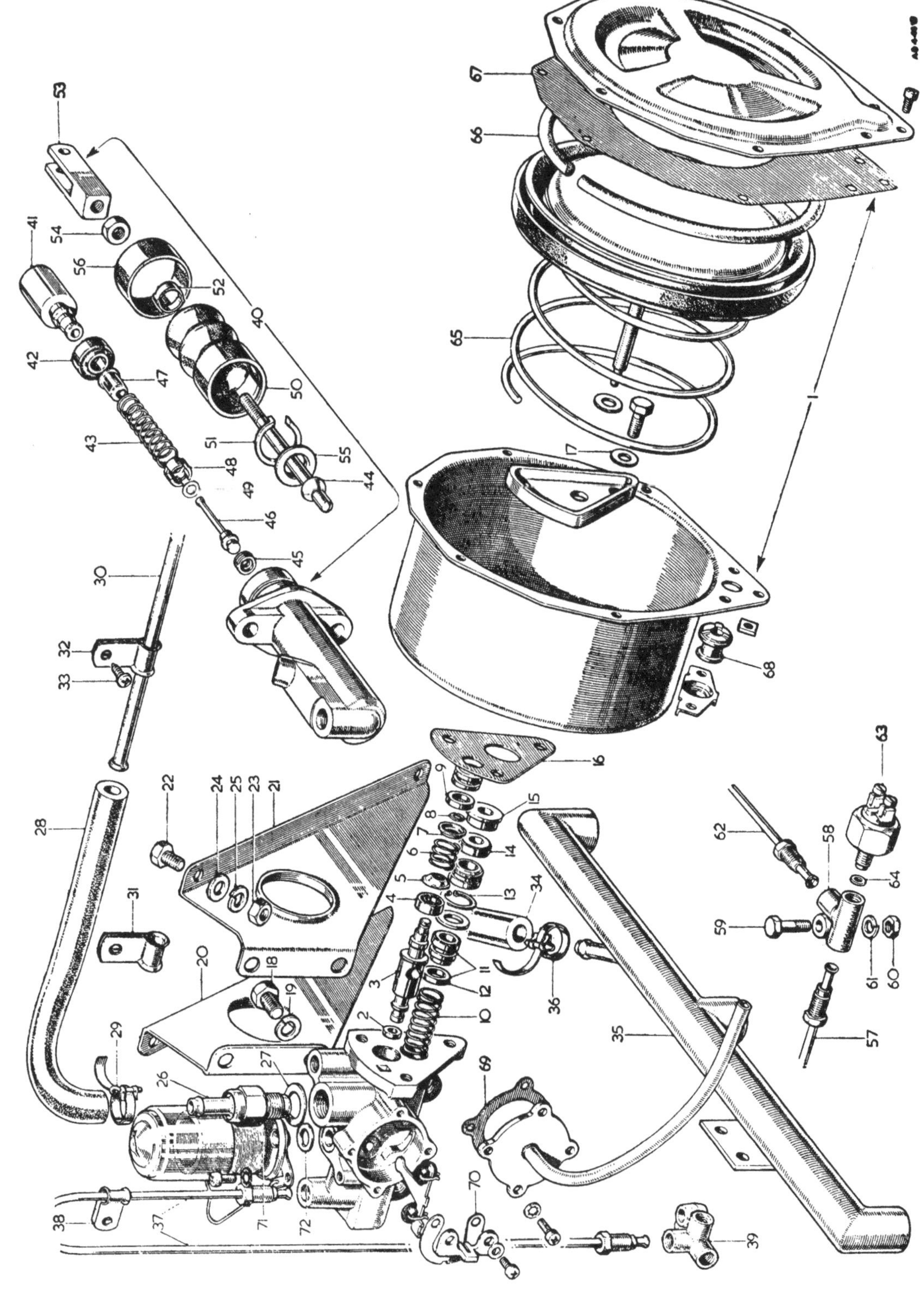

KEY TO THE COMPONENTS OF THE SERVO ASSISTED BRAKING SYSTEM

No.	Description	No.	Description	No.	Description
1.	*Vacuum servo unit.*	25.	*Washer (spring).*	51.	*Circlip.*
2.	*Valve operating piston seal (small).*	26.	*Non-return valve.*	52.	*Dust cover clip.*
3.	*Valve operating piston.*	27.	*Valve to servo unit gasket.*	53.	*Fork-end.*
4.	*Valve operating piston seal (large).*	28.	*Valve to vacuum pipe hose.*	54.	*Locknut.*
5.	*Valve operating piston spring abutment.*	29.	*Hose clip.*	55.	*Retaining washer.*
6.	*Valve operating piston return spring.*	30.	*Vacuum pipe.*	56.	*Retaining band.*
7.	*Return spring retainer.*	31.	*Pipe to pedal box top clip.*	57.	*Master cylinder to 3-way connection pipe.*
8.	*Return spring retainer circlip.*	32.	*Pipe to dash panel clip.*	58.	*3-way connection.*
9.	*Cylinder plug seal.*	33.	*Screw.*	59.	*3-way connection to wheelarch bolt.*
10.	*Hydraulic piston return spring.*	34.	*Vacuum pipe to balance pipe hose.*	60.	*Nut.*
11.	*Hydraulic piston assembly.*	35.	*Balance pipe.*	61.	*Washer (spring).*
12.	*Hydraulic piston tapper seal.*	36.	*Hose clip.*	62.	*3-way connection to servo unit pipe.*
13.	*Sealing spacer retaining circlip.*	37.	*Servo unit to 4-way connection pipe.*	63.	*Stop light switch.*
14.	*Vacuum cylinder piston gland seal.*	38.	*Pipe clip.*	64.	*Switch to 3-way connection gasket.*
15.	*Vacuum cylinder piston guide bush.*	39.	*4-way connection.*	65.	*Vacuum cylinder piston return spring*
16.	*Vacuum cylinder to body gasket.*	40.	*Brake master cylinder assembly.*	66.	*Locking plate seal.*
17.	*Washer (copper).*	41.	*Plunger.*	67.	*Vacuum cylinder end cover gasket.*
18.	*Servo unit to mounting bracket screw.*	42.	*Plunger seal.*	68.	*Vacuum pipe to cylinder sleeve (rubber).*
19.	*Washer (spring).*	43.	*Spring.*	69.	*Vacuum pipe to body gasket.*
20.	*R/H servo unit mounting bracket.*	44.	*Push-rod.*	70.	*Control valve return spring.*
21.	*L/H servo unit mounting bracket.*	45.	*Valve seal.*	71.	*Air filter element.*
22.	*Bracket to pedal box screw.*	46.	*Valve stem.*	72.	*Filter to body sealing washer (rubber).*
23.	*Nut.*	47.	*Spring retainer.*		
24.	*Washer (plain).*	48.	*Valve spacer.*		
		49.	*Washer (spring).*		
		50.	*Dust cover.*		

With the front of the car jacked up and working underneath the front wing, unscrew the hydraulic outlet pipe union from the body of the servo unit.

Seal the open ends of the pipes to prevent dirt entering and to avoid unnecessary loss of fluid.

Remove the three set bolts holding the servo unit to the mounting bracket under the front wing. The servo with the hydraulic inlet pipe and vacuum hose attached may now be pulled downwards and removed from the car.

Replacing

Replacement is a reversal of the above instructions. The braking system must then be bled as described in Section M.2.

SECTION N

ELECTRICAL SYSTEM

SERIES BN4

GENERAL DESCRIPTION

The 12-volt electrical equipment incorporates compensated voltage control for the charging circuit. The positive earth system of wiring is employed.

Battery details may be found in **"General Data".**

The generator is mounted on the right of the cylinder block and driven by an endless belt from the crankshaft pulley. A rotatable mounting enables the bolt tension to be adjusted.

The voltage control unit adjustment is sealed and should not normally require attention. The fuses are carried in external holders mounted in an accessible position on the right-hand side of the engine compartment together with spare fuses.

The starter motor is mounted on the flywheel housing on the right-hand side of the engine unit and operates on the flywheel through the usual sliding pinion device.

The headlamps employ the double-filament dipping system. Both lamps are fitted with double-filament bulbs, both dipping according to the regulations existing in the countries concerned.

NOTE:—Whenever booster charging of the battery or electric welding of the body is carried out, the battery earth lead must be disconnected to prevent damage to the electric system.

Section N.1

BATTERY MAINTENANCE

In order to keep the battery in good condition a periodical inspection should be made and the following carried out:

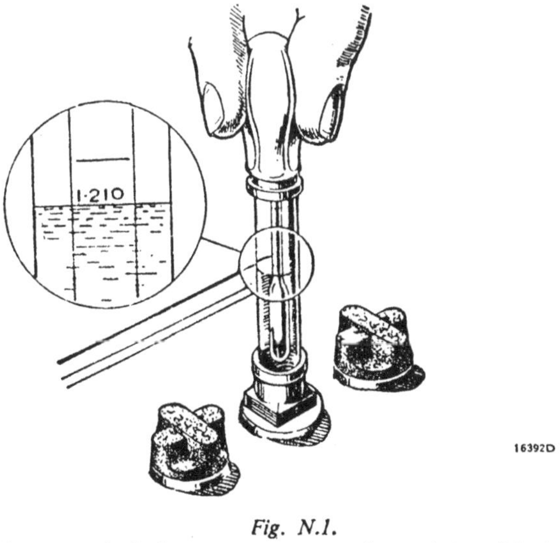

Fig. N.1.
When using the hydrometer to ascertain the condition of the battery take the reading at eye-level, ensuring that the float is free.

(*1*) Topping up

Remove the filler plug from each cell weekly and examine the level of the electrolyte. Add distilled water as required to bring the level of the electrolyte just above the top of the perforated plate covering the separators.

NOTE.—Do not use tap-water and do not use a naked light when examining the condition of the cells. Wipe away all dirt and moisture from the top of the battery.

(*2*) Testing the condition of the battery

At the recommended mileage examine the condition of the battery by taking hydrometer readings.

If the level of the electrolyte is too low for hydrometer readings to be taken top up with distilled water and recharge the battery for at least thirty minutes (to ensure thorough mixing of the electrolyte and newly added water) before taking hydrometer readings.

The specific gravity readings and their indications are as follows:

	Home Trade and climates below 27° C. (80° F.)	*Climates frequently above 27° C. (80° F.)*
Cell fully charged	1·270 to 1·290	1·210 to 1·230
Cell about half-discharged ..	1·190 to 1·210	1·130 to 1·150
Cell completely discharged ..	1·110 to 1·130	1·050 to 1·070

The figures given in the table are corrected to an electrolyte temperature of 16° C. (60° F.) and the hydrometer readings obtained must also be corrected to suit the temperature of the electrolyte.

For every 3° C. (5° F.) above 16° C. (60° F.) add ·002.

For every 3° C. (5° F.) below 16° C. (60° F.) subtract ·002.

The readings of all cells should be approximately the same. If one cell gives a reading which differs from the remainder by forty points (·040 S.G.) or more, an internal fault in that cell is indicated. The battery should then be checked by the nearest battery specialist. Should the battery be in a low state of charge, it should be recharged by taking the car for a long daytime run or by charging from external source of D.C. supply at a current rate of 5 amperes until the cells are gassing freely.

After examining the battery check the vent plugs, making sure that the air passages are clear, and screw the plugs into position. Wipe the top of the battery to remove all dirt and moisture.

STORAGE

If a battery is to be out of use for any length of time it should first be fully charged and then given a freshening charge about every fortnight.

A battery must never be allowed to remain in a discharged condition as this will cause the plates to become sulphated.

Section N.2

INITIAL FILLING AND CHARGING (Dry Uncharged Batteries)

1. Preparation of an electrolyte

When a new battery has been supplied dry it is necessary to fill the cells with electrolyte of the correct specific gravity.

The specific gravity of the filling-in solution should be the same as that required at the end of the charge (i.e. 1·260 for climates below 27°C. (80°F.) and 1·210 for climates frequently above 27°C. (80°F.)).

The electrolyte is prepared by mixing distilled water and concentrated sulphuric acid 1·840 S.G. The mixing must be carried out in a lead-lined tank or a suitable glass or earthenware vessel. Steel or iron containers must **not** be used. The acid must be added slowly to the water while the mixture is stirred with a glass rod. **Never add the water to the acid,** as the resulting chemical reaction may have dangerous consequences.

To produce electrolyte of the correct specific gravity use the following proportions of acid and distilled water:

To obtain specific gravity (corrected to 16° C. (60° F.)	*Add 1 part by volume of 1·840 S.G. acid to distilled water by volume as below*
1·260	3·2 parts
1·210	4·3 parts

Heat is produced by the mixture of acid and water and the mixture should therefore be allowed to cool before it is poured into the battery, otherwise the plates, separators, and moulded container may be damaged.

2. Filling in and soaking

The temperature of the filling-in solution, battery, and charging room should be between 16° and 38°C. (60° and 100°F.).

Carefully break the seals in the filling holes and half-fill each cell in the battery with electrolyte of the appropriate specific gravity. The quantity of electrolyte required to half-fill a two-volt cell is $\frac{1}{2}$ pint (·28 litre).

Allow to stand for at least six to twelve hours then complete the filling of the cells by the addition of more electrolyte of the same specific gravity as before until the level reaches the bottom of the filling holes, and allow the battery to stand for at least another two hours before commencing the first charge.

3. Duration and rate of initial charge

Charge at a constant current of 3·5 amps. (4·5 amps. for 11-plate battery) until the voltage and temperature-corrected specific gravity readings show no increase over five successive hourly readings. This period is dependent upon the length of time the battery has been stored since manufacture, and will be from 40 to 80 hours.

Throughout the charge the electrolyte in each cell must be maintained level with the top of the perforated plate covering the separators by the addition of electrolyte of the same specific gravity as the original filling-in solution.

If, during charge, the temperature of the acid in any cell of the battery reaches the maximum permissible temperature, i.e. 38° C. (100° F.) in climates ordinarily below 27° C. (80° F.), 49° C. (120° F.) in climates frequently above 27° C. (80° F.), the charge must be interrupted and the battery temperature allowed to fall at least 5·5° C. (10° F.) before charging is resumed.

At the end of the charge carefully check the specific gravity in each cell to ensure that, when corrected to 16°C. (60°F.), it lies between the specified limits. If any cell requires adjustment some of the electrolyte must be siphoned off and replaced either by distilled water or by electrolyte as originally used for filling in, depending on whether the specific gravity is too high or too low. Continue the charge for an hour or so to ensure adequate mixing of the electrolyte and again check the specific gravity readings. If necessary, repeat the adjustment process until the desired reading is obtained in each cell. Finally, allow the battery to cool, and siphon off any electrolyte over the tops of the separator guards.

Section N.3

INITIAL FILLING AND CHARGING (Dry-charged Batteries)

Dry-charged batteries are supplied without electrolyte but with the plates in a charged condition. When required for service it is only necessary to fill each cell with sulphuric acid of the correct specific gravity. No initial charging is required.

Preparing the electrolyte

Prepare the electrolyte as detailed in Section N.2.

Filling the battery

Remove the tapes securing the filler plugs, unscrew the filler plugs, and fill each cell with electrolyte to the top of the separators, **in one operation.** The temperature of the filling room, battery, and electrolyte should be maintained between 16 and 38°C. (60 and 100°F.). If the battery has been stored in a cool place it should be allowed to warm up to room temperature before filling.

Putting into use

Measure the temperature and specific gravity of the electrolyte in each of the cells. Allow to stand for 20 minutes and then re-check. The battery is ready for service unless the electrolyte temperature has risen by more than 5·5°C., (10°F.), or the specific gravity has fallen by more than 10 points (·010 S.G.). In this event, re-charge the battery at the normal re-charge rate until the specific gravity remains constant for three successive hourly readings and all cells are gassing freely. During the charge the electrolyte must be kept level with the top of the separator guard by the addition of distilled water.

Section N.4

DYNAMO

Description

The dynamo is a shunt-wound two-pole two-brush machine, arranged to work in conjunction with a compensated voltage control regulator unit. A fan, integral with the driving pulley, draws cooling air through the generator, inlet and outlet holes being provided in the end brackets of the unit.

The output of the dynamo is controlled by the regulator and is dependent on the state of charge of the battery and the loading of the electrical equipment in use. When the battery is in a low state of charge, the generator gives a high output, whereas if the battery is fully charged, the dynamo gives only sufficient output to keep the battery in good condition without any possibility of overcharging. In addition, an increase in output is given to balance the current taken by lamps and other accessories when in use. Further, a high boosting charge is given for a few minutes immediately after starting up, thus quickly restoring to the battery the energy taken from it by the electric starting motor.

Section N.5

DYNAMO MAINTENANCE

Lubrication

At the recommended mileage inject a few drops of oil to Ref. D (page Q.1) into the hole marked "oil" at the end of the bearing housing.

Inspection of Brushgear and Commutator

Periodically inspect the brush-gear and commutator. Access to the brushgear on earlier dynamos is gained by removing the metal band cover from around the yoke. Some dynamos are now produced without brushgear inspection windows in the yoke and it is necessary to unscrew the two through bolts and withdraw the commutator end bracket before access to the brushgear can be gained.

Check that the brushes move freely in their holders by holding back the brush springs and pulling gently on the flexible connectors. If a brush is inclined to stick, remove it from its holder and clean its sides with a petrol-moistened cloth. Be careful to replace brushes in their original positions in order to retain the "bedding". Brushes which have worn so that they will not "bed" properly on the commutator must be renewed.

The commutator should be clean, free from oil or dirt, and should have a polished appearance. If it is dirty, clean it by pressing a fine dry cloth against it while the engine is slowly turned over by hand. If the commutator is very dirty, moisten the cloth with petrol.

Belt Adjustment

Occasionally inspect the dynamo driving belt and adjust if necessary to take up any undue slackness by turning the dynamo on its mounting. Care should be taken to avoid overtightening the belt, which should have sufficient tension only to drive without slipping.

See that the generator is properly aligned, otherwise undue strain will be thrown on the bearings.

Section N.6

TESTING IN POSITION TO LOCATE FAULT IN CHARGING CIRCUIT

In the event of a fault in the charging circuit, adopt the following procedure to locate the cause of the trouble.

(1) Inspect the driving belt and adjust if necessary.

(2) Check that the dynamo and control box are connected correctly. The larger dynamo terminal must be connected to control box terminal "D", and the smaller dynamo terminal to control box terminal "F". Check the control box terminal "E" and associated earthing cable for tightness.

(3) Switch off all lights and accessories, disconnect the cables from the dynamo terminals and connect the two terminals with a short length of wire.

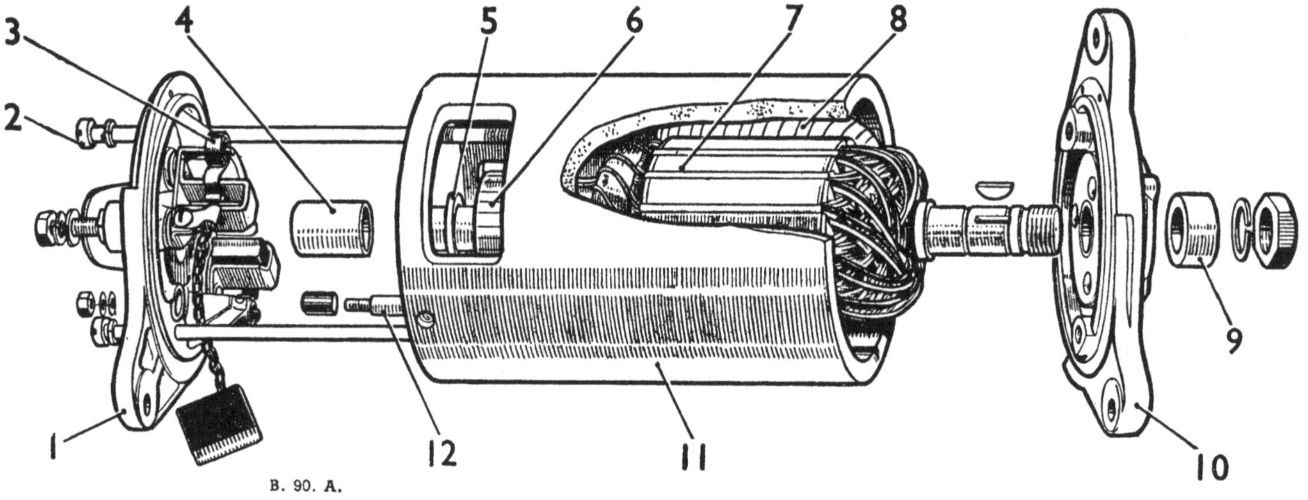

Fig. N.2. The dynamo exploded.

1. *Commutator end bracket.*
2. *Through bolt.*
3. *Brush spring.*
4. *Brush.*
5. *Thrust collar.*
6. *Commutator.*
7. *Armature.*
8. *Field coil.*
9. *Distance collar.*
10. *Drive end bracket.*
11. *Yoke.*
12. *Field terminal post.*

(4) Start the engine and set to run at normal idling speed.

(5) Clip the negative lead of a moving coil voltmeter, calibrated 0 to 20 volts, to one dynamo terminal, and the other lead to a good earthing point on the yoke.

(6) Gradually increase the engine speed, when the voltmeter reading should rise rapidly without fluctuation. Do not allow the voltmeter reading to reach 20 volts and do not race the engine in an attempt to increase the voltage. It is sufficient to run the dynamo up to a speed of 1,000 r.p.m. If there is no reading, check the brushgear as described in (7) following.

If there is a low reading of approximately ½ to 1 volt, the field winding may be at fault (see "Field Coils"). If there is a reading of 4 to 5 volts, the armature winding may be at fault (see "Armature").

NOTE.—Excessive sparking at the commutator in the above test indicates a defective armature which should be replaced.

(7) Remove the cover band (when fitted) and examine the brushes and commutator. Hold back each of the brush springs and move the brush by pulling gently on its flexible connector. If the movement is sluggish, remove the brush from its holder and ease the sides by lightly polishing on a smooth file. Always replace brushes in their original position.

If the brushes are badly worn, new brushes must be fitted and bedded to the commutator. The minimum permissible length of brush is $\frac{7}{16}$ in.

Test the brush spring tension with a spring scale. The tension of the springs when new is 36 to 44 oz. (1020 to 1247 gr.) In service it is permissible for this value to fall to 30 oz. (840 gr.) before performance may be affected. Fit new springs if the tension is low. If the commutator is blackened or dirty, clean it by holding a petrol moistened cloth against it while the engine is turned slowly by hand cranking. Re-test the dynamo as in (6); if there is still no reading on the voltmeter there is an internal fault and the complete unit, if a spare is available, should be replaced. Otherwise the unit must be dismantled for internal examination.

When reassembling a "windowless" yoke dynamo, the brushes must first be held clear of the commutator in the usual way, i.e., by partially withdrawing the brushes from their brush-boxes until each brush is trapped in position by the side pressure of its spring. The brushes can be released on to the commutator with a small screwdriver or similar tool when the end bracket is assembled to within about half-an-inch of the yoke. Before closing the gap between end bracket and yoke, see that the springs are in correct contact with the brushes.

(8) If the dynamo is in good order, remove the link from between the terminals and restore the original connections, taking care to connect the larger dynamo terminal to control box terminal "D", and the smaller terminal to control box terminal "F".

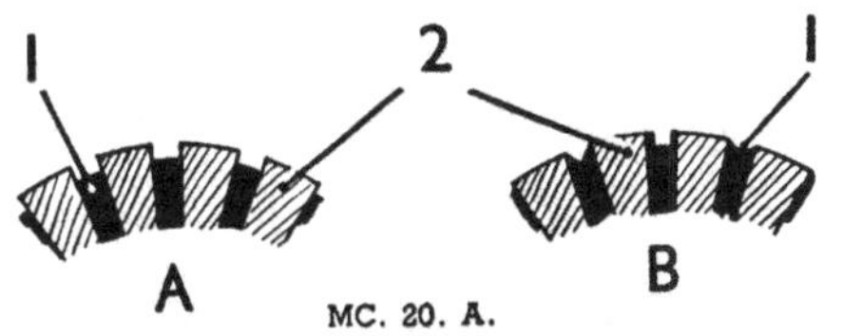

Fig. N.4. Undercutting the commutator.
'A' is the correct and 'B' is the incorrect method.
1. Insulation. 2. Segments.

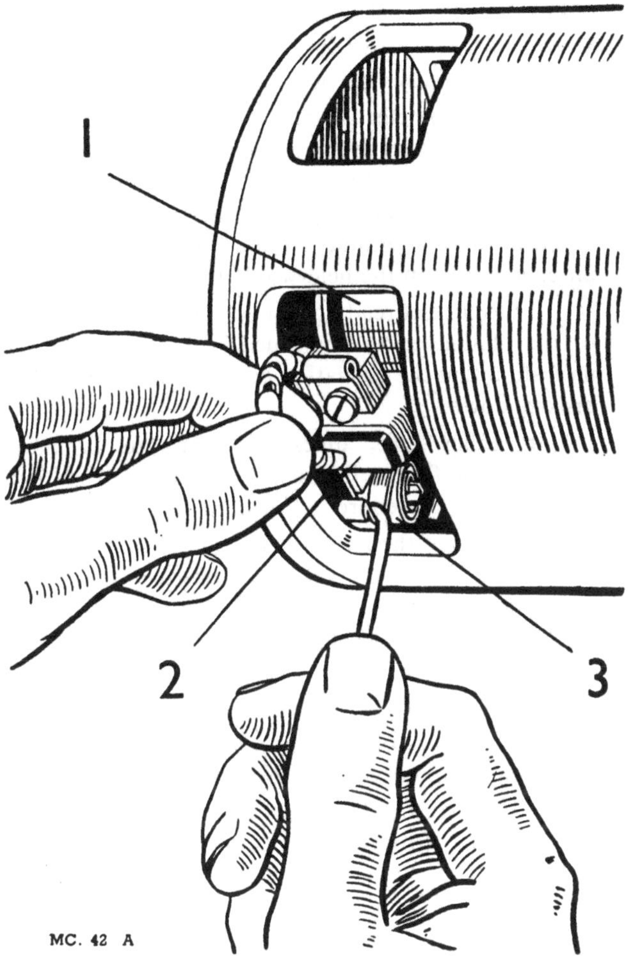

Fig. N.3. Removing a dynamo brush from its holder.
1. Commutator. 2. Brush. 3. Brush spring.

Section N.7

DYNAMO ASSEMBLY

To remove

(1) Disconnect the two leads to the dynamo.

(2) Disconnect the high tension lead and the two low tension leads to the coil.

(3) Slacken the nut securing the sliding link and the two bolts holding the dynamo to its mounting bracket.

(4) Push the dynamo downwards to slacken the fan belt so that the latter can then be removed.

(5) Remove the setpin from the upper end of the sliding link and take out the nuts and bolts from the mounting bracket.

(6) Lift the dynamo clear of the engine.

(7) Unscrew the two nuts securing the coil to its bracket on the dynamo and remove the coil.

To Dismantle

(1) Take off the driving pulley.

(2) On earlier type dynamos, remove the cover band, hold back the brush springs and remove the brushes from their holders.

(3) Unscrew and withdraw the two through bolts.

(4) The commutator end bracket can now be withdrawn from the dynamo yoke.

(5) The driving end bracket together with the armature can now be lifted out of the yoke.

(6) The driving end bracket, which on removal from the yoke has withdrawn with it the armature and armature shaft ball bearing, need not be separated from the shaft unless the bearing is suspected and requires examination, or the armature is to be replaced; in this event the armature should be removed from the end bracket by means of a hand press.

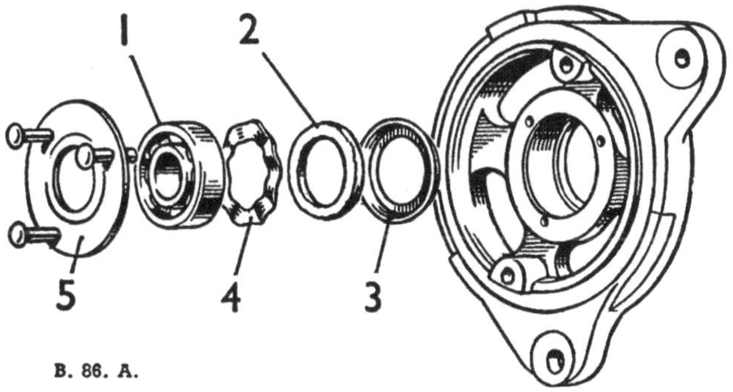

Fig. N.5. Dynamo drive end bracket.
1. Bearing. 3. Oil retaining washer.
2. Felt washer. 4. Corrugated washer.
5. Bearing retaining plate.

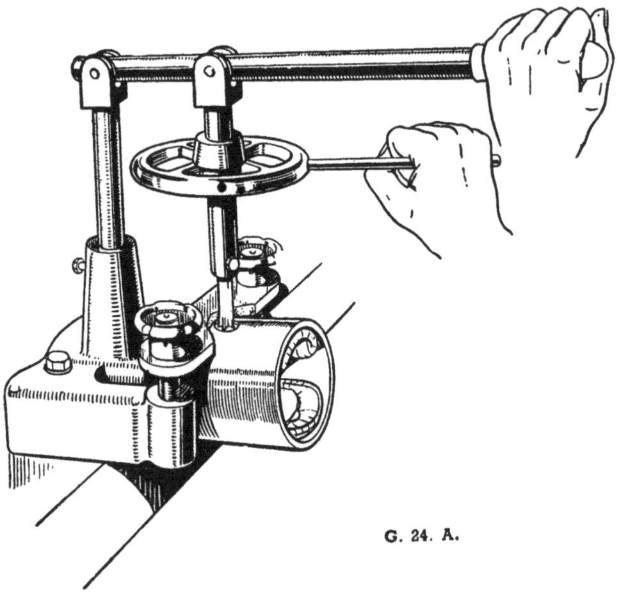

G. 24. A.

Fig. N.6. Using the wheel operated screwdriver.

Section N.8

INSPECTION AND OVERHAUL

Commutator

A commutator in good condition will be smooth and free from pits or burned spots. Clean the commutator with a petrol-moistened cloth. If this is ineffective, carefully polish with a strip of fine glass paper while rotating the armature.

To remedy a badly worn commutator mount the armature, with or without the drive end bracket, in a lathe, then rotate at high speed and take a light cut with a very sharp tool. Do not remove more metal than is necessary. Polish the commutator with very fine glass paper. Undercut the insulators between the segments to a depth of $\frac{1}{32}$ in. (·8 mm.) with a hacksaw blade ground to the thickness of the insulator.

The most common armature faults are open or short-circuited windings. Indication of an open-circuited armature winding is given by burnt commutator segments. A short-circuited armature winding is easily identified by discolouration of the overheated windings and badly burned commutator segments.

If armature testing facilities are not available, an armature can be checked by substitution.

To remove the armature shaft from the drive end bracket and bearing, support the bearing retaining plate firmly and press the shaft out of the drive end bracket. When fitting the new armature, support the inner journal of the ball bearing, using a mild steel tube of suitable diameter, whilst pressing the armature shaft firmly home. The mild steel tubes should be approximately 4 in. long and $\frac{1}{8}$ in. thick, the internal diameter being $\frac{11}{16}$ in.

Do not use the drive end bracket as a support for the bearing whilst fitting an armature.

Field Coils

Measure the resistance of the field coils, without removing them from the dynamo yoke, by means of an ohmmeter connected between the field terminal and the yoke.

The ohmmeter should read 6 ohms approximately.

If an ohmmeter is not available, connect a 12-volt d.c. supply with an ammeter in series between the field terminal and dynamo yoke. The ammeter reading should be approximately 2 amperes. Zero on the ammeter or an "Infinity" ohmmeter reading indicates an open-circuit in the field winding.

If the current reading is much more than 2 amperes, or the ohmmeter reading much below 6 ohms, it is an indication that the insulation of one of the field coils has broken down.

In either case, unless a substitute dynamo is available, the field coils must be replaced. To do this, carry out the procedure outlined below:—

(1) Drill out the rivet securing the field coil terminal assembly to the yoke, and unsolder the field coil connections.

(2) Remove the insulation piece which is provided to prevent the junction of the field coils from contacting with the yoke.

(3) Mark the yoke and pole shoes in order that they can be refitted in their original positions.

(4) Unscrew the two pole shoe retaining screws by means of the wheel-operated screwdriver.

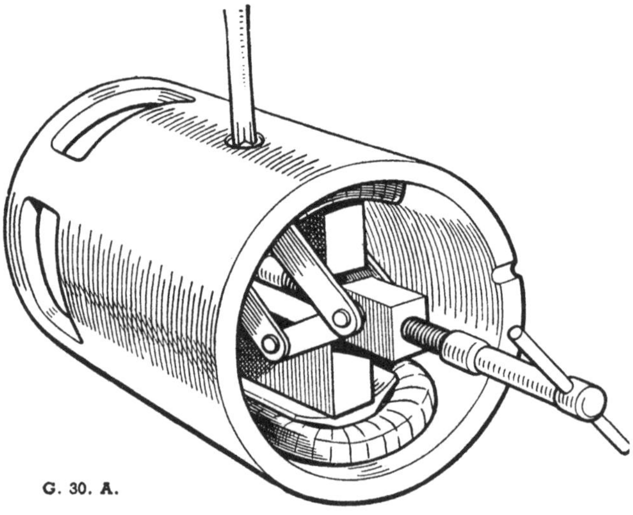

G. 30. A.

Fig. N. Showing the pole shoe expander in position.

(5) Draw the pole shoes and coils out of the yoke and lift off the coils.

(6) Fit the new field coils over the pole show and place them in position inside the yoke. Take care to ensure that the taping of the field coils is not trapped between the pole shoes and the yoke.

(7) Locate the pole shoes and field coils by lightly tightening the fixing screw.

(8) Fully tighten the screws by means of the wheel-operated screwdriver and lock them by caulking.

(9) Replace the insulation piece between the field coil connections and the yoke.

(10) Re-solder the field connections to the field coil terminal tags and re-rivet the terminal assembly to the yoke.

Bearings

Bearings which have worn to such an extent that they will allow side movemeut of the armature shaft must be replaced.

To replace the bearing bush in a commutator end bracket, proceed as follows:—

(1) Remove the old bearing bush from the end bracket. The bearing can be withdrawn with a suitable extractor or by screwing a tap into the bush for a few turns and pulling out the bush with the tap. Use an $\frac{11}{16}$ in. tap. Screw the tap squarely into the bush to avoid damaging the bracket.

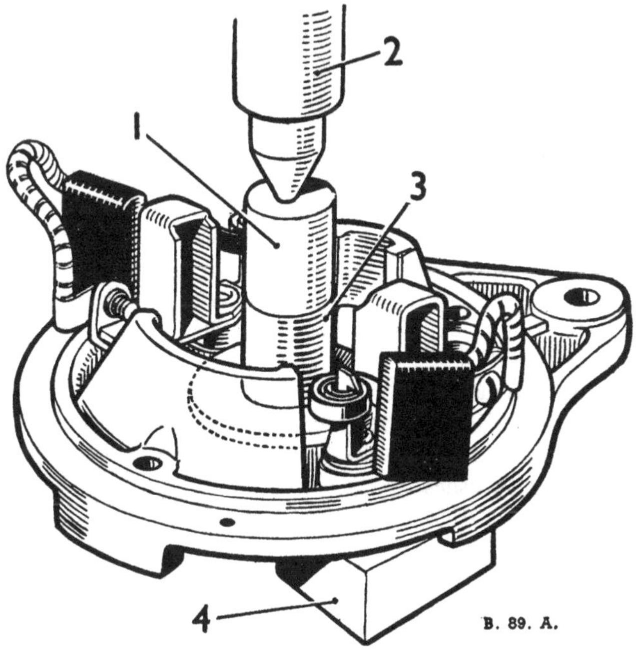

Fig. N.8. Fitting a new bush to the dynamo commutator end bracket.
1. Shouldered mandrel. 2. Hand press. 3. Bearing bush. 4. Supporting block

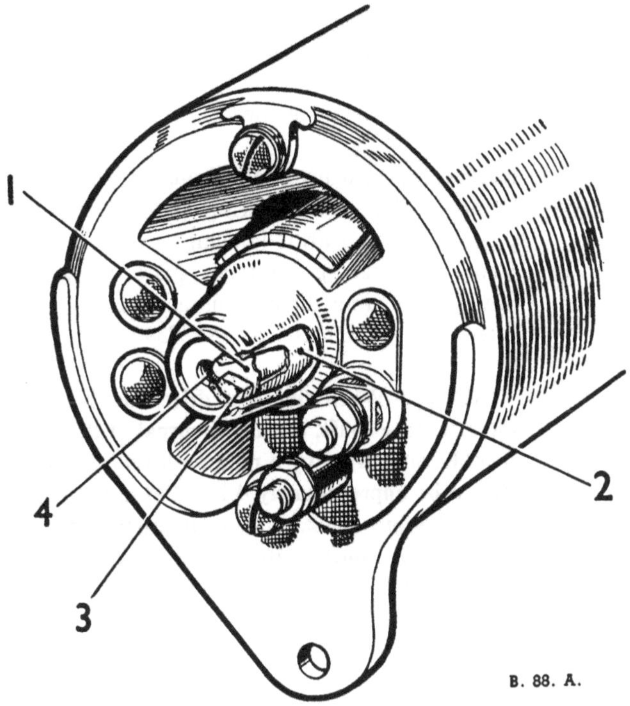

Fig. N.9. Dynamo lubrication.
1. Aluminium disc. 2. Porous bronze bush. 3. Felt ring. 4. Oil hole.

(2) Insert the felt ring and aluminium disc in the bearing housing, then press the new bearing bush into the end bracket, using a shouldered, highly polished mandrel of the same diameter as the shaft which is to fit in the bearing, until the visible end of the bearing is flush with the inner face of the bracket. Porous bronze bushes must not be opened out after fitting, or the porosity of the bush may be impaired.

NOTE.—Before fitting the new bearing bush it should be allowed to stand for 24 hours completely immersed in thin (S.A.E.20) engine oil; this will allow the pores of the bush to be filled with lubricant. In cases of extreme urgency, this period may be shortened by heating the oil to 100°C. (212°F.), for two hours then allowing to cool before removing the bearing bush.

The ball bearing at the driving end is replaced as follows:—

(1) Drill out the rivets which secure the bearing retaining plate to the end bracket and remove the plate.

(2) Press the bearing out of the end bracket and remove the corrugated washer, felt washer and oil retaining washer.

(3) Before fitting the replacement bearing see that it is clean and pack it with high melting point grease.
(4) Place the oil retaining washer, felt washer and corrugated washer in the bearing housing in the end bracket.
(5) Locate the bearing in the housing and press it home. The outer bearing journal is a light push-fit in the bearing housing.
(6) Refit the bearing retaining plate using rivets having the same dimensions as those originally fitted. Use Lucas No. 188739 for the end brackets.

NOTE.—When fitting a drive end bracket to the armature shaft, the inner journal of the bearing must be supported by a mild steel tube. This tube should be approximately 4 in. (100 mm.) long and $\frac{1}{8}$ in. (3 mm.) thick. with an internal diameter of $\frac{11}{16}$ in. (17 mm.). Do not use the drive end bracket as a support for the bearing when fitting an armature.

Section N.9

ASSEMBLING AND REPLACING

In the main the reassembly of the dynamo is a reversal of the dismantling procedure. Before refitting the dynamo, however, inject S.A.E.30 oil into the commutator end bracket as previously described. The replacement is the reverse of the procedure "To Remove" in Section N.7. Check the fan belt adjustment as described in Section C.

Section N.10

THE STARTER

To Test on Vehicle

(1) Switch on the lamps and operate the starter control. If the lights go dim, but the starter is not heard to operate, an indication is given that current is flowing through the starter windings but that the starter pinion is meshed permanently with the geared ring on the flywheel. This was probably caused by the starter being operated while the engine was still running. In this case the starter must be removed from the engine for examination.
(2) Should the lamps retain their full brilliance when the starter switch is operated, check that the switch is functioning. If the switch is in order, examine the connections at the battery, starter switch and starter, and also check the wiring between these units. Continued failure of the starter to operate indicates an internal fault, and the starter must be removed from the engine for examination.

Sluggish action of the starter is usually caused by a poor connection in the wiring which produces a high resistance in the starter circuit. Check as described above.

Damage to the starter drive is indicated if the starter is heard to operate but does not crank the engine.

Section N.11

SERVICING THE STARTER

To Remove and Replace

Release the starter cable from the terminal and unscrew the two starter securing bolts. Manœuvre the starter forwards below the oil filter and lift clear of the engine.

Examination of Commutator and Brush Gear

(1) Remove the starter cover band and examine the brushes and the commutator.
(2) Hold back each of the brush springs and move the brush by pulling gently on its flexible connector. If the movement is sluggish remove the brush from its holder and ease the sides by lightly polishing on a smooth file. Always replace brushes in their original positions. If the brushes are worn so that they no longer bear on the commutator, or if the brush flexible lead has become exposed on the running face, they must be renewed.
(3) If the commutator is blackened or dirty, clean it by holding a petrol-moistened cloth against it while the armature is rotated.
(4) Secure the body of the starter in a vice and test by connecting it with heavy-gauge cables to a 12-volt battery. One cable must be connected to the starter terminal and the other held against the starter body or end bracket. Under these light load conditions the starter should run at a very high speed.

If the operation of the starter is still unsatisfactory, it should be dismantled for detailed inspection and testing.

To Dismantle

(1) Take off the cover band at the commutator end, hold back the brush springs and take out the brushes.
(2) Extract the split pin at the driving end and remove the nut (left-hand thread), spring, washer, pinion and sleeve, restraining spring and collar and spring sleeve.
(3) Remove the terminal nuts and washers from the terminal post and screw out the two through-bolts.
(4) Remove the commutator end bracket, the attachment bracket and the armature.

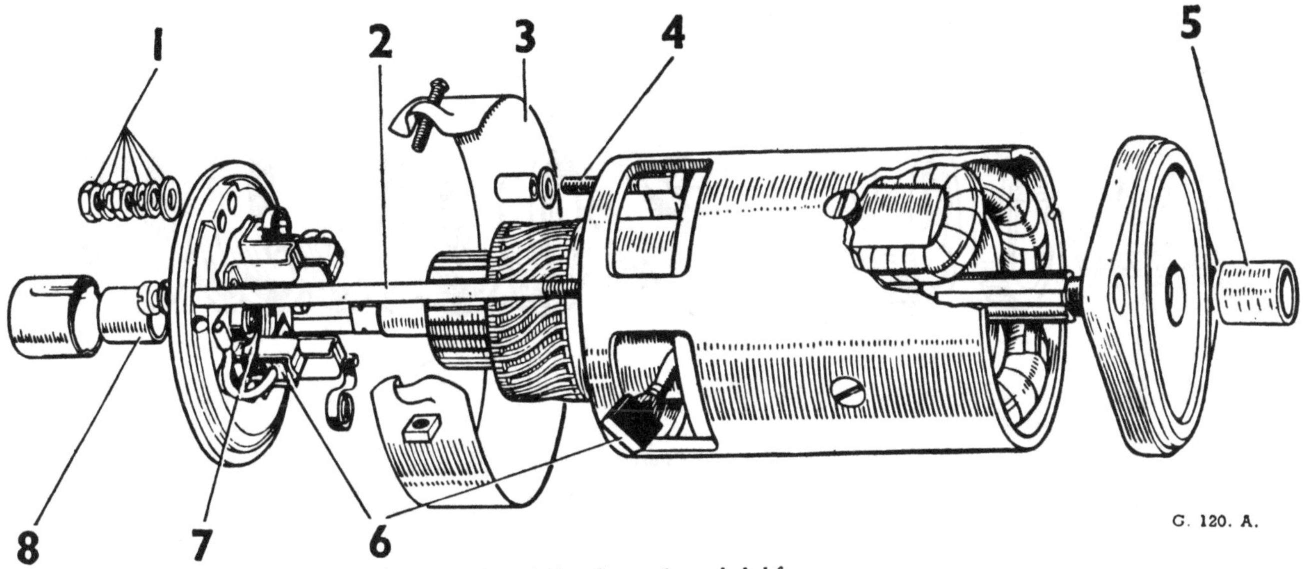

Fig. N.10. Starter in exploded form.

1. Terminal nuts and washers.
2. Through bolt.
3. Cover band.
4. Terminal post.
5. Bearing bush.
6. Brushes.
7. Brush spring.
8. Bearing bush.

Brushes

(1) Test the brush springs with a spring balance. The correct tension is 30 to 40 ozs. (850 to 1134 gm.). Fit a new spring if the tension is low.

(2) If the brushes are worn so that they no longer bear on the commutator, or if the flexible connector has become exposed on the running face, they must be renewed. Two of the brushes are connected to terminal eyelets attached to the brush boxes on the commutator end bracket. The other two brushes (Fig. N.12) are connected to tappings on the field coils.

The flexible connectors must be removed by unsoldering and the connectors of the new brushes secured in their place by soldering. The brushes are pre-formed, so that bedding of the working face to the commutator is unnecessary.

Drive

(1) If the pinion is tight on the screwed sleeve, wash away any dirt with paraffin (kerosene).

(2) If any parts are worn or damaged they must be renewed.

(3) Remove the cotter pin from the shaft nut at the end of the starter drive. Hold the squared starter shaft extension at the commutator end by means of a spanner and unscrew the square shaft nut. Lift off the main spring, washer, screwed sleeve with pinion, collar, pinion restraining spring and spring restraining sleeve.

Commutator

A commutator in good condition will be smooth and free from pits and burned spots. Clean the commutator with a cloth moistened with petrol (gasoline). If this is ineffective, carefully polish with a strip of fine glass-paper, while rotating the armature. To remedy a badly worn commutator, dismantle the starter drive as described above and remove the armature from the end bracket. Now mount the armature in a lathe, rotate it at a high speed and take a light cut with a very sharp tool. Do not remove any more metal than is absolutely necessary, and finally polish with very fine glass-paper.

The mica on the starter commutator **must not be undercut.**

Field Coils

The field coils can be tested for an open circuit by connecting a 12-volt battery, having a 12-volt bulb in one of the leads, to the tapping point of the field coils to which the brushes are connected and the field terminal post. If the lamp does not light, there is an open circuit in the wiring of the field coils.

Lighting of the lamp does not necessarily mean that the field coils are in order, as it is possible that one of them may be earthed to a pole shoe or to the yoke. This may be checked by removing the lead from the brush connector and holding it on a clean part of the starter yoke. Should the bulb now light it indicates that the field coils are earthed.

Should the above tests indicate that the fault lies in the field coils, they must be renewed When renewing field coils carry out the procedure detailed in **Section N.8.**

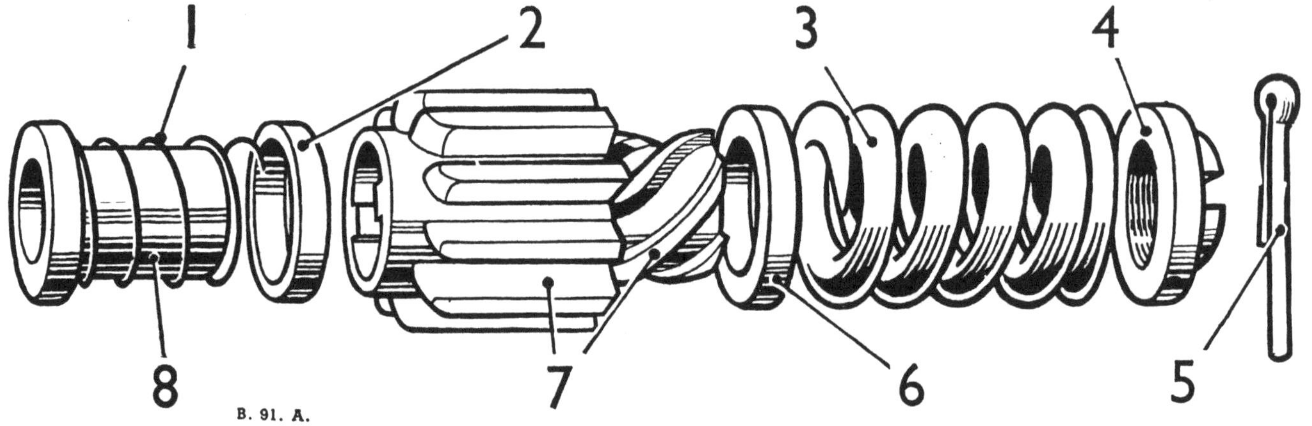

Fig. N.11. The starter pinion assembly.

1. *Restraining spring.*
2. *Collar.*
3. *Main spring.*
4. *Shaft nut.*
5. *Split pin.*
6. *Washer.*
7. *Screwed sleeve and pinion.*
8. *Restraining spring sleeve.*

Armature

Examination of the armature will in many cases reveal the cause of failure, *e.g.* conductors lifted from the commutator due to the starter being engaged while the engine is running and causing the armature to be rotated at an excessive speed. A damaged armature must in all cases be renewed—no attempt should be made to machine the armature core or to true a distorted armature shaft.

Bearings (Commutator End)

Bearings which are worn to such an extent that they will allow excessive sideplay of the armature shaft must be renewed. To renew the bearing bush, proceed as follows:—

Press the new bearing bush into the end bracket, using a shouldered mandrel of the same diameter as the shaft which is to fit in the bearing.

NOTE.—The bearing bush is of the porous phosphor-bronze type, and before fitting, new bushes should be allowed to stand completely immersed for twenty-four hours in thin engine oil in order to fill the pores of the bush with lubricant.

Reassembly

The reassembly of the starter is a reversal of the operations described in this section.

NOTE.—When reassembling the starter drive the locating nut must be re-caulked to the armature shaft.

Section N.12

CONTROL BOX

This unit contains the cut-out and voltage regulator. The regulator controls the generator output in accordance with the load on the battery and its state of charge. When the battery is discharged, the generator gives a high output so that the battery receives a quick recharge, which brings it back to its normal state in the minimum time.

On the other hand, if the battery is fully charged the generator will give a trickle charge only, which is sufficient to keep the battery in good condition without over charging, thus avoiding damage to the plates.

The regulator also causes the generator to give a controlled boosting charge immediately after starting up, which quickly restores to the battery the energy taken from it when starting. After about 30 minutes

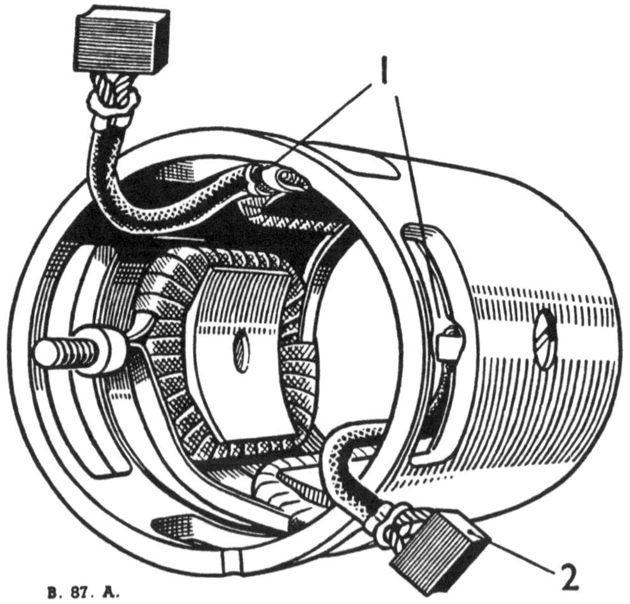

Fig. N.12 Starter yoke.

1. *Brushes.*
2. *Tapping on field coil.*

running, the output of the generator falls to a steady rate, best suited to the particular state of charge of the battery.

The cut-out is an automatic switch for connecting and disconnecting the battery with the generator. This is necessary because the battery would otherwise discharge through the generator with the engine stopped or running at low speed.

Regulator Adjustment

The regulator is carefully set during manufacture, and in general it should not be necessary to make any further adjustment. If however, the battery does not keep in a charged condition, or if the generator output does not fall when the battery is fully charged, the setting should be checked, and if necessary corrected.

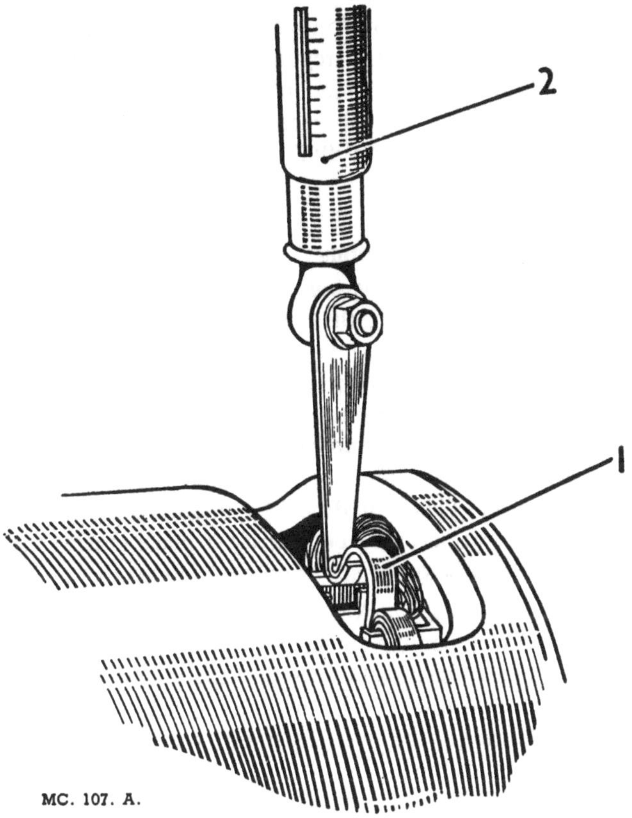

Fig. N.13. Checking brush spring tension.

1. Brush spring. 2. Spring scale.

It is important, before altering the regulator setting when the battery is in a low state of charge, to check that its condition is not due to a battery defect or to slipping of the generator belt.

Checking and Adjusting the Electrical Setting

The electrical setting can be checked without removing the cover from the control box.

(1) Withdraw the cables from the terminals marked "A" and "A.1" at the control box and join them

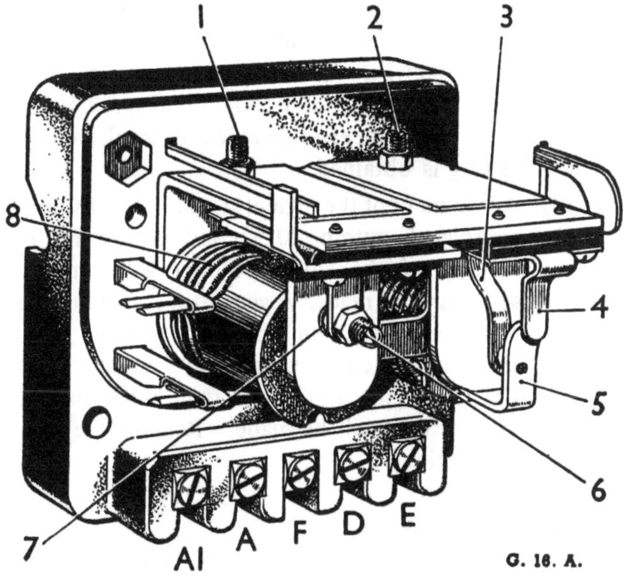

Fig. N.14. Control box.

1. Regulator adjusting screw.
2. Cut-out adjusting screw.
3. Fixed contact blade.
4. Stop arm.
5. Armature tongue and moving contact.
6. Regulator moving contact.
7. Fixed contact.
8. Regulator series windings.

together. Connect the negative lead of a moving coil (0 to 20 volts) voltmeter, to control box terminal "D" and connect the other lead to terminal "E".

(2) Slowly increase the speed of the engine until the voltmeter needle "flicks" and then steadies. This should occur at a voltmeter reading between the limits given for the appropriate temperature of the regulator. If the voltage at which the reading becomes steady is outside these limits the regulator must be adjusted.

(3) Shut off the engine and remove the control box cover. Release the locknut securing the regulator

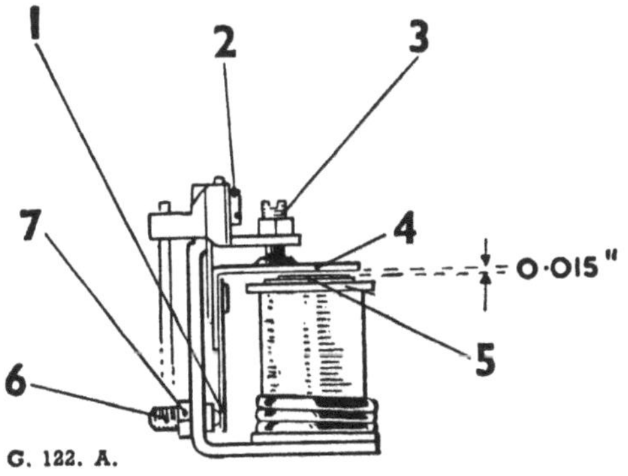

Fig. N.15. Regulator mechanical setting.

1. Armature tension spring.
2. Armature securing screws.
3. Fixed contact adjustment screw.
4. Armature.
5. Core face and shim.
6. Voltage adjusting screw.
7. Lock nut.

adjusting screw and turn the adjusting screw in a clockwise direction to raise the setting, or in an anti-clockwise direction to lower the setting. Turn the screw a fraction at a time and tighten the locknut. Repeat this procedure until the desired setting is obtained.

(4) Adjustment of the regulator open circuit should be completed within 30 seconds otherwise over-heating of the shunt winding will cause false settings to be made. A generator run at high speed on open circuit will build up a high voltage, therefore when adjusting the regulator do not run the engine up to more than half throttle or a false setting will be made. Remake the original connections.

Mechanical Setting

The mechanical settings of the regulator, shown in Fig. N.15 are accurately adjusted before leaving the factory, and provided that the armature carrying the moving contact is not removed, these settings should not be tampered with. If however, the armature has been removed, the regulator will have to be reset. To do this, proceed as follows :—

(1) Slacken the fixed contact locking nut (3), and unscrew the contact until it is well clear of the armature moving contact. Slacken the voltage adjusting screw locking nut (7) and unscrew the adjuster until it is well clear of the armature tension spring. Slacken the two armature assembly securing screws (2).

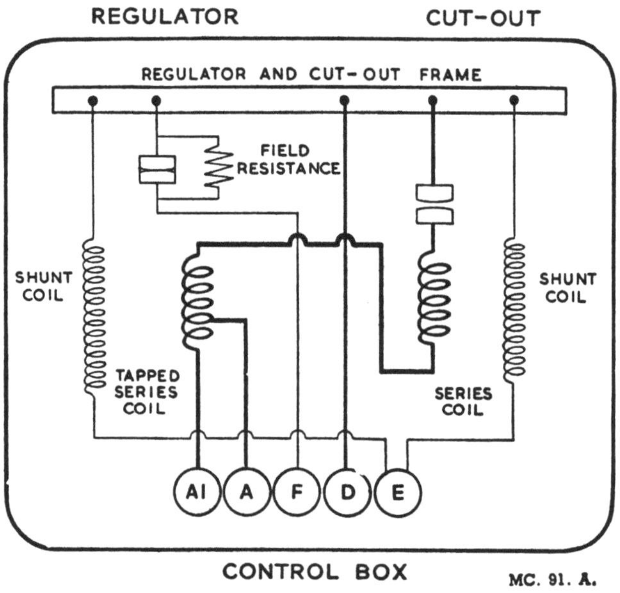

Fig. N.16. Internal connections of the control box.

(2) Insert a ·015 in. (·38 mm.) feeler gauge (which should be wide enough to completely cover the core face), between the armature and the core shim. Take care not to turn up, or damage the edge of the shim. Press the armature squarely down against the gauge and re-tighten the two armature securing screws.

(3) With the gauge still in position, screw the adjustable contact down until it just touches the armature contact. Tighten the locking nut and remove the feeler gauge. Reset the voltage adjusting screw as described under "Electrical Setting".

Cleaning Regulator Contacts

After periods of long service it may be found necessary to clean the regulator contacts. Fine carborundum stone or fine emery cloth may be used. Carefully wipe away all traces of dust or other foreign matter, using a clean fluffless cloth moistened with methylated spirits.

Cut-Out Electrical Setting

If the regulator is correctly set but the battery is still not being charged the cut-out may be out of adjustment. To check the voltage at which the cut-out operates remove the control box cover and connect the voltmeter between the terminals "D" and "E". Start the engine and slowly increase its speed until the cut-out contacts are seen to close, noting the voltage at which this occurs. This should be 12·7 to 13·3 volts.

If operation of the cut-out takes place outside these limits, it will be necessary to adjust. To do this :—

(1) Slacken the locknut of (2) Fig. N.14, securing the cut-out adjusting screw and turn the screw in a clockwise direction to raise the voltage setting, or in an anti-clockwise direction to reduce the setting. Turn the screw a fraction at a time and then tighten the locknut.

(2) Test after each adjustment by increasing the engine speed and noting the voltmeter readings at the instant of contact closure. Electrical settings of the cut-out, like the regulator, must be made as quickly as possible because of temperature rise effects. Tighten the locknut after making the adjustment.

(3) Adjustment of the drop-off voltage (8·5 to 11 volts) is effected by carefully bending the fixed contact blade. If the cut-out does not operate there may be an open circuit in the wiring of the cut-out and regulator unit in which case the unit should be removed for examination or renewal.

Cut-out Mechanical Setting

If for any reason the cut-out armature has to be removed from the frame, care must be taken to obtain

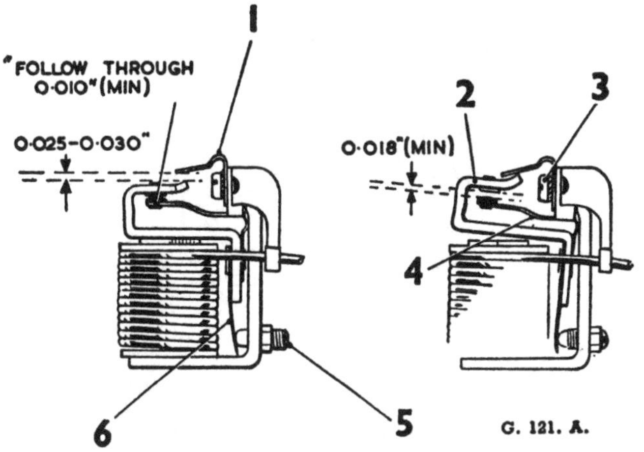

Fig. N.17. Cut-out mechanical setting.

1. *Stop arm.*
2. *Armature tongue and moving contact.*
3. *Armature securing screw.*
4. *Fixed contact blade.*
5. *Cut-out adjusting screw.*
6. *Armature tension spring.*

the correct air gap settings on reassembly. These can be obtained as follows :—

(1) Slacken the adjusting screw locking nut Fig. N.17, and unscrew the adjusting screw (5) until it is well clear of the armature tension spring (6). Slacken the two armature assembly securing screws (3).

(2) Press the armature squarely down against the copper sprayed core face and re-tighten the two armature assembly securing screws.

(3) Using a pair of round-nosed pliers, adjust the gap between the armature stop-arm and the armature tongue by bending the stop-arm. The gap must be ·025 in. to ·030 in. (·6 mm. to ·7 mm.) when the armature is pressed squarely down on the core face.

(4) Similarly, the insulated contact blade must be bent so that, when the armature is pressed squarely down against the core face, there is a minimum "follow through", or contact deflection of ·010 in. (·25 mm.) The contact gap, when the armature is in the free position must be ·018 in. (·45 mm.) minimum. Reset the cut-out adjusting screw as described under "Cut-out Electrical Setting".

Cleaning Cut-out Contacts

If the contacts appear rough or burnt, place a strip of fine glass paper between, and with them closed by hand, draw the paper through. This should be done two or three times with the rough side towards each contact. Wipe away all dust or other foreign matter, using a clean fluffless cloth moistened with methylated spirits.

Do not use emery cloth or carborundum stone for cleaning the cut-out contacts.

Section N.13

FUSE UNIT

Description

The fuse unit, which is mounted on the bulkhead under the engine cowl, is an open insulated moulding carrying two single-pole cartridge-type fuses which are held by spring clips between grub-screw-type terminal blocks. Two spare fuses are carried in recesses in the fuse unit base and are positioned by a common retaining spring. The fuse which bridges the terminal blocks (A1—A2) is to protect general auxiliary circuits, *e.g.* the horn and interior lamps, which are independent of the ignition switch. The other fuse, bridging terminal blocks (A3—A4), is to protect ignition auxiliary circuits, *e.g.* the fuel gauge, windscreen wiper motor and flashing indicators, which only operate when the ignition is switched on.

To Remove

(1) Disconnect the cables from the battery.
(2) Remove the two nuts securing the fuse unit to the bulkhead.
(3) Slacken the terminal grub screws and withdraw the cables to release the fuse unit.

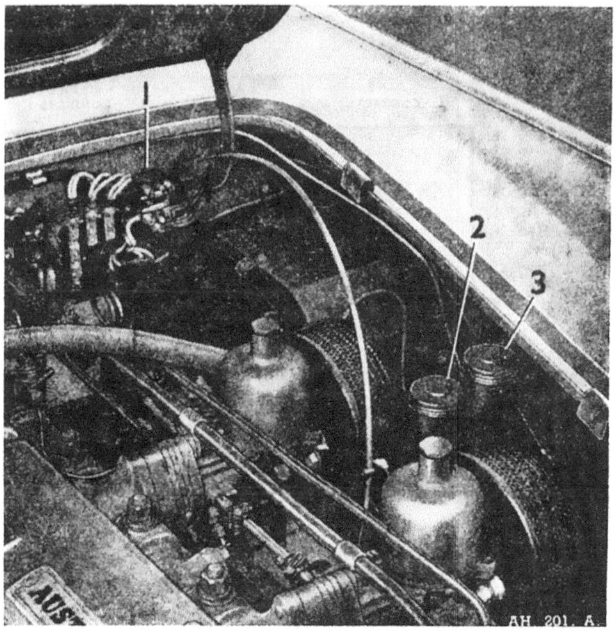

Fig. N.18. Location of fuse unit.

1. *Fuse unit.* 2. *Brake fluid reservoir.* 3. *Clutch fluid reservoir.*

To Replace

(1) Ensure all fuses are serviceable.
(2) Reconnect the cables to the appropriate terminals on the fuse unit in accordance with the colour code given in the wiring diagram.
(3) Secure the fuse unit to the bulkhead.
(4) Reconnect the battery cables and test the circuits concerned.

Section N.14

THE FLASHER UNIT

Description

The Lucas flasher unit is situated in the engine compartment and is operated by a self-cancelling steering column direction switch, a warning lamp being provided in the centre of the facia panel.

The unit is contained in a small cylindrical metal container, one end of which is rolled over on to an insulated plate carrying the mechanism and three terminals. The unit depends for its operation on the linear expansion of a length of wire which becomes heated by an electric current flowing through it. This actuating wire controls the movement of a spring-loaded armature attached to a central steel core and carrying a moving contact—the sequence of operation being as follows:—

When the direction-indicator switch is turned either to left or right, current flows through the actuating wire,

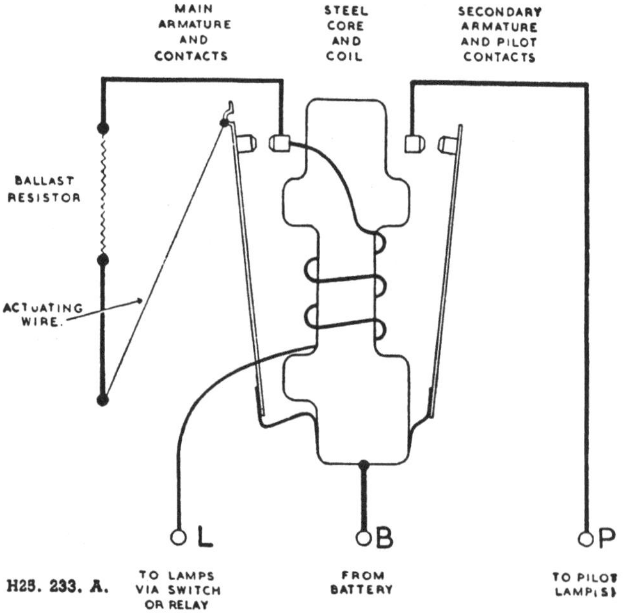

Fig. N.19. Diagram of flasher unit model FL5.

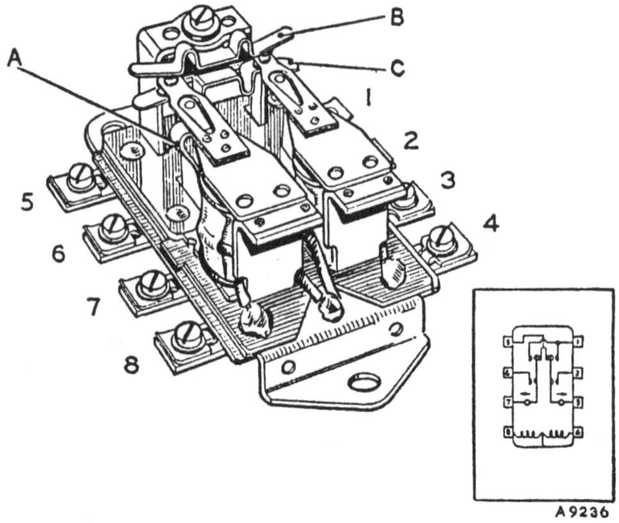

Fig. N.20. Brake switch overriding relay, model DB10, with cover removed and (inset) internal connections.

A. *Inner contacts.* B. *Outer upper contacts.* C. *Outer lower contacts.*

ballast resistor and a coil wound on the central core and thence to earth via the flasher lamp filaments. This current is limited by the ballast resistor to a value which will ensure that the flasher lamp filaments do not light at this stage. The actuating wire grows in length under the heating influence of the current and allows the armature to move inwards to its alternative position, thereby closing a pair of contacts in the supply circuit to the flasher lamps and, at the same time, short-circuiting the actuating wire. The increased electro-magnetic attraction of the armature to the core, due to the full lamp current now flowing through the coil, serves to hold the closed contacts firmly together. At the same time a secondary spring-loaded armature is attracted to the core and closes a pilot warning lamp circuit so that now both flasher lamps and warning lamp are illuminated.

Since, however, heating current no longer flows through the short-circuited wire, the latter cools down and consequently contracts in length. The main armature is therefore pulled away from the core, the contacts opened and the light signals extinguished. The consequent reduction of electro-magnetism in the core allows the secondary armature to return to its original position and so extinguish the pilot warning light. The above sequence of operations continues to be repeated until the indicator switch is returned to the off position. A diagram of the flasher unit is shown in Fig. N.19.

Functions of Warning Lamp

The warning lamp not only serves to indicate that

the flasher unit is functioning correctly but also gives warning of any bulb failure occurring in the external direction-indicator lamps—since a reduction in bulb current flowing through the coil reduces the electromagnetic effect acting on the secondary armature and so prevents closure of the pilot light contacts.

The Brake Switch Overriding Relay

When stop-light filaments are used also as direction lights, it is essential that responses to the flasher unit should override simultaneous applications of the brake switch. In the event of simultanous applications being made, the relay shown in Fig. N.20 allows the appropriate stop-light filament to flash and the other to remain steadily illuminated as long as the brake pedal is depressed.

Operation of the direction-indicator switch to right or left first energises the appropriate relay operating coil which effects movement of its associated armature in the direction shown by the arrow (Fig. N.20 inset). By this means, flasher unit terminal 'L' is connected to relay terminals '2' and '3' (or '6' and '7') and, thus, to the indicating lamps. As long as the relay coil remains energised, connection to the brake switch on the corresponding side is interrupted.

Checking Faulty Operation

In the event of trouble occurring with a flashing light direction-indicator system, the following procedure should be followed:—

(1) Check the bulbs for broken filaments.
(2) Refer to vehicle wiring diagram and check all flasher circuit connections.
(3) Switch on the ignition.
(4) Check with a voltmeter that flasher unit terminal 'B' is a battery voltage with respect to earth.
(5) Connect together flasher unit terminals 'B' (or 'X') and 'L' and operate the direction-indicator switch. If the flasher lamps now light, the flasher unit is defective and must be replaced.
(6) If the lamps do not light in test (5), check the brake switch overriding relay as follows:—
    (*a*) Temporarily link relay terminal 'I' to terminals '2' and '3'.
    The left-hand lamps should now flash.
    (*b*) Temporarily link relay terminal 'I' to terminals '6' and '7'
    The right-hand lamps should now flash.
    (*c*) If the lamps do flash in test (6), the relay is defective and requires either re-setting, see "Checking and Re-setting Air Gaps".
    (*d*) Direction-indicator switches are best checked by substitution.

Fig. N.21. The location of the brake switch overriding relay

Maintenance

Flasher units cannot be dismantled for subsequent reassembly. A defective unit must therefore be replaced, care being taken to reconnect as the original.

The cover of the brake switch overriding relays can be withdrawn for checking air-gap settings. No further dismantling is possible. In the event of defective coils or contacts occurring, relays must be replaced as complete units, care being taken to reconnect as the original.

Similarly, defective direction-indicator switches are normally replaceable only as complete units.

Replacement of Flasher Unit

When replacing a flasher unit or installing a flashing light system, it is advisable to test the circuits before connections to flasher terminals 'L', 'B' and 'P' are made. When testing, join the cables normally connected to these terminals together and operate the direction-indicator switch. In the event of a wrong connection having been made, the ignition auxiliaries fuse will blow but no damage will be done to the flasher unit.

Flasher units must be handled with care. Factory-made settings, though good for conditions of normal automobile duty, can be thrown off balance by rough handling.

Checking and Re-setting Relay Air-Gaps

Prise off the relay cover, noting the non-reversible locating slot between terminals '6' and '7'.

Each armature controls three pairs of contacts, two pairs being normally open and one pair normally closed. For setting purposes three contacts can be identified as follows:

Inner pairs, adjacent to bobbins, normally open.
Outer lower pairs, normally open.
Outer upper pairs, normally closed.

When an inner pair of contacts is just touching, a relay in correct adjustment will have an armature-to-bobbin core gap of 0·010 in. to 0·015 in. (·25 mm. to ·38 mm.). In addition, when these contacts are separated by a 0·007 in. to 0·013 in. (·2 mm. to ·3 mm.) gap, the outer lower contacts must be separated by 0·012 in. to 0·018 in. (·3 mm. to ·4 mm.) gap. If the gaps are not within these limits, the relay must be re-set.

Adjustments are made by bending the fixed contact carriers with a suitably slotted bending tool. Setting is effected in three stages, as follows :—

(1) Insert a 0·010 in. (·25 mm.) gauge between one of the armatures and its bobbin core.
(2) Press down the armature.
(3) Adjust the height of the inner contact carrier until the inner pair of contacts is just touching.
(4) Remove the gauge.
(5) Insert the gauge between the inner pair of contacts and lightly press down the armature.
(6) Adjust the outer lower contact carrier until the outer lower contacts are just touching.
(7) Remove the gauge.
(8) With the outer lower contacts just touching, adjust the upper contact carrier until a 0·015 in. (·38 mm.) gauge is a sliding fit between the outer upper contacts.
(9) Remove the gauge and refit the cover.

Section N.15

WINDSCREEN WIPERS

Maintenance

(1) An inspection should be made of the rubber wiping elements which after long service become worn and should be renewed.
(2) The rubber grommet or washer around the wheel-box spindle should be lubricated with a few drops of glycerine.
(3) Methylated spirits (de-natured alcohol) should be used to remove oil, tar spots and other stains from the windscreen. It has been found that the use of some silicone and wax-based polishes for this purpose can be detrimental to the rubber wiping elements.
(4) The gearbox and cable rack are packed with grease during manufacture and need no further lubrication.

Checking Switching Mechanism

If the wiper fails to park or parks unsatisfactorily, the limit switch in the gearbox cover should be checked. Unless the limit switch is correctly set, it is possible for the wiper motor to overrun the open circuit position and continue to draw current.

Resetting the Limit Switch

Slacken the four screws securing the gearbox cover and observe the projection near the rim of the limit switch. Position the projection in line with the groove in the gearbox cover. Turn the limit switch 25° in an anti-clockwise direction and tighten the four securing screws. If the wiping blades are required to park on the opposite side of the screen, the limit switch should be turned back 180° in a clockwise direction.

Checking Current Consumption

If the wiper fails to operate, or operates unsatisfactorily, switch on the wiper and note the current being supplied to the motor. The normal running current should be 2·3 to 3·1 amps. Use a 0 to 15 amp. moving coil ammeter connected in the wiper circuit, then proceed as follows :—

Wiper takes no Current

Examine the fuse protecting the wiper circuit. If the fuse has "blown", examine the wiring of the motor circuit and of all other circuits protected by that fuse. Replace any cables which are badly worn or chafed, if necessary fitting protective sleeving over the cables to prevent a recurrence of the fault.

If the external wiring is found to be in order, replace the fuse with one of the recommended size. Then proceed as for the wiper taking an abnormally high current.

If the fuse is intact, examine the wiring of the motor circuit for breaks and ensure that the wiper control

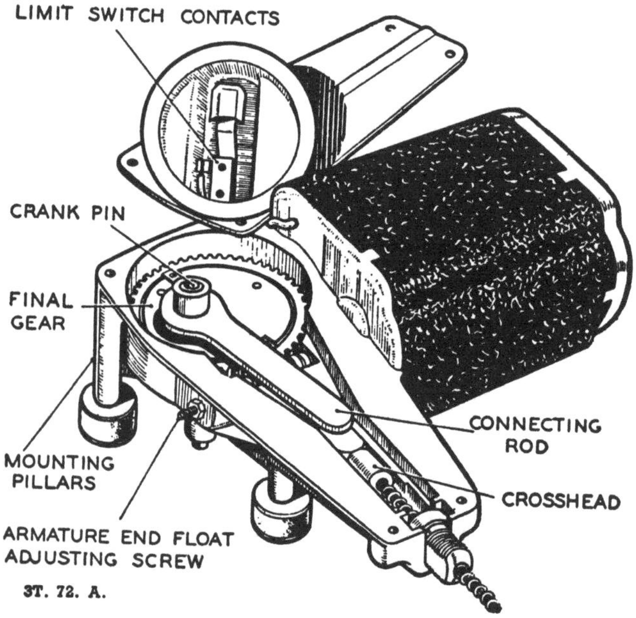

Fig. N.22. Wiper motor gearbox cover removed (Model D.R.2)

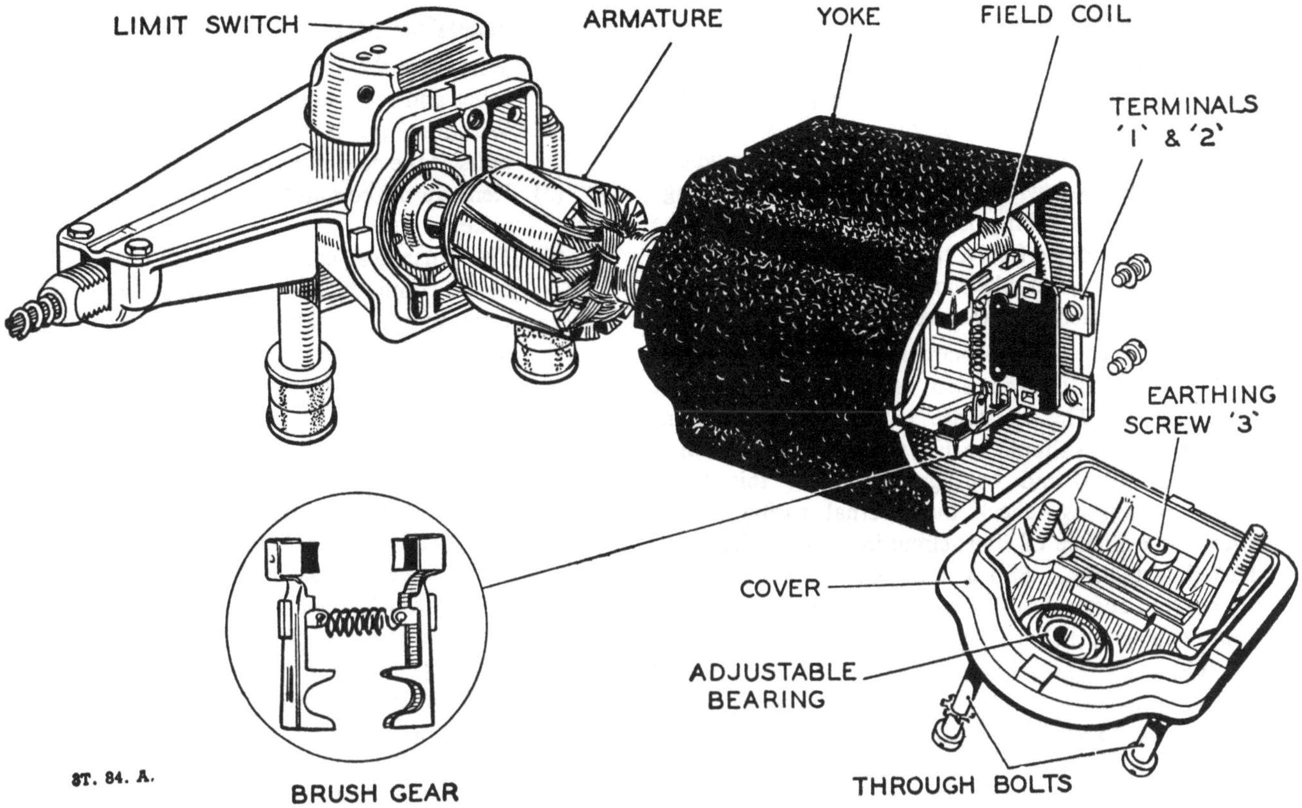

Fig. N.23. Wiper motor exploded.

switch is operating correctly.

When a current-operated thermostat is fitted, test it by connecting an ohmmeter across its terminals in place of the two cables. If a closed circuit is indicated, the thermostat is in order, and the cables must be refitted. An open circuit means that the thermostat has operated but not reset. Check the thermostat by substitution. Adjustment of the thermostat must not be attempted.

If the thermostat is in order, proceed as for the wiper taking an abnormally high current.

Wiper takes Abnormally Low Current

Check that the battery is fully charged. The performance of the motor is dependent on the condition of the battery.

Remove the commutator end bracket and examine the brush gear, ensuring that it bears firmly on the commutator. The tension spring must be renewed if the brushes do not bear firmly on the commutator. Brush levers must move freely on the pivots. If these levers are stiff they should be freed by working them backwards and forwards by hand.

Examine the commutator and, if necessary, clean with a petrol-moistened cloth. A suspected armature should be checked by substitution.

Wiper takes Abnormally High Current

If an abnormally high current is shown on the ammeter, this may be due to excessive load on the driving shaft. The stall current of the motor cold is 14 amp., and hot is 8 amp.

If there is no obvious reason for this, such as a sticking wiper blade, a check should be made at the gearbox.

Remove the gearbox cover and examine the gear assembly, checking that a blow on the gearbox end bracket has not reduced the armature end float. The armature end float adjusting screw must be set to give an armature end play of 0·008 in. (·20 mm.) to 0·012 in. (·30 mm.).

Sluggish operation with excessive current consumption may be caused through frictional losses in badly positioned or defective connecting tubes. The connecting tubes can be checked, using a cable gauge. (Details of this gauge can be obtained from any Lucas Agent.) The gauge cable is similar in appearance to the driving rack but is 0·010 in. (·26 mm.) larger in diameter and is less flexible. The gauge will not easily pass through connecting tubes having less than the minimum permissible curvature.

To check the tubing using the gauge, it is necessary to remove the inner rack. Insert the gauge into the connecting tube as far as the first wheelbox and then withdraw it. Remove the tubing connecting the wheelboxes. Insert and withdraw the gauge. If the gauge moves freely, the tubing is correctly installed. If the gauge does not move freely, the tubing must be checked for sharp bends and obstructions. Check the wheelboxes for alignment and then re-assemble.

Pieces of carbon short-circuiting adjacent segments of the commutator will also cause excessive current consumption. The resistance between adjacent commutator segments should be 0·34 to 0·41 ohms. Cleaning the commutator and brushgear removes this fault. When dismantling, check the internal wiring of the motor for evidence of short-circuiting due to chafed or charred insulation. Slip a new piece of sleeving over any charred connections, and arrange them so that they do not rub against sharp edges.

While the motor is dismantled check the value of the field resistance. If it is found to be much lower than 12·8 to 14 ohms, a short-circuit in the windings is indicated and a new field coil must be fitted. Other evidence of a short circuit will be given by charred leads from the field coil.

To Remove the Rack and Motor Unit

Release the wiping arms from the spindles, see Fig. N.24, disconnect the union on the Bundy tube at the gearbox, and remove the nuts from the motor mounting bolts. Withdraw the motor and cable rack clear of the Bundy tube.

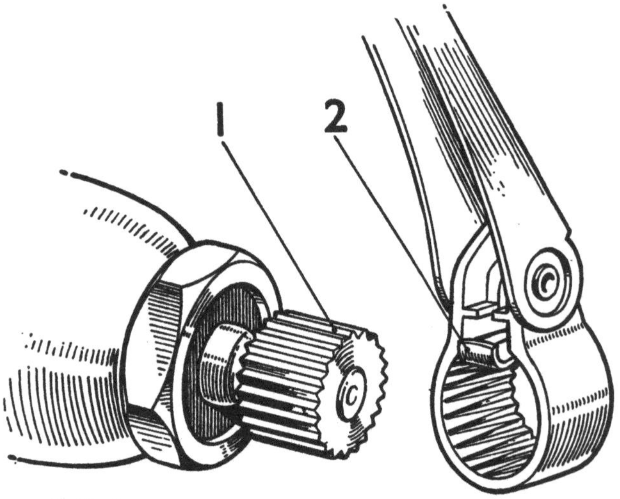

G. 17. A.

Fig. N.24. Windscreen wiper arm assembly.
1. Splined driving drum. 2. Retaining clip.

To Dismantle the Motor

(1) Withdraw the four screws securing the gearbox cover and remove the cover.
(2) Withdraw the terminal screws and the through bolts at the commutator end bracket.
(3) Remove the commutator end bracket clear of the yoke.
(4) The brush gear can be removed by lifting it clear of the commutator and withdrawing it as a unit. Care should be taken at this point to note the particular side occupied by each brush so that each may be replaced in its original setting on the commutator.
(5) Access to armature and field coils can be gained by withdrawing the yoke.
(6) If it is necessary to remove the field coil, unscrew the two screws securing the pole piece to the yoke. These screws should be marked so that they can be replaced in their original holes.
(7) Press out the pole pieces complete with field coil, marking the pole piece so that it can be replaced in its correct position inside the yoke. The pole piece can now be pressed out of the field coil.

To Dismantle the Gearbox Unit

Remove the circlip and washer from the crosshead connecting link pin and lift off the crosshead and cable rack assembly. Then remove the circlip and washer from the final gear shaft located underneath the gearbox unit. Remove any burr from the circlip groove before lifting out the final gear. The armature and worm drive can now be withdrawn from the gearbox. All gear teeth should be examined for signs of damage or wear and, if necessary, new gears fitted.

Reassembly

Reassembly is a reversal of the above procedures. When reassembling, the following components should be lubricated, using the lubricants recommended:—

(1) **Armature bearings.** These should be lubricated with S.A.E.20 engine oil—the self-aligning bearing being immersed in this for 24 hours before assembly.
(2) **Armature shaft (commutator end).** Apply S.A.E.20 engine oil sparingly.
(3) **Felt lubricator in gearbox.** Apply S.A.E.20 engine oil carefully.
(4) **Worm wheel bearings, crosshead, guide channel, connecting rod, crank pin, eccentric coupling assembly, worm and final gear shaft.** Grease liberally with grease.
(5) **Cable rack and wheelboxes.** Grease liberally with grease.

Testing

Switch on the ignition and the wiper control. The two wiper areas should be approximately symmetrical on the windscreen.

Fitting a Blade to a Wiper Arm

Pull the wiper arm away from the windscreen and insert the curved "wrist" of the arm into slotted spring fastening of the blade. Swivel the two components into engagement.

Fitting a Wiper Arm to Driving Spindle

(1) First ensure that the wiper spindles are in the correct parking position by switching on the ignition and turning the wiper control on and then off.
(2) To fit the arms, press the headpieces on to the spindles at the correct parking angle until the retaining clip is heard to snap over the end of the spindle drum.
(3) Switch off the wiper control. The arms should come to rest in the correct parking position.

Adjusting

Correct operation can be obtained by adjusting the position of the arms relative to the spindles. If necessary the position of the arms may be adjusted by removing and re-engaging them with the spined driving spindles the angular pitch of the splines being 5°.

Do not attempt to turn the arms whilst in position, but press back the retaining clip (Fig. N.24) in the headpieces and withdraw the arms from the driving spindles. Refit in the desired position. The above adjustment may affect the self-parking position. If so, it may be corrected by adjustment of the limit switch position, as described above.

If the arms and blades are required to come to rest on the opposite side, the limit switch should be turned through 180°. It should be noted that the switch cover is designed for turning through a sector only and not through 360°. This feature prevents unnecessary twisting of the external flexible connections.

Section N.16

IGNITION SWITCH

Description

The ignition switch is a rotary barrel-type Yale lock located centrally on the facia panel. Operation of the switch is carried out by inserting the ignition key into the lock and turning it in a clockwise direction. In addition to controlling the primary circuit of the ignition coil the switch also operates as a master switch for the ignition, fuel gauge and pump, and the flashing indicators.

To Test in the Vehicle

Test the ignition switch in the manner described in Section B.

To Remove

(1) Unscrew the locknut securing the switch to the instrument panel and release the electrical connections at the rear of the switch.
(2) Remove the switch from the instrument panel.

Section N.17

DIRECTION INDICATOR WARNING LAMP

To Remove and Dismantle

(1) Pull out the bulb holder with bulb from the rear of the warning lamp.
(2) Unscrew the bulb.
(3) To release the green lens unscrew the chrome retaining ring situated on the front of the facia panel.

To Reassemble and Install

The reassembly and installation is the reversal of the procedure "To Remove and Dismantle".

Section N.18

PANEL LIGHT BULBS

To Remove

(1) Pull out the bulb holder with bulb from the rear of the warning lamp.
(2) Unscrew the bulb.

To Replace

The replacement of a panel light bulb is a reversal of the removal procedure.

Section N.19

HEADLAMP MAIN BEAM WARNING LIGHT BULB

For details see Section N.18.

Section N.20

IGNITION WARNING LIGHT BULB

For details see Section N.18.

Section N.21

FUEL GAUGE

To Remove

(1) Release the "T" and "B" terminals from behind the gauge.

(2) Unscrew the centrally placed knurled securing nut.
(3) Withdraw the gauge forwards of the instrument panel.

To Replace

The replacement is the reversal of the procedure "To Remove", noting the following point :—

(1) Connect the cables in accordance with the colour code given in the wiring diagram.

Section N.22

OVERDRIVE SWITCH

To Remove

(1) Unscrew the locknut securing the switch to the instrument panel.
(2) Withdraw the switch from its locating hole in the panel.
(3) Disconnect the cables from the instrument panel.

To Replace

The replacement of the overdrive switch is a reversal of the procedure "To Remove". Reconnect the cables in accordance with the colour code in the wiring diagram.

Section N.23

PANEL LAMPS SWITCH

To remove

(1) Unscrew the two screws securing the panel lamps switch to the underside of the instrument panel.
(2) Disconnect the cables from the switch terminals.
(3) Withdraw the switch from the instrument panel.

To Replace

The replacement of the panel lamps switch is a reversal of the procedure "To Remove". Reconnect the cables in accordance with the colour code given in the wiring diagram.

Section N.24

WINDSCREEN WIPER SWITCH

To Remove

(1) Disconnect the switch knob from the switch by pushing in the retaining plunger and pulling the knob away from the instrument panel.
(2) Unscrew the nut on the outside of the switch body.
(3) Disconnect the cables from the switch terminals.
(4) Withdraw the switch from the instrument panel.

To Install

The installation of the windscreen wiper switch is a reversal of the procedure "To Remove". Reconnect the cables in accordance with the colour code given in the wiring diagram.

Section N.25

LIGHTING SWITCH

The removal and installation of this switch is tne same as that described for the windscreen wiper switch.

Section N.26

HEADLIGHT BULBS

To Remove

(1) Unscrew the screw securing the front rim and remove the rim from the headlight unit.
(2) Remove the dust-excluding cover to expose the three spring loaded adjustment screws.
(3) Press the light unit inwards against the tension of the adjusting screw springs and turn it in an anti-clockwise direction until the heads of the screws can be disengaged through the slotted holes in the light unit rim.

NOTE.—Do not disturb the screws as this will alter the lamp setting.

(4) Twist the adaptor in an anti-clockwise direction and pull it off.
(5) Remove the bulb.

To Replace

(1) Install the replacement bulb in the holder, taking care to locate it correctly.
(2) Engage the projections on the inside of the adaptor with the slots in the holder, press on and secure by twisting in a clockwise direction.
(3) Position the light unit so that the heads of the adjustment screws protrude through the slotted holes in the flange, press the unit in and turn in a clockwise direction.
(4) Replace the dust-excluding cover and refit the front rim.

Section N.27

HEADLAMP BEAM SETTING

The lamps should be set so that the main driving beams are straight ahead and parallel to one another, and parallel to the road surface. If adjustment to the setting is required, first remove the front rim and rubber as previously described. Set each lamp to the correct position in the vertical plane by means of the vertical adjustment screw at the top of the reflector unit. Turn the screw in a clockwise direction to raise the beam and in an anti-clockwise direction to lower it. Horizontal adjustment can be altered by turning the adjustment screws on each side of the light unit.

The setting of the lamps can best be carried out by placing the car in front of a blank wall at the greatest possible distance, taking care that the surface on which the car is standing is level and not sloping relative to the wall.

It will be found an advantage to cover one lamp while setting the other.

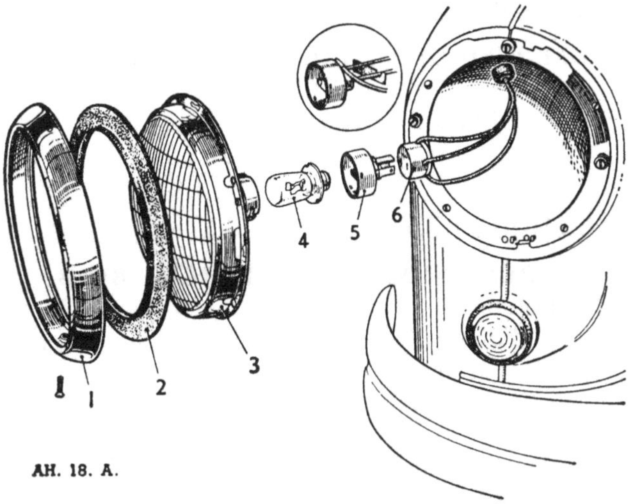

Fig. N.25. Headlamp exploded.

1. Front rim. 2. Rubber seal. 3. Glass and reflector. 4. Bulb. 5. Bulb holder (U.S.A.) 6. Three pin socket.
Inset shows bulb holder for all models except U.S.A.

Section N.28

REPLACING A LIGHT UNIT

In the event of damage to either the front lens or reflector, a replacement light unit must be fitted as follows :—

(1) Remove the light unit as already described.
(2) Withdraw the three screws from the unit rim and remove the seating rim and unit rim from the light unit.
(3) Position the replacement light unit between the unit rim and setting rim, taking care to see that the die cast projection at the edge of the light unit fits into the slot in the seating rim, and also check that the seating rim is correctly positioned. Finally secure in position by means of the three fixing screws.

NOTE.—**In order to comply with the lighting regulations in certain States of America, a sealed beam unit must be fitted in place of the Lucas light unit.**

Cars intended for the American market are fitted with special headlamp bulb adaptors and Ward and Goldstone sockets.

When replacing a Lucas Light Unit by a Sealed Beam Unit, it is only necessary when connecting up to withdraw the Lucas adaptor from the Ward and Goldstone socket. The socket can then be plugged directly to the Sealed Beam Unit.

Section N.29

HEADLAMP DIPPING SWITCH

To Remove

The switch is foot-operated and is mounted on a bracket welded to the floor assembly.

(1) Remove the two screws securing the switch to the bracket and withdraw the switch.
(2) Disconnect the three cables from the connectors.
(3) Check the operation of the dip switch. Lightly smear the mechanism with petroleum jelly. A faulty switch must be renewed as a complete unit.

To Replace

The installation of the switch is a reversal of the procedure "To Remove". Reconnect the cables in accordance with the colour code given in the wiring diagram.

Section N.30

HORN-PUSH AND DIRECTION INDICATOR SWITCH

Description

The combined horn-push and direction indicator switch is mounted on the steering wheel hub and comprises a spring-metal push covering the hub with the indicator switch lever positioned in its centre. The switch cables pass through a long tube down the steering column shaft secured by an olive in the base of the

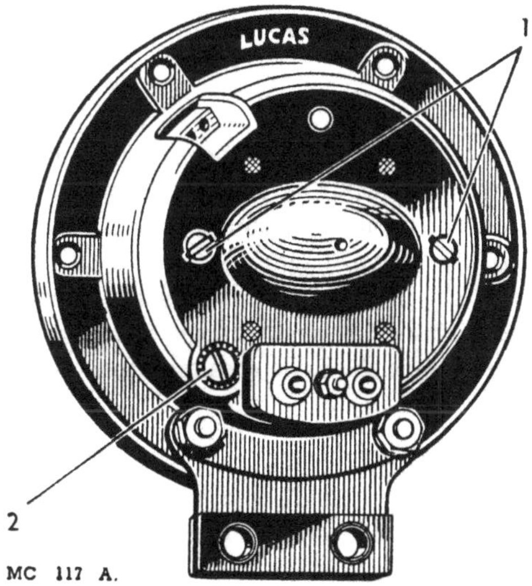

Fig. N.26. Rear view of the horn.
1. Coil securing screws. 2. Adjustment screw.

steering box. With an adjustable column the stator tube is in two parts, the shorter piece being attached to the horn quadrant. When the push is depressed, it is earthed through the steering column, thus completing the circuit.

To Remove

(1) Disconnect, at the nearest snap connections, the horn and flasher light cables protruding from the end of the stator tube.
(2) Where an adjustable type column is fitted remove the three grub screws in the steering-wheel hub and withdraw the quadrant, together with the short stator tube and cables. The long part of the stator tube remains in the steering column.

NOTE.—**The short stator tube has an indentation in it which fits in a slot in the long stator tube. The horn quadrant must be withdrawn without any twisting motion to avoid enlarging the slot in the long stator tube. Any enlargement of this slot will result in excessive movement of the horn quadrant after replacement.**

(3) When a non-adjustable type column is fitted, remove the nut and olive at the bottom end of the steering box. This will free the stator tube, which in this case is in one piece. The horn quadrant can then be withdrawn into the interior of the car, complete with stator tube and cables.
Plug the hole left in the bottom of the steering box to prevent the oil draining out.
(4) Clean and examine the switch assembly, this can only be renewed as a complete assembly.
(5) Renew any cables which are damaged.

To Replace

The replacement of the horn button and indicator switch is the reversal of the procedure "To Remove", reconnect the cables in accordance with the colour code given in the wiring diagram.

NOTE.—**In order to facilitate the threading of the horn and flasher cables through the long tube it is advisable to tape the cable ends together.**

Section N.31

COMBINED SIDE AND FLASHER LIGHTS BULB

To Remove

(1) Prise back the rubber lip and insert a screwdriver blade under the glass retaining collar.
(2) Lever the collar out from the lamp body.
(3) Remove the lamp glass and unscrew the bulb.

To Replace

The installation of a side lamp bulb is a reversal of the procedure "To Remove".

Section N.32

COMBINED STOP, TAIL AND FLASHER LIGHT BULBS

For details see section N.31.

Section N.33

REAR NUMBER PLATE LIGHT BULB

To Remove

(1) Unscrew the screw securing the lamp cover and lift off the cover.
(2) Remove the bulb or bulbs as fitted.

To Replace

The installation of a stop/tail light bulb is a reversal of the procedure "To Remove".

Section N.34

MODIFIED EUROPEAN LIGHT UNIT

Cars exported to Europe are now fitted with the new European-type headlamps. These lamp units are fitted

Fig. N.27. Combined side and flasher light in exploded form.
1. Light rim. 2. Glass. 3. Bulb.

with special bulbs and front lenses giving an asymmetrical beam to the right-hand or left-hand side according to the regulations prevailing in the country concerned.

Access to the bulb is gained in the same way as described in Section N.26. The bulb, however, is released from the reflector by withdrawing the three-pin socket and pinching the two ends of the wire retaining clip to clear the bulb flange (see Fig. N.28).

When replacing the bulb care must be taken to see that the rectangular pip on the bulb flange engages the slot in the reflector seating for the bulb.

Replace the spring clip with its coils resting in the base of the bulb flange and engaging in the two retaining lugs on the reflector seating

The appropriate replacement bulbs are listed in Section N.35. They are not interchangeable with those used in conjunction with the Continental-type headlamps previously fitted.

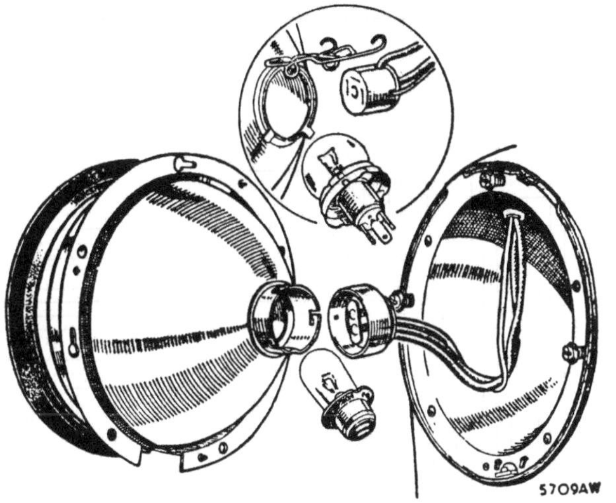

Fig. N.28

The headlamp light unit, with the European-type lamp bulb arrangement inset

Section N.35

REPLACEMENT BULBS

	Watts	B.M.C. Part No.
Headlamp R.H.D.	50/40	BFS 414
Headlamp L.H.D. (not Europe)	50/40	BFS 415
Headlamp L.H.D. (Europe, except France—early cars)	45/50	BFS 370
Headlamp R.H.D. and L.H.D. (Europe except France—European light units)	45/40	BFS 410
Heaplamp L.H.D. (France only—yellow—European light units)	45/50	BFS 411
Pilot lamps (combined flashing indicators)	6/21	BFS 380
Stop/tail lamp	6/21	BFS 380
Number plate lamp	6	BFS 989
Warning and panel lights	2 2	BFS 987

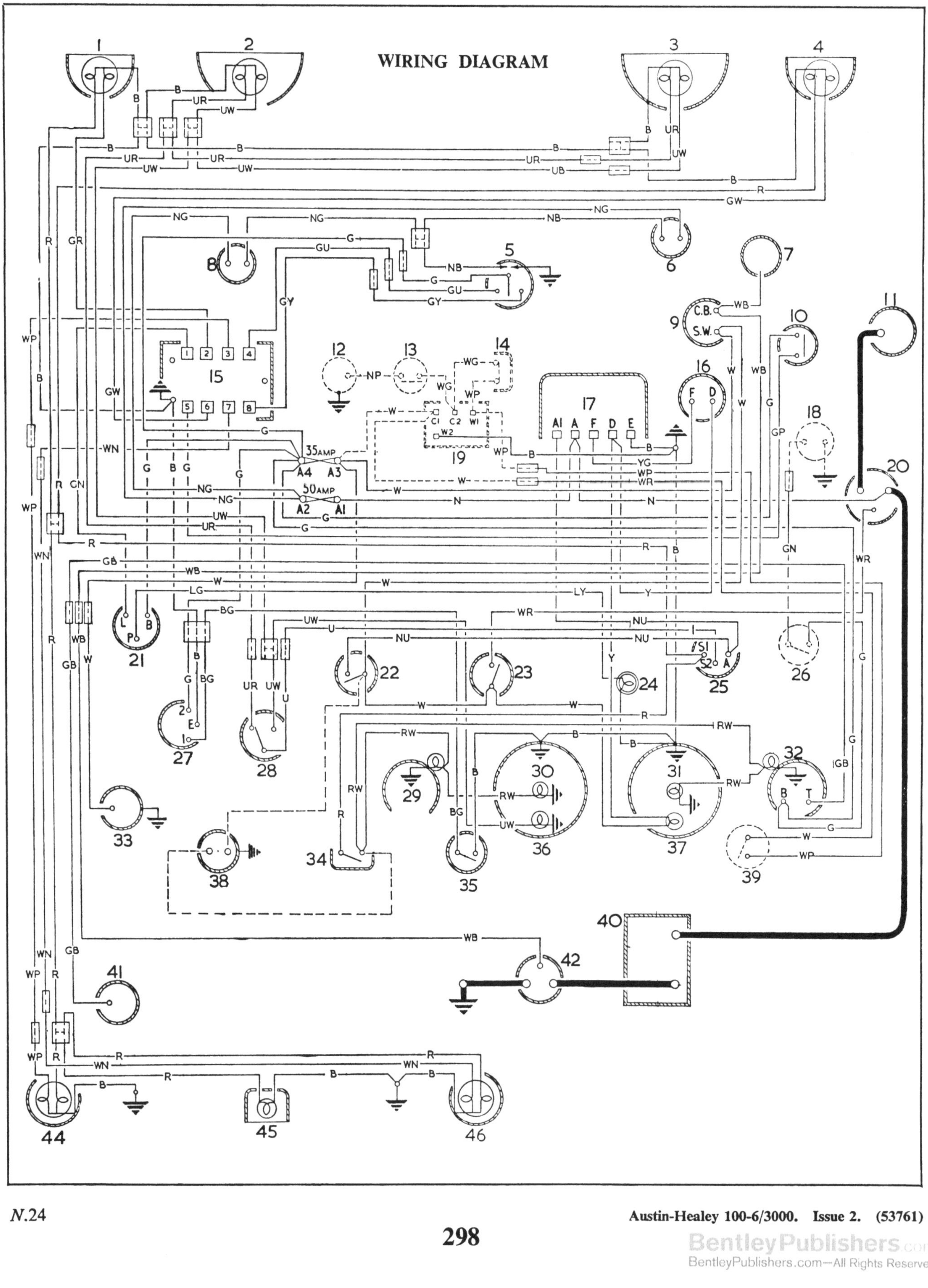
WIRING DIAGRAM
35AMP
50AMP
A1
A2
A3
A4
C.B.
S.W.
F
D
E
A
L
P
B
T

KEY TO THE WIRING DIAGRAM

No.	*Description*
1.	L.H. pilot lamp and front flasher
2.	L.H. head lamp
3.	R.H. head lamp
4.	R.H. pilot lamp and front flasher
5.	Direction switch and horn push
6.	R.H. horn
7.	Distributor
8.	L.H. horn
9.	Ignition coil
10.	Stop lamp switch
11.	Starter
12.	Overdrive solenoid (when fitted)
13.	Gearbox switch
14.	Throttle switch (when fitted)
15.	Flasher relay
16.	Dynamo
17.	Control box
18.	Heater motor (when fitted)
19.	Overdrive relay (when fitted)
20.	Starter solenoid
21.	Flasher unit
22.	Ignition switch
23.	Starter switch
24.	Flasher warning lamp
25.	Lighting switch
26.	Heater switch (when fitted)
27.	Wiper motor
28.	Dip switch
29.	Panel light
30.	Panel light
31.	Panel light
32.	Fuel gauge
33.	Petrol pump
34.	Panel lamps switch
35.	Wiper motor switch
36.	High beam warning lamp
37.	No charge warning lamp
38.	Cigar lighter (when fitted)
39.	Overdrive switch (when fitted)
40.	12 volt battery
41.	Tank unit
42.	Battery master switch
44.	L.H. stop and tail lamp and rear flasher
45.	Number plate lamp
46.	R.H. stop and tail lamp and rear flasher

COLOUR CODE			
U	BLUE	GN	GREEN with BROWN
UR	BLUE with RED	GB	GREEN with BLACK
UW	BLUE with WHITE	Y	YELLOW
W	WHITE	YG	YELLOW with GREEN
WR	WHITE with RED	N	BROWN
WG	WHITE with GREEN	NU	BROWN with BLUE
WP	WHITE with PURPLE	NG	BROWN with GREEN
WN	WHITE with BROWN	NP	BROWN with PURPLE
WB	WHITE with BLACK	NB	BROWN with BLACK
G	GREEN	R	RED
GR	GREEN with RED	RW	RED with WHITE
GY	GREEN with YELLOW	B	BLACK
GU	GREEN with BLUE	BG	BLACK with GREEN
GW	GREEN with WHITE	LG	LIGHT GREEN
GP	GREEN with PURPLE	WU	WHITE with BLUE

SECTION NN

ELECTRICAL SYSTEM

SERIES BN6

Section No. NN.1 Battery

NOTE

This information should be used in conjunction with that contained in Section N.

Section NN.1

BATTERY

The Austin-Healey (Series BN.6) model is equipped with two 6 volt batteries mounted beneath the floor; both batteries are accessible for servicing from inside the car. Partly withdraw the spare wheel into the luggage compartment and raise the hinged lid located in the spare wheel floor behind the driver's and passenger's seats.

Servicing details are the same as given in Section N.

Fig. NN.1
Pull the two "Tenax" fasteners to release the straps securing the hinged lid to gain access to the batteries

SECTION NNN

ELECTRICAL SYSTEM

Mk. I and II (SERIES BN7 and BT7)
AND Mk. II and Mk. III (SERIES BJ7 and BJ8)

NOTE

For details of the electrical system refer to Section N. The battery arrangement of the BN7 is described in Section NN.

Section NNN.1

SEALED BEAM LIGHT UNITS

Cars exported to the U.S.A. are now fitted at the Works with headlamps incorporating sealed beam light units.

To change a sealed beam light unit, remove the headlamp rim and slacken the three retaining screws securing the light unit rim. Rotate the rim anti-clockwise to disengage the slotted holes from the heads of the retaining screws. Pull the light unit forward and disconnect the three-pin socket to release it from the back-shell.

Section NNN.2

THE C42 DYNAMO

Testing

To test on vehicle when dynamo is not charging

(1) Make sure that belt slip is not the cause of the trouble. It should be possible to deflect the belt approximately ½ in. (13 mm.) at the centre of its longest run between two pulleys with moderate hand pressure.

(2) Check the Lucar connections on the commutator end bracket. The large connector carries the main dynamo output and should be connected to the control box terminal 'D'. The smaller connector carries the field current and should be connected to the control box terminal 'F'. Check the earth connections of the control box.

(3) After switching off all lights and accessories pull off the connectors from the terminal blades of the dynamo and connect the two blades with a short length of wire.

(4) Start the engine and set to run at normal idling speed.

(5) Clip the negative lead of a moving-coil-type voltmeter, calibrated 0–20 volts, to one dynamo terminal and the positive lead to a good earthing point on the yoke.

(6) Gradually increase the engine speed, when the voltmeter reading should rise rapidly and without fluctuation. Do not allow the voltmeter reading to reach 20 volts and do not race the engine in an attempt to increase the voltage. It is sufficient to run the dynamo up to a speed of 1,000 r.p.m. If the voltage does not rise rapidly and without fluctuation the unit must be dismantled for internal examination.

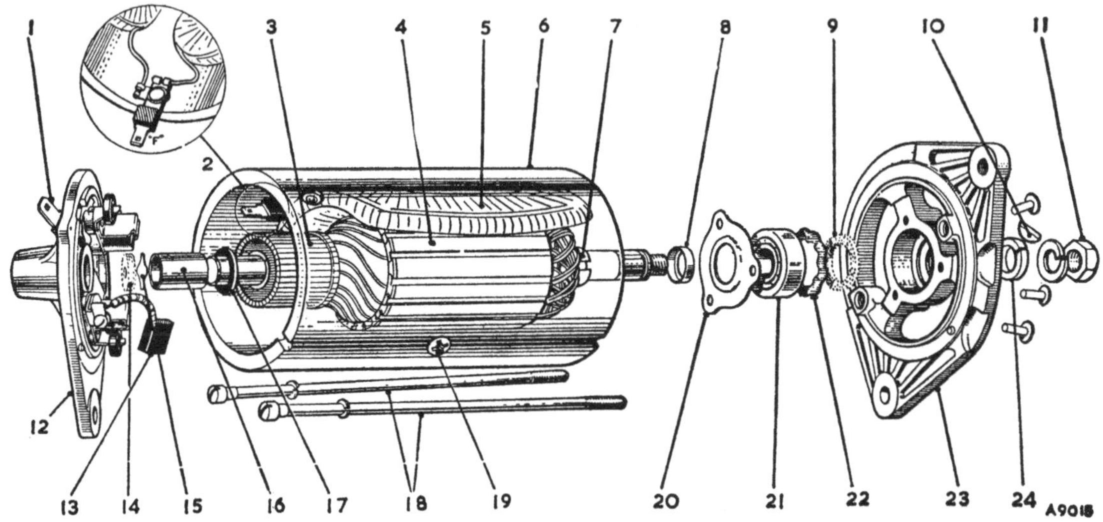

Fig. NNN.1. Components of the dynamo

1. *Output terminal 'D'.*
2. *Field terminal connections.*
3. *Commutator.*
4. *Armature.*
5. *Field coils.*
6. *Yoke.*
7. *Shaft collar.*
8. *Shaft collar retaining cup.*
9. *Felt ring.*
10. *Shaft key.*
11. *Shaft Nut.*
12. *Commutator end bracket.*
13. *Brushes.*
14. *Felt ring.*
15. *Felt ring retainer.*
16. *Porous bronze bush.*
17. *Fibre thrust washer.*
18. *Through-bolts.*
19. *Pole-shoe securing screws.*
20. *Bearing retaining plate.*
21. *Ball bearing.*
22. *Corrugated washer.*
23. *Drive end bracket.*
24. *Pulley spacer.*

Excessive sparking at the commutator during this test indicates a defective armature.

If a radio suppression capacitor is fitted between the output terminal and earth, disconnect the capacitor and retest the dynamo before dismantling. If a reading is now given on the voltmeter the capacitor is defective and must be renewed.

Servicing

Remove the dynamo from the car and dismantle it, using the methods given in Section N.8.

Brush gear

Having removed the commutator end bracket from the yoke, hold back each of the brush springs and move each brush by pulling gently on its flexible connector. If the movement is sluggish, remove the brush from its holder and ease the sides by lightly polishing on a smooth file. Always refit brushes in their original positions. If the brushes are badly worn, new brushes must be fitted and bedded to the commutator. The minimum permissible length of brush is $\frac{1}{4}$ in. (6·35 mm.).

Test the brush spring pressures, using a spring balance held radially to the commutator. The correct pressures are 33 oz. (935 gm.) maximum when exerted on a new brush and 16 oz. (454 gm.) minimum on a brush worn to $\frac{1}{4}$ in. (6·35 mm.). Fit new springs if the tension is low.

Commutator

A commutator in good condition will be smooth and free from pits or burned spots.

The commutator is of fabricated construction and can be re-skimmed during service, but care must be taken to ensure that the finished diameter is not less than 1·187 in. (30·16 mm.). The process of re-skimming consists of rough-turning, undercutting, and diamond-turning—in this order. Whether or not rough-turning is carried out depends upon the severity and unevenness of the wear which may have taken place. If a commutator cannot be completely cleaned up without going below the specified diameter the armature must be renewed. The width of undercut slots must not exceed ·060 to ·065 in. (1·52 to 1·65 mm.) with a depth of $\frac{1}{32}$ in. (·8 mm.). It is important to see that the insulating material is cleaned from the sides of each slot to a minimum depth of ·015 in. (·381 mm.).

If a non-diamond-tipped tool is used for machining, the commutator must afterwards be lightly polished with very fine glass-paper. Do not use emery-cloth.

Armature

Indication of an open-circuited armature winding will be given by burnt commutator segments. If armature-testing facilities are not available, check the armature by substitution. To separate the armature shaft from the drive end bracket press the shaft out of the bearing. When fitting a new armature support the inner race of the ball bearing, using a mild-steel tube of suitable diameter, while pressing the armature shaft firmly home.

Field coils

Measure the resistance of the field coils, without removing them from the dynamo yoke, by means of an ohmmeter connected between the field terminal and the yoke. A reading of 4·5 ohms should be obtained.

If an ohmmeter is not available connect a 12-volt D.C. supply with an ammeter in series between the field terminal and the dynamo yoke. The ammeter reading should be approximately 2·5 amps. A zero reading on theam meter or an infinity ohmmeter reading indicates an open circuit in the field winding. If the current reading is much more than 2·5 amperes, or the ohmmeter reading is much below 4·5 ohms, it is an indication that the insulation of one of the field coils has broken down.

In either case, unless a substitute dynamo is available the field coils must be renewed in accordance with the following procedure:

(1) Drill out the rivet securing the field coil terminal assembly to the yoke and remove the insulating sleeve from the terminal blade to protect it from the heat of soldering.
(2) Unsolder the terminal blade and earthing eyelet.
(3) Remove the insulation piece which is provided to prevent the junction of the field coils from contacting the yoke.
(4) Mark the yoke and pole-shoes so that the latter can be refitted in their original positions.
(5) Unscrew the two pole-shoe retaining screws by means of a wheel-operated screwdriver (see Fig. N.6).
(6) Draw the pole-shoes and coils out of the yoke and lift off the coils.
(7) Fit the new field coils over the pole-shoes and place them in position inside the yoke. Take care to ensure that the taping of the field coils is not trapped between the pole-shoes and the yoke.
(8) Locate the pole-shoes and field coils by lightly tightening the fixing screws.
(9) Fully tighten the screws by means of the wheel-operated screwdriver.
(10) Solder the original terminal blade and earthing eyelet to the appropriate coil ends.
(11) Refit the insulating sleeve and re-rivet the terminal assembly to the yoke.
(12) Refit the insulation piece behind the junction of the two coils.

Bearings

Bearings which are worn to such an extent that they will allow side-movement of the armature shaft must be renewed.

To fit a new bearing at the commutator end of the dynamo proceed as follows:

(1) Remove the old bearing brush from the end bracket. The bearing can be withdrawn with a suitable extractor or by screwing a $\frac{5}{8}$ in. tap into the bush a few turns and pulling out the bush with the tap. Screw the tap squarely into the bush to avoid damaging the bracket.

(2) Withdraw and clean the felt ring retainer and felt ring.

(3) Insert the felt ring and felt ring retainer in the bearing housing. Press the new bearing bush into the end bracket until it is flush with the inner face of the bracket, using a shouldered mandrel of the same diameter as the shaft which is to fit in the bearing. Porous bronze bushes must not be opened out after fitting or the porosity of the bush may be impaired.

NOTE.—Before fitting the new bearing bush allow it to stand completely immersed in S.A.E. 30 engine oil for 24 hours to fill the pores of the bush with oil.

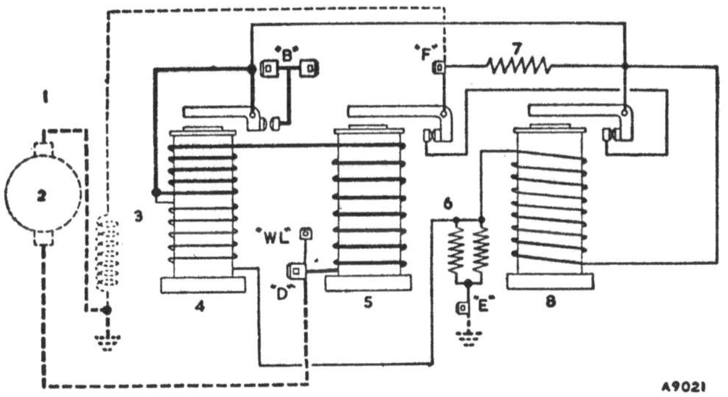

Fig. NN.2. The charging circuit

1. *Dynamo.*
2. *Armature.*
3. *Field.*
4. *Cut-out relay.*
5. *Current regulator.*
6. *Swamp resistor.*
7. *Field resistor.*
8. *Voltage regulator.*

The ball bearing at the driving end is renewed as follows:

(1) Drill out the rivets which secure the bearing retaining plate to the end bracket and remove the plate.

(2) Press the bearing out of the end bracket.

(3) Remove and clean the corrugated washer and felt ring.

(4) Before fitting the replacement bearing see that it is clean and pack it with a high-melting-point grease.

(5) Place the felt ring and corrugated washer in the bearing housing in the end bracket.

(6) Locate the bearing in the housing and press it home.

(7) Fit the bearing retaining plate. Insert the new rivets from the pulley side of the end bracket and open them over the plate by means of a punch to secure the plate rigidly in position.

Reassemble the dynamo, following the instructions given in section N.8.

Section NNN.3

CONTROL BOX MODEL RB340

General description

The RB340 control box operates on the current-voltage system of dynamo output regulation. Three units are housed in the control box: two separate vibrating armature-type single-contact regulators and a cut-out relay. One regulator is responsive to changes in current and the other to changes in voltage.

The voltage regulator and cut-out relay are temperature-compensated to allow for operating changes in the circuit resistance and for climatic variations in battery voltage. The effect of temperature fluctuation on control box settings is further minimized by the use of a swamp resistor connected in series with the shunt coils of the voltage regulator and cut-out relay.

For adjustment purposes toothed cams are carried on the front limb of each magnet frame to enable voltage and current settings to be made with a special tool (see Fig. NNN.3).

The control box settings are accurately adjusted during manufacture and the cover should not be removed unnecessarily.

Preliminary checking of charging circuits.

Before disturbing any electrical adjustments examine the items mentioned below to ensure that the fault does not lie outside the control box.

Check the battery by substitution or with a hydrometer and a heavy discharge tester. Inspect the dynamo driving belt. This should be just taut enough to drive without slipping. Check the dynamo by substitution, or by disconnecting the cables from the dynamo terminals and linking the large terminal 'D' to the small terminal 'F', connecting a voltmeter between this link and earth, and running the dynamo up to about 1,000 r.p.m., when a rising voltage should be shown.

Inspect the wiring of the charging circuit and carry out continuity tests between the dynamo, control box, and ammeter.

Check the earth connections, particularly that of the control box.

In the event of reported under-charging, ascertain that this is not due to low mileage.

Voltage regulator

Method of adjustment

Checking and adjusting should be completed as rapidly as possible to avoid errors due to heating of the operating coil. Withdraw the cables from the control box terminal blades 'B'. To enable the engine to be started it will be necessary to join the ignition and battery feeds together with a suitable lead. Connect a first-grade 0–20 moving-coil voltmeter between control box terminal 'D' and a good earthing point. A convenient method of making this connection is to withdraw the ignition warning light feed from the control box terminal 'WL' and to clip the voltmeter lead of appropriate polarity to the small terminal blade thus exposed, this terminal being electrically common with terminal 'D'. Start the engine and run the dynamo at 4,500 r.p.m. The voltmeter reading should be steady and lie between the appropriate limits shown below according to the temperature.

Ambient temperature	*Voltage setting*
10° C. (50° F.)	14·9 to 15·5
20° C. (68° F.)	14·7 to 15·3
30° C. (86° F.)	14·5 to 15·1
40° C. (104° F.)	14·3 to 14·9

An unsteady reading may be due to dirty contacts. If the reading is steady but occurs outside the appropriate limits an adjustment must be made. Proceed as follows.

Stop the engine and remove the control box cover. Restart the engine and run the dynamo at 4,500 r.p.m. Using a suitable tool (see Fig. NNN.3), turn the voltage adjustment cam until the correct setting is obtained. Turn the tool clockwise to raise the setting or anti-clockwise to lower it. Check the setting by stopping the engine and then again raising the generator speed to 4,500 r.p.m. Restore the original connections and refit the cover.

Current regulator

On-load setting

The current regulator on-load setting is equal to the maximum rated output of the dynamo, which is 30 amperes.

Method of adjustment

The dynamo must be made to develop its maximum rated output, whatever the state of charge of the battery might be at the time of setting. The voltage regulator must therefore be rendered inoperative, and to achieve this the bulldog clip shown in Fig. NNN.3 is used to keep the voltage regulator contacts together.

Remove the control box cover and, using a bulldog clip, short out the contacts of the voltage regulator.

Withdraw the cables from the control box terminal blades 'B' and connect a first-grade 0–40 moving-coil ammeter between these cables and one of the terminal blades 'B'. It is important that terminal 'B' should carry only this one connection. All other load connections, including the ignition coil feed, must be made to the battery side of the ammeter. Switch on all lights and accessories, start the engine, and run the dynamo at 4,500 r.p.m. The ammeter pointer should be steady and indicate a current of 29–31 amperes. An unsteady reading may be due to dirty contacts. If the reading is too high or too low an adjustment must be made. Proceed as follows.

Using a suitable tool (see Fig. NNN.3), turn the current adjustment cam until the correct setting is obtained. Turn the tool clockwise to raise the setting or anti-clockwise to lower it. Stop the engine, restore the original connections, and refit the control box cover.

Cut-out relay

Cut-in adjustment

Checking and adjusting should be completed as rapidly as possible to avoid errors to heating of the operating coil.

Connect a first-grade 0–20 moving-coil voltmeter between control box terminal 'D' and a good earthing point. A convenient method of making this connection is to withdraw the ignition warning light feed from control box terminal 'WL' and to clip the voltmeter lead of appropriate polarity to the small terminal blade thus exposed, this terminal being electrically common with terminal 'D'. Switch on an electrical load such as the headlamps, start the engine, and gradually increase the engine speed. Observe the voltmeter pointer. The voltage should rise steadily and then drop slightly at the instant of contact closure. The cut-in voltage is that which is indicated immediately before the pointer drops back and should be within the limits 12·7 to 13·3 volts. If the cut-in occurs outside these limits an adjustment must be made. In this event proceed as follows.

Remove the control box cover. Using a suitable tool (see Fig. NNN.3), turn the cut-out relay adjustment cam until the correct setting is obtained. Turn the tool clockwise to raise the setting or anti-clockwise to lower it. Stop the engine, restore the original connections, and refit the cover.

Drop-off adjustment

Withdraw the cables from control box terminal blades 'B'. To enable the engine to be started it will be necessary to join the ignition and battery feeds together with a suitable lead. Connect a first-grade 0–20 moving-coil voltmeter between control box terminal 'B' and a good earthing point. Start the engine and run up to approximately 4,500 r.p.m.

Slowly decelerate and observe the voltmeter pointer. Opening of the contacts, indicated by the voltmeter pointer dropping to zero, should occur between 9·5 and 11 volts. If the drop-off occurs outside these limits an adjustment must be made. Proceed as follows:

Stop the engine and remove the control box cover. Adjust the drop-off voltage by carefully bending the fixed contact bracket. Reducing the contact gap will raise the drop-off voltage and increasing the gap will lower it. Retest, and if necessary readjust until the correct drop-off setting is obtained. This should result in a contact 'follow through' of blade deflection of 010 to ·020 in. (·254 to ·508 mm.). Restore the original connections and refit the cover.

Adjustment of air gap settings

Air gap settings are accurately adjusted during manufacture and should require no further attention. If the original settings have been disturbed, it will be necessary to make adjustments in the manner described below

Armature-to-bobbin core gaps of voltage and current regulators

Disconnect the battery. Using a suitable tool (see Fig. NNN.3) turn the adjustment cam of the regulator being adjusted, to the point given minimum lift to the armature tensioning spring (by turning the tool to the fullest extent anti-clockwise). Slacken the appropriate contact locking nut and unscrew the contact. Insert a feeler gauge of ·056 in. (1·42 mm.) thickness between the armature and the regulator head, as far back as the two rivets heads on the underside of the armature. With the gauge in position press squarely down on the armature, screw in the contact until it just touches the armature contact. Tighten the locknut and withdraw the gauge. Repeat this procedure on the remaining regulator.

Note:- On earlier type regulators having a copper sh im on the regulator head, the air gap setting is ·045 in. (1 15 mm.) and care must be taken not to damage the copper shim.

Carry out the electrical setting procedure.

Contact 'follow through' and armature-to-bobbin core gap of cut-out relay

Press the armature squarely down against the copper separation on the core face. Adjust the fixed contact bracket to give a 'follow through' or blade deflection of the moving contact of ·010 to ·020 in. (·254 to ·508 mm.).

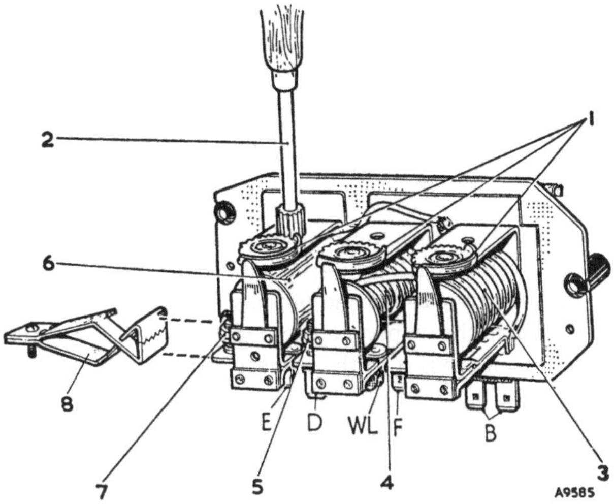

Fig. NNN.3. The control box with cover removed

1. *Adjustment cams.*
2. *Setting tool.*
3. *Cut-out relay.*
4. *Current regulator.*
5. *Current regulator contacts.*
6. *Voltage regulator.*
7. *Voltage regulator contacts.*
8. *Bulldog clip.*

Adjust the armature back stop to give a core gap of ·035 to ·045 in. (·899 to 1·147 mm.).

Check the cut-in and drop-off voltage settings.

Cleaning contacts

Regulator contacts

To clean the voltage or current regulator contacts use fine carborundum stone or silicon-carbide paper, followed by methylated spirits (denatured alcohol).

Cut-out relay contacts

To clean the cut-out relay contacts use a strip of fine glass-paper—never carborundum stone or emery-cloth.

Section NNN.4

WINDSCREEN WIPER

Apart from the renewal of perished wiper blades, the windscreen wiper requires no periodic maintenance. Efficient operation of the wiper blades is, however, dependent upon the cleanliness of the windscreen, and oil and tar spots should be removed, using methylated spirits (denatured alcohol). Do not use silicone- or wax-based-polishes for this purpose.

Inspect the rubber wiping elements, which after long service become worn and should be renewed.

Lubricate the rubber grommet or washer around the wheelbox spindle with a few drops of glycerine.

The gearbox and cable racks are packed with grease during manufacture and need no further lubrication.

Fitting a blade to a wiper arm

Pull the wiper arm away from the windscreen and insert the end of the arm into the slotted spring fastening of the blade. Push home until the raised portion of the arm engages the hole in the blade.

Fitting a wiper arm to the driving spindle

First ensure that the wiper spindles are in the correct parking position by switching on the ignition and turning the wiper control on and then off.

To fit the arms, press the headpieces onto the spindles at the correct parking angle until the retaining clip is heard to snap over the end of the spindle drum.

Switch the wiper control on and off. The arms should come to rest in the correct parking position.

Resetting the limit switch

If the wiper fails to park or parks unsatisfactorily, the limit switch in the gearbox cover should be checked. Unless the limit switch is correctly set, it is possible for the wiper motor to overrun the open-circuit position and continue to draw current.

Slacken the four screws securing the gearbox cover and observe the projection near the rim of the limit switch. Position the projection in line with the groove in the gearbox cover. Turn the limit switch 25° in an anti-clockwise direction and tighten the four securing screws. If the wiper blades are required to park on the opposite side of the windscreen, the limit switch should be turned back 180° in a clockwise direction

Testing on a vehicle

If the wiper fails to operate, or operates unsatisfactorily, connect a 0–15 moving-coil ammeter in the wiper circuit, switch on the wiper, and note the current being supplied to the motor. The normal running current should be 2·7 to 3·4 amps.

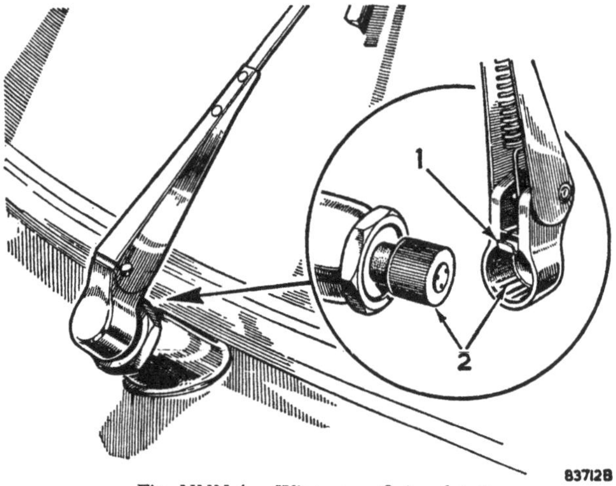

Fig. NNN.4. Wiper arm fixing details
1. *Retaining clip.* 2. *Splined drive.*

Wiper takes no current

Examine the fuse protecting the wiper circuit. If the fuse has blown examine the wiring of the motor circuit and of all other circuits protected by that fuse. Renew, if necessary, any cables which are badly worn or chafed, fitting protective sleeving over the cables to prevent a recurrence of the fault.

If the external wiring is found to be in order, replace the fuse with one of the recommended rating. Then proceed as for the wiper taking an abnormally high current.

If the fuse is intact, examine the wiring of the motor circuit for breaks and ensure that the wiper control switch is operating correctly.

When a current-operated thermostat is fitted test it by connecting an ohmmeter across its terminals in place of the two cables. If a closed circuit is indicated the thermostat is in order and the cables must be refitted. An open circuit means that the thermostat has operated but not reset. Check the thermostat by substitution. Adjustment of the thermostat must not be attempted.

If the thermostat is in order, proceed as for the wiper taking an abnormally high current.

Wiper takes abnormally low current

Check that the battery is fully charged. The performance of the motor is dependent on the condition of the battery.

Remove the commutator end bracket and examine the brush gear, ensuring that it bears firmly on the commutator. The tension spring must be renewed if the brushes do not bear firmly on the commutator. Brush levers must move freely on the pivots. If these levers are stiff they should be freed by working them backwards and forwards by hand.

Examine the commutator and, if necessary, clean with a fuel-moistened cloth. A suspected armature should be checked by substitution.

Wiper takes abnormally high current

If an abnormally high current is shown on the ammeter, this may be due to excessive load on the driving shaft. The stall current of the motor when cold is 14 amps. and when hot is 8 amps.

If there is no obvious reason for this, such as a sticking wiper blade, a check should be made at the gearbox.

Remove the gearbox cover and examine the gear assembly, checking that a blow on the gearbox end bracket has not reduced the armature end-float. The armature end-float adjusting screw must be set to give an armature end-play of ·008 to ·012 in. (·20 to ·30 mm.).

Sluggish operation with excessive current consumption may be caused through frictional losses in badly positioned or defective connecting tubes. The connecting tubes can be checked, using a cable gauge. (Details of this gauge can be obtained from any Lucas Agent.) The gauge cable is similar in appearance to the driving rack but is ·010 in. (·25 mm.) larger in diameter and is less flexible. The gauge will not easily pass through connecting tubes having less than the minimum permissible curvature.

To check the tubing remove the motor and inner rack. Insert the gauge into the connecting tube as far as the first wheelbox and then withdraw it. Remove the tubing connecting the wheelboxes. Insert and withdraw the gauge. If the gauge moves freely the tubing is correctly installed. If the gauge does not move freely the tubing must be checked for sharp bends and obstructions. Check the wheelboxes for alignment and then re-assemble.

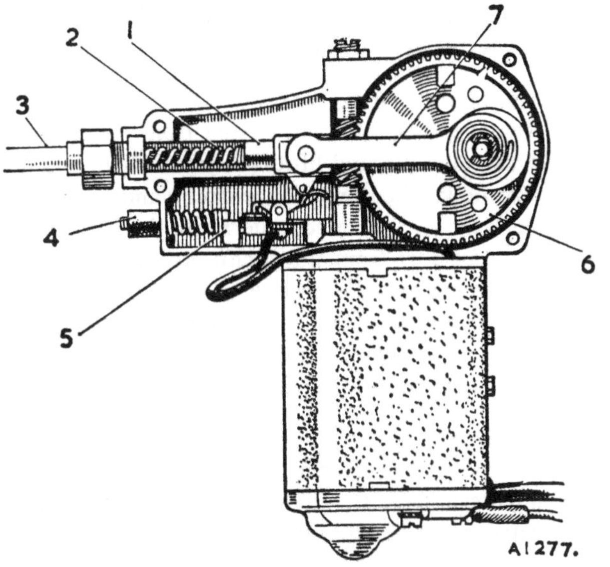

Fig. NN.5

1. *Cross-head.*
2. *Cable rack.*
3. *Outer casing.*
4. *Adjusting nut*
5. *Park switch.*
6. *Final gear.*
7. *Connecting rod.*

Removing the motor, gearbox, and wheelboxes

The motor and gearbox is located beneath the passenger's side of the fascia panel and is mounted on a bracket secured to the bulkhead panel by three screws.

The cable rack connected to the cross-head in the gearbox passes through outer casings which connect the gearbox to the first wheelbox and the first wheelbox to the second wheelbox.

Disconnect the wiper arms, the electrical connections from the motor, and the outer cable from the gearbox housing. Remove the three screws securing the bracket to the bulkhead panel and withdraw the motor, bracket, and cable rack from beneath the fascia panel.

Slacken the cover screws in each wheelbox and remove the cable rack outer casings.

Remove the nut, front bush, and washer from the front of each wheelbox and remove the wheelbox together with the rear bush and spindle tube from beneath the fascia panel.

Replacement is a reversal of the removal sequence, but care must be taken to ensure that the wheelboxes are correctly lined up and that the cable rack engages the gear and spindle assemblies.

Dismantling the motor

Withdraw the four screws securing the gearbox cover and remove the cover.

Withdraw the connectors and through-bolts at the commutator end bracket.

Remove the commutator end bracket clear of the yoke.

The brush gear can be removed by lifting it clear of the commutator and withdrawing it as a unit. Care should be taken at this point to note the particular side occupied by each brush so that each may be replaced in its original setting on the commutator.

Access to the armature and field coils can be gained by withdrawing the yoke.

If it is necessary to remove the field coil, unscrew the two screws securing the pole-piece to the yoke. These screws should be marked so that they can be returned to their original holes.

Press out the pole-piece complete with field coil, marking the pole-piece so that it can be replaced in its correct position inside the yoke. The pole-piece can now be pressed out of the field coil.

Pieces of carbon short-circuiting adjacent segments of the commutator will also cause excessive current consumption. The resistance between adjacent commutator segments should be ·34 to ·41 ohm. Cleaning the commutator and brush gear removes this fault. When dismantling, check the internal wiring of the motor for evidence of short-circuiting due to chafed or charred insulation. Slip a new piece of sleeving over any charred connections, and arrange them so that they do not rub against sharp edges.

While the motor is dismantled check the value of the field resistance. If it is found to be lower than 12·8 to 14 ohms a short circuit in the windings is indicated and a new field coil must be fitted. Other evidence of a short circuit will be given by charred leads from the field coil.

Dismantling the gearbox unit

Remove the circlip and washer from the cross-head connecting link pin and lift off the cross-head and cable rack assembly. Then remove the circlip and washer from the final gear shaft located underneath the gearbox unit. Remove any burr from the circlip groove before lifting out the final gear. The armature and worm drive can now be withdrawn from the gearbox. All gear teeth should be examined for signs of damage or wear and, if necessary, new gears fitted.

Reassembling

Reassembly is a reversal of the above procedures. When reassembling, the following components should be lubricated, using the lubricants recommended.

Armature bearings

These should be lubricated with S.A.E. 20 engine oil, the self-aligning bearing being immersed in this for 24 hours before assembly.

Armature shaft (commutator end)

Apply S.A.E. 20 engine oil.

Felt lubricator in gearbox

Apply S.A.E. 20 engine oil.

Worm wheel bearings, cross-head, guide channel, connecting rod, crankpin, eccentric coupling assembly, worm, and final gear shaft

Grease liberally.

Switch on the ignition and the wiper control. The two wiper areas should be approximately symmetrical on the windshield.

Section NNN.5

SIDE, STOP, TAIL AND FLASHER LAMPS

To remove

Unscrew the two screws, lift off the lens, and withdraw the bulb.

To replace

Replacement is the reverse of the above procedure.

The pilot and direction indicator lamps have single filament bulbs and may be fitted either way round.

The tail and stop lamp bulbs have twin filaments, and an offset bayonet fixing to ensure correct replacement.

Section NNN.6

HEADLAMPS (Later type)

To remove

Unscrew the retaining screw at the bottom of the lamp rim and withdraw the rim.

Unscrew the three Phillips screws securing the light unit retaining plate. Remove the plate, lift the light unit forward, and pull off the three-pin plug from the back of the light unit.

To replace

Reverse the removal procedure, but ensure that the lugs moulded on the back of the lens engage in the slots of the back shell.

Section NNN.8

FUSES

An additional 10 amp fuse protects the number plate lamp circuit.

The fuse is located in a nylon tube situated in the boot wiring loom to the right of the boot floor catch.

To renew the fuse, twist and release the end of the tube and withdraw the fuse.

Section NNN.7

REPLACEMENT BULBS

	Watts	BMC Part No.
Pilot lamps		
Except Germany and Sweden	6	BFS 207
Germany and Sweden	5	—
Stop and tail lamps		
Except Germany and Sweden	21/6	BFS 380
Germany and Sweden	18/4	—
Direction indicator lamps		
Except Germany and Sweden	21	BFS 382
Germany and Sweden	18	—
Number plate illumination lamp		
Except Germany and Sweden	6	BFS 207
Germany and Sweden	4	—

Wiring Diagram
(Early Cars)

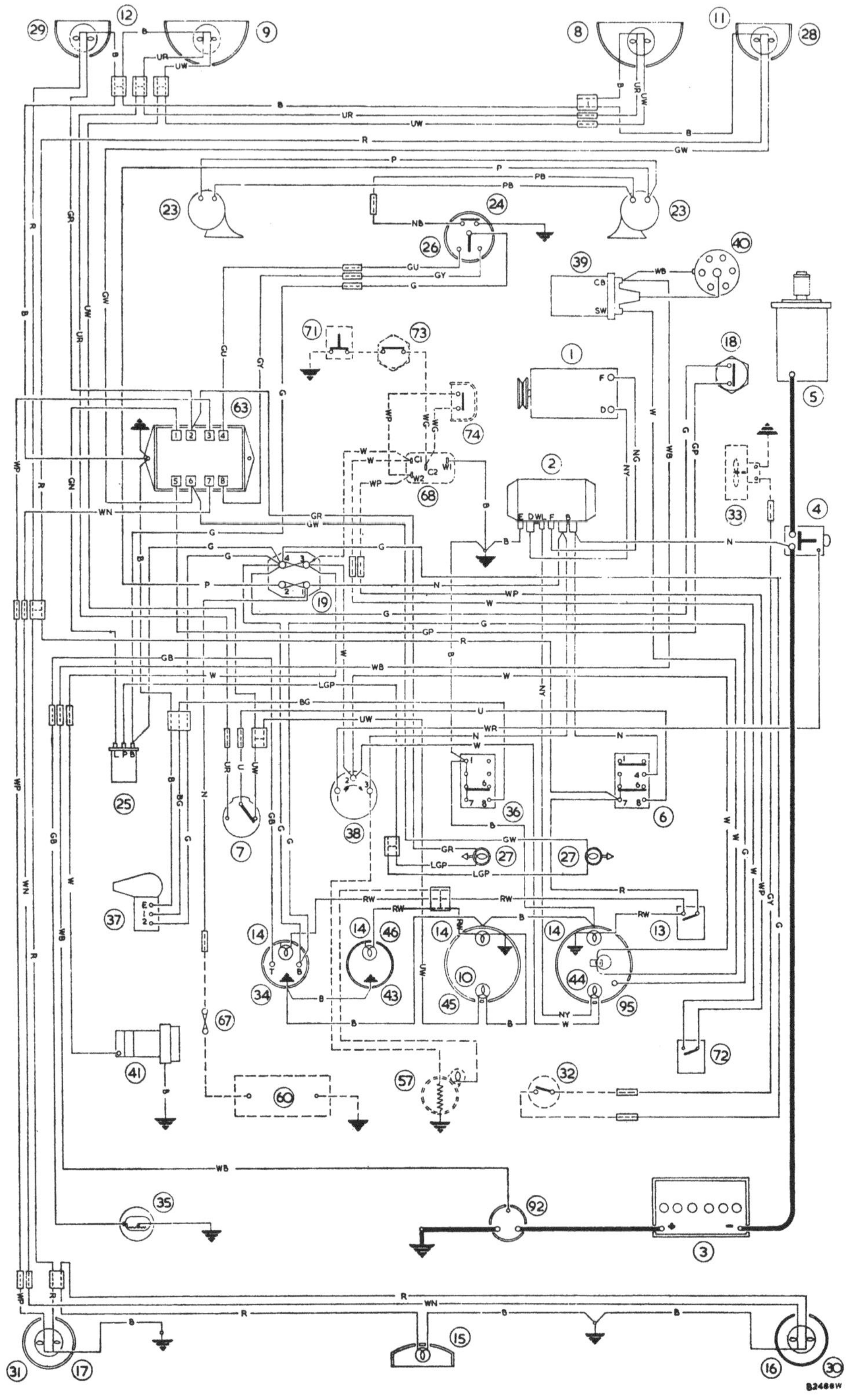

KEY TO THE WIRING DIAGRAM
(Early cars)

No.	*Description*
1.	Dynamo
2.	Control Box
3.	Battery
4.	Starter solenoid
5.	Starter motor
6.	Lighting switch
7.	Headlight dip switch
8.	R.H. Headlamp
9.	L.H. Headlamp
10.	Main-beam warning light
11.	R.H. pilot lamp
12.	L.H. pilot lamp
13.	Panel light switch
14.	Panel lights
15.	Number-plate illumination lamp
16.	R.H. Stop and tail lamp
17.	L.H. Stop and tail lamp
18.	Stop light switch
19.	Fuse unit (50 amps. 1–2, 35 amps. 3–4)
23.	Horns
24.	Horn-push
25.	Flasher unit
26.	Direction indicator switch
27.	Direction indicator warning lights
28.	R.H. Front flasher lamp
29.	L.H. Front flasher lamp
30.	L.H. Rear flasher lamp
31.	L.H. Rear flasher lamp
32.	Heater motor switch
33.	Heater motor*
34.	Fuel gauge
35.	Fuel gauge tank unit
36.	Windscreen wiper motor switch
37.	Windscreen wiper motor
38.	Ignition/starter switch
39.	Ignition coil
40.	Distributor
41.	Fuel pump
43.	Oil pressure gauge
44.	Ignition warning light
45.	Speedometer
46.	Water temperature gauge
57.	Cigar lighter*
60.	Radio*
63.	Flasher relay
67.	Line fuse*
68.	Overdrive relay (25-amp. fuse)*
71.	Overdrive solenoid*
72.	Overdrive manual control switch
73.	Overdrive gear switch*
74.	Overdrive throttle switch*
92.	Battery cut off switch
95.	Revolution counter

CABLE COLOUR CODE

B	BLACK	R	RED
U	BLUE	W	WHITE
N	BROWN	Y	YELLOW
G	GREEN	L.G	LIGHT GREEN
P	PURPLE		

When a cable has two colour code letters the first denotes the main colour and the second denotes the tracer colour.

All items marked thus * fitted as optional extra, circuits shown dotted.

WIRING DIAGRAM
(Later cars)

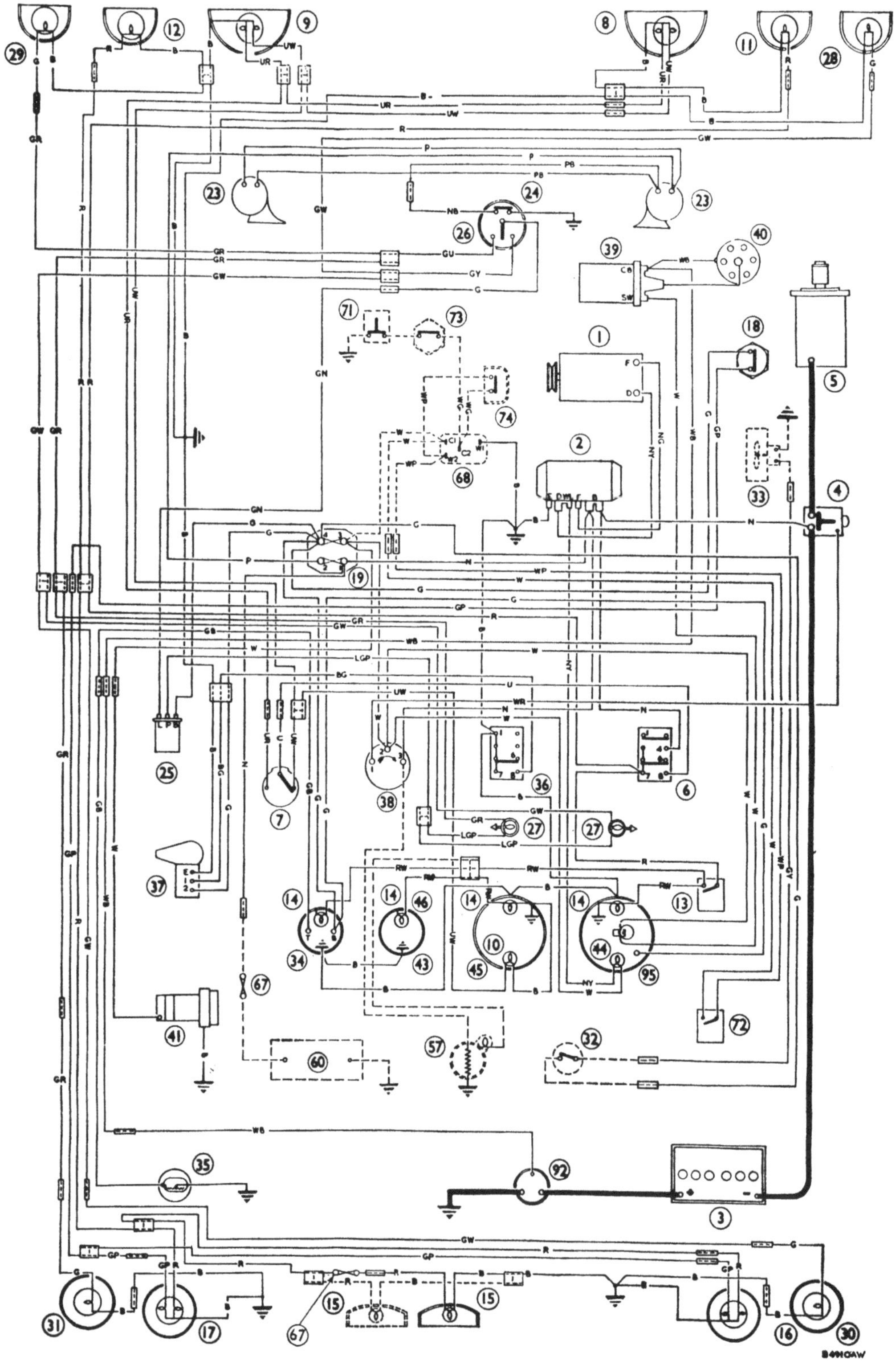

KEY TO WIRING DIAGRAM

(Later Cars)

No.	*Description*
1.	Dynamo.
2.	Control box.
3.	Battery.
4.	Starter solenoid.
5.	Starter motor.
6.	Lighting switch.
7.	Headlamp dip switch.
8.	R.H. headlamp.
9.	L.H. headlamp.
10.	Main beam warning lamp.
11.	R.H. pilot lamp.
12.	L.H. pilot lamp.
13.	Panel lamp switch.
14.	Panel lamps.
15.	Number plate illumination lamp.
16.	R.H. stop and tail lamp.
17.	L.H. stop and tail lamp.
18.	Stop lamp switch.
19.	Fuse unit (50 amps. 1-2, 35 amps. 3-4).
23.	Horns.
24.	Horn push.
25.	Direction indicator flasher unit.
26.	Direction indicator switch.
27.	Direction indicator warning lamps.
28.	R.H. front direction indicator lamp.
29.	L.H. front direction indicator lamp.
30.	R.H. rear direction indicator lamp.
31.	L.H. rear direction indicator lamp.
32.	Heater motor switch*.
33.	Heater motor*.
34.	Fuel gauge.
35.	Fuel gauge tank unit.
36.	Windscreen wiper switch.
37.	Windscreen wiper motor.
38.	Ignition/starter switch.
39.	Ignition coil.
40.	Distributor.
41.	Fuel pump.
43.	Oil pressure gauge.
44.	Ignition warning light.
45.	Speedometer.
46.	Water temperature gauge.
57.	Cigar lighter.
60.	Radio.*
67.	Line fuse.
68.	Overdrive relay unit.*
71.	Overdrive solenoid.*
72.	Overdrive manual control switch.
73.	Overdrive gear switch.*
74.	Overdrive throttle switch.
92.	Battery cut-off switch.
95.	Revolution counter.

CABLE COLOUR CODE

N.	Brown	P.	Purple	W.	White
U.	Blue	G.	Green	Y.	Yellow
R.	Red	L.G.	Light Green	B.	Black

When a cable has two colour code letters the first denotes the main colour and the second denotes the tracer colour.

* Optional extra, circuit shown dotted.

SECTION O

WHEELS AND TYRES

Section O.1

DESCRIPTION

The Austin-Healey 100 'Six' is fitted with 5·90—15 tubeless tyres upon 15 × 4J ventilated steel disc wheels, or, as an optional alternative, with 5·90—15 Dunlop Road Speed tyres upon 15 × 4J wire spoked wheels with knock-on type hub caps.

Section O.2

ADJUSTMENTS IN THE VEHICLE

The purpose of the following adjustments is to obtain the best performance from the wheels and tyres. Proceed as detailed below. Other faults should be diagnosed after consulting Section O.5.

(1) When the car is new: After the first long run or after 50 miles (80 km.) of short runs, jack up the wheels and hammer the nuts to make sure they are tight. Always jack up the wheel before using the hammer.

(2) Tighten the road wheel nuts (or knock-on hub caps). In the case of disc wheels tighten the nuts to a torque wrench reading of 60 to 62·5 lb. ft. (8·3 to 8·64 kg.m.). **Do not overtighten.**

(3) Check the tyre pressures regularly with a gauge and inflate them to the recommended pressures.

Section O.3

WHEEL AND TYRE ASSEMBLIES (Tubeless)

To Remove

(1) Apply the handbrake and scotch one of the wheels.

(2) Jack up the vehicle sufficiently to ensure that the wheel with a fully inflated tyre can be removed or installed.

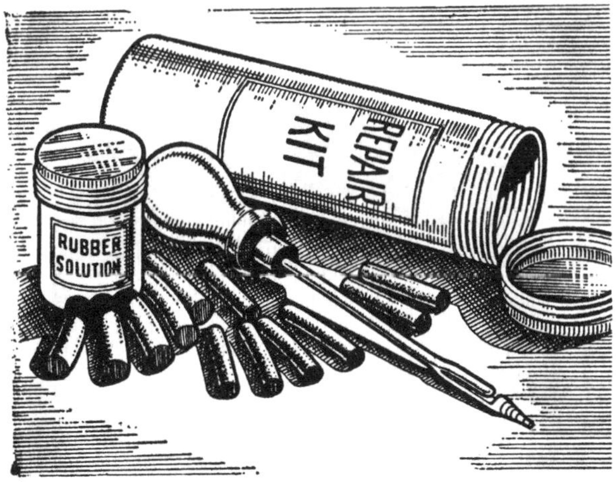

Fig. O.1. Tyre Repair Kit.

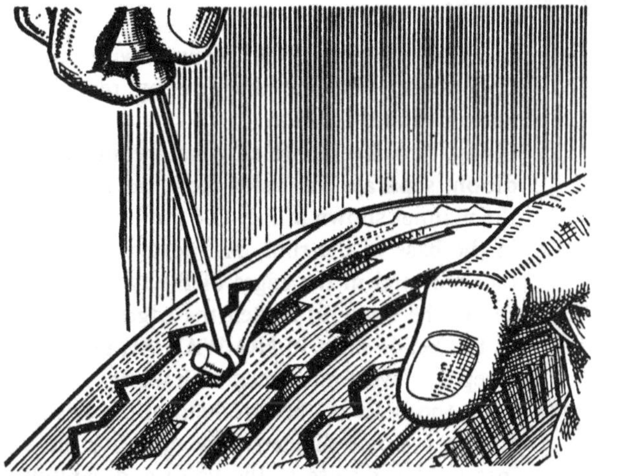

Fig. O.2. Inserting the plug and needle through the hole in the tyre.

(3) Remove the hub cover.

(4) Remove the five wheel nuts, which have right-hand threads, *i.e.* turn anti-clockwise to loosen. Lift off the wheel.

To Repair Simple Tyre Penetrations

Normally a tubeless tyre will not leak as the result of penetration by a nail or other normal puncturing objects provided that it is left in the tyre, but a repair should be effected at the earliest convenient time.

In the case of a nail penetrating the tyre, a repair can be carried out externally without removing the tyre from the rim, providing the special repair kit is available. If the hole fails to seal, mark the spot and extract the nail, taking note of the direction of penetration. If the tyre is leaking and the puncturing object cannot be located by sight, immerse the wheel and tyre in water.

Repair the tyre as follows :—

(1) Insert the needle of the repair kit through the hole in the tyre in the same direction as the penetration to free it from road grit. Dip the needle into the rubber solution and re-insert it through the hole, repeating this operation until the hole is well lubricated with the solution.

(2) Select a repair plug of about twice the diameter of the puncturing object, stretch and roll it into the eye of the needle, about ¼ in. (6 mm.) from its end. Dip the plug into the rubber solution and insert the needle through the hole in the tyre so that the end of the rubber plug passes through the hole into the interior of the tyre. Withdraw the needle, leaving the plug in the tyre, and cut off the plug about ⅛ in. (3 mm.) from the surface of the tread.

(3) Inflate the tyre (see "**To inflate the tyre**").

To Dismantle

(1) Lay the wheel on the ground, with the valve uppermost. Deflate the tyre by removing the valve cap and interior.

(2) Using tyre levers, which must be in good condition, separate the beads from the rim flange in the manner shown in Fig. O.6 until both beads are in the base of the rim. As inextensible wires are incorporated in the edges of the tyre, no attempt should be made to stretch the edges over the rim as the beads must in **NO WAY BE DAMAGED.** Keep the levers moistened with water.

(3) With the bead of the tyre held in the base of the rim at a point diametrically opposite the valve, insert a lever close to the valve and carefully lift the tyre over the rim. Using two levers at intervals of about 6 in. (15 cm.) apart, continue to lift the tyre bead over the rim until it is entirely free.

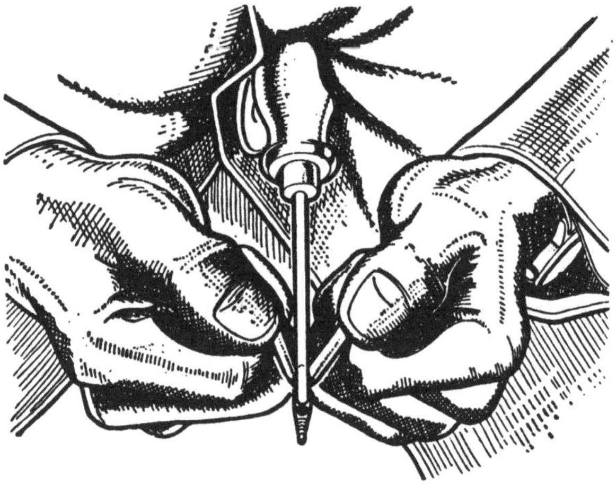

Fig. O.3. Rolling the plug into the needle eye.

(4) Stand the tyre and wheel upright, keeping the bead in the base of the rim. Lever the bead over the rim flange, and at the same time push the wheel away from the tyre with the other hand to completely remove the tyre off the wheel.

To Repair Severe Tyre Penetrations

Severe penetrations which are outside the scope of the small repair kit can be repaired in a similar manner to conventional covers which will necessitate the removal of the tyre (see above).

Repair the tyre as follows:—

(1) Inspect the tyre for damage and remove any puncturing objects.

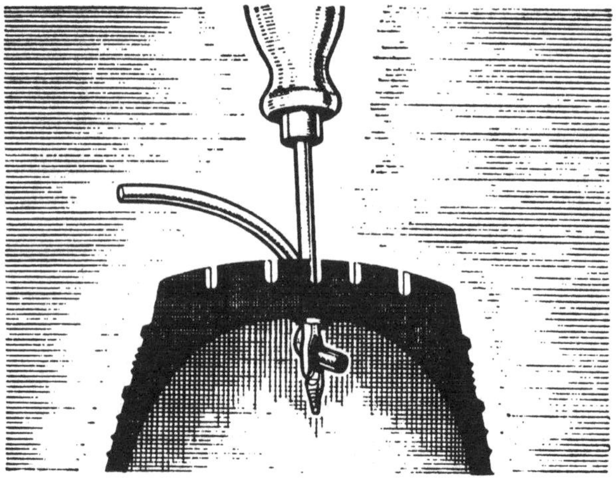

Fig. O.4. The inserted plug prior to withdrawing the needle.

(2) Clean the area around the hole on the inside of the tyre, roughen with a scratchbrush and apply a rubber solution to the surface to receive an ordinary tube patch such as the Dunlop "Vulcafix" patch, or preferably use an uncured rubber patch and vulcanise it in position.

(3) In the event of more serious damage, the tubeless tyre can undergo a major vulcanised repair in the same way as a normal tyre. The tubeless tyre can also be re-treaded.

To Prepare the Rim before Fitting the Tyre

(1) Examine the condition of the wheel and renew if cracked, or if the attachment holes are elongated.

(2) Check for loose rivets in the base of the wheel rim and fit oversize rivets if necessary, ensuring that they do not protrude beyond the height of the original rivets. An airtight seat must be maintained.

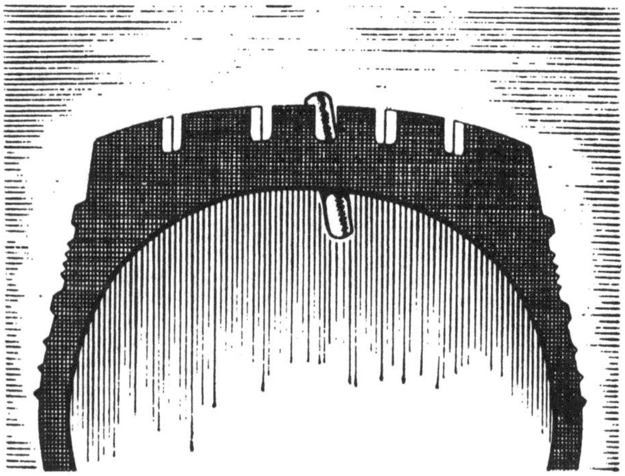

Fig. O.5. Plug inserted in the tyre and cut-off.

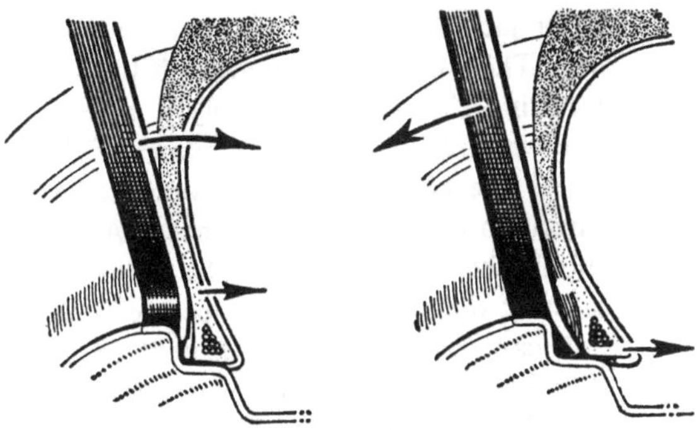

Fig. O.6.
Removing a tyre bead off the rim.

(3) Clean off all rust, rubber, etc., from the wheel flange and rim seat, using steel wool, emery or similar cleaning medium. In extreme cases of rusty rims it may be necessary to use a wire brush or even a file.

(4) Remove any dents in the flange by hammering out carefully to maintain an airtight seat.

(5) High spots in the welded joint of the rim must be filed or buffed away.

To Renew a Valve

A valve must never be refitted once it has been removed from the rim, and it must be renewed every time a new tyre is fitted.

Cut out or pull outwardly the old valve from the rim.

Lubricate a new valve with soap solution and pull it through the rim hole from the inside. The valve should be pulled until the flange on the rubber base of the valve is in full contact with the inner rim surface. If the valve is pulled too far, the base will be damaged and another new valve will have to be fitted.

The use of the Schrader valve mounting Tool No. 553 is recommended, so as to avoid damage.

Fig. O.7.
The use of a tourniquet to seal the beads.

To Reassemble

When replacing the tubeless tyre a similar technique has to be employed to that used for removal, first fitting the tyre into the base of the rim at a point opposite the valve. Make sure the valve interior is removed and that the balance spots near the tyre bead are at the valve position. Wipe clean and moisten the beads of the tyre, rim flanges and tyre levers with clean water. Do not use petrol. Carry out the final fitting of the tyre, using levers which are in good condition and free from burrs. Take small "bites" with the levers.

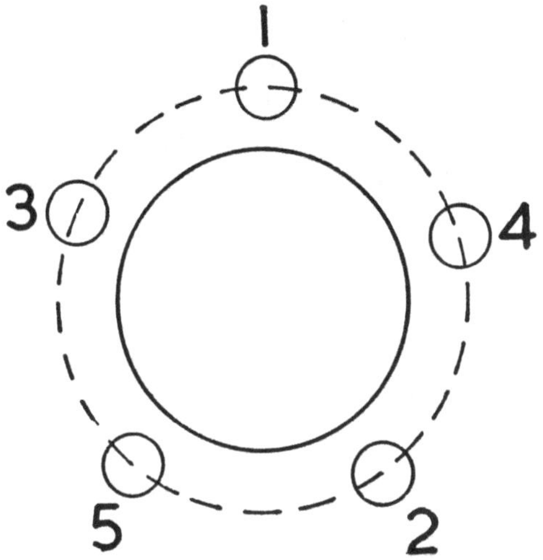

Fig. O.8.
Order of tightening the wheel nuts (disc wheels).

To Inflate the Tyre

(1) Before inflating the tyre, bounce the crown of the tyre on the ground at various points round its circumference, to snap the beads home against the rim. This will provide a partial seal.

(2) Connect an air line to the valve, with its interior plunger omitted, and inflate the tyre with the wheel in an upright position. If a seal cannot be effected by the first rush of air, bounce the tyre on its crown at various points round the circumference with the air line still attached. In cases of difficulty apply the special tourniquet consisting of a strap incorporating a lever, but a suitably strong cord or rope around the circumference of the tread and a twisting bar or stick will also serve. If no air line is available and the tyre has to be inflated by a foot- or hand-pump, then the use of a tourniquet is essential to force the beads outwards against the rim flanges to effect a seal. Remove the air line, insert the valve interior and re-inflate, for test purposes, to 40 lb./sq. in. (2·8 kg./cm.^{2}).

(3) Allow the tyre to stand for a few minutes so that any free air trapped between the flange and the bead clinch can escape. Test the complete assembly in a water tank to check for leaks, special attention being paid to the areas at the beads, valve and wheel rivets. Should leakage occur at the valve base, this can only be rectified by renewing the valve. Loss of air around the bead seat and flange is generally due to a high spot on the rim (foreign matter, rust, weld, etc.) and in most cases this can be cured by holding the tyre bead away from the rim, with the tyre deflated, in order to effect further cleaning of the rim bead seat with emery, steel wool, etc. Air leakage at the rivets can be remedied by peening over the rivet head with a ball-pane hammer. The rivet should be backed up with another, and preferably larger, hammer. In extreme cases where major leaks occur at the flanges or rivets, mark off the position of the leaks on the tyre and the rim before removing the tyre for inspection and rectification.

(4) When satisfied that there are no air leaks, and that the tyre is correctly fitted, adjust the tyre to the recommended working pressure.

To Replace

(1) Install the wheel and tyre assembly on the hub.

(2) Screw on the wheel nuts, ensuring that the chamfered end is engaged in the conical seat in the wheel, and tighten them in the sequence shown in Fig. O.8.

(3) Release the jack and scotch; check the wheel nuts again with a torque wrench set to 60 to 62·5 lb. ft. (8·3 to 8·64 kg.m.).

(4) Place the rim of the hub cover over two of the buttons on the wheel centre and give the outer face a sharp blow of the fist over the third button.

Section O.4

WHEEL AND TYRE ASSEMBLIES (Spoke Wheels)

Wheel Changing

(1) First loosen the "knock-on" hub cap, then jack up the car. If it is a front wheel which is to be changed the lip on the platform of the screw type jack must project into the recess in the spring plate, whilst the platform should be across the outer rim of the spring plate, the flat end between the lower wishbone links.

(2) For lifting the rear wheels, place the lifting platform across the lowest spring leaf, to the rear of the axle, with the lipped end on the outside of the spring and up against the spring "U" bolt, this avoids any turning movement.

After jacking, the hub cap can be screwed right off. The wheel is then pulled off the splined hub.

(3) Refitting the wheel is simply a reversal of this removal procedure, but the splines of the hub and wheel are so fine that the operator should be careful not to jam the splines. A little grease should be smeared upon the splines and cone faces of the hub and wheel before refitting. The hub cap threads will also benefit from an occasional application of grease.

Remember that hub caps fitted to right-hand side hubs have left-hand threads, left-hand hubs have right-hand threads, however, the direction for turning is clearly marked on each cap. Caps should be finally tightened with a mallet.

Once a year remove the wheels for examination and regreasing.

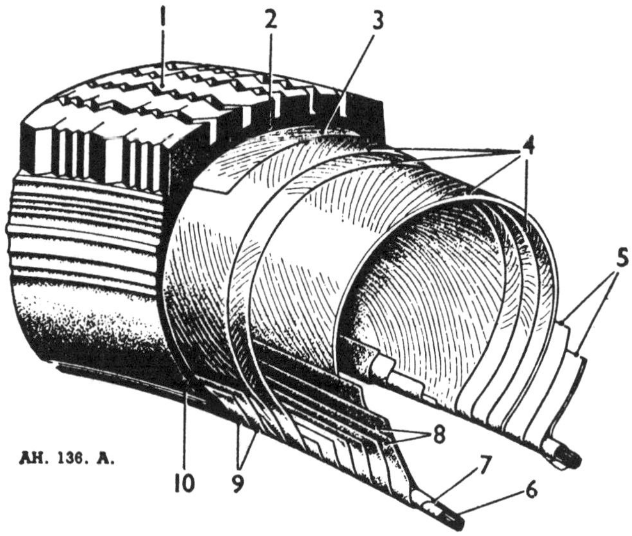

Fig. O.9. Diagrammatic illustration of tyre construction.

1. Tread.
2. Soft cushion rubber.
3. Breaker.
4. Casing plies.
5. Fillers.
6. Bead wires.
7. Bead wrapping.
8. Fillers.
9. Chafers.
10. Wall rubber.

Tyre Removal

(1) Remove all valve parts to deflate the tyre and push both tyre beads off the rim seats.

(2) Commence to remove the bead on the valve side of the cover. Insert a lever at the valve position and, while pulling on this lever, push the bead into the well of the rim diametrically opposite the valve.

(3) Insert a second lever about 2 in. away from the first lever and gradually prise the bead over the rim flange. Remove both beads one at a time, over the flange on the side of the rim which has the narrower bead seat. If the tyre is removed with the tyre lying flat, the narrower bead seat should be upwards.

(4) Continue with one lever while holding the removed portion of the bead with the other lever. The tube can then be removed.
(5) Stand the cover upright with the wheel in front.
(6) Insert a lever from the front between the bead and the flange and pull the cover back over the flange.
(7) If difficult to remove, keep the strain on the bead with the lever and tap off with a rubber mallet.

Tyre Replacement

(1) Place the cover on top of the wheel and push as much as possible of the lower bead by hand into the well of the rim. Insert a lever to prise the remaining portion of the lower bead over the rim flanges.
(2) Slightly inflate the tube until it begins to round out and insert it in the cover with the valve through the hole in the rim. Take care that the valve, which is fitted in the side of the tube, is on the correct side of the rim and that the tube and spot markings coincide ; a point of balance already described.
(3) Commence to fit the second bead by pushing it into the well of the rim diametrically opposite the valve.

Fig. O.10. Excessive tyre distortion from persistent underinflation causes rapid wear on the shoulders and leaves the centre standing proud. If the effects of underinflation are aggravated by other factors, such as camber and excessive braking, the irregular and rapid wear is more pronounced.

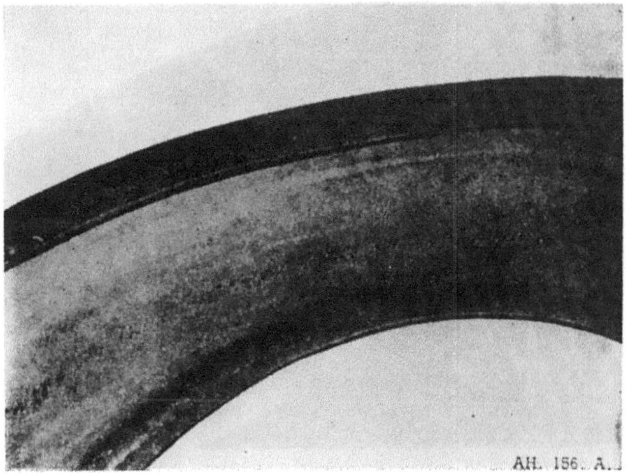

Fig. O.11. This casing is breaking up due to over-flexing and heat generation.

(4) Lever the bead over the flange either side of this position, finishing at the valve, when the bead will be completely fitted.
(5) Ease the valve in the rim hole and push upwards by hand to enable the beads to seat correctly and then pull the valve firmly back into position.
(6) Inflate the tyre and see that the beads are seated evenly round the rim: check by the line on the cover.

During fitting, check pressure frequently to make sure that 40 lbs./sq. in. (2·8 kg./cm^2.) air pressure is not exceeded as there can be a risk of severe bead damage. If 40 lbs./sq. in. (2·8 kg./cm^2.) pressure will not seat beads properly, deflate, lubricate, centralise the tyre and re-inflate.

IMPORTANT:—Lock wheels down when using a fitting machine. Stand clear of the tyre when inflating and do not lean over it until the beads are properly seated.

Note: Water on levers considerably eases the fitting and removing of beads.

Section O.5

FACTORS AFFECTING TYRE LIFE AND PERFORMANCE

Radial-ply tyres

Radial ply tyres (Dunlop SP41) should only be fitted in sets of four, although in certain circumstances, it is permissible to fit a pair on the rear wheels, tyres of different construction must not be used on the same axle. A pair must never be fitted to the front wheels with conventional tyres at the rear.

Inflation Pressures

Pressures which are higher than those recommended for the car reduce comfort. They may also reduce tread

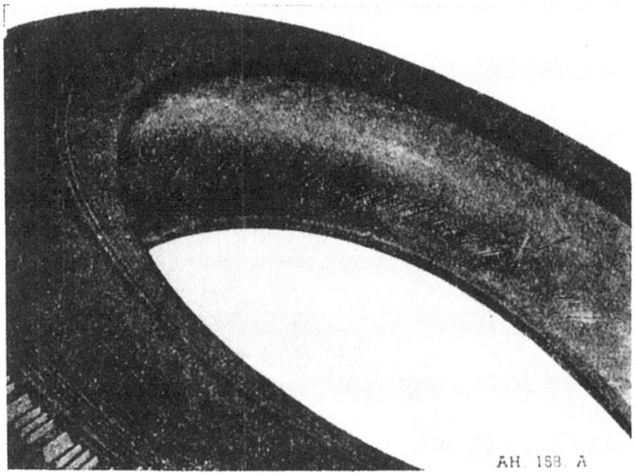

Fig. O.12. Running deflated has destroyed this tyre.

life due to a concentration of the load and wear on a smaller area of tread, aggravated by increased wheel bounce on uneven road surfaces. Excessive pressures overstrain the casing cords, in addition to causing rapid wear, and the tyres are more susceptible to impact fractures and cuts.

Effect of Temperature

Pressures in warm tyres should not be reduced to standard pressure for cold tyres. "Bleeding" the tyres increases their deflections and causes their temperatures to climb still higher. The tyres will also be underinflated when they have cooled.

Speed

High speed is expensive and the rate of tread wear may be twice as fast at 50 m.p.h. (80 km./hr.) as at 30 m.h.p. (48 km./hr.).

High speed involves :—

(1) Increased tyre temperatures due to more deflections per minute and a faster rate of deflection and recovery. The resistance of the tread to abrasion decreases with increase of temperature.

(2) Fierce acceleration and braking.

(3) More tyre distortion and slip when negotiating bends and corners.

(4) More "thrash" and "scuffing" from road surface irregularities.

Braking

"Driving on the brakes" increases the rate of tyre wear, apart from being generally undesirable. It is not necessary for wheels to be locked for an abnormal amount of tread rubber to be worn away.

Other braking factors not directly connected with the method of driving can affect tyre wear, for instance correct balance and lining clearances, and freedom from binding, are very important. Braking may vary between one wheel position and another due to oil or foreign matter on the shoes even when the brake mechanism is free and correctly balanced.

Brakes should be relined and drums reconditioned in complete sets. Tyre wear may be affected if shoes are relined with non-standard material having suitable characteristics or dimensions, especially if the linings differ between one wheel position and another in such a way as to upset the brake balance. Front tyres, and particularly near front tyres, are very sensitive to any condition which adds to the severity of front braking in relation to the rear.

"Picking-up" of shoe lining leading edges can cause grab and reduce tyre life. Local "pulling-up" or flats on the tread pattern can often be traced to brake drum eccentricity, fig. O.13. The braking varies during each wheel revolution as the minor and major axis of the eccentric drum pass alternately over the shoes. Drums should be free from excessive scoring and be true when mounted on their hubs with the road wheels attached.

Climatic Conditions

The rate of tread wear during a reasonably dry and warm summer can be twice as great as during an average winter.

Water is a rubber lubricant and tread abrasion is much less on wet roads than on dry roads. In addition resistance of the tread to abrasion decreases with increase in temperature.

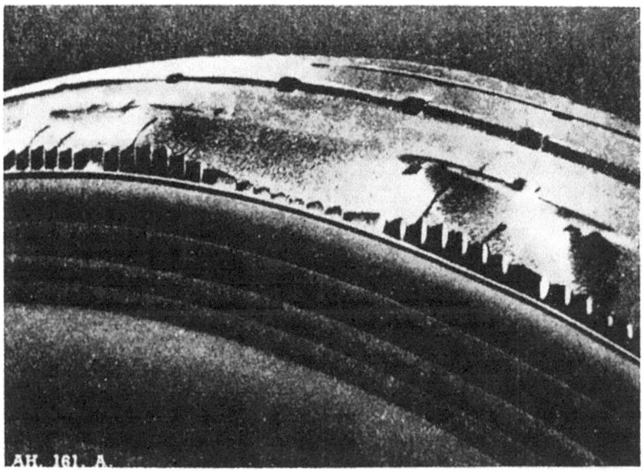

Fig. O.13. Local excessive wear due to brake drum eccentricity.

When a tyre is new its thickness and pattern depth are at their greatest. It follows that heat generation and pattern distortion due to flexing, cornering, driving and braking are greater than when the tyre is part worn.

Higher tread mileages will usually be obtained if new tyres are fitted in the autumn or winter rather than in the spring or summer. This practice also tends to reduce the risk of road delays because tyres are more easily cut and penetrated when they are wet than when they are dry. It is, therefore, advantageous to have maximum tread thickness during wet seasons of the year.

Road Surface

Present day roads generally have better non-skid surfaces than formerly. This factor, combined with improved car performance, has tended to cause faster tyre wear, although developments in tread compounds and patterns have done much to offset the full effects.

Road surfaces vary widely between one part of the country and another, often due to surfacing with local material. In some areas the surface dressing is coarser than others; the material may be comparatively harmless rounded gravel, or more abrasive crushed granite, or knife-edged flint. Examples of surfaces producing very slow tyre wear are smooth stone setts and wood blocks, but their non-skid properties are poor.

Bends and corners are severe on tyres because a car can be steered only by misaligning its wheels relative to the direction of the car. This condition applies to the rear tyres as well as the front tyres. The resulting tyre slip and distortion increase the rate of wear according to speed, load, road camber and other factors, fig. O.14.

The effect of hills, causing increased driving and braking torques with which the tyres must cope, needs no elaboration.

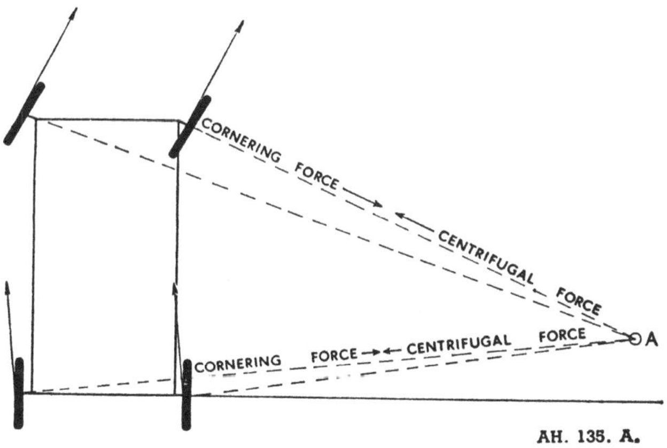

Fig. O.14. Slip when cornering causes increased tyre wear.

Impact Fractures

In order to provide adequate strength, resistance to wear, stability, road grip and other necessary qualities, a tyre has a certain thickness and stiffness. Excessive and sudden local distortion, such as may result from striking a kerb, a large stone or brick, an upstanding manhole cover, or a deep pothole may fracture the casing cords, figs. O.15 and O.16.

Impact fractures often puzzle the car owner because the tyre and road spring may have absorbed the impact without his being aware of anything unusual. Only one or two casing cords may be fractured by the blow and the weakened tyre fails some time later. Generally there is no clear evidence on the outside of the tyre unless the object has been sufficiently sharp to cut it.

This damage is not associated solely with speed and care should be exercised at all times, particularly when drawing up to a kerb.

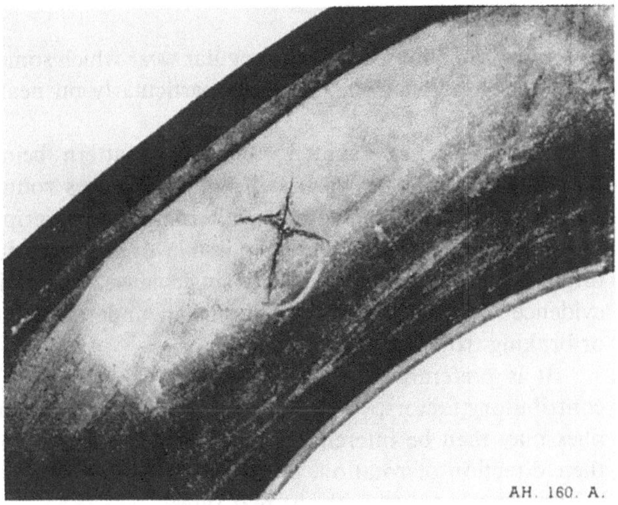

Fig. O.15. Severe impact has fractured the casing.

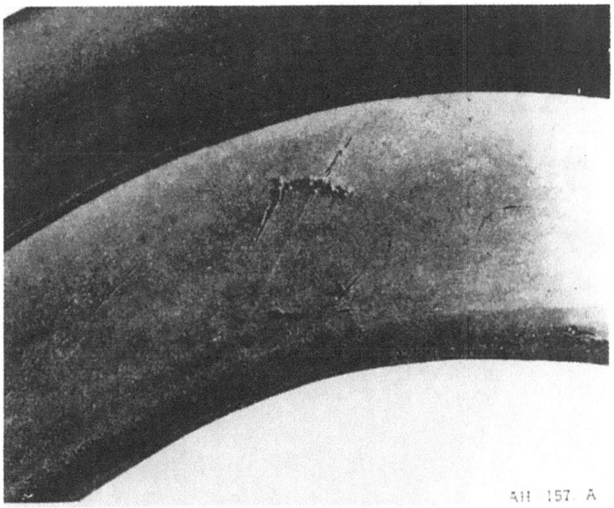

Fig. O.16. A double fracture caused by the tyre being crushed between the rim and an obstacle, such as the edge of a kerb.

"Spotty Wear"

Fig. O.17 shows a type of irregular wear which sometimes develops on front tyres and particularly on nearside front tyres.

The nature of "spotty" wear—the pattern being much worn and little worn at irregular spacings round the circumference—indicates an alternating "slip grip" phenomenon, but it is seldom possible to associate its origin and development with any single cause. There is evidence of camber wear, misalignment, underinflation, or braking troubles.

It is preferable to check all points which may be contributory factors. The front tyres and wheel assemblies may then be interchanged, which will also reverse their direction of rotation, or better still the front tyres may be interchanged with the rear tyres.

Points for checking are :—

(1) Inflation pressures and the consistency with which the pressures are maintained.
(2) Brake freedom and balance, shoe settings, lining condition, drum condition and truth.
(3) Wheel alignment.

Fig. O.17. Irregular "Spotty" wear, to which a variety of causes may contribute.

Fig. O.18. Fins or feathers caused by severe misalignment. With minor misalignment, probably aggravated by road camber, the ribs may have sharp edges instead of upstanding fins. These conditions will usually be accompanied by heel and toe wear across the tread due to its being distorted and worn away laterally instead of in a true rolling direction.

(4) Camber and similarity of camber of the front wheels.
(5) Play in hub bearings, swivel pin bearings, suspension bearings, and steering joints.
(6) Wheel concentricity at the tyre bead seats.
(7) Balance of the wheel and tyre assemblies.
(8) Conditions of road springs and shock absorbers.

Corrections which may follow a check of these points will not always effect a complete cure and it may be necessary to continue to interchange wheel positions and reverse directions of rotation at suitable intervals.

Irregular wear may be inherent in the local road conditions such as from a combination of steep camber, abrasive surfaces, and frequent hills and bends. Driving methods may also be involved. Irregular wear is likely to be more prevalent in summer than in winter, particularly on new or little worn tyres.

Wheel Alignment and Road Camber

It is very important that correct wheel alignment should be maintained. Misalignment causes a tyre tread

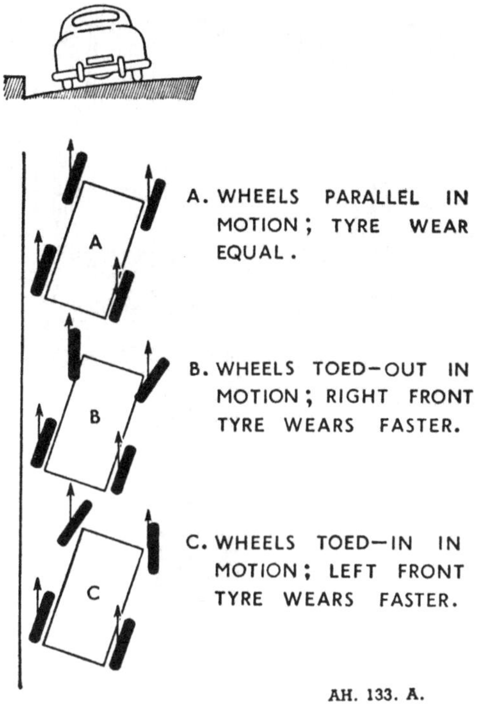

Fig. O.19.
Exaggerated diagram of the way in which road camber affects a car's progress.

to be scrubbed off laterally because the natural direction of the wheel differs from that of the car.

An upstanding fin on the edge of each pattern rib is a sure sign of misalignment and it is possible to determine from the position of the "fins" whether the wheels are toed in or toed out, see fig. O.18. Fins on the inside edges of the pattern ribs—nearest to the car—and particularly on the off-side tyre, indicate toe-out.

With minor misalignment the evidence is less noticeable and sharp pattern edges may be caused by road camber even when wheel alignment is correct. In such cases it is better to make sure by checking with an alignment gauge.

Road camber affects the direction of the car by imposing a side thrust and if left to follow its natural course the car will drift to the near side. This is instinctively corrected by steering towards the road centre. As a result the car runs crab-wise. Fig. O.19 shows, in exaggerated form, the effect this has upon the tyres.

The near front tyre sometimes persists in wearing faster and more unevenly than the other tyres even when the mechanical condition of the car and tyre maintenance are satisfactory. The more severe the average road camber the more marked will this tendency be. This is an additional reason for the regular interchange of tyres.

Camber Angle

This angle normally requires no attention unless disturbed by a severe impact, however, it is always advisable to check this angle if steering irregularities develop, see L6.

Wheel camber usually combined with road camber, causes a wheel to try to turn in the direction of lean, due to one side of the tread attempting to make more revolutions per mile than the other side. The resulting increased tread shuffle on the road and the off centre tyre loading tend to cause rapid and one sided wear. If wheel camber is excessive for any reason the rapid and one sided tyre wear will be correspondingly greater. Unequal cambers introduce unbalanced forces which try to steer the car one way or the other. This must be countered by steering in the opposite direction which results in faster tread wear.

Section O.6

TYRE AND WHEEL BALANCE

Static Balance

Unbalance in wheel and tyre assemblies may be responsible for various effects such as wheel wobble, abnormal wear of tyres and suspension parts, vibration in the steering or, in extreme cases, in the whole car. If any of these faults develop for which no other cause can be found, wheel and tyre balance should be checked and corrected according to instructions supplied by the manufacturer of the balancing machine.

When wheels are to be re-balanced it is essential that the weight of the car be removed from the tyres as soon as possible after a run so that temporary flat spots do not form on the tyres. Nylon tyres are particularly prone to this and re-balancing with the tyres in this condition is pointless.

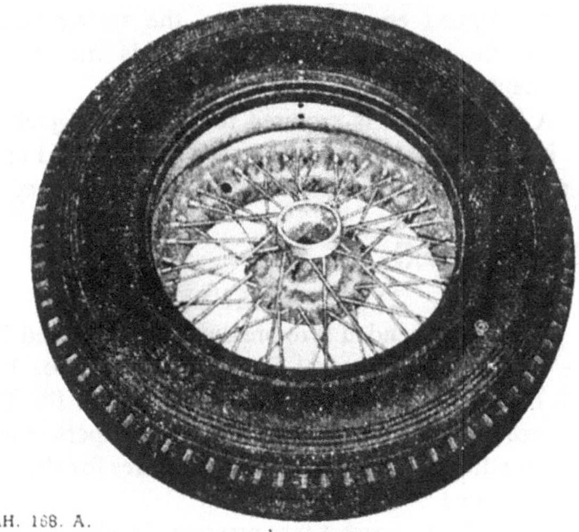

AH. 168. A.

Fig. O.20.
Correct fitting relationship of Dunlop covers and tubes.

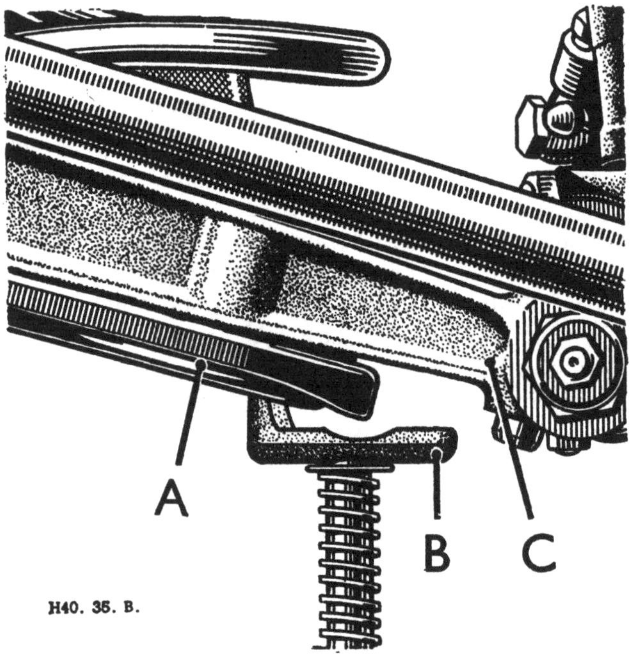

Fig. O.21. Front position for screw jack.
A. Front suspension spring plate. B. Jack platform.
C. Front suspension lower wishbone arm.

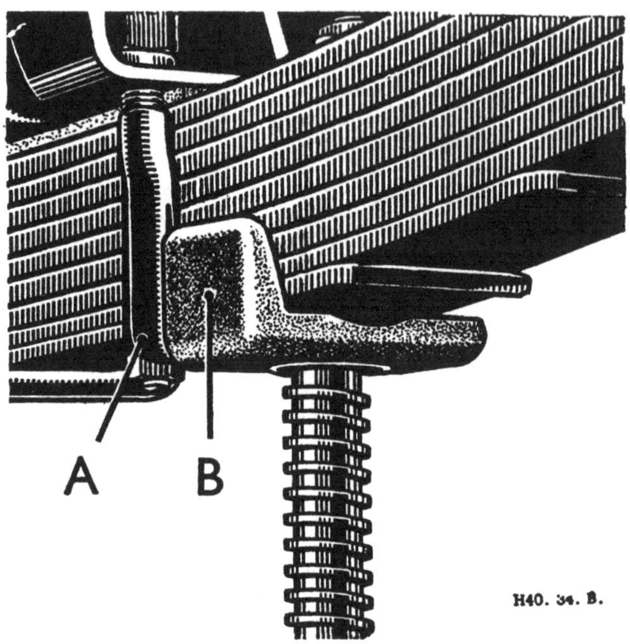

Fig. O.22. Rear position for screw jack.
A. Rear spring "U" Bolt. B. Jack platform lip.

Section O.7

LIFTING GEAR

The Under-Axle Type of Screw Jack

This type of jack is used for all models. For the front wheels the lifting platform of the jack should be placed across the outer rim of the spring recess in the spring plate.

For lifting the rear wheels, place the lifting platform across the lowest spring leaf, to the rear of the axle, with the lipped end on the outside of the lower plate, so that the flat end is between the bottom wishbone links and the lipped end projects into the spring and up against the spring U-bolt; this avoids any turning movement.

A long handle is required to operate the jack and this is obtained by joining together the provided extension and tommy bars, the latter being the turning medium.

Section O.8

TYRE PRESSURES

The recommended tyre pressures for BN4 and BN6 cars are given on General Data page 8, and for BN7, BT7, BJ7 and BJ8 cars on General Data page 18. Pressures are quoted for normal and maximum performance speed conditions. The increase in pressures for the latter must be noted.

Maintain the correct tyre pressures by checking with an accurate tyre gauge at least once a week, and inflating if necessary.

Any unusual pressure loss must be investigated. Under-inflation causes rapid tyre wear, and even more serious is the possibility of damage to the fabric of the tyre owing to the excessive flexing of the tyre walls.

At Car Nos. 10300 (BN7) and 10299 (BT7), Road Speed tyres with an improved tread pattern were introduced. The new tyres can only be fitted to earlier cars in complete sets, and the bolts securing the rear bump rubber brackets to the wheel arches must be reduced in length by 0·5 in. (12.7 mm.) to prevent fouling between the bolts and tyres.

Section O.9

CARE OF TYRES

To obtain the best mileage occasionally interchange the front and rear wheels and bring the spare into use.

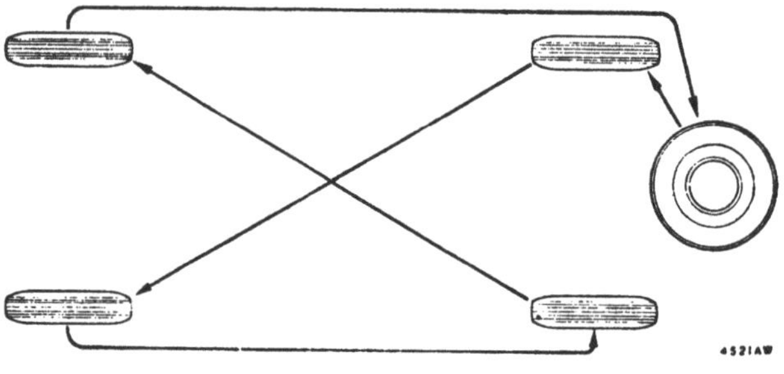

Fig. O.23.
System of tyre changing to regularize tyre wear.

SECTION P

BODYWORK

SERIES BN4

Section P.1

MAINTENANCE OF BODYWORK

Coachwork

Regular care of the body finish is necessary if the new appearance of the car exterior is to be maintained against the effects of air pollution, rain, and mud.

Wash the bodywork frequently, using a soft sponge and plenty of water containing a mild detergent. Large deposits of mud must be softened with water before using the sponge. Smears should be removed by a second wash in clean water, and with the sponge if necessary. When dry, clean the surface of the car with a damp chamois-leather. In addition to the regular maintenance, special attention is required if the car is driven in extreme conditions such as sea spray, or on salted roads. In these conditions and with other forms of severe contamination an additional washing operation is necessary which should include underbody hosing. Any damaged areas should be immediately covered with paint and a complete repair effected as soon as possible. Before touching-in light scratches and abrasions with paint, thoroughly clean the surface. Use petrol/white spirit (gasoline/hydrocarbon solvent) to remove spots of tar or grease.

The application of B.M.C. Car Polish is all that is required to remove traffic film and to ensure the retention of the new appearance.

Bright trim

Never use an abrasive on stainless, chromium, aluminium, or plastic bright parts and on no account clean them with metal polish. Remove spots of grease or tar with petrol/white spirit (gasoline/hydrocarbon solvent) and wash frequently with water containing a mild detergent. When the dirt has been removed polish with a clean cloth or chamois-leather until bright. Any slight tarnish found on stainless or plated parts which have not received regular washing may be removed with B.M.C. Chrome Cleaner. An occasional application of mineral light oil or grease will help to preserve the finish, particularly during winter, when salt may be used on the roads, but these protectives must not be applied to plastic finishes.

Windshield

If windshield smearing has occurred it can be removed with B.M.C. Screen Cleaner.

Interior

Clean the carpets with a stiff brush or vacuum cleaner preferably before washing the outside of the car. The most satisfactory way to give the carpets a thorough cleaning is to apply BMC 2-way Cleaner with a semi-stiff brush, brush vigorously, and remove the surplus with a damp cloth or sponge. Carpets should not be cleaned by the 'Dry Clean' process. The upholstery may be treated with BMC 2-way Cleaner applied with a damp cloth and light rubbing action.

A razor blade will remove transfers from the window glass.

Section P.2

DISMANTLING

Bonnet Top

The bonnet at its rear edge, has two brackets which form part of the hinges. A leg from the bulkhead is secured to each bracket by two nuts and bolts, therefore removal of the bonnet top is achieved by withdrawing the bolts from each bracket then lifting off the panel.

Grille

The grille is secured at eight points, all easily accessible from beneath the car. Three ¼in. U.N.F. set pins will be found at the top and bottom of the grille, while one $\frac{3}{16}$in. U.N.F. setpin is located each side.

Bumper Bars

Rear: The rear bumper is best dismantled by releasing the two nuts immediately behind the bumper bar at the junction of bracket and bumper. The brackets supporting the bumper are secured by two setpins to each chassis frame side member accessible within the luggage compartment.

Fig. P.1. Rear bumper fixing.
1. Fixing bolts. 2. Supporting brackets.

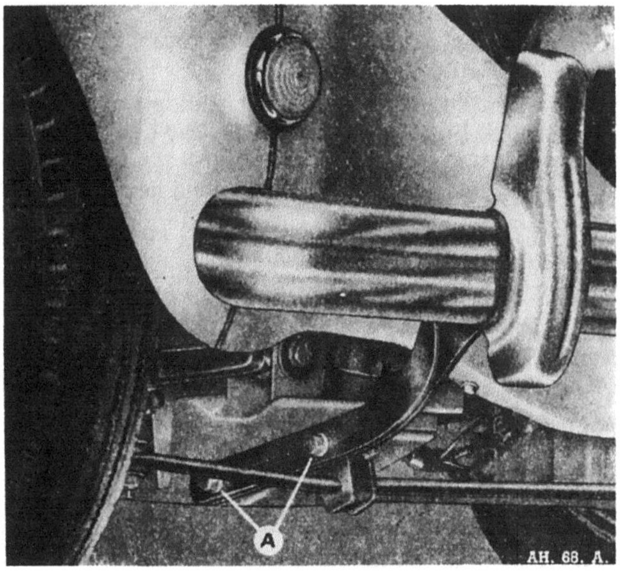

Fig. P.2. Front bumper fixing.
A. Bracket to chassis setpins.

Front: In a manner similar to that employed for its counterpart at the rear, the front bumper bar is held by two nuts to the supporting brackets which are secured by two setpins to each chassis side member.

Front Apron

Once the front bumper has been released at its forward end, the apron can be readily removed from the bodywork.

Sidescreens and Sockets

The sidescreens each have one locating dowel at their base. These dowels are a snug push fit into sockets let into the top of each door.

If necessary the sidescreen sockets can be screwed out of the door using either a broad blade screwdriver or, preferably, a special tool which incorporates a pilot, see Fig. P.3.

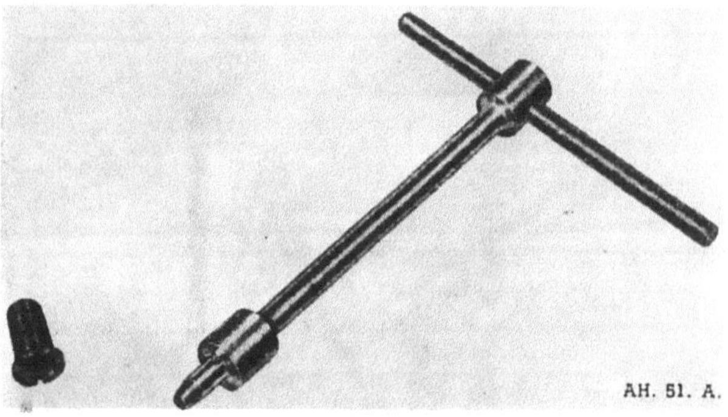

Fig. P.3.
Sidescreen socket and extractor tool.

Doors

Hinges and Door Removal: Both the upper and lower hinge of each door is secured to the door post by four cross-head screws plus one hexagon head setpin. At the door frame each hinge is fixed by four cross-head screws.

There is a check strap fitted to each door which must be released when dismantling a door from the bodywork. This check strap can be released by withdrawing the two setpins from the coupling bracket in the door pillar. Thus with the door wide open, the hinges can readily be uncoupled from the door and the door removed.

Fig. P.4.
The door top moulding is held in place by three cross-head screws A.

Casing: Each door casing, complete with its trimming, can be removed from the door shell after its sixteen securing cross-head screws have been withdrawn from around the casing perimeter.

Door Top Moulding: The aluminium moulding at each door top edge is held in place by three cross-head screws.

Outer Handles: Each outer door handle is secured by a nut accessible from the inside of the door and a Phillip screw accessible from the outside when the handle is raised.

Inner Handles: To remove the door operating handle, the chrome cap behind the handle concerned must be pushed inwards against the spring pressure. When pressure to the cup is applied a dowel pin is visible, passing through the handle stem. Withdraw the pin when the handle and cup can be removed.

Windscreen

The windscreen frame is secured to the scuttle at each side by two nuts and bolts and a single setpin. Each nut and setpin head are accessible within the cockpit behind the fascia. The bolt heads can be seen at the door pillars when the doors are open.

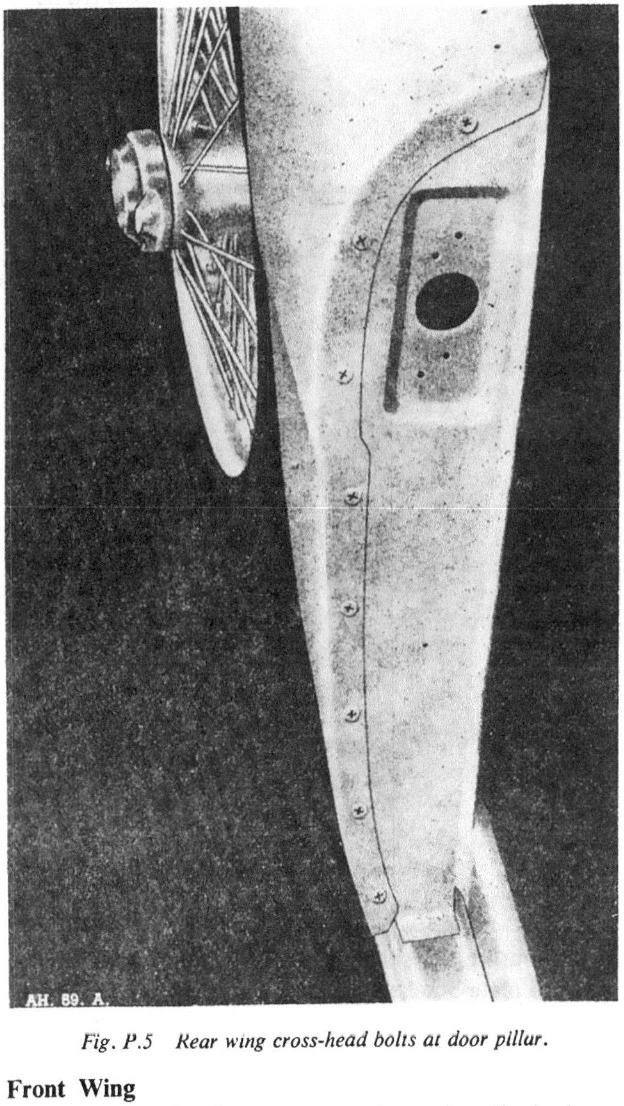

Fig. P.5 Rear wing cross-head bolts at door pillar.

Front Wing

Remove the door concerned as described above. The next operation in dismantling a front mud-wing is to remove both the head and sidelight concerned; details of lens and reflector unit removal procedures are given in Section N of this manual. The outer case of the headlamp

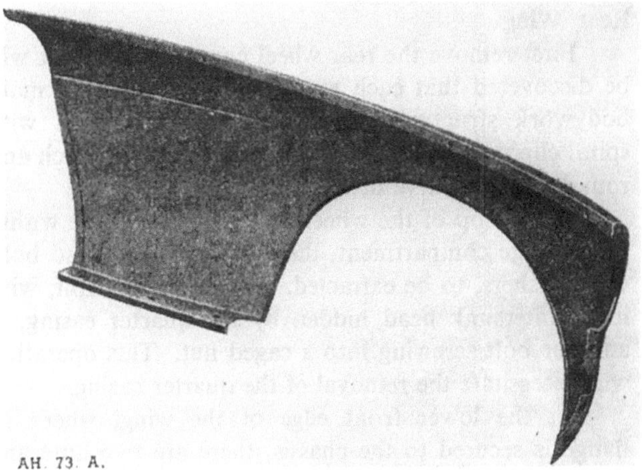

Fig. P.6. The front wing. Securing holes may be seen at top, bottom and front flanges.

is held by four bolts with brass nuts accessible beneath the wing. The sidelamps are secured by three cross-head bolts and nuts.

Beneath the headlight aperture there are three bolts which secure the wing to the cowl centre. These bolts screw into spring-clip type nuts. Along the top edge of the wing flange and forward of the scuttle, four bolts, screwing into clip nuts, clamp the wing to the bonnet surround.

In the cockpit, behind the fascia, there are a further three setpins that screw into clip nuts on the wing. These ¼in. nuts and bolts secure the lower flange of the wing to the underside of the scuttle.

Before the wing can be finally removed there are a number of metal thread screws to be extracted that fix the wing, on the inside of the door pillar, also extract the two rivets securing the rubber water channel to the rear section of the wing.

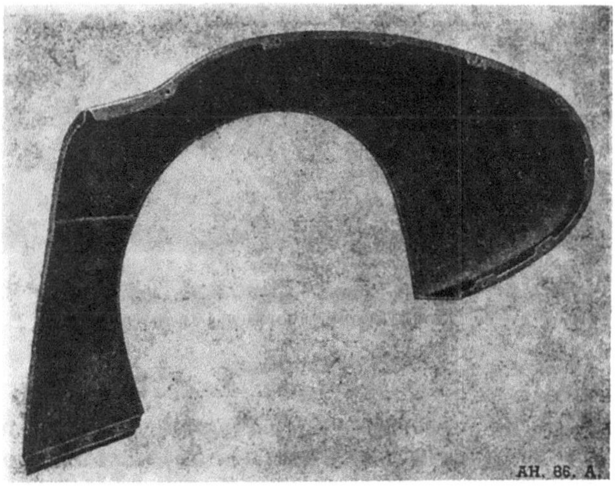

Fig. P.7. Rear wing securing flanges.

Rear Wing

First remove the rear wheel concerned when it will be discovered that each rear wing is fixed to the main bodywork structure by six square thread bolts with spiral clip nuts which are located over the wheelarch and round the rear curve of the wing.

At the top of the wheelarch, head accessible within the luggage compartment, there is a plain nut and bolt, with washers, to be extracted. Within the cockpit, with its countersunk head hidden by the quarter casing, is another bolt screwing into a caged nut. This operation will necessitate the removal of the quarter casing.

At the lower front edge of the wing, where its flange is secured to the chassis, there are two nuts and bolts and a vertical drive screw.

To complete the wing dismantling extract the eight $\frac{3}{16}$in. counter sunk cross-head nuts and bolts with their plain and spring washers that fix the wing leading edge to the door pillar.

Hood Frame

The hood frame is secured at each side to the rear quarter panel, immediately behind the seats, by two bolts, see Fig. P.8.

With these bolts withdrawn the hood frame complete with its fabric can be removed from the bodywork.

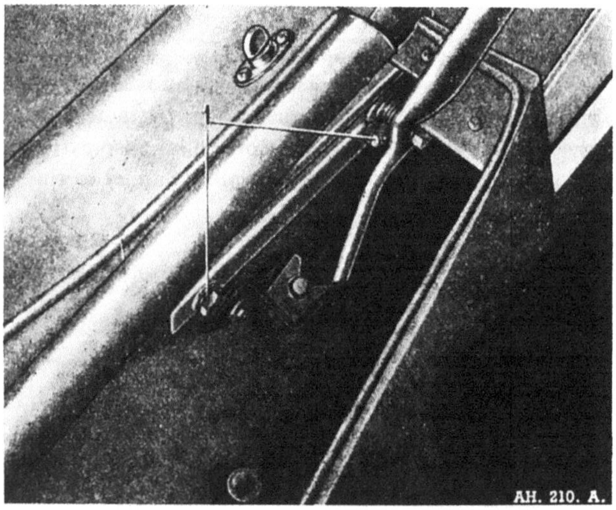

Fig. P.8. Remove the two bolts "1" in order to release the hood frame.

Shroud

The shroud is not removed for normal maintenance work, however, if it should become necessary to remove the shroud due to damage the following fixing points must be made free.

Each outer wing half should be dismantled, see "Front Wing" also the front bumper, grille, apron, windscreen and driving mirror complete with bracket, also the cockpit moulding. In addition the bonnet top must be removed thus giving access to the fixing points around the perimeter of the opening to the engine compartment.

At the rear of the opening there are five drive screws holding the shroud to the scuttle. At the front of the bonnet opening three cross-head bolts and nuts secure the shroud to the front cross bracing of the bodywork. Still working within the opening, at each side, there are two countersunk cross-headed bolts with nuts that fix the shroud to upright braces from the chassis frame.

Fig. P.9. Shroud upright brace with two cross-head bolts at A.

Securing the shroud to each wheelarch panel there are two plate brackets from which the two nuts and bolts must be extracted. The cowl, which is part of the shroud, has two brackets that secure the body member to the frame dumb-irons. From these brackets extract the two nuts and bolts.

Finally free the rear end of the shroud. This is secured to the scuttle just above the fascia by five "pop" rivets, with a further "pop" rivet and two soft rivets at each side fixing the shroud to the scuttle.

The complete shroud can now be lifted clear of the frame and remainder of the bodywork.

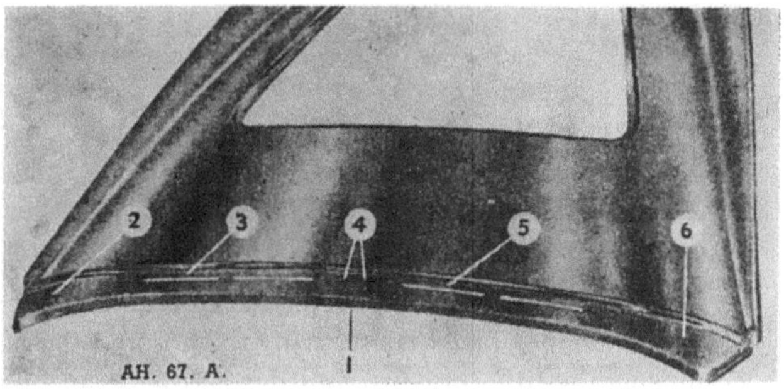

Fig. P.10. Shroud rear fixing.
1. *Five "pop" rivet holes on lip.*
2 *and* 6. *Holes for tonneau cover studs.*
3 *and* 5. *Demister ducts.*
4. *Fixing holes for driving mirror.*

Gearbox Cover

The gearbox cover, or tunnel, is secured at each side flange by six metal thread screws to the floor boards. The heads of these screws are hidden from view until the carpet is peeled back.

Immediately before the tunnel there is a carpet covered bulkhead plate which can be removed for further access to gearbox and clutch housing. This is fixed to the bulkhead by six self-tapping screws.

Fascia Panel

The first operation for removing the fascia is to drop the heater controls temporarily out of the way. To do this unscrew the two round-headed bolts and nuts securing the controls to the fascia.

Remove the steering wheel as described in Section K.

There are five screws, the heads of which are under the fascia, passing through the fascia panel into tapped holes of brackets behind the fascia. There is also one screw adjacent to the ignition switch. By extracting these screws the instrument panel can be brought forward into the cockpit thus giving access to the rear of each instrument.

Grab Handle

The passenger grab handle is fixed to the fascia panel by two round head screws the heads of which are situated behind the panel.

Seats

To adjust or remove the passenger seat the cushion must be lifted whereupon the heads of four setpins are revealed. These setpins (two each side of the seat frame) secure the seat to the body floor. On their extraction the seat may be removed or repositioned, there being four alternative holes for adjustment each side of the seat frame. For the driver an adjustable driving seat is provided for forward or rearward positioning by pushing the lever, beneath the seat, toward the runner then moving the seat to the required setting and releasing the lever.

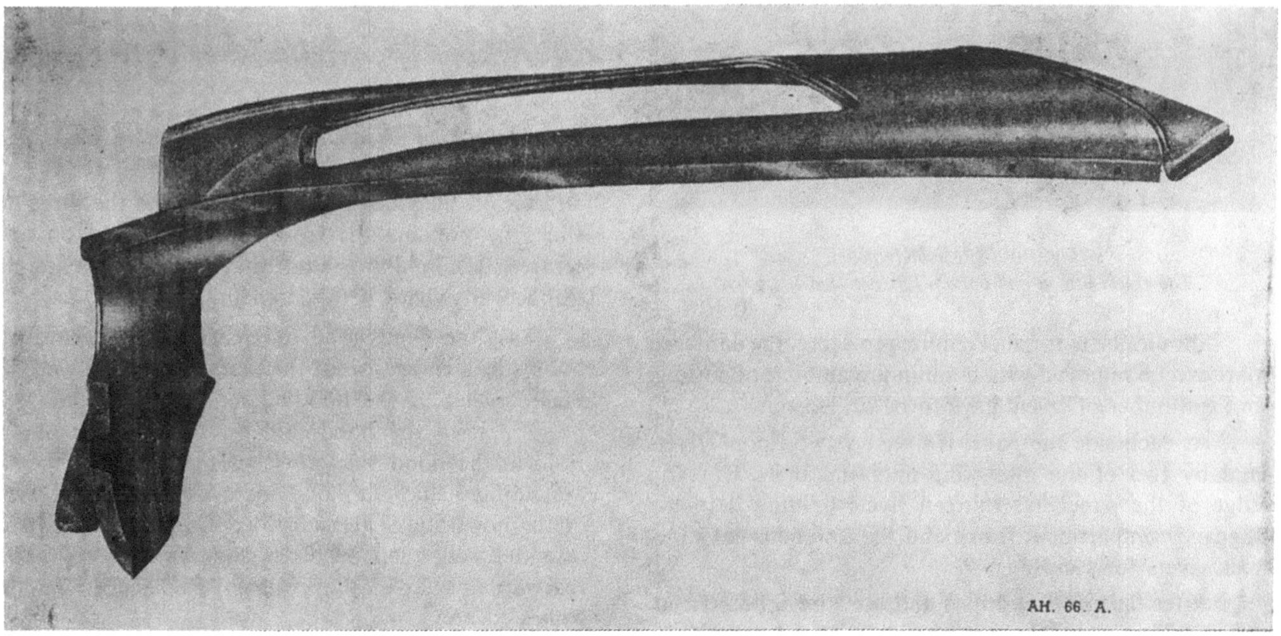

Fig. P.11. The complete shroud, removed from the bodywork, showing the side fixings.

Lift out the seat cushion to gain access to the six nuts securing the seat to the runners.

The seat runners, with their packing pieces, are bolted to the floor, each runner having three bolts with nuts accessible beneath the floor.

Boot Lid

Each hinge may be disconnected from the boot lid by releasing the two hinge nuts from the underside of the lid.

The limit cable must be released from its connection under the lid.

The locking handle can be withdrawn through lock and compartment lid after its two securing screws are extracted. The round heads of these screws are visible on the underside of the lid.

Four cross-headed screws with nuts and washers fix the lock assembly to a supporting bracket riveted and welded to the lid. There is little that can go wrong with this lock, however it does benefit from the occasional application of oil particularly to the spring that is partially visible at the closing edge of the boot lid.

Rear Body Panel

This panel, which forms the lower rear part of the luggage compartment, is the most likely panel to suffer damage in the event of a rear collision.

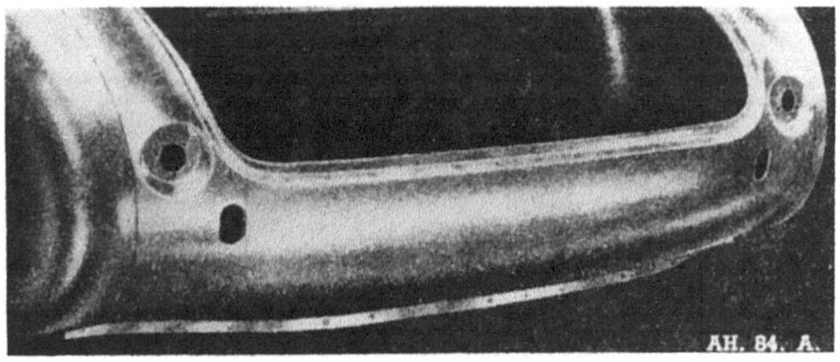

Fig. P.12. Rear body panel.
The rivets at top and bottom lips are clearly shown.

Should the rear panel require replacing, the damaged part can be removed with a minimum amount of cutting, but a number of rivets have to be drilled out.

At each side the panel is fixed by two "pop" rivets and by two of the mud-wing nuts and bolts. The top edge of the panel has thirteen rivets holding it to the luggage compartment frame and the lower lip has nineteen rivets for securing.

After the rivets and nuts and bolts have been freed, the rear light should be removed allowing the panel to be cut along the welded seam at each side which can be felt inside the compartment.

Naturally, the new panel must be re-welded along the same lines and secured by the requisite number of rivets and nuts and bolts.

Section P.3

HOOD (First type to Car No. BN4 68959)

To lower the hood first remove the rear seat squab. Release the tenex fasteners securing the hood to the body commencing with the turn buttons at the rear of the doors and working in towards the body centre. Release the toggle clamps at each end of the windshield top rail

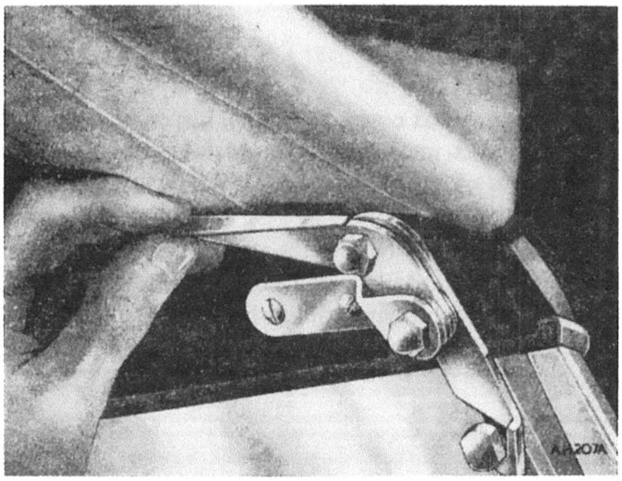

Fig. P.13.
Releasing the toggle clamp on windshield top rail.

and lift off the front hood rail. Collapse the three hood sticks by pressing the front and rear sticks on to the centre stick, making sure that no hood material is trapped anywhere between them.

Pass the front hood rail rearwards underneath the hood sticks and rest it at the rear of the hood well. Fold the rear end of the hood under the rear window and position it in the hood well on top of the front rail. Fold in the hood sides towards the centre of the well, ascertaining that the hood material is completely clear of the hood sticks at each side. Lower the hood sticks into the well, then push the sticks and hood material rearwards as far as they will go. Replace the rear seat squab.

Raising the hood is an exact reversal of the above order.

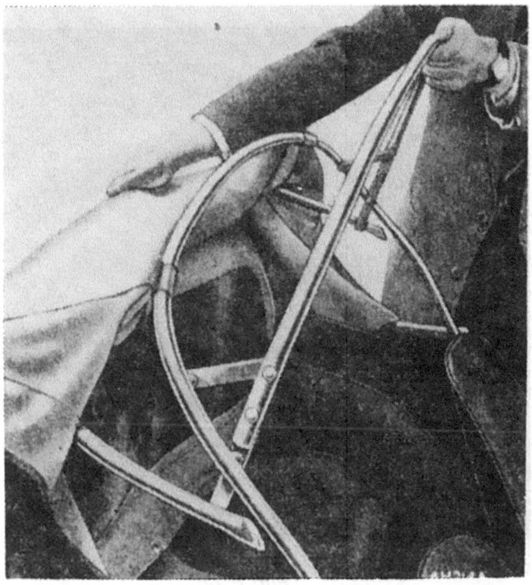

Fig. P.14. Collapsing the hood sticks together.

Fig. P.15. Passing the hood rail rearwards underneath the hood sticks.

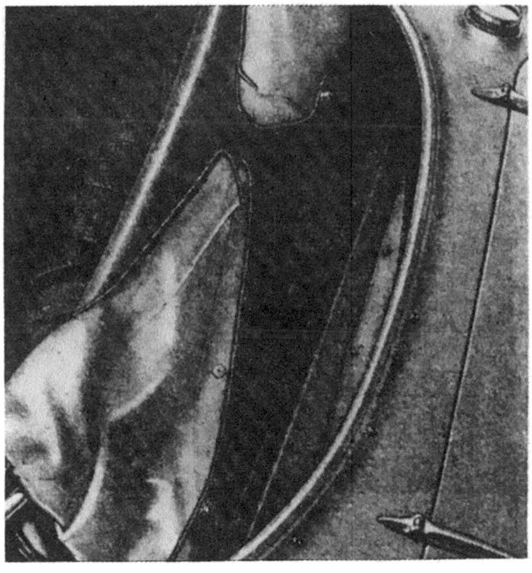

Fig. P.16. Hood side flaps folded into centre of well, leaving hood sticks clear of material at each side.

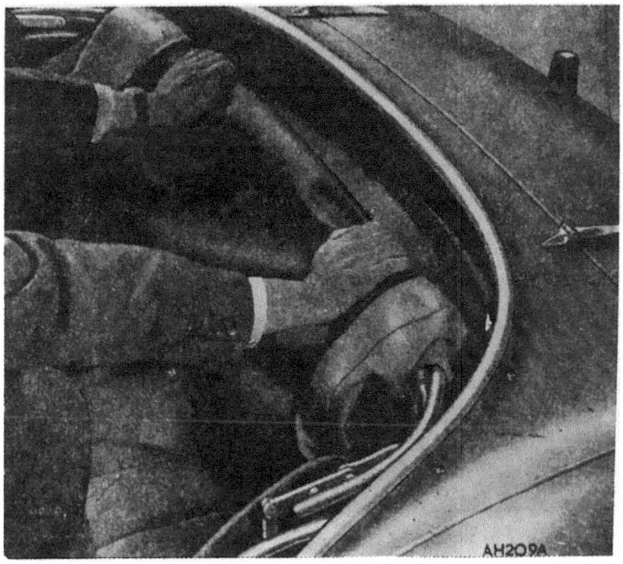

Fig. P.17. Hood assembly in forward position.

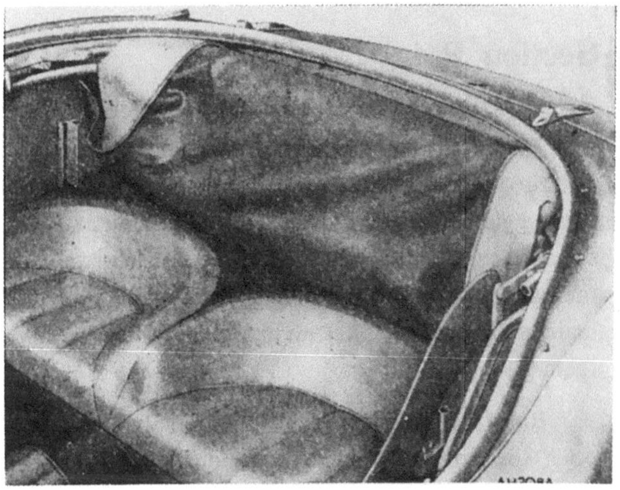

Fig. P.18. Hood assembly pushed back into well.

Fig. P.19. Rear seat squab hinged forward to show slide '1' and securing clip '2'.

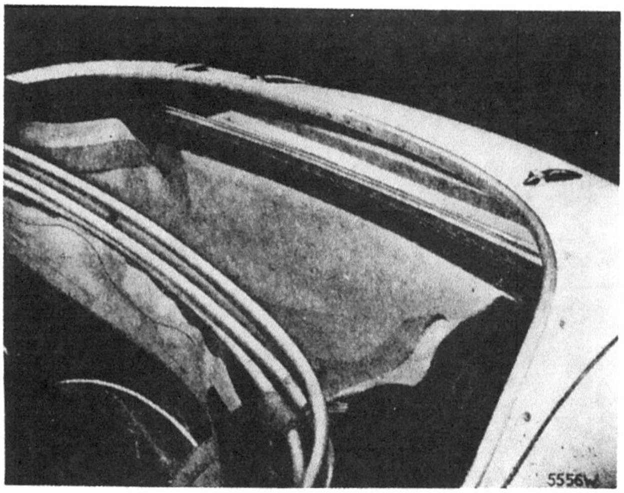

Fig. P.20.
The hood laid out preparatory to stowage with the front rail located under the rear deck panel.

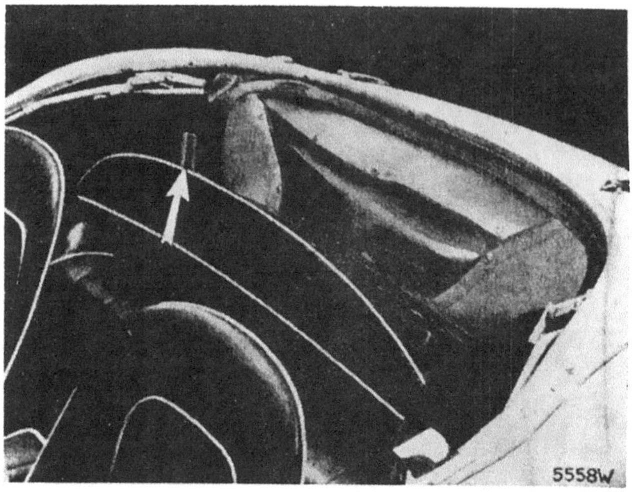

Fig. P.22.
Push the hood frame under the rear deck panel. Make certain that the hood material is clear of the rear seat squab channels.

Section P.4

HOOD (Second type from Car No. BN4 68960)

To lower the hood undo the fasteners securing the hood to the body side and rear deck panel. Release the two toggle catches and lift the front hood rail away from the windshield, pulling the hood material forward over the hood frame. Fold the hinged hood sticks up to the main stick, ensuring that no material is trapped between them.

Raise the rear seat squab and pull it forward to open the hood stowage compartment. Lift the hood assembly from the body sockets and lay it down in position preparatory to stowing (see Fig. P.20).

Take the front rail and stow it under the rear deck panel. Fold the side flaps inwards with the under side of the fasteners towards the window; do not fold with the heads of the fasteners in this position as they may damage the window panel. Lift up the hood material from the well of the stowage compartment and push the frame into position under the rear deck panel. Fold the rear window into the compartment arranging the hood material clear of the squab locating channels.

Fig. P.21.
Fold the side flaps inwards, and lift up the hood material from the well of the stowage compartment.

Fig. P.23.
The hood stowed away and the rear seat squab replaced in the locating channels.

Section P.5

PAINT REFINISHING INSTRUCTIONS

Operation	*Material*	*Thinning*	*Drying times*	*Application*	*Instructions*
Stripping original paint	Water-soluble paint remover, e.g. Sunbeam Anti-corrosives 'Stripolene 799'	—	—	Brush	Remove the original finish with a scraper after allowing paint-strip 10 minutes to react (repeat if necessary). Wash off thoroughly with cold water, rubbing with wire wool. Dry. Blow out crevices with compressed air. Strip a small area at a time to enable correct neutralizing of the stripper
Metal abrading	Emery-cloth, e.g. Howarth Blue Twill, grade $1\frac{1}{2}$ M	—	—	Hand or disc	Paper thoroughly to ensure satisfactory key. Wipe with cleaner solvent or white spirits
Acid etching	Apply Deoxidine 125 (I.C.I.)	1 part Deoxidine, 1 part water	—	Brush	Apply solution generously and rub in with wire wool. Do not allow Deoxidine solution to dry off before the wash-off operation. Allow approximately five minutes to complete reaction. Wash thoroughly with cold water to remove all traces of Deoxidine solution, followed by a hot rinse. Thoroughly dry surfaces with a clean cloth and blow out crevices with compressed air
Priming	Synthetic primer G.I.P. No. S3178 or Grey cellulose primer G.I.P. C3971 MOD	6 to 1 with Z1048 50/50 with 2045M	$\frac{1}{2}$-hour to 4 hours $\frac{1}{4}$-hour	Spray Spray	Apply one thin coat of synthetic primer (recommended for superior adhesion) or one thin coat of cellulose primer (recommended for good adhesion). The use of a primer coat enhances adhesion and gives the system a much greater safety factor
Applying stopper	Stopper Grey G.I.P. 824D or Stopper Brown G.I.P. 1543	—	6–8 hours, or overnight if possible	Glazing knife	Apply stopper in thin layers, allowing 15–20 minutes' drying between applications. Heavy layers result in insufficient drying, with subsequent risk of cracking
Filling	Primer Filler Grey G.I.P. C3663M	50/50 with 2045M	3–4 hours	Spray	Apply two or three full coats, allowing 15–25 minutes' drying time between coats

Wet-sanding	Abrasive paper 280 grade	—	—	—	Rub down wet until smooth; a guide coat (a weak contrasting colour) may be used to ensure that the whole surface is rubbed level. Wash off thoroughly with water, sponge all sludge, wash off, dry with clean sponge. Dry off. Minimum of paint should be removed consistent with a satisfactory surface. Film thickness after rubbing should be ·0025 in. (·06 mm.) min.
Applying sealer or undercoat	Sealer Grey or Sealer White or Red undercoat (see B.M.C. Paint Scheme schedule)	50/50 with 2045M	15–20 minutes	Spray	Apply one coat, flash off
Dry-sanding or de-nibbing as required	320 grade paper	—	—	—	De-nib or dry-sand with 320 paper. Clean with white spirit. The grade of paper quoted is from the 3M Company (Minnesota Mining and Mfg. Co. Ltd.); the grade of paper may vary according to manufacture
Applying colour coats	B.M.C. body finishes (see B.M.C. Paint Scheme schedule)	50/50 with 2045M	5–10 minutes' flash between coats. Overnight dry	Spray	Apply two double coats with a 5–10-minute flash between coats. Overnight dry
Flatting colour coat	Flatting 320 or 400 paper (dependent on conditions)	—	—	Hand	Flat with 320 or 400 paper, dependent on conditions
Applying final colour coat	B.M.C. body finishes (see B.M.C. Paint Scheme schedule)	50/50 with 2045M	Overnight dry	Spray	Spray final double colour coat
Polishing	Cut and polish (see B.M.C. Paint Scheme schedule)	—	—	Hand or machine	The colour coat must be thoroughly dry before polishing. After cutting, burnish to a high gloss with a clean mop, and finally clean with a liquid polish, e.g. Apollo liquid polish

NOTE.—(1) For faster drying of undercoats or local repairs G.I.P. thinners 1523 may be used.
(2) Under extreme circumstances of heat and or humidity, retarder G.I.P. Z1694 can be used added to the 2045M thinners.

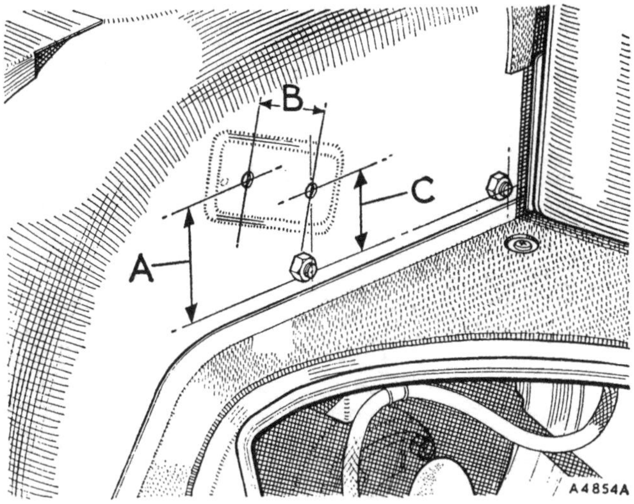

Fig. P.24. The rear wheel arch anchorage point location.
A. *2·5 in. (63·5 mm.).* B. *1·187 in. (30·16 mm.).*
C. *1·875 in. (47·63 mm.).*

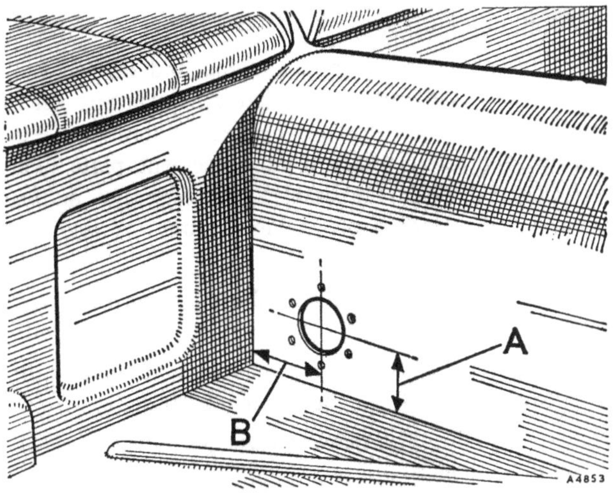

Fig. P.25. The propeller shaft tunnel anchorage point location.
A. *1·562 in. (39·69 mm.).* B. *2 in. (50·8 mm.) BN6 and BN7*
7 in. (177·8 mm.) BN4 and BT7.

Section P.6

SEAT BELT FITTING (Early models)

To enable seat belts to be fitted to the BN4, BN6, BN7, and BT7 models prior to the Healey 3000 Mk. II, it is necessary to modify the body to incorporate the requisite attachment points for the belts.

A seat belt kit Part No. AHB9141 is available from B.M.C. Service Ltd. for this purpose.

The anchorage points are on the rear wheelarches, the rear floor, and the sides of the propeller shaft tunnel. The procedure for modifying the body at these points is as follows.

Rear Wheelarch

Remove the hood and hoodsticks, and the carpet trim from the platform adjacent to the hoodstick socket. Detach the corner finisher from the forward end of the platform, remove the screw in the centre of the wood block, and the retaining screw at the approximate centre of the platform.

Remove the four screws retaining the rear seat to the seat pan and detach the seat.

Gently ease the side trim pad from the rear retainer and the retainer on the front wall of the seat pan. Swing the trim pad carefully forward to expose the wheelarch. Mark off and drill two holes of $\frac{11}{32}$ in. (8·73 mm.) diameter at the positions shown in Fig. P.24, using the centres of the two existing weld nuts as a datum. The area of the wheelarch around the holes must be raised to receive the reinforcement plate attached to the two weld-bolts. The plate assembly must lie flush with the outer face of the wheelarch when fitted.

Refit the trim pad, wood block, finisher and carpet trim. Punch two holes in the side trim pad to correspond with the holes in the wheelarch, and fit the reinforcement plate and bolt assembly from the outside of the wheel-arch. Attach the belt bracket, plain washers, spring washer and capped nuts as described in Section PPP.2.

Propeller shaft tunnel

Remove the carpet and felt, mark out and cut a $1\frac{1}{2}$ in. (38·10 mm.) hole in the position shown in Fig. P.25 and drill six $^{13}/_{64}$ in. (5·16 mm.) equally spaced holes on a pitch circle of 2 in. (50·80 mm.) diameter. The mounting bracket can be used as a template for this purpose.

Fit the mounting bracket from the underside of the tunnel with its projection facing into the car, and secure the bracket with the six No. 10 U.N.F. pan head screws, the nuts and spring washers to face into the car interior.

Cut a 1 in. (25·4 mm.) hole in the carpet and felt to clear the projection of the mounting bracket and replace the felt and carpet.

Assemble the plain washer and then the anti-rattle washer to the hexagon headed set screw. Fit the distance piece to the belt bracket on the short belt so that the large diameter of the distance piece faces onto the mounting bracket and the set in the belt bracket faces away from the tunnel. Assemble the setscrew and washers, and the belt bracket and distance piece to the mounting bracket. Fit the nut and spring washer from the underside of the tunnel.

Floor

Remove the carpets and felt and drill two $\frac{11}{32}$ in. (8·73 mm.) diameter holes at the positions shown in Fig. P.26. Fit the hexagon headed setscrews to the quick-release bracket, place a reinforcement plate over the holes and place the bracket and setscrews in position with the setscrews projecting through the underside of the floor.

The head of the quick-release pin must face towards the door sill.

Fit the other reinforcement plate retaining nuts and spring washers from the underside of the floor.

NOTE.—The quick release pin is not used as such on these models.

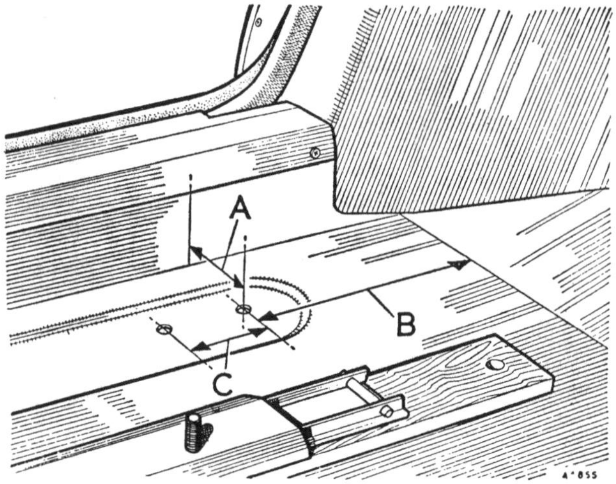

Fig. P.26. The floor anchorage point location.
A. *1·375 in. (34·93 mm.).* B. *3·375 in. (85·73 mm.)* C. *2 in. (50·8 mm.).*

SECTION PP

BODYWORK

SERIES BN6

Section No. PP.1 Hood

NOTE

This information should be used in conjunction with that contained in Section P.

Section PP.1

HOOD

The main hood stick of the Austin-Healey (Series BN.6) is located in sockets on the top of the rear quarter trim panel. Two die-cast retainers fasten the hood to the rear deck panel, and two toggle catches secure the front of the hood to the windshield. Folding, removal, or stowage of the hood should be undertaken by the following method.

1. Pull the rear floating stick forward to the main stick using the loop provided.
2. Release the two turn buttons and the Tenax fasteners securing the hood side flaps to the outside of the body.
3. Remove the rear end of the hood by disengaging the securing plate from the two die castings on the rear deck panel.
4. Release the toggle catches from the top corners of the windshield.
5. Break the cantrail hinge links and pull the front rail up to the main hood stick, ensuring that the hood covering is not trapped between the sticks.
6. Fold the hood onto the hood sticks with the window at the rear and the seam of the hood running along the top of the sticks. The side flaps should be folded inwards with the underneath part of the fasteners against the rear window to avoid damage.
7. Lift the folded hood and sticks from the body sockets and stow with the sticks passing through the leather covered stirrup brackets and resting in the carpet covered recesses near the floor. Ensure that the cantrail hinge links are retained by the stirrup brackets to prevent chafing of the back of the seats when they are pushed back in the rear position.

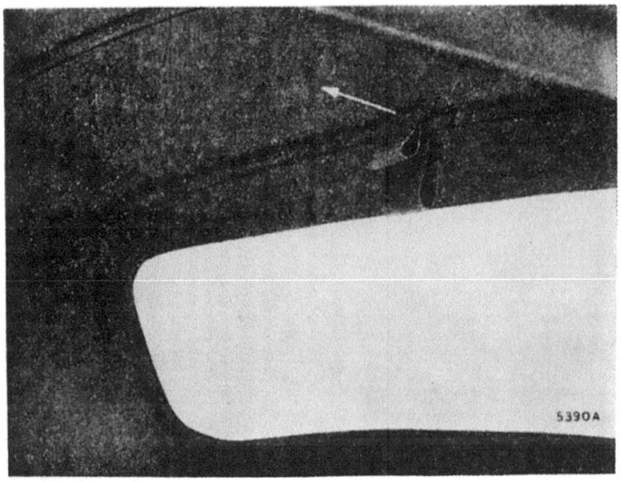

Fig. PP.1
Pull the rear floating stick forward, using the finger loop provided.

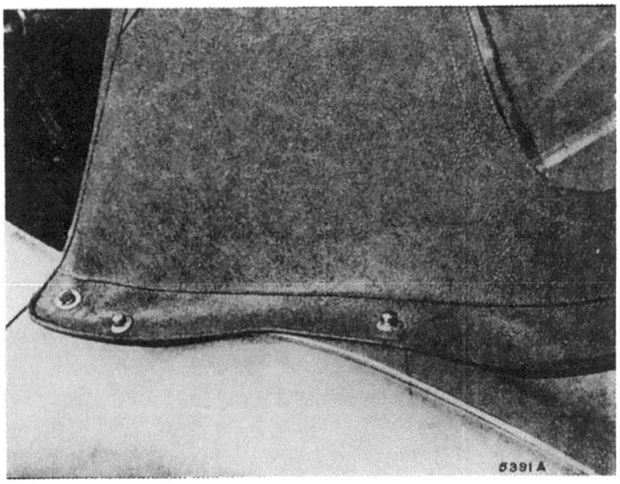

Fig. PP.2
Release the two turn buttons and Tenax fasteners securing the hood flaps to the body.

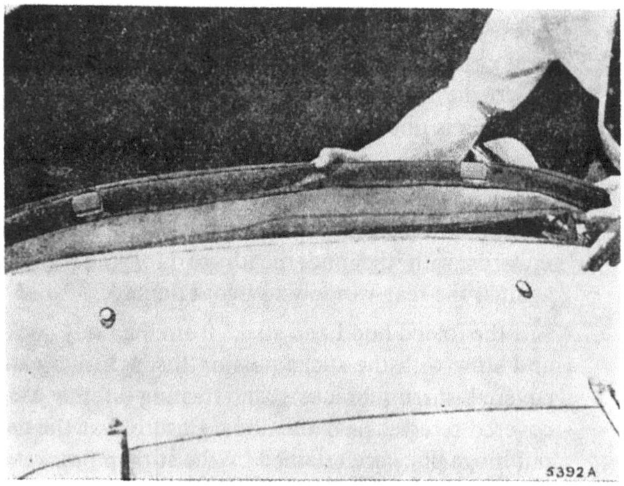

Fig. PP.3
Disengage the hood plate from the rear deck panel.

Fig. PP.4
Release the toggle clamp from the windshield top rail.

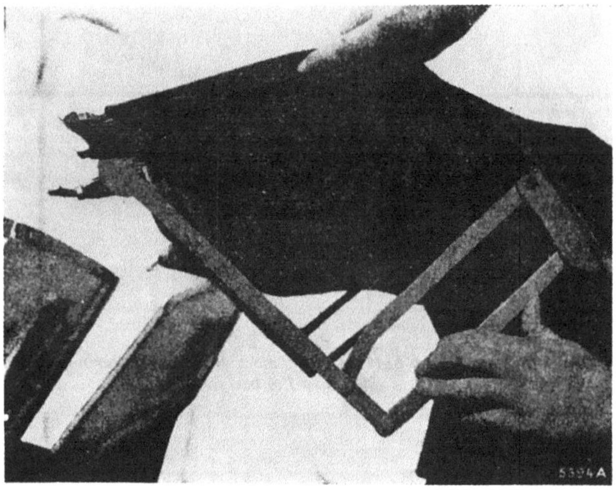

Fig. PP.5
Break the cantrail hinge points and pull the front rail up to the main hood stick.

Fig. PP.6
Fold the hood onto the upright hood sticks with the seam of the hood running along the top and the window to the rear.

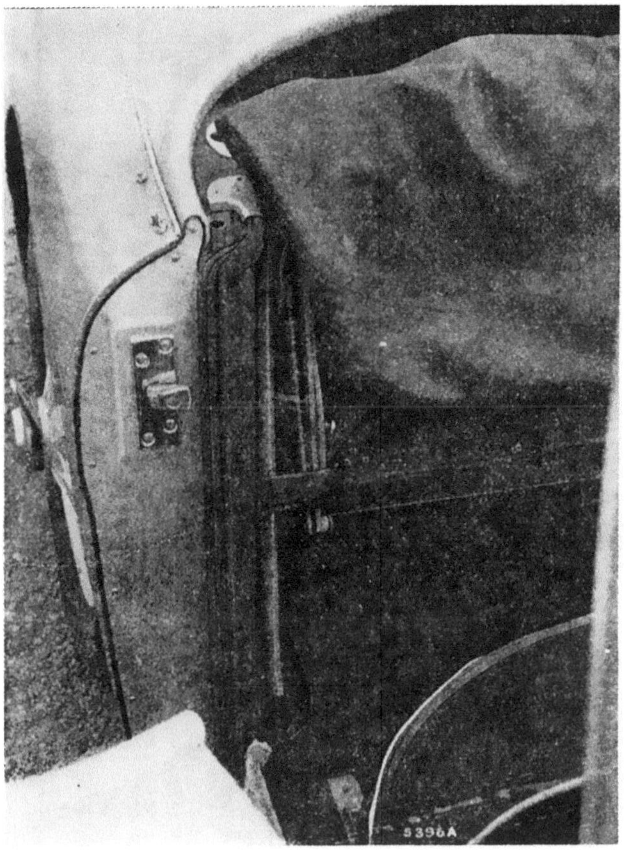

Fig. PP 8
The hood stowed away and the seats pushed back into their most rearward position.

Fig. PP.7
Lift the hood assembly from the body sockets and stow away, passing the sticks through the retaining stirrups into the recess near the floor.

SECTION PPP

BODYWORK

Mk. I and II (SERIES BN7 and BT7)
AND Mk. II and Mk. III (SERIES BJ7 and BJ8)

NOTE

This Section should be used in conjunction with Sections P.1, P.2, and. P.5, The hood of the BN7 is dealt with in Section PP.1.

Section PPP.1

HOOD (Series BT7)

To lower, fold, and stow the hood proceed as follows:—

(1) Break the rear stick hinge links, by upward pressure on their tabs and pull the stick forward to the main support by using the loops provided.

(2) Lift the rear seat squab and open it forward onto the seats.

(3) Release the two buttons securing the side flaps and the fasteners along the rear deck panel (see Fig. PP.2).

(4) Release the toggle catches from the top corners of the windshield (see Fig. P.13).

(5) Remove the hood from the sticks and place the front rail under the rear deck panel. Fold the sticks together and stow under the rear deck panel and over the front rail (see Fig. PPP.1).

(6) Make the first fold in the hood at the seam above the window, fold again and fold the side flaps on top.

(7) Loosely fold again, making sure the hood material is well clear of the seat channels. Finally replace the rear seat squab.

Fig. PPP.2.
Fold the hood side flaps well over after making the second fold in the hood.

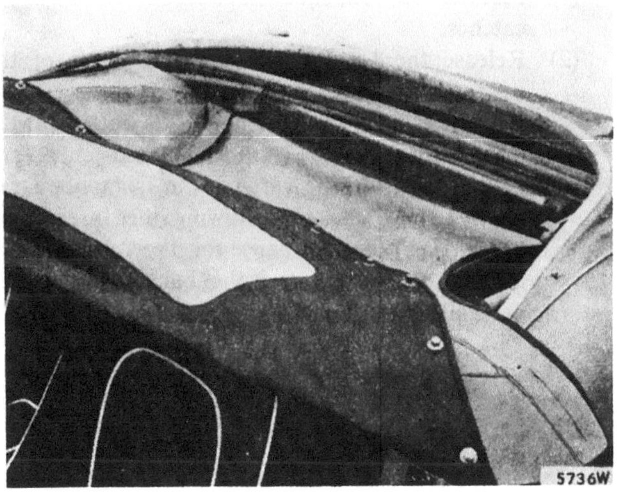

Fig. PPP.1.
Hood sticks and front rail in position under the rear deck panel with hood inverted and the first fold made.

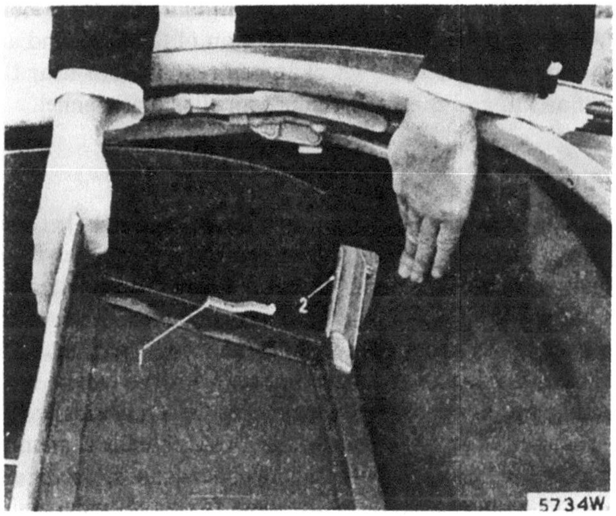

Fig. PPP.3.
Keep the hood material clear of the rear seat squab slides "1" and securing clips "2".

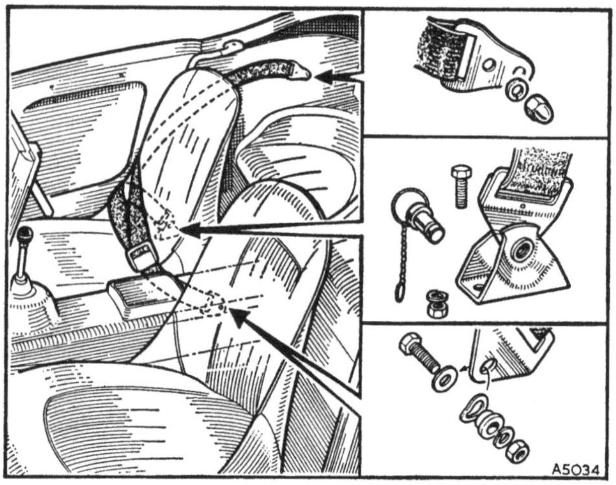

Fig. PPP.4.
Showing the fixing points on the body and the order of assembly of the fixing components.

Section PPP.2

SEAT BELTS (Healey 3000 Mk II and Mk III)

Seat belts for the front seats can be supplied by B.M.C. Service Ltd. Attachment points for these belts are incorporated in the construction of the body and are located on the rear wheel arch, the rear floor near the door sill, and the sides of the propeller shaft tunnel.

Fitting

The two holes for the floor fixing are located between the rear of the door sill and seat, and can be felt from the underside of the floor. Cut the carpet and underfelts to allow a complete metal to metal contact between the base of the channel bracket and the floor. Attach the channel bracket to the belt bracket with the clevis pin so that the angle of the belt bracket faces inwards and the head of the pin faces outwards. Assemble the channel bracket to the floor with the two hexagon set screws and fit the nuts and spring washers from the underside of the floor. Ensure no carpet or felt is trapped under the bracket.

Remove the chromium plated capped nuts and plain washers visible on the rear wheel arch. Fit the other belt bracket over the two studs, ensuring that the belt is not twisted. Refit the plain washers, spring washers and capped nuts. The belt bracket is fitted on top of the trimming at this attachment point.

Lift the carpet at the rear end of the propeller shaft drive tunnel and remove the rubber plug from the hole in the tunnel near its junction with the floor. Cut a 1 in. (25·4 mm.) diameter hole in the carpet centred around the hole in the tunnel.

The short belt is fitted with the angle of the attachment bracket facing away from the tunnel and the large diameter of the distance piece bearing against the tunnel. Assemble the anti-rattle washer on the small diameter of the distance piece with its concave face away from the shoulder. Fit the belt bracket over the small diameter of the distance piece with the angle of the bracket facing away from the large diameter.

Assemble the plain washer to the hexagon set screw and pass the screw through the belt bracket, distance piece and tunnel fixing hole. Make sure that the shouldered distance piece makes a complete metal to metal contact with the tunnel, and fit the nut and spring washer to the set screw from the inside of the tunnel.

Note

The short belt must be fitted to the same side of the tunnel as the seat for which the belt is to be used.

Section PPP.3

LOWERING AND RAISING THE HOOD OF THE CONVERTIBLE MODEL (Mk. II and Mk. III)

To lower the hood proceed as follows:

(1) Pull the rear seat squab forwards from its rubber catches.
(2) Release the toggle catches at each end of the windscreen top rail. Pull the top of the catches downwards and then disengage the bottom of the catches from the hooks on the windscreen pillars.
(3) Standing beside the car, lift the front hood rail from the windscreen allowing the three hood supporting rails to collapse together. Ensure that the hood material and the back window fold naturally without buckling or creasing.

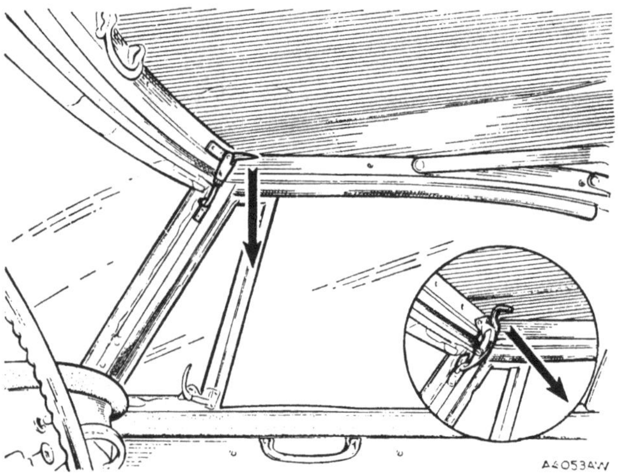

Fig. PPP.5.
Release the toggle catches at each end of the windshield top rail. Earlier type shown inset.

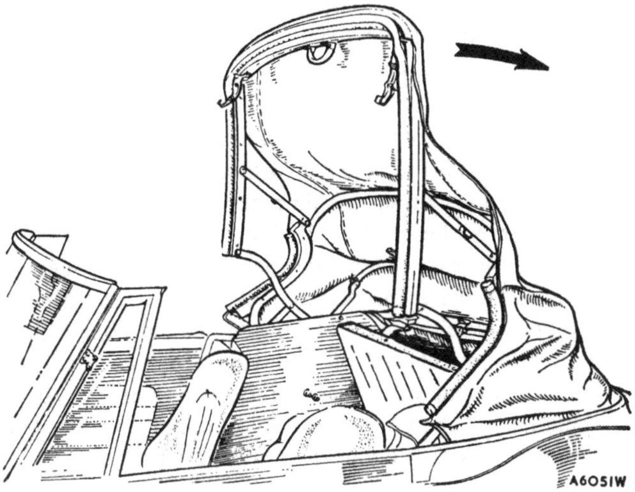

Fig. PPP.6.
Lift the front hood rail from the windshield allowing the hood supporting rails to collapse together.

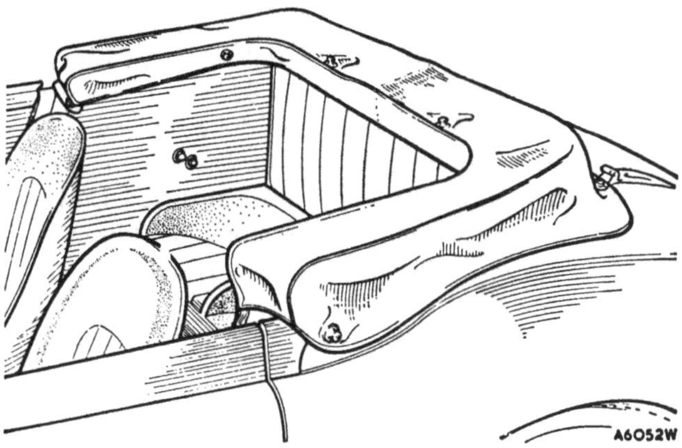

Fig. PPP.8.
Fit the hood cover using the fasteners provided.

(4) Push the front section of the hood backwards, inverting it in the process, and ease the hood and supporting rails into the well behind the rear seat.

(5) Press the front hood rail section downwards on each side of the car in turn, making sure that the hood framework does not damage the interior trim panels.

(6) Push the rear seat squab back into its retaining catches.

(7) Fit the hood cover, securing first the two Tenax fasteners on each side of the car adjacent to the doors and then the fasteners on each interior trim panel. Press home the two fasteners on the tonneau panel and then secure the hood cover to the top of the rear seat squab with the two fasteners provided.

Raising the hood is a reversal of the above sequence. Having lifted the hood from the well behind the rear seat, sit inside the car and use the handle at the centre of the front hood rail to pull the hood down towards the top rail of the windshield. Engage the bottom of the toggle catches in the hooks on the windshield pillars and press the top of the catches firmly forwards and upwards towards the hood.

Having raised the hood and fastened the catches, make sure that the lower edge of the hood material immediately behind the door openings is inserted into the cockpit drain channel on each side of the car.

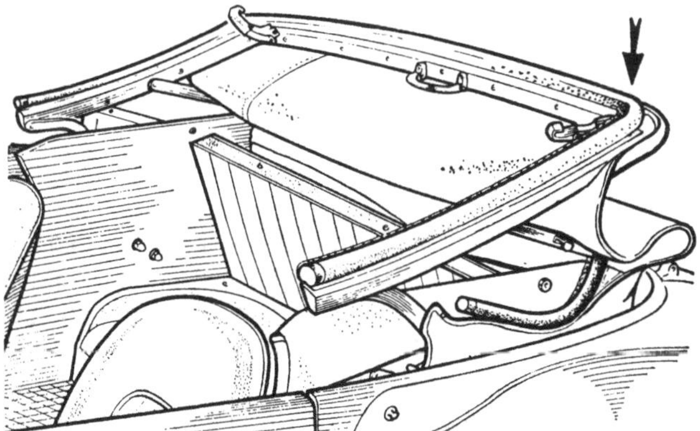

Fig. PPP.7.
Press the front hood rail section downwards.

Fig. PPP.9.

Rear window

The rear window may be folded down when extra ventilation is required with the hood in the raised position. Undo the two clips at the top of the window and release the window frame from the three metal clips on the hood frame. Fold the window panel down, avoiding creasing or buckling the transparent window material.

A zip fastener holds the rear window in place on later models.

Section PPP.4

REMOVING AND REPLACING THE HOOD ASSEMBLY (CONVERTIBLE MODEL Mk. II and III)

Removal

To remove the complete hood and hood frame assembly from the car proceed with the following:

Undo the four screws which hold each rear seat pan in position and lift out the rear seats. This will give access to the nuts securing the rear seat squab hinges. Remove the screws and nuts from the two hinges and lift out the seat squab.

Release the door draught excluder from the top edge of the rear interior trim panels on each side of the car. Unscrew the plated dome nuts from the seat belt anchorage points and the self-tapping screws from each panel. Remove both trim panels from the car.

With the hood in the raised position to minimise the tension on the hood framework helper springs, detach the springs at their top ends from the hood mechanism.

Remove the three hood material inner cappings from inside the car. Remove the self-tapping screws to release central capping rail from tonneau panel. Undo the window clips and remove the complete rear window panel from the hood. Remove the self-tapping screws to release the two side capping rails from the body. These will remain attached to the hood material.

The hood mechanism is attached to the body at three points adjacent to the door openings on each side of the car. With the hood still in the raised position, remove the nut from the rearmost of the three fixing points. Lower the hood and remove the two nuts and bolts on each side from the remaining attachment points, noting that at these points packing washers are fitted between the hood frame pivot bracket and the body.

Lift the complete hood from the car.

Replacement

Replacement is a reversal of the removal sequence, but attention must be given to the following points.

When placing the hood assembly on the car, locate the rear hole of each pivot bracket over the rear stud of the body attachment point, with the hood in the collapsed position. With the packing washers in place, fit the two nuts and bolts on each side making sure that the bolts with tapered heads are fitted in the holes nearest the sides of the car. The thin end of the taper must face towards the back of the car. Before tightening the nuts and bolts raise the hood and fit the nut and washer to the body attachment stud on each side of the car. Clip the hood to the top rail of the windscreen to ensure correct alignment of the hood assembly in relation to the windscreen. Fully tighten all the nuts and bolts.

Fit the helper springs to the hood mechanism with the hood in the raised position.

Release the hood from the windscreen and partially fold it back to enable the material side cappings to be screwed to the body. Ensure at this stage that the hood material immediately behind the door openings is correctly located in the cockpit drain channel.

Clip the top of the rear window panel in position and screw the central capping rail into place, noting that the two longest screws must be fitted at the extreme ends of this rail.

Section PPP.5

REMOVING AND REPLACING THE COCKPIT DRAIN TUBE HOSES (CONVERTIBLE MODEL Mk. II and III)

Rubber extension hoses are fitted to the three tubes leading from the cockpit drain channel around the lower edge of the hood. One drain tube is located at the back of the cockpit in the centre of the car. The other two tubes are at the side of the cockpit adjacent to the door openings taking water from the drain channel into each rear wheel arch.

To remove the rear drain tube rubber hose, undo the screws which hold the rear seat pans in position and lift out the rear seats. This will give access to the nuts securing the rear seat squab hinges. Remove the screws and nuts from the two hinges and lift out the seat squab. The rubber hose may now be pulled off its tube and removed from beneath the cockpit trim carpet.

To gain access to a side drain tube rubber hose, remove the two rear seats, the rear seat squab, and the trim panel on the appropriate side of the car.

Replacement is a reversal of the above procedures.

Section PPP.6

REMOVING AND REPLACING THE WINDSHIELD FRAME ASSEMBLY (CONVERTIBLE MODEL Mk. II and III)

Remove the windshield wiper arms from their driving spindles.

Remove the two self-tapping screws from each corner of the fascia panel and the two screws securing the driving mirror to the scuttle. It will now be possible to lift the scuttle top liner assembly from the scuttle, noting the six locating holes in the scuttle for the demister duct bezel pegs.

Unscrew the set bolt holding the bracket at the centre of the windshield bottom rail to the scuttle top.

Remove the four screws from the rear of each windshield pillar. Pull the windshield and frame assembly forwards from the pillars and lift it from the car.

Before replacing the windshield assembly ensure that the aperture at the base of each pillar is sealed with Glasticon Compound or Dum-Dum putty.

A rubber weather strip is located in a channel in the windshield lower frame member to effect a seal between the windshield and the scuttle top. This should be inspected for signs of perishing or splitting, and must be renewed if necessary.

Fit all the windshield screws and the centre-fixing set bolt before fully tightening any of the pillar screws.

Section PPP.7

REMOVING AND REPLACING THE WINDSHIELD GLASS (CONVERTIBLE MODEL Mk. II and III)

Remove the windshield frame assembly from the car (see Section PPP.6).

To dismantle the windshield frame from the glass, remove the two screws from each corner of the frame (top and bottom). The four parts of the frame can now be separated.

It will be seen that the sealing of the windshield glass is performed by a glazing rubber in two sections. One section is in the frame top member, and the other is in the frame lower member and the two side members.

Ensure that the glazing rubber is in good condition without any signs of perishing or cracking. No additional sealing compound is necessary when assembling the glass to the windshield frame with the glazing rubber.

The rubber strip behind the glazing rubber must not overlap the corner brackets.

Insert the shorter section of the glazing rubber into the frame top member and position the longer section centrally in the lower member. Assemble these frame members to the windshield glass. Insert the glazing rubber protruding from the lower member of the frame into the two side members and assemble them to the glass. Refit the screws at both ends of the top and lower frame members.

Section PPP.8

REMOVING AND REPLACING A WINDSHIELD PILLAR (CONVERTIBLE MODEL Mk. II and III)

Remove the windshield frame assembly from the car (see Section PPP.6).

Ease the draught excluder from the pillar which is to be removed, drilling out the rivet holding the excluder at the top of the pillar.

Unscrew the four nuts and bolts holding the lower end of the pillar to the scuttle. Pull the pillar upwards from the body. Retain the packing piece located between it and the body, and the sealing compound from around the base of the pillar.

Reverse the above procedure when replacing the pillar. Select packing pieces to attain the correct fit of the pillar without straining the windshield and frame assembly and fitting all the windshield frame screws before fully tightening the pillar nuts and bolts.

Section PPP.9

REMOVING AND REPLACING THE DOOR COMPONENTS (CONVERTIBLE MODEL Mk. II and III)

Interior Door Handles

Push the handle escutcheon away from the handle to be removed to expose the retaining pin. Push out the pin and withdraw the handle, escutcheon, and spring. When removing the door lock remote control handle it may be necessary to turn the escutcheon until its slots are aligned with the retaining pin. This will allow the escutcheon to be pushed away from the handle.

Door Interior Trim

Remove the interior door handles (see above).

Unscrew the self-tapping screws and remove the plated door pull.

Unscrew the self-tapping screws around the edge of the outer trim panel and remove the panel. Remove the screws around the edge of the inner trim panel and lift off the panel.

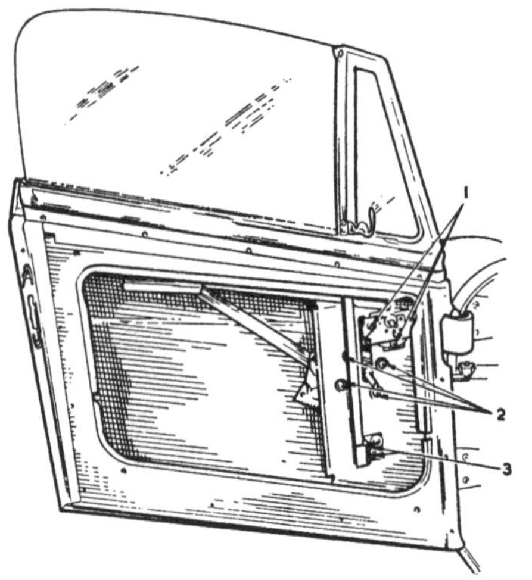

Fig. PPP.9.
A door with the trim panels removed.

1. Door lock remote control screws.
2. Window regulator screws.
3. Window guide channel nuts.

Removal of the outer trim panel will have revealed the self-tapping screws which hold the top trim pad in position.

Door Glass and Ventilator

Remove the door interior trim (see overleaf.)

Unclip the door glass inner weatherstrip and the door waist plated moulding complete with the attached rubber seal.

Remove the four screws retaining the ventilator assembly to the door upper panel. Unscrew the nuts near the window regulator mechanism to release the lower end of the window guide channel.

Lower the glass inside the door and release the regulator arm from its guide channel in the glass frame.

Pull the ventilator assembly and door glass upwards together from the door.

When replacing the ventilator and door glass make sure that the glass is correctly located in the guide channel before assembly to the door.

Window Regulator

Remove the door glass and ventilator.

Unscrew the four screws from the regulator and withdraw the assembly complete with its arm from the door.

Door Lock Remote Control

Remove the window regulator (see above).

Remove the three screws retaining the door lock remote control assembly, and the three screws holding the door lock and guide plate to the door.

Push the catch of the lock down into the locked position and push it through the aperture in the door. Drop the remote control downwards and withdraw the complete assembly from the door.

Door Outer Handle

Remove the door lock remote control (see above).

From inside the door, remove the nut and washer from the handle stud. Pull the door handle up and remove the screw revealed, holding and retrieving the nut and washer from inside the door. The handle assembly with its seating washer may then be withdrawn from the door.

Door lock (later cars)

Remove the door interior trim, see overleaf.

Remove the three screws retaining the door lock and guide plate to the door, and the three screws retaining the door lock remote control assembly.

Remove the two inside screws from the window channel.

Remove the four screws and remove the door lock.

Section PPP.10

SEAT BELTS

The following instructions refer to fitting the approved 'Kangol Magnet' seat belt to the fixing points incorporated in the body structure.

Rear wheel arch

(1) Remove the plastic cap from the fixing boss.

(2) Place the small bracket of the long belt on the short $\frac{7}{16}$ in. bolt followed by the waved washer and distance piece (the small diameter of the distance piece towards the bolt head).

(3) Secure the bracket to the fixing boss.

Side member

(4) Locate the fixing point and cut the carpet to expose the fixing boss.

(5) Place the large washer on the square headed adaptor and secure to the fixing boss.

(6) If the adaptor is not square with the sill when fitted, remove the adaptor, add the small shim washer and repeat item 5.

(7) Place the belt bracket on the $\frac{7}{16}$ in. bolt, followed by the waved washer and distance piece (the small diameter towards the bolt head).

(8) Secure the bracket to the adaptor (from the sill side).

Drive shaft tunnel

(9) Place the bracket of the short belt on the remaining $\frac{7}{16}$ in. bolt followed by the waved washer and distance piece (the small diameter towards the bolt head).

(10) Secure the bracket to the exposed boss on the same side as the seat for which the belt is being fitted.

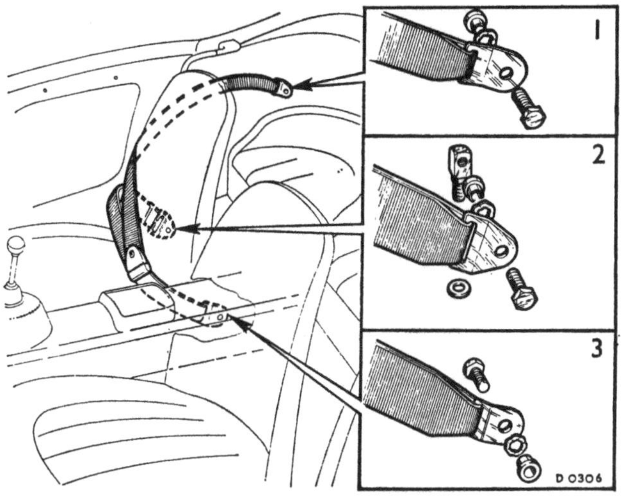

Fig. PPP.10

Seat belt fixings

1. Rear wheel arch. 2. Side member. 3. Drive shaft tunnel.

SECTION Q

LUBRICATION

RECOMMENDED LUBRICANTS

RECOMMENDED LUBRICANTS

BP	CASTROL	DUCKHAM'S	ESSO	FILTRATE	MOBIL	SHELL	STERNOL
ENGINE AND GEARBOX							
All conditions down to 5°C. (41°F.)							
Energol SAE 40 or Super Visco-Static 20W/50.	Castrol XL	Q20/50	Esso Motor Oil 40/50. Esso Motor Oil 40. Esso Extra Motor Oil 20/40.	Filtrate Heavy Filtrate 20W/50.	Mobiloil AF Mobiloil Special 20W/40.	Shell Super Motor Oil Shell X-100 40. Shell X-100 Multigrade 20W/40 or 20W/50.	Sternol WW 40 or WW Multi-grade 20W/50
All conditions between 5°C. (41°F.) and —12°C. (10°F.)							
Energol SAE 20W. Super Visco-Static 10W/40 or Visco Static.	Castrolite or Castrol XL.	Q20/50 or Q5500	Esso Motor Oil 20, 20W/30 or Esso Extra Motor Oil 10W/30	Filtrate Zero or Filtrate10W/30	Mobiloil Arctic or Mobiloil Special 10W/30 Mobiloil Special 10W/40	Shell Super Motor Oil Shell X-100 20W. Shell X-100 Multigrade 10W/30 or 20W/40 or 20W/50	Sternol WW 20 or WW Multi-grade 10W/40 or 20W/50
All conditions below —12°C. (10°F.)							
Energol SAE 10W or Super Visco-Static 10W/40 or Visco-Static	Castrol Z or Castrolite	Q5500	Esso Motor Oil 10W Esso Extra Motor Oil 10W/30	Filtrate Sub-Zero 10W or Filtrate 10W/30	Mobiloil Special 10W/30 Mobiloil 10W or Mobiloil Super 10W/40	Shell Super Motor Oil or Shell X-100 10W or Shell X-100 Multigrade 10W/30	Sternol WW 10 or WW Multi-grade 10W/40
REAR AXLE AND STEERING GEAR							
All conditions down to —12°C. (10°F.)							
BP Gear Oil SAE 90 EP	Castrol Hypoy	Duckham's Hypoid 90	Esso Gear Oil GP 90/140 or GP 90	Filtrate EP Gear 90	Mobilube GX 90	Spirax 90 EP	Ambroleum EP 90
All conditions below —12°C. (10°F.)							
BP Gear Oil SAE 80 EP	Castrol Hypoy Light	Duckham's Hypoid 80	Esso Gear Oil GP 80	Filtrate EP Gear 80	Mobilube GX 80	Spirax 80 EP	Ambroleum EP 80
GREASE POINTS (EXCEPT FRONT HUB BEARINGS)							
Energrease L2	Castrolease LM	Duckham's LB 10 Grease	Esso Multi-purpose Grease H	Filtrate Super Lithium Grease	Mobilgrease MP	Shell Retinax A	Ambroline LHT
FRONT HUB BEARINGS							
Energrease B2	Castrolease BNS Grease	Duckham's DB 500 Grease	Esso Bearing Grease B2	Filtrate Bentonite Grease	—	Shell Retinax DX	Ambroline HTB Grease
OIL CAN AND CARBURETTER							
Visco-Static or Super Visco-Static 10W/40	Castrolite	Q5500	Esso Extra Motor Oil 10W/30	Filtrate 10W/30 Multigrade	Mobiloil Special 10W/30 or Mobiloil Super 10W/40	Shell Super Motor Oil	Sternol WW Multigrade 10W/40
UPPER CYLINDER LUBRICANT							
BP Upper Cylinder Lubricant	Castrollo	Duckham's Adcoid Liquid	Esso Upper Cylinder Lubricant	Filtrate Petroyle	Mobil Upperlube	Shell Upper Cylinder Lubricant	Sternol Magikoyl

Fig. Q.1.
The engine oil filler cap.

Section Q.1

ENGINE (A)

Check the oil level in the engine and refill if necessary to the 'MAX' mark on the dipstick and never let it fall below the 'MIN' mark. The oil filler cap is on the forward end of the rocker cover and is released by turning it anti-clockwise.

ENGINE OIL CHANGE (A)

The sump should be drained and refilled with the appropriate grade of lubricant. The sump plug should be removed after a journey when the oil is still warm and fluid.

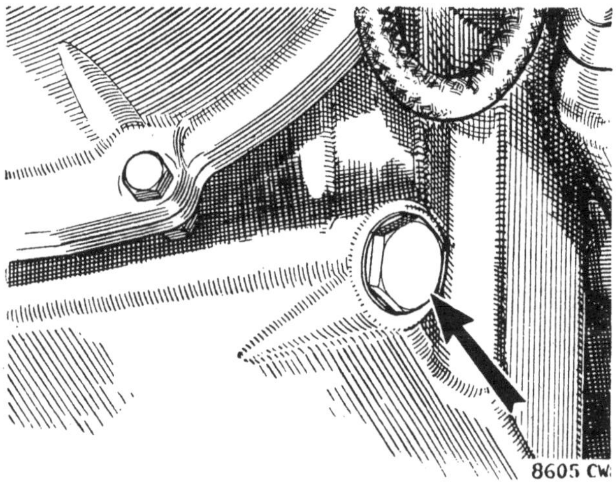

Fig. Q.2.
The engine sump drain plug.

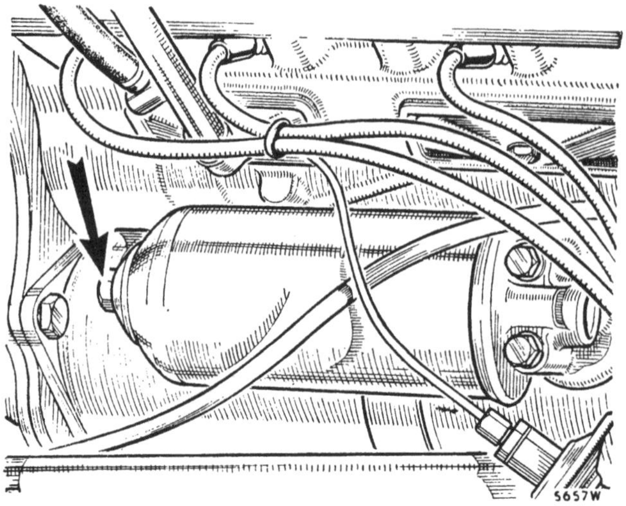

Fig. Q.3.
The engine oil filter with its centre-securing bolt.

ENGINE OIL FILTER

Fit a new engine oil filter element.

The oil filter is of the full-flow renewable element type and the bowl must be removed and washed in petrol (gasoline). The filter is released by unscrewing the central bolt securing the filter to the filter head, Place a suitable container under the assembly to catch the oil that will be released when the seal between the bowl and the cylinder block is broken.

When refitting ensure that the seating washer for the filter body is correctly positioned, clean and serviceable. Ensure that the washers below the element inside the bowl are fitted correctly. The small felt washer must be positioned between the element pressure plate and the metal washer above the pressure spring. It is essential for efficient oil filtration that the felt washer should be in good condition and be a snug fit on the centre-securing bolt.

Run the engine and make certain that there are no oil leaks.

DYNAMO BEARING (D)

Apply a few drops of oil to the commutator end dynamo bearing via the oil hole provided in the bearing housing.

Do not over-oil.

AIR CLEANERS (A)

Remove, clean in petrol (gasoline), drain and moisten the air cleaners with oil. In exceptionally dusty conditions this attention may be required at more frequent intervals.

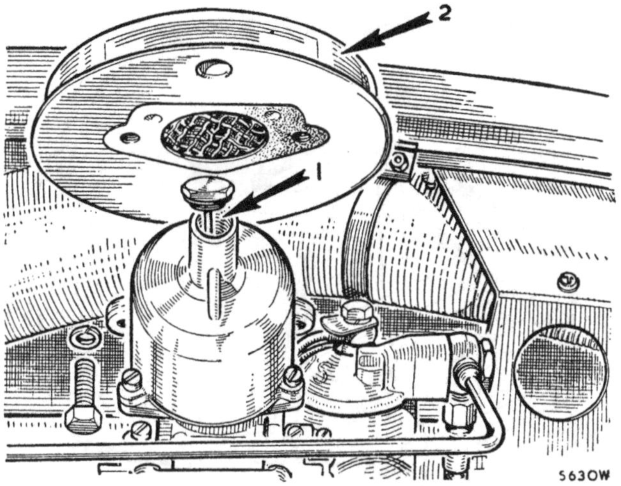

Fig. Q.4. A carburetter damper reservoir (1) and an air cleaner (2)

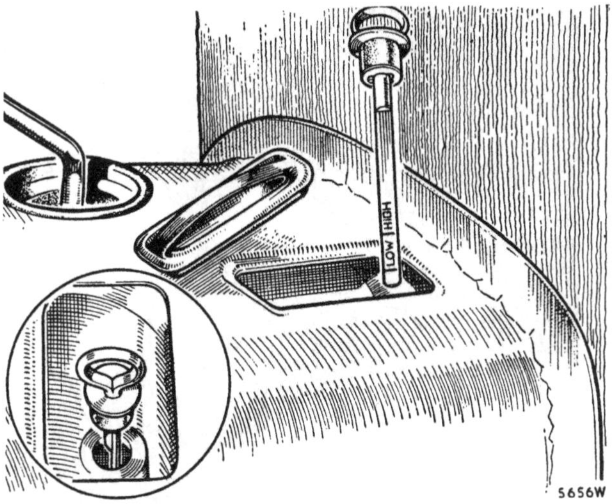

Fig. Q.6. The gearbox combined filler plug and dipstick.

CARBURETTER DAMPERS (D)

Unscrew the top from each suction chamber, pour in a small quantity of oil to bring the oil level to within ½ in. (13 mm.) of the top of the hollow piston rod, and replace the cap. In no circumstances should a heavy-bodied lubricant be used. Failure to lubricate the piston dampers will cause the pistons to flutter and reduce acceleration.

WATER PUMP (C)

Remove the water pump lubrication plug on the water pump casing and add a small quantity of grease. The greasing of the pump must be done very sparingly, otherwise grease will run past the bearings on to the face of the carbon sealing ring impairing its efficiency.

Section Q.2

GEARBOX (A)

GEARBOX AND OVERDRIVE (where applicable)

Check the oil level and top up if necessary. For access lift the floor covering and take out the inspection panel in the top right-hand side of the gearbox cover when the filler plug will be accessible.

On the later central gear change gearboxes, the combined dipstick and filler plug is located under an access panel on the left-hand side of the gearbox cover in front of the gear lever.

Remove the combined dipstick and filler plug, and fill to the correct level with oil.

The capacity of the gearbox and overdrive unit, which are connected by oilways is given in 'General Data'.

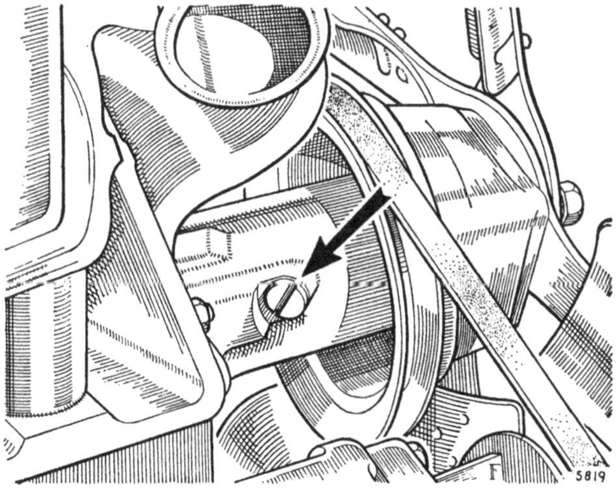

Fig. Q.5. The water pump plug.

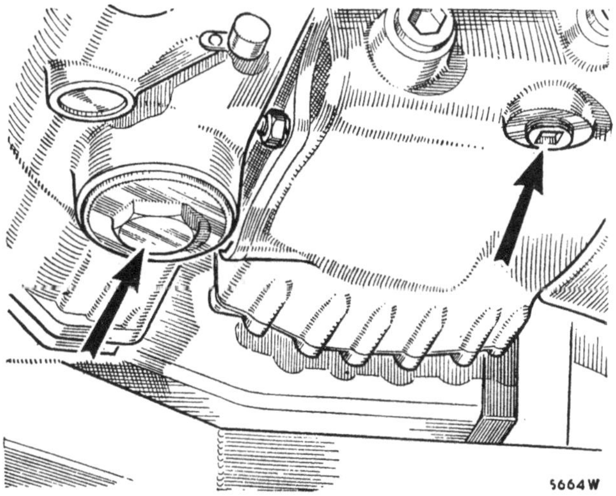

Fig. Q.7. The overdrive and gearbox drain plugs.

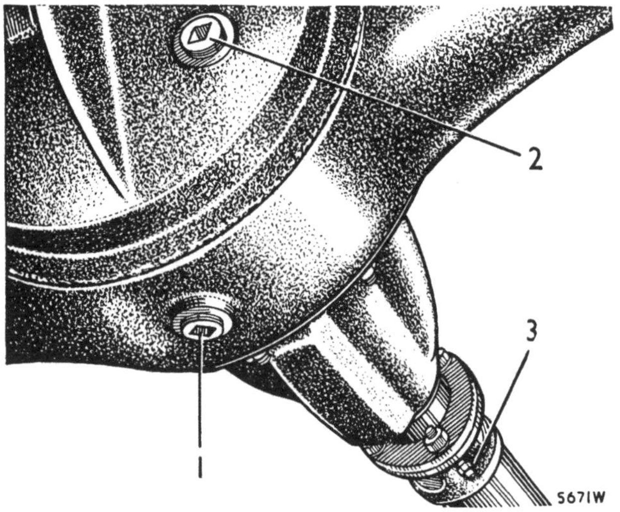

Fig. Q.8. The rear axle drain plug (1), filler plug (2) and the propeller shaft rear universal joint lubrication point (3).

Section Q.3

REAR AXLE (B)

The combined filler and level plug situated on the rear of the axle casing is reached from underneath the car. Use a special key to remove the plug and replenish up the filler plug hole if necessary with oil. Wipe away excess oil from the casing.

NOTE.—It is essential that only hypoid oil is used in the rear axle.

Section Q.4

STEERING (B)

STEERING GEARBOX (B)

Check the oil level, and top up if necessary with oil

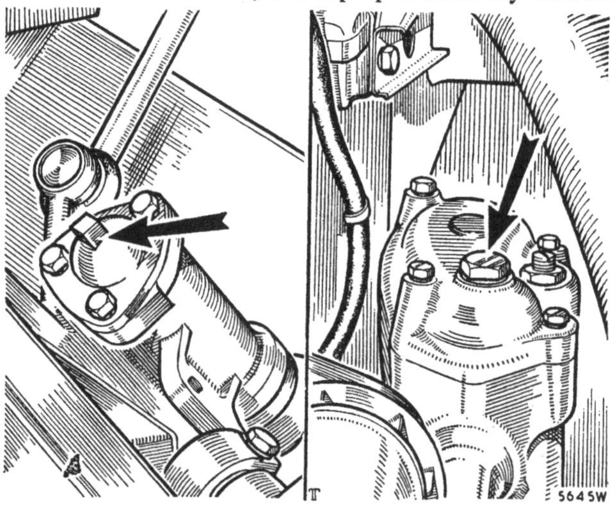

Fig. Q.9. The steering idler and steering gearbox filler plugs. NOTE.—On earlier cars the steering gearbox filler is located on the side of the unit and access to it is gained from under the wheelarch.

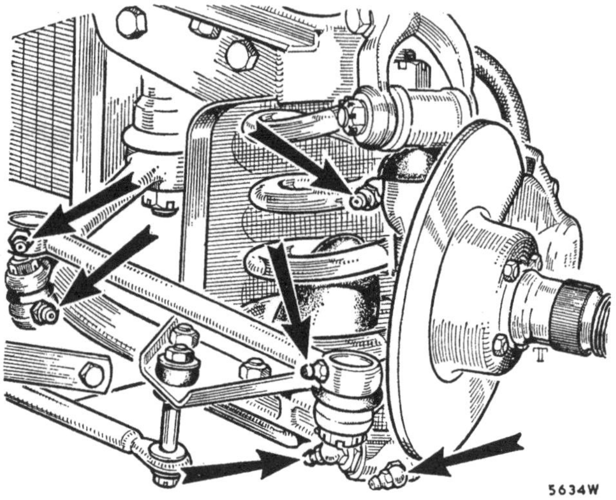

Fig. Q.10 The steering and front suspension lubrication points.

The correct level is flush with the bottom of the filler hole.

Ensure that no dirt enters the steering gearbox when removing or replacing the filler plug.

STEERING IDLER (B)

Remove the square-headed plug from the top of the steering idler and replenish if necessary with oil to just below the screw thread.

NOTE.—Steering idler lubrication is extremely important as, if neglected, severe loading will be imposed on the steering gearbox.

Section Q.5

GREASE POINTS (C)

Grease nipples are situated at the points listed below and should receive three or four strokes of the grease gun.

(1) Front suspension lower outer fulcrum pins (one nipple each side).

(2) Swivel pins (two nipples each side). It is better to grease the swivel axle pins when the weight of the car has been taken off the suspension with a jack or sling. This will allow the lubricant to penetrate around the bushes more effectively.

(3) Steering ball joint connections (three nipples each side).

(4) Rear spring shackle pins (one nipple located at the rear end of each spring shackle).

(5) Propeller shaft universal joints (one nipple each) and the sliding yoke (one nipple) at the gearbox end. The sliding yoke and the front universal

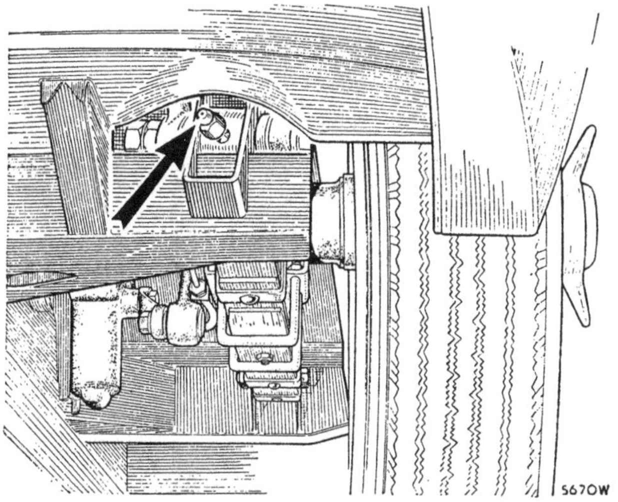

Fig. Q.11. The rear spring rear shackle lubrication point.

joint are best attended to from within the car after the short section at the rear of the gearbox cover has been removed. On later cars with gearbox covers of fibreglass material, a rubber inspection panel on the left-hand side of the rear of the cover may be removed for access to the front universal joint.

The rear joint may be lubricated from below or through the hinged panel behind the seats (2-seater models). Move the car to bring the nipples to the required position.

(6) Handbrake cable (one nipple) and the balance lever (one nipple). These are accessible from underneath the car at the rear axle.

Section Q.6

DISTRIBUTOR (C and D)

Remove the distributor cap and lubricate the following points:

Cam Bearing (D)

Lubricate the distributor camshaft bearing by withdrawing the rotor arm from the top of the distributor spindle and carefully adding a few drops of oil round the screw exposed to view. Take care to refit the rotor arm correctly by pushing it on to the shaft and turning until the key is properly located.

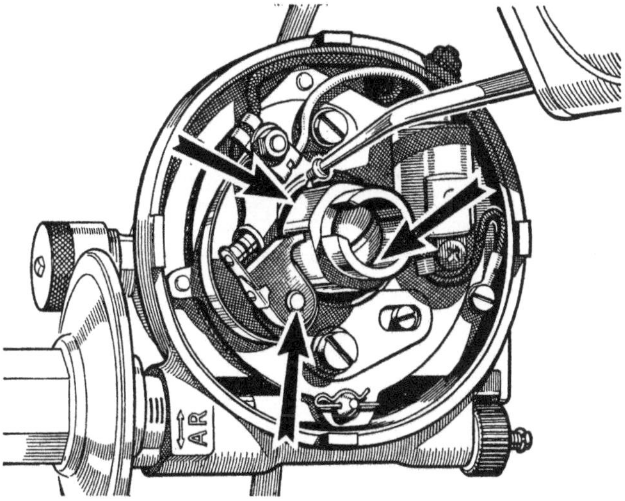

Fig. Q.12. The distributor oiling points.
The arrows indicate the automatic timing control, the cam bearing and the contact breaker pivot oiling points.

Cam (C)

Lightly smear the cam with a very small amount of grease or, if this is not available, clean engine oil may be used.

Automatic Timing Control (D)

Carefully add a few drops of oil through the hole in the contact breaker base-plate through which the cam passes. Do not allow the oil to get on or near the contacts. Do not over-oil.

Contact Breaker Pivot (D)

Add a spot of oil to the moving contact pivot pin.

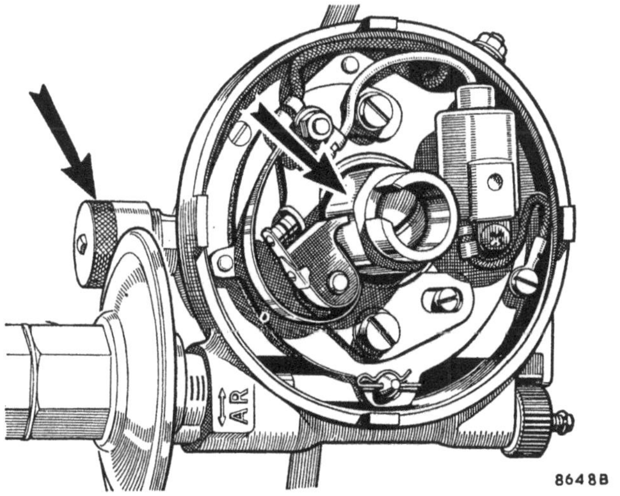

Fig. Q.13. The distributor greasing points.
The arrows indicate the drive shaft greaser and the cam.

SECTION R

CHASSIS FRAME

CHASSIS HORIZONTAL ALIGNMENT CHECK

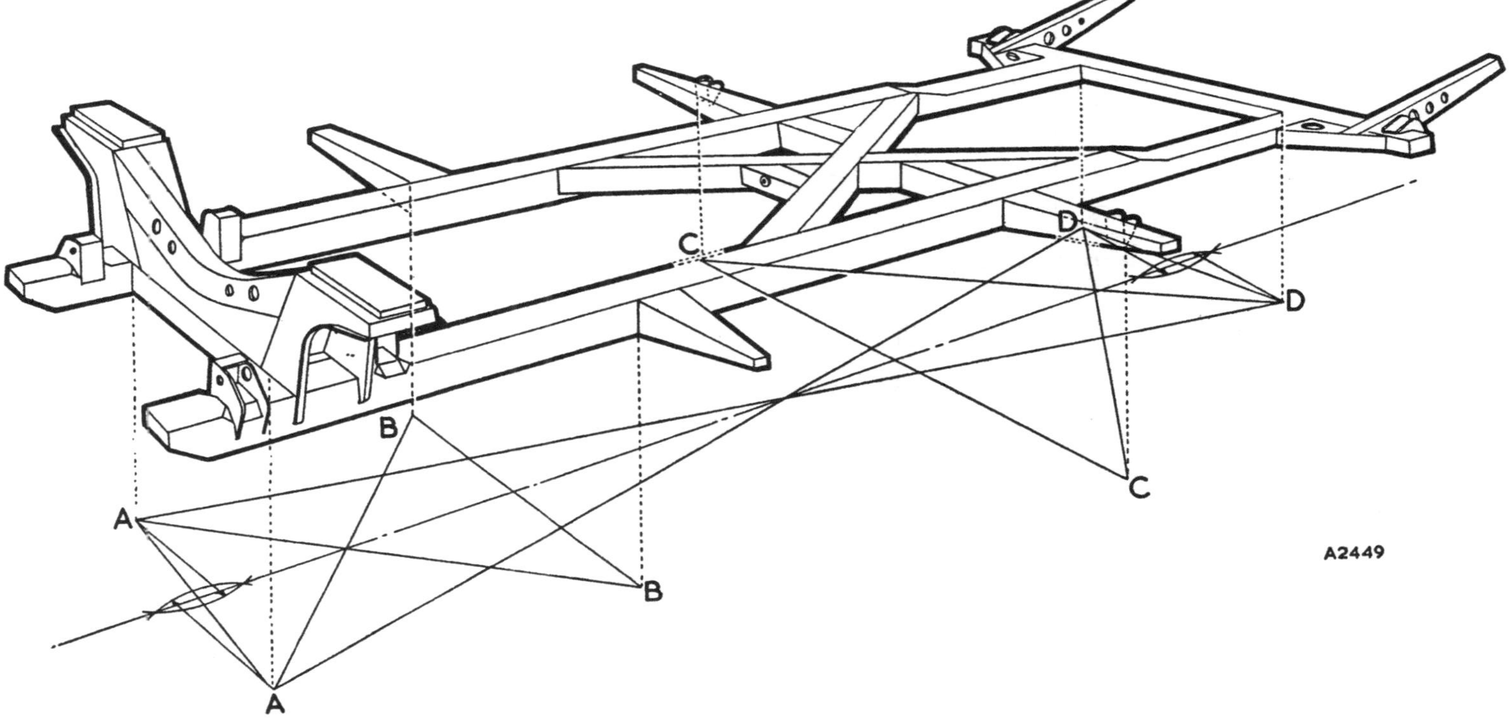

A preliminary check of the alignment can best be carried out by the system of diagonal and measurement checks from points projected on to a level floor by means of a plumb-bob.

A centre-line can then be established by means of a large pair of compasses, and any deviation from correct alignment will be made evident by the failure of the diagonals to intersect on the centre-line or by considerable deviations in the measurements.

A—A	B—B	C—C	D—D
Inside frame width 17 in. (43.2 cm.)	Outside frame width 23 in. (58.4 cm.)	Outside of rear spring front brackets $40\frac{11}{16}$ in. (103.4 cm.)	Inside frame width 17 in. (43.2 cm.)

CHASSIS ALIGNMENT DIAGRAM

(series BN4, BN6, BN7, BT7 and BJ7 cars)

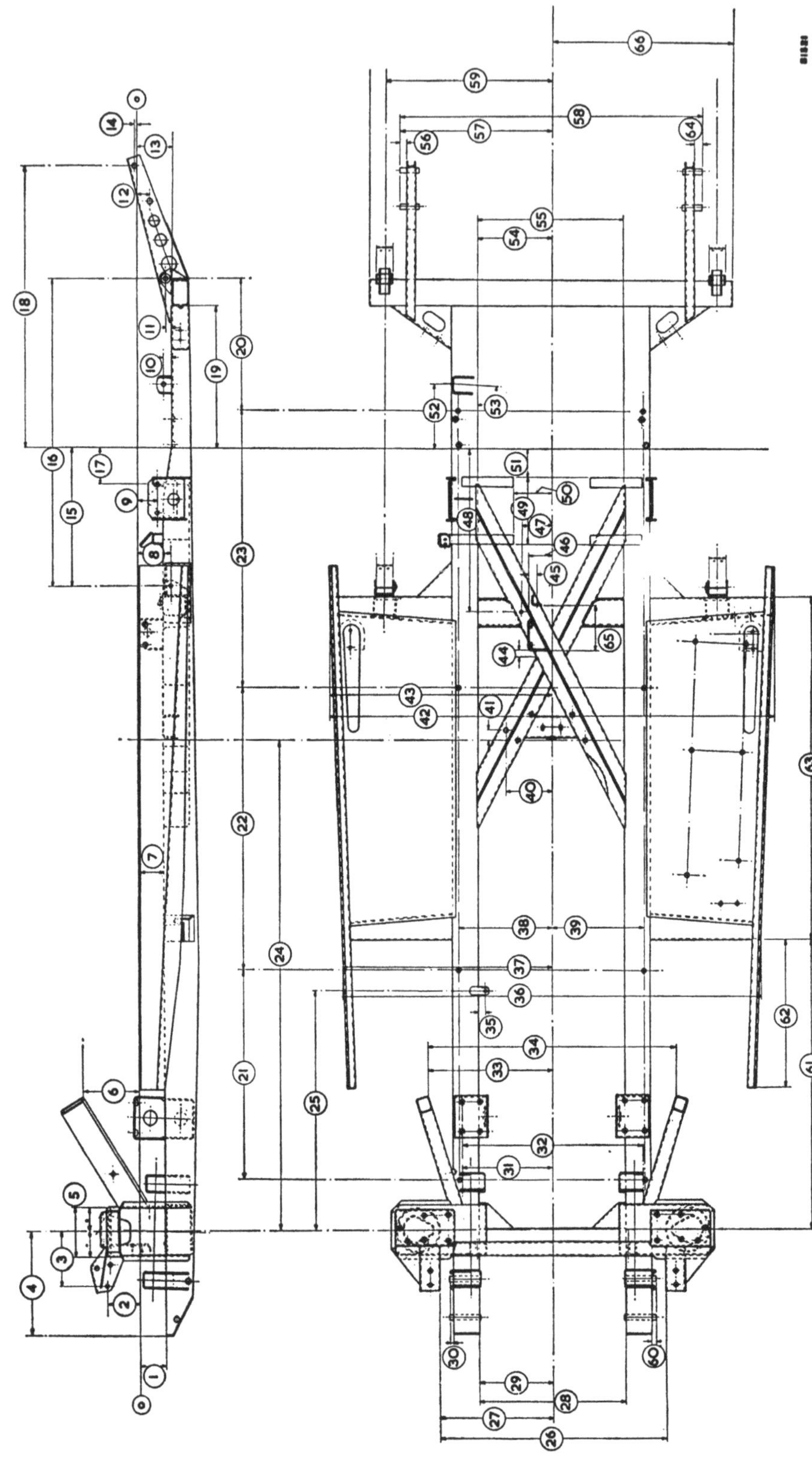

KEY TO THE CHASSIS ALIGNMENT DIAGRAM

(Series BN4, BN6, BN7, BT7 and BJ7 cars)

No.	*Measurement*
1.	$3\frac{1}{4}$ in. (8·26 cm.) parallel.
2.	3·723 to 3·754 in. (9·46 to 9·54 cm.)
3.	6·196 to 6·226 in. (15·74 to 15·81 cm.).
4.	12 in. (30·48 cm.).
5.	$5\frac{3}{4}$ in. (14·6 cm.).
6.	$6\frac{1}{2}$ in. (16·51 cm.).
7.	3 in. (7·62 cm.).
8.	$4\frac{1}{8}$ in. (10·48 cm.)
9.	$2\frac{1}{2}$ in. (6·35 cm.).
10.	1 in. (2·54 cm.).
11.	$\frac{3}{4}$ in. (1·91 cm.).
12.	1·55 in. (3·94 cm.).
13.	$4\frac{1}{4}$ in. (10·8 cm.).
14.	$\frac{11}{32}$ in. (·873 cm.).
15.	16 in. (40·64 cm.).
16.	$35\frac{1}{2}$ in. (90·13 cm.).
17.	$4\frac{1}{8}$ in. (10·48 cm.).
18.	$42\frac{11}{32}$ in. (82.15 cm.).
19.	$16\frac{3}{8}$ in. (41·59 cm.).
20.	15 in. (38·1 cm.).
21.	24 in. (61 cm.).
22.	$32\frac{1}{2}$ in. (82·55 cm.).
23.	32 in. (76·2 cm.).
24.	$56\frac{9}{32}$ in. (143 cm.)
25.	$27\frac{1}{2}$ in. (69·89 cm.).
26.	$26\frac{1}{4}$ in. (66·72 cm.).
27.	$13\frac{1}{8}$ in. (33·36 cm.).
28.	17 in. (43·18 cm.).
29.	$8\frac{1}{2}$ in. (21·59 cm.).
30.	$\frac{1}{2}$ in. (1·27 cm.).
31.	$10\frac{1}{2}$ in. (26·67 cm.).
32.	21 in. (53·34 cm.).
33.	$14\frac{13}{32}$ in. (36·6 cm.).

No.	*Measurement*	
34.	$28\frac{13}{16}$ in. (73·2 cm.).	
35.	$\frac{3}{4}$ in. (1.91 cm.).	
36.	$48\frac{3}{4} \pm \frac{1}{16}$ in. (123·9 ± ·16 cm.).	
37.	$24\frac{3}{8} \pm \frac{1}{32}$ in. (61·95 ± ·08 cm.).	
38.	$10\frac{3}{4}$ in. (27·31 cm.).	
39.	$10\frac{3}{4}$ in. (27·31 cm.).	
40.	$5\frac{1}{4}$ in. (13·34 cm.).	
41.	$1\frac{1}{4}$ in. (3·18 cm.).	
42.	$51\frac{1}{2} \pm \frac{1}{16}$ in. (130·89 ± ·16 cm.).	
43.	$25\frac{3}{4} \pm \frac{1}{32}$ in. (65·45 ± ·08 cm.).	
44.	$\frac{3}{4}$ in. (1·91 cm.).	
45.	$\frac{7}{8}$ in. (2·22 cm.).	
46.	$2\frac{7}{8}$ in. (7·3 cm.).	
47.	$3\frac{1}{2}$ in. (8·89 cm.).	
48.	$18\frac{3}{4}$ in. (47·61 cm.).	
49.	$7\frac{13}{16}$ in. (19·84 cm.).	Series BN6 and BN7 cars only.
50.	$4\frac{1}{2}$ in. (11·43 cm.).	Series BN6 and BN7 cars only.
51.	$3\frac{5}{16}$ in. (8·41 cm.).	Series BN6 and BN7 cars only.
52.	$7\frac{3}{8}$ in. (18·73 cm.).	
53.	87°	
54.	$8\frac{1}{2}$ in. (21·59 cm.).	
55.	17 in. (43·18 cm.).	
56.	$\frac{7}{8}$ in. (2·22 cm.).	
57.	$17\frac{9}{16}$ in. (44·61 cm.).	
58.	$35\frac{1}{8}$ in. (89·22 cm.).	
59.	$19\frac{1}{4}$ in. (48·88 cm.).	
60.	$\frac{1}{2}$ in. (1·27 cm.).	
61.	$33\frac{3}{8}$ in. (84·77 cm.).	
62.	17 in. (43·18 cm.).	
63.	$39\frac{7}{16}$ in. (100·13 cm.).	
64.	$\frac{7}{8}$ in. (2·22 cm.).	
65.	$5\frac{5}{16}$ in. (13·49 cm.).	
66.	21 in. (53·32 cm.).	

Section R.1

CHASSIS ALIGNMENT

To check the chassis alignment of a car which has been damaged a system of diagonal and measurement checks from points projected from the underframe on to a level floor is used (see pages R.1 and R.2).

To ensure that the alignment check is performed accurately the chassis must first be raised so that its datum line (o) (see page R.2) is parallel with the floor. Use the comparative measurements given on page R.2 to achieve this condition. Elevate the rear of the chassis to a convenient working height, and then adjust the height of the front of the chassis until the points given on page R.2 for the front and rear on both sides of the frame are in the correct vertical position **relative to each other;** for example, if the rear point is 36 in. (91·4 cm.) from the floor and is quoted as 2 in. (5·1 cm.) above the chassis datum line, the front point, quoted as 1 in. (2·54 cm.) below the datum line, must be 33 in. (83·8 cm.) above the floor.

At the same time, it will be helpful to check the **relative** heights of all the intermediate points given on page R.19 so that any distortion of the car in the vertical plane will be ascertained.

Chalk over the area of the floor directly below the points shown on page R.1. Using a plumb-line, project the points from the chassis on to the floor, marking the positions with a pencilled cross. The centre between each pair of points can be established by means of a large pair of compasses and the central points marked on the floor. In addition, diagonals can be determined between any two pair of points and the points of intersection marked on the floor. At this stage a length of thin cord covered with chalk can be held by two operators in such a position that it passes through as many of the central points and intersections marked as possible. While the cord is held taut a third operator raises the centre of it and then allows it to spring back smartly to the floor. If the resulting white line passes through all the points the chassis alignment is satisfactory. Any points through which the white line does not pass will be in a position where the underframe is out of alignment.

Considerable deviations in the transverse and longitudinal measurements given on page R.2 confirm chassis misalignment. It must be understood that allowance must be made for normal manufacturing tolerances and that a reasonable departure from nominal dimensions can be permitted without detriment to road performance.

SECTION S

HEATER

Section S.1

HEATER KITS

Heater kits were fitted as standard equipment to the BN4 models commencing at car No. 68960 and to BN6 models commencing at car No. 501. The kits are available under Part No. 8G9048.

Section S.2

HEATER FITTING INSTRUCTIONS

(1) Drain the cooling system.

(2) Remove the fresh air intake control and panel, discard the panel and fit the control cable and knob (13) to the new lever control panel. Remove the heater cover which is secured to the dash fitting the heater unit. Remove the grommets in the dash panel to enable the inlet and outlet pipes to pass through.

(3) Fit the heater temperature control cable to the tap (1) on the heater unit. Attach the long rubber inlet pipe (2) and the short outlet pipe (3) to the heater top with the clips supplied. The flexible hose (4) from heater to blower unit should also be fitted to the heater unit at this stsge. Offer up the heater unit crossing the inlet pipe over the outlet pipe before pushing them through the dash panel. Secure the heater unit to the dash panel with the discarded cover fixings but do not tighten the fixings until the flexible hose (heater to blower) has been passed through the dash panel and under the front wing. Push the two new grommets (5) on to the inlet and outlet pipes and fit them into the holes in the dash panel.

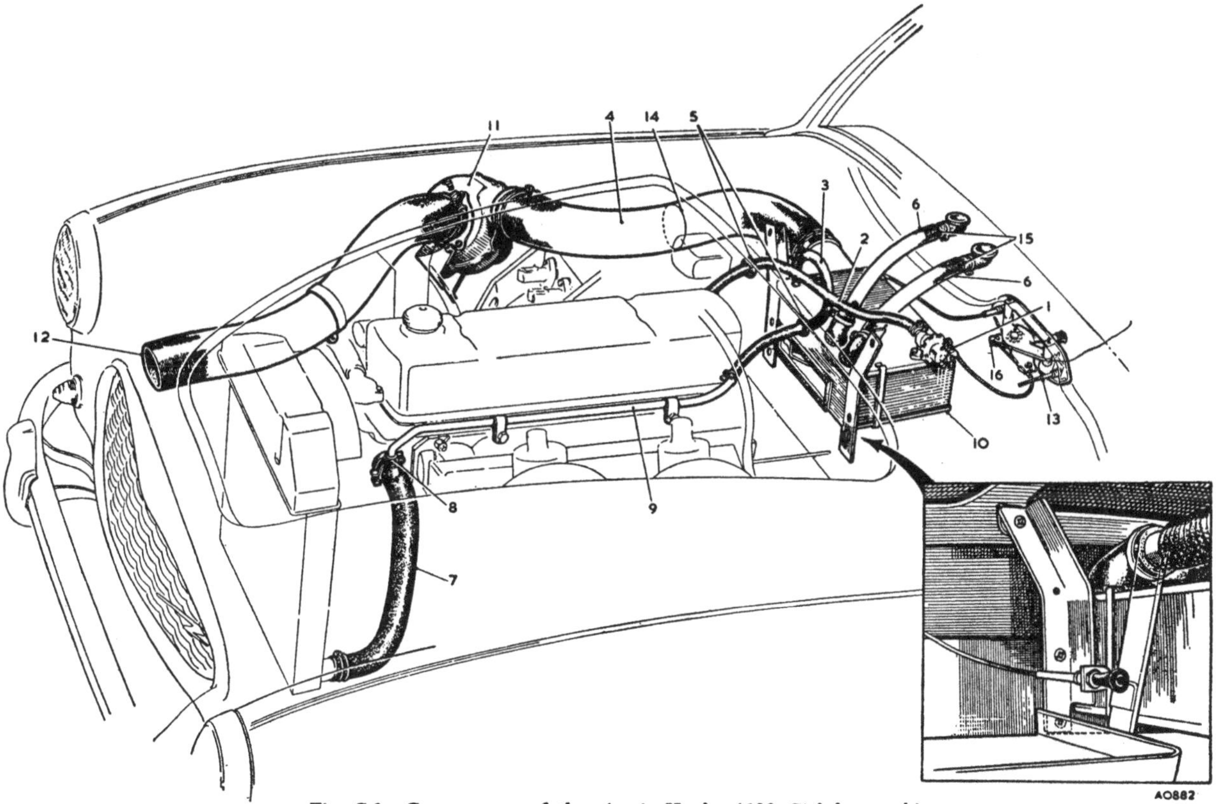

Fig. S.1 Components of the Austin-Healey '100 Six' heater kit fitted in position. The fresh air intake (standard fitment) is not shown.

panel with four setscrews, nuts and spring washers, and the round cover plate (14) for the fresh air intake pipe at the end of the dash panel. Disconnect the choke control (inset) bracket from the parcel shelf (where applicable) and the fascia panel support stay, and move them to one side to allow more freedom of movement when

(4) To fit the demister hoses (6), first remove the aluminium pipes (15) from the elbows and fit them to the long and short flexible demister hoses leaving approximately ¾ in. of aluminium pipe showing. Fit the two rubber elbows into the demister vent holes behind the fascia panel; slip on the securing clips and connect the aluminium

pipes which are already connected to the flexible hoses and secure them with clips. Connect the other ends to the heater unit.

(5) Fit the heater control cable push-pull assembly to the lever (16) on the control panel and thread the cold air control cable (13) through the dash panel; connect the cable to the valve on the cold air intake pipe. Secure the switch panel to the fascia with the existing screws. Refit the choke control bracket and the fascia panel stay.

(6) Remove the bottom radiator hose and fit the new hose (7) supplied in the kit. Fit the copper outlet pipe to the adaptor (8) on the bottom radiator hose then secure the copper pipe (9) to the engine using the carburetter holding bolts through the clips already on the pipe. Secure the rubber outlet pipe to the copper outlet pipe with a clip. Push the rubber inlet pipe on to the water tap on the right-hand side of the cylinder head and secure with a clip.

(7) Place the blower unit (11) in position on the inside of the front wing, thread the flexible air intake pipe behind the wing stays, slip on the securing clip and then push the end of the pipe into the large hole in the front panel (12). Fit the other end on to the blower and secure with a clip. Now connect the free end of the large flexible hose to the blower unit with two clips. Secure the blower unit to the flitch plate with four screws, nuts and washers. Connect the earth wire to the body and the remaining wire to the main harness (green to brown).

(8) Connect the switch wire to the two existing leads which will be found under the fascia panel. Fill the radiator, switch on and test for water leaks.

SECTION T

SERVICE TOOLS

Description	Part No.	BN4 BN6	BN7 BT7
Engine and Clutch			
Crankshaft gear and pulley remover	18G2	*	*
Crankshaft pulley replacer	18G16		*
Crankshaft gear replacer	18G16	*	*
Valve rocker bush remover and replacer	18G21		*
Valve seat cutter and pilot handle	18G27	*	*
Valve seat cutting tool fibre box	18G27B	*	*
Valve seat finishing cutter—exhaust	18G28		*
Valve seat glaze breaker—exhaust	18G28A	*	*
Valve seat narrowing cutter—top—exhaust	18G28B	*	
Valve seat narrowing cutter—bottom—inlet and exhaust	18G28C	*	
Grinding in tool	18G29	*	*
Suction pad for grinding in tool	18G29A	*	*
Valve seat finishing cutter—inlet	18G30		*
Valve seat glaze breaker—inlet	18G30A		*
Valve seat narrowing cutter—top—inlet	18G30B		*
Valve seat narrowing cutter—bottom—inlet and exhaust	18G30C		*
Main bearing cap remover (basic tool)	18G42A	*	*
Main bearing cap remover adaptor	18G42A	*	*
Piston ring compressor	18G55A	*	*
Camshaft gear remover	18G58	*	*
Water pump bearing remover and replacer	18G61	*	*
Oil pump release valve grinding in tool	18G69	*	*
Clutch centralizer	18G79	*	*
Clutch assembly gauging fixture	18G99A	*	*
Valve spring compressor	18G106	*	*
Camshaft liner reamer (basic tool)	18G123A	*	*
Camshaft liner reamer	18G123C	*	*
Camshaft liner reamer	18G123D	*	*
Camshaft liner reamer	18G123E	*	*
Camshaft liner reamer	18G123F	*	*
Camshaft liner reamer	18G123L	*	*
Camshaft liner reamer pilot	18G123T	*	*
Camshaft liner reamer pilot	18G123AA	*	*
Camshaft liner reamer pilot	18G123AB	*	*
Camshaft liner reamer pilot	18G123AM	*	*
Camshaft liner remover and replacer (basic tool)	18G124A	*	*
Camshaft liner remover and replacer adaptor	18G124C	*	*
Camshaft liner remover and replacer adaptor	18G124D	*	*
Camshaft liner remover and replacer adaptor	18G124E	*	*
Camshaft liner remover and replacer adaptor	18G124F	*	*
Camshaft liner remover and replacer adaptor	18G124H	*	*
Camshaft liner remover and replacer adaptor	18G124L	*	*
Valve seat narrowing cutter—bottom—exhaust	18G174C		*
Valve seat cutter pilot	18G174D	*	*
Valve seat finishing cutter—inlet	18G373	*	
Valve seat glaze breaker—inlet	18G373A	*	
Starting nut spanner	18G391	*	*
Cylinder head spanner	18G545	*	*

Description	Part No.	BN4 BN6	BN7 BT7
Gearbox			
Bearing and oil seal replacer (basic tool)	18G134	*	*
Rear oil seal replacer adaptor	18G134N		*
Synchromesh unit assembly ring 1st and 2nd	18G262	*	*
Synchromesh unit assembly ring 3rd and 4th	18G263	*	*
Rear oil seal remover (basic tool)	18G389	*	*
Rear oil seal remover adaptor	18G389B		*
Overdrive			
Oil seal replacer	18G177	*	
Roller clutch assembly spring	18G178	*	
Accumulator housing piston ring compressor	18G179	*	*
Operating piston ring compressor	18G180	*	*
Accumulator housing ring guide	18G181	*	*
Accumulator housing remover	18G182	*	*
Oil pump body remover	18G183	*	*
Oil pump body replacer	18G184	*	*
Dummy mainshaft	18G185	*	*
Mainshaft bearing replacer	18G186	*	*
Hydraulic pressure gauge	18G251	*	*
Rear oil seal remover adaptor	18G389D	*	*
Front Suspension			
Hub assembly remover, wire wheels (basic tool)	18G8	*	*
Hub remover adaptor L.H. wire wheels	18G8H	*	*
Hub remover centre screw extension, wire wheels	18G8J	*	*
Hub remover adaptor R.H. wire wheels	18G8K	*	*
Hub bearing inner race remover adaptor	18G8L	*	*
Front suspension spring compressor	18G37	*	*
Front suspension checking plate	18G56		*
Swivel axle bush reamer—top	18G64	*	*
Swivel axle bush reamer—bottom	18G65	*	*
Swivel axle bush reamer wrench	18G68	*	*
Swivel axle bush remover and replacer	18G85	*	*
Front suspension assembly fixture	18G89	*	*
Hub remover disc wheels	18G220	*	*
Hub remover adaptor, disc wheels	18G220A	*	
Hub remover adaptor ring, disc wheels	18G220D	*	
Hub remover pads, disc wheels	18G220E	*	
Hub bearing remover and replacer (basic tool)	18G260		*
Hub bearing outer race remover and replacer (inner bearing)	18G260J		*
Hub bearing outer race remover and replacer (outer bearing)	18G260K		*
Hub remover disc, wheels	18G304		*
Hub remover adaptors, disc wheels	18G304B		*
Rear Axle and Rear Suspension			
Bevel pinion flange wrench	18G34A	*	*
Differential bearing remover (basic tool)	18G47C	*	*
Differential bearing remover adaptor	18G47AD	*	*
Bearing oil seal replacer (basic tool)	18G134	*	*
Rear hub bearing remover adaptor	18G134K		*
Rear hub oil seal replacer adaptor	18G134AQ	*	

Description	Part No.	BN4 BN6	BN7 BT7
Rear Axle and Rear Suspension—*continued*			
Bevel pinion setting gauge	18G191	*	*
Differential bearing setting gauge	18G191A	*	*
Bevel pinion bearing pre-load gauge	18G207	*	*
Rear hub remover, disc wheels	18G220	*	
Rear hub remover adaptor, disc wheels	18G220A	*	
Rear hub remover adaptor ring	18G220D	*	
Rear hub remover disc pads, disc wheels	18G220E	*	
Rear hub nut spanner	18G258	*	*
Bevel pinion bearing outer race remover	18G264	*	*
Bevel pinion bearing outer race remover adaptor	18G264D	*	*
Bevel pinion bearing outer race remover adaptor	18G264H	*	*
Bevel pinion inner race remover and replacer	18G285		*
Rear hub remover	18G304		*
Rear hub remover bolt adaptors (3)	18G304B		*
Rear hub remover thrust pad	18G304K		*
Propeller Shaft			
Propeller shaft flange remover	18G75A	*	*
Steering			
Steering arm remover	18G75A	*	*
Steering ball joint separator	18G125	*	*
Steering wheel nut spanner	18G512	*	
Miscellaneous			
Radiator reverse flush adaptor	18G187	*	*
Torque wrench 30 to 140 lb.ft. (4.15 to 5.53 kg.m.)	18G372	*	*
Torque wrench 2 to 8 lb.ft. (.28 to 1.11 kg.m.)	18G536	*	*
Torque wrench 5 to 30 lb.ft. (.69 to 4.15 kg.m.)	18G537	*	*

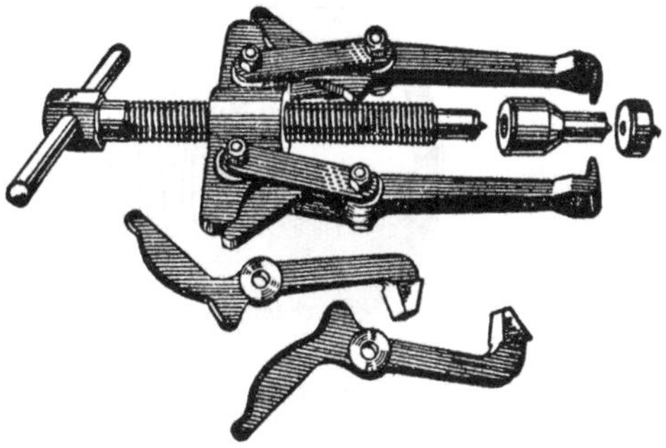

18G2

18G2. Extractor for Driving Flange

A multipurpose tool consisting of:
(1) Extractor (basic tool).
(2) Alternative pair of legs (for pulleys).
(3) Short thrust pad.
(4) Long thrust pad.

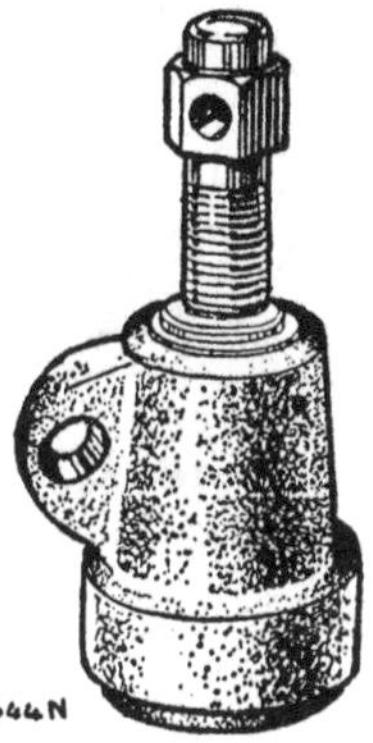

18G8

18H8. Hub Assembly Remover (Basic Tool)

Internally threaded for attachment to the hub and the end of the centre screw provided with a hardened steel ball to reduce friction when engaging the stub axle. Provision is also made for the use of a tommy-bar to prevent the tool turning in operation.

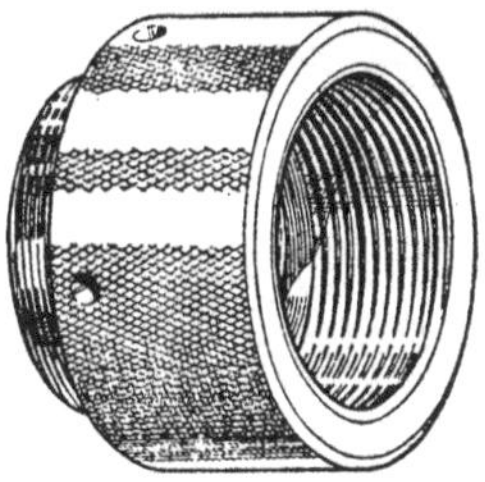

18G8H

18G8H. Front Hub Remover Adaptor L.H. (Wire Wheels only).

Internally threaded to suit left-hand hubs, and must be used in conjunction with remover 18G8. The centre screw extension 18H8J must be used to pilot in the adaptor, thus ensuring a straight pull on the centre screw.

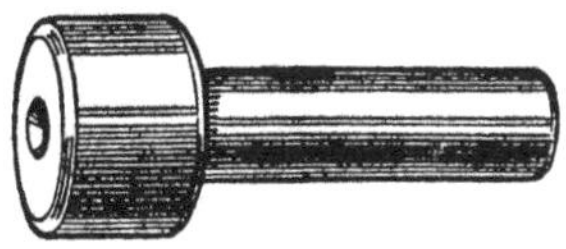

18G8J

18G8J. Front Hub Remover Centre Screw Extension (Wire Wheels only)

Must be used to pilot in the adaptors 18G8H and 18G8K to ensure a straight pull on the centre screw.

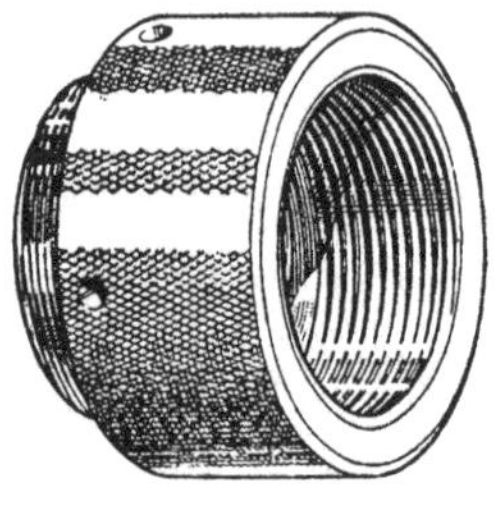

18G8K

18G8K. Front Hub Remover Adaptor R.H. (Wire Wheels only)

Internally threaded to suit the right-hand hub of the vehicle. This adaptor is used with the remover 18G8 and pilot 18G8J.

18G8L

18G8L. Front Hub Bearing Inner Race Remover Adaptor

Use with hub remover 18H8; a peg and chain is provided to retain the locking ring and half-ring when not in use.

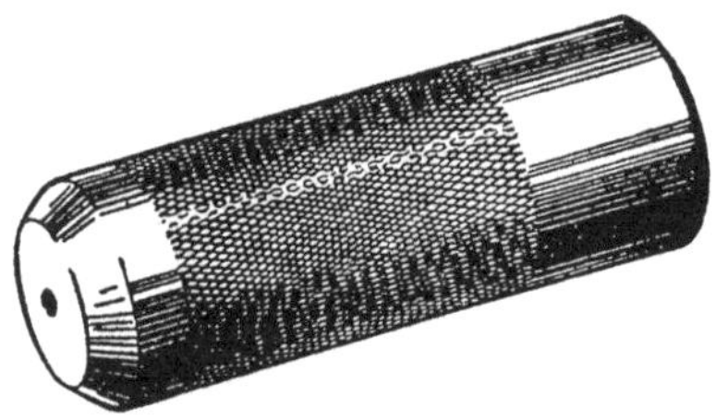

18G16

18G16. Crankshaft Gear and Pulley Replacer

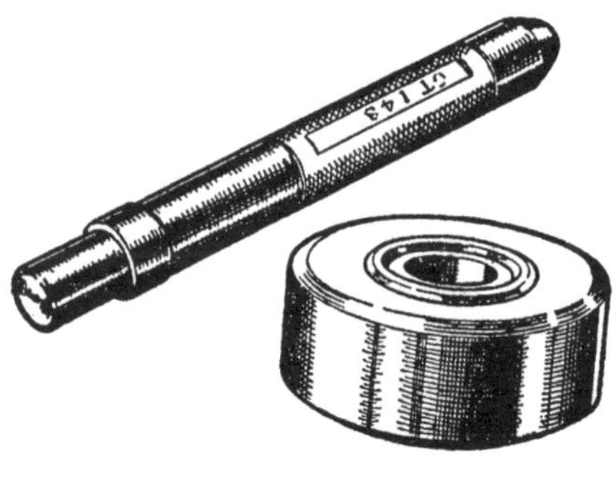

18G21

18G21. Valve Rocker Bush Remover and Replacer

The flange of the driver is recessed to prevent the split bush from opening when being driven into position; the anvil is also recessed to retain the rocker during the operation. Use of a light press is desirable when using this tool; alternatively a vice or copper-faced hammer may be used.

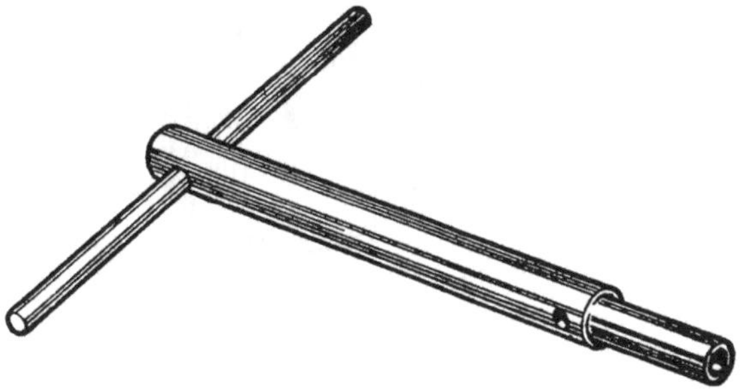

18G27

18G27. Valve Seat Cutter and Pilot Handle

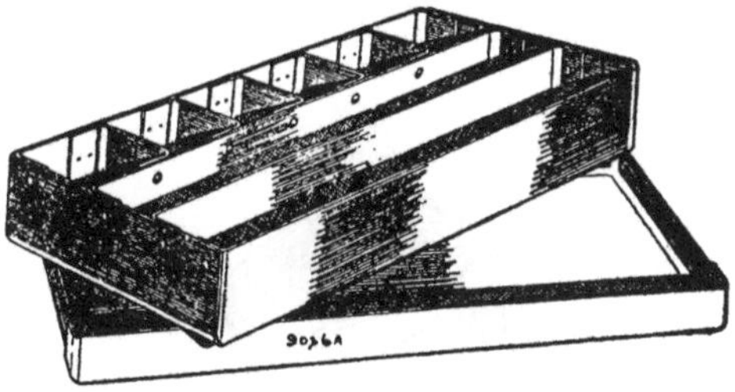

18G27B

18G27B. Valve Seat Cutting Tool Fibre Box

A fibre box for the storage of valve seat cutting tools. Partitioned to protect the machined edge of the cutters.

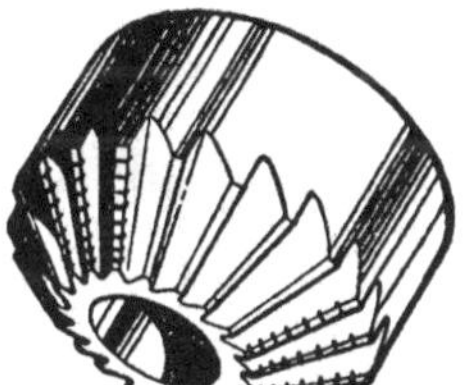

18G28

18G28. Exhaust Valve Seat Finishing Cutter

For use with handle 18G27 and pilot 18G174D.

18G28A

18G28A. Exhaust Valve Seat Glaze Breaker

For use with pilot 18G174D and handle 18G27.

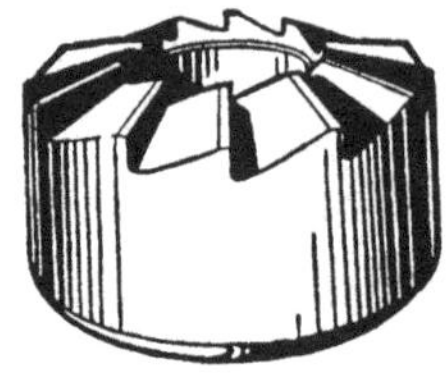

18G28B

18G28B. Exhaust Valve Seat Narrowing Cutter—Top

Use with pilot 18G174D and handle 18G27.

18G29. Valve Suction Grinder

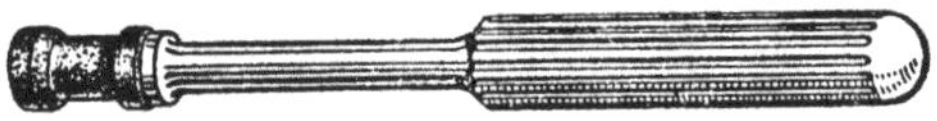

18G29

A handle complete with suction pad 18G29A; a further suction pad 18G29B is available for models fitted with smaller valves.

18G30

18G30. Valve Seat Finishing Cutter

Use with pilot 18G174D and handle 18G27.

18G30A

18G30A. Valve Seat Glazed Breaker

Use with pilot 18G174D and handle 18G27.

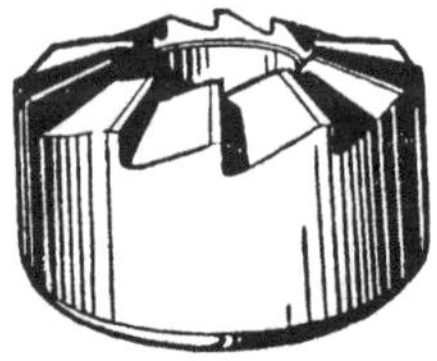

18G30B

18G30B. Valve Seat Narrowing Cutter—Top

Use with pilot 11G174D and handle 18G27.

18G30C

18G30C. Valve Seat Narrowing Cutter—Bottom

For use with pilot 18G174D and handle 18G27.

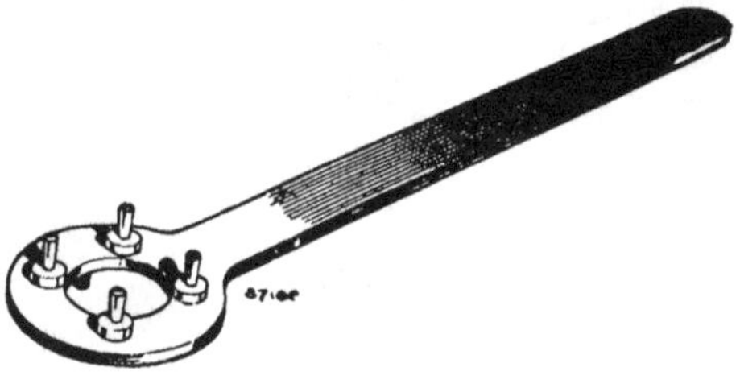

18G34A

18G34A. Bevel Pinion Flange Wrench

The pegs of the wrench engage the holes in the bevel pinion flange to prevent rotation when releasing or tightening the flange out.

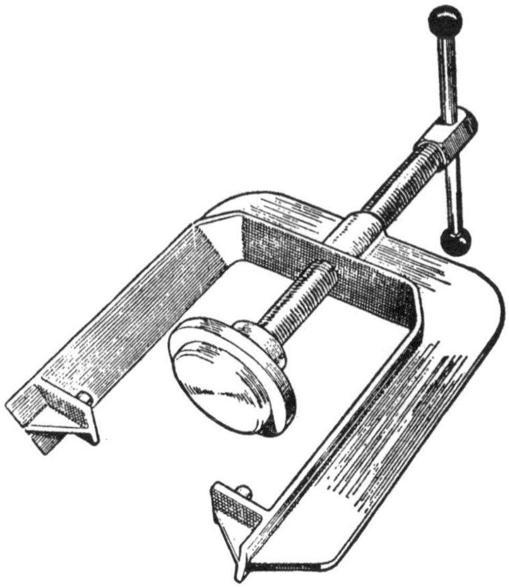

18G37

18G37. Front Suspension Spring Compressor

The spring compressor thrust pad is ball-mounted to assist in lining up the spring and spring seat.

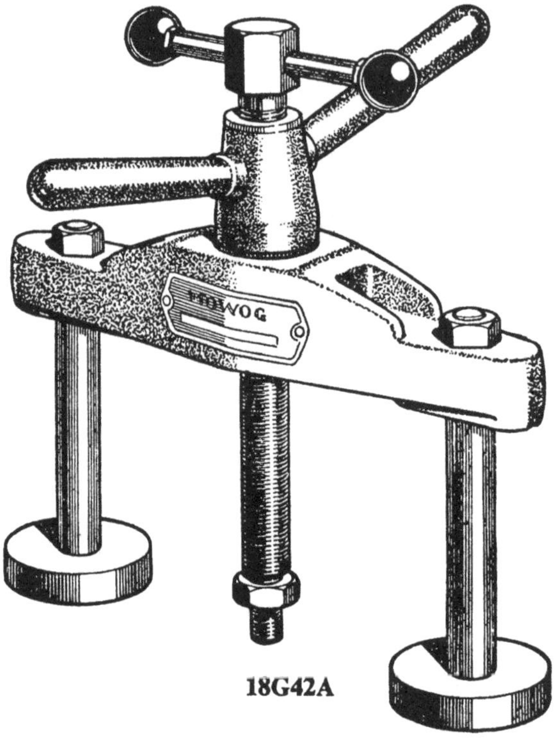

18G42 A Extractor for Main Bearing Caps

18G42A

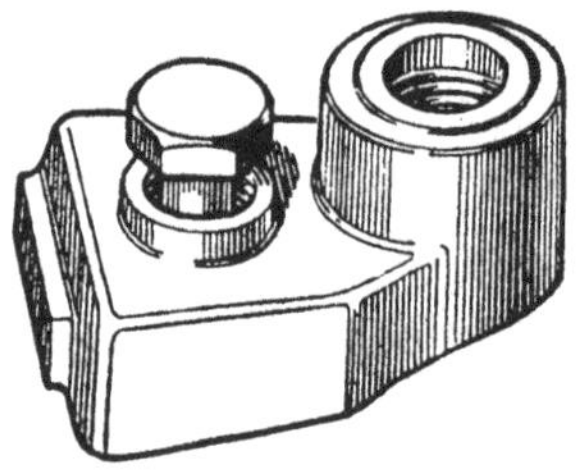

18G42C. Adaptor

The frame has feet suitably spaced to locate on the crankcase flanges. The appropriate adaptor is screwed first onto the drive screw and then into the main bearing cap.

18G42C

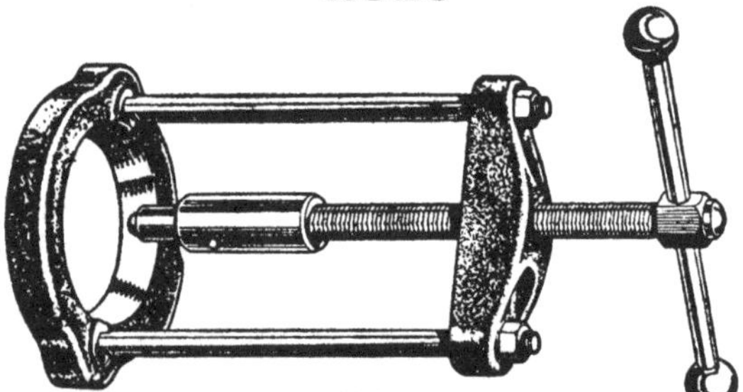

18G47C. Extractor for Differential Bearings

18G47C

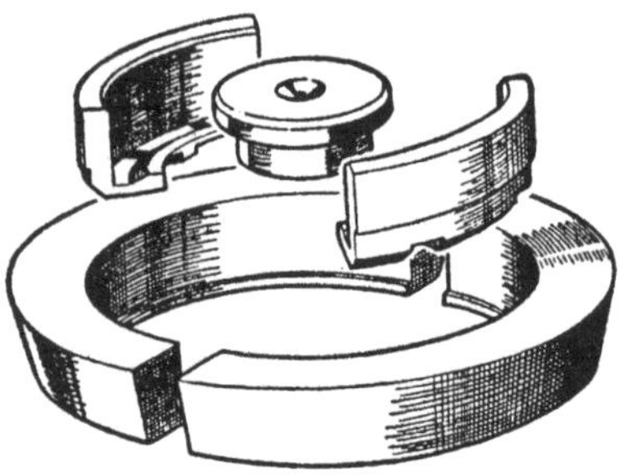

18G47AD. Adaptor

18G47AD

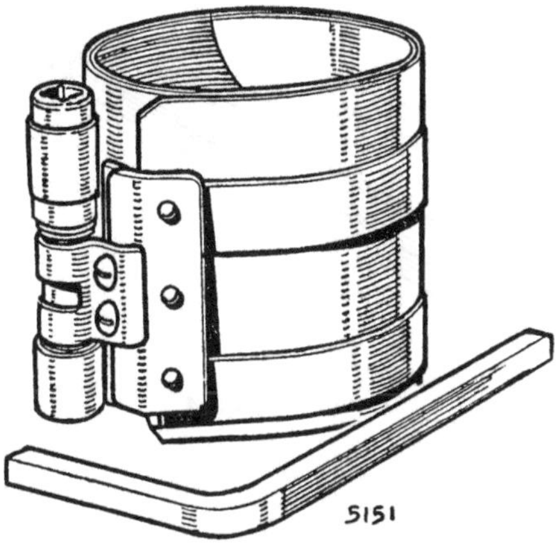

18G55A

18G55A. Piston Ring Compressor

Designed to cover a wide range of pistons, it is easy to operate and will compress the strongest piston ring, making assembly to the bore a quick and easy operation.

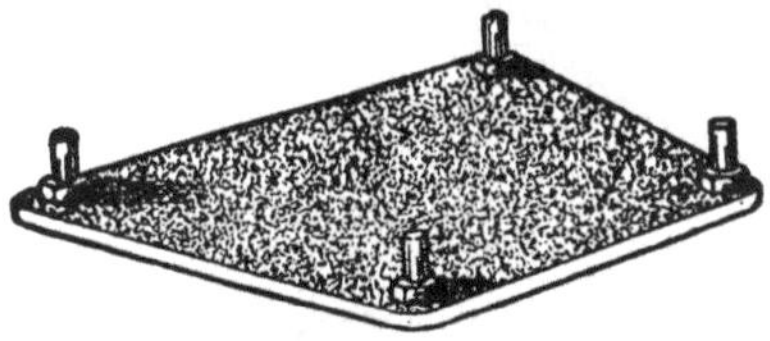

18G56

18G56. Front Suspension Checking Plate

The checking plate has four pegs accurately positioned to enable the lower wishbone arms to be correctly aligned when assembling the front suspension unit.

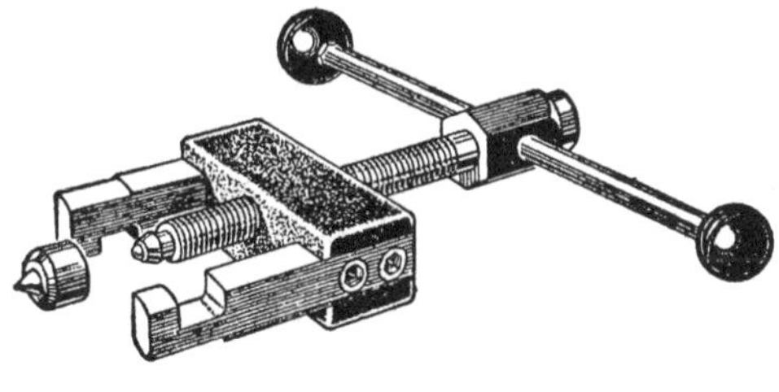

18G58

18G58. Camshaft Gear Remover

This tool will remove the tightest camshaft gear quickly and without damage. The centre drive screw incorporates a hardened steel ball to reduce friction; a separate thrust pad is provided for insertion between the camshaft and the ball to protect the end of the camshaft.

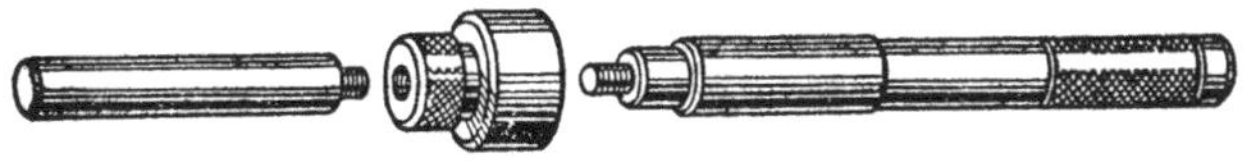

18G61

18G61. Water Pump Bearing Remover and Replacer

This tool, consisting of three parts—a driver, dummy front bearing, and a pilot—will ensure that the rear bearing is removed without damage to the pump body.

18G64. Swivel Axle Bush Reamer—Top

This reamer is designed to pilot in one bush while the other is being reamed, thus ensuring that the holes are directly in line.

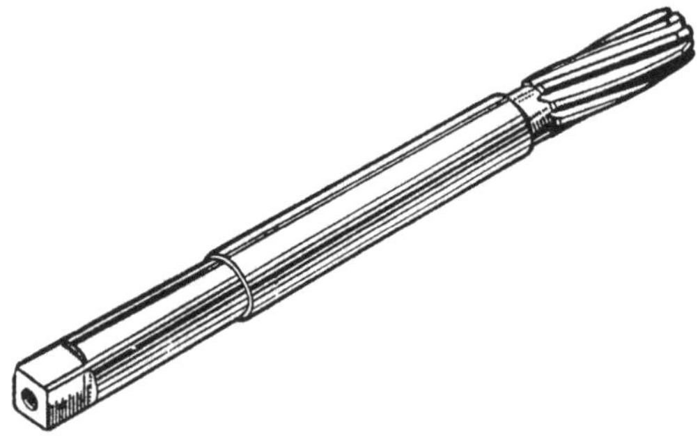

18G64

18G65. Swivel Axle Bush Reamer—Bottom

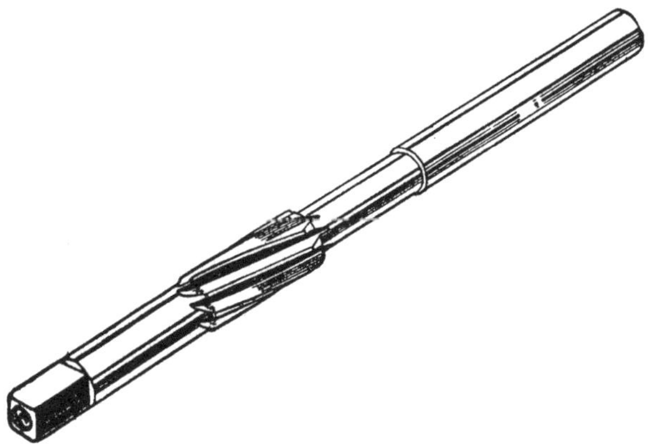

18G65

For use with reamers 18G64 and 18G65. The weight has been carefully determined to ensure that the reamer will pass through the bushes during rotation without any undue pressure by the operator.

18G68

18G69. Oil Pump Release Valve Grinding-in Tool

Designed to facilitate the removal and grinding-in of the engine oil release valve. Tightening the set screw expands the rubber plunger, which ensures that the tool is a tight fit when inserted into the hollow oil release valve.

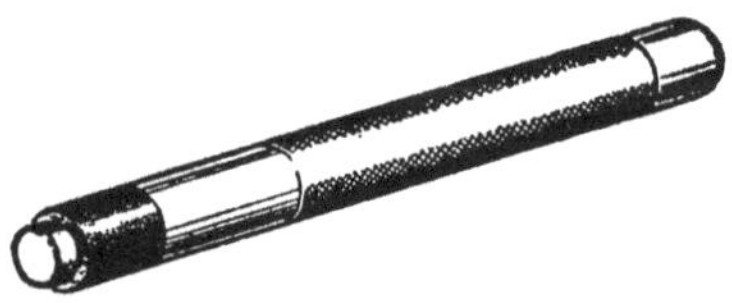

18G69

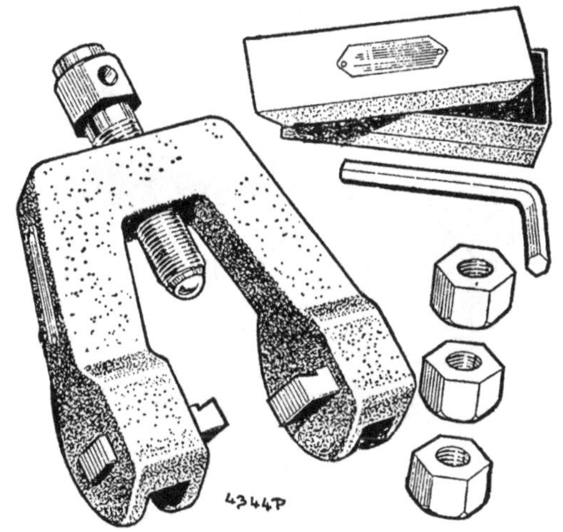

18G75A

18G75A. Steering Arm Remover

Thread protectors 18G75C, to suit BSF-threaded shafts, are included with this tool. The steel jaws are adjustable and an Allen key is provided to release and secure the jaw screws.

18G79

18G79. Clutch Centraliser

18G85

18G85. Swivel Axle Bush Remover and Replacer

This tool enables swivel axle bushes to be removed and fitted without the distortion which would occur if an improvised drift were used. The shoulder of the driver is recessed to prevent the split bushes from opening when being pressed into position. The tool should be used with a press.

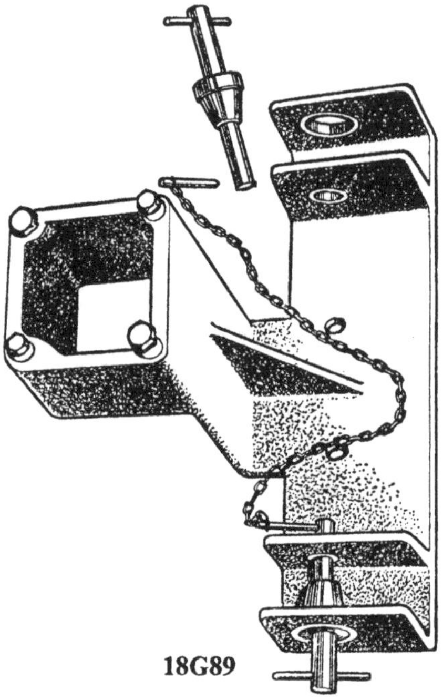

18G89. Front Suspension Assembly Fixture

Designed to serve the dual purpose of accurately assembling a front suspension unit for fitting as a replacement to a vehicle, and also as a means of checking a unit suspected of damage or misalignment. This tool must be used in conjunction with checking plate 18G56.

18G89

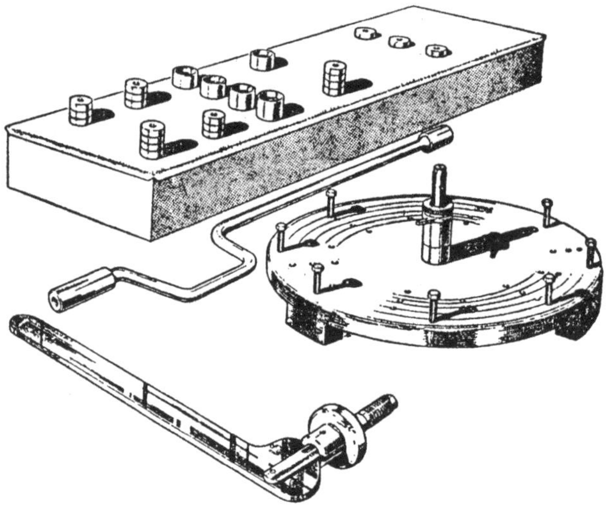

18G99A. Clutch Assembly Gauging Fixture

With the use of this tool a clutch assembly can be quickly dismantled, rebuilt, and finally adjusted with a high degree of accuracy. This is a universal tool for clutch assembly from 6¼ in. to 11 in. diameter.

18G99A

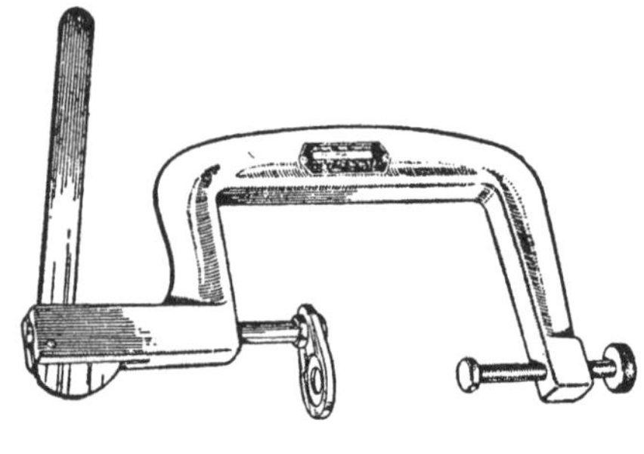

18G106. Valve Spring Compressor

This tool is designed for overhead-valve engines. It has a cam and lever action, and screw adjustment. The adaptor ring is shaped to facilitate the fitting of cotters.

18G106

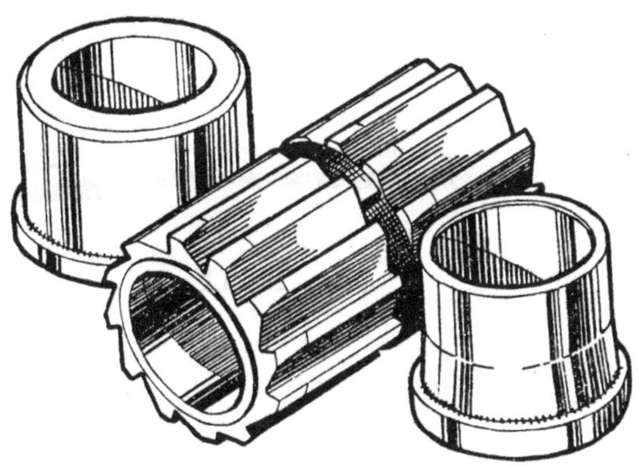

18G123A. Camshaft Liner Reamer (basic tool)

Cutters: 18G123C, 18G123D, 18G123E, 18G123F.
Pilots: 18G123AB, 18G123L, 18G123T, 18G123AA, 18G123AM.

This equipment is essential when reconditioning cylinder blocks, otherwise camshaft liners cannot be reamed in line and in consequence the clearance between the camshaft journal and the liner will be incorrect. The basic tool must be used with the cutters and pilots supplied separately. Operating instructions are supplied with each basic tool.

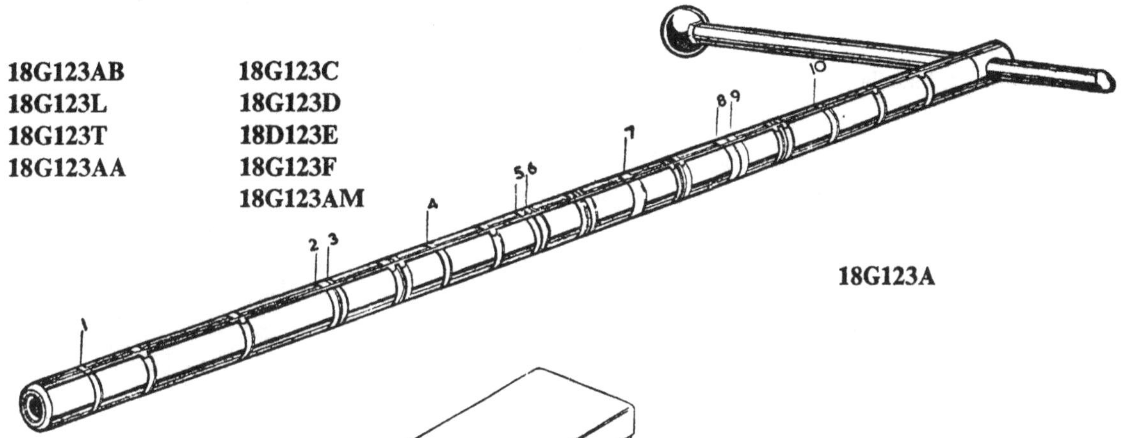

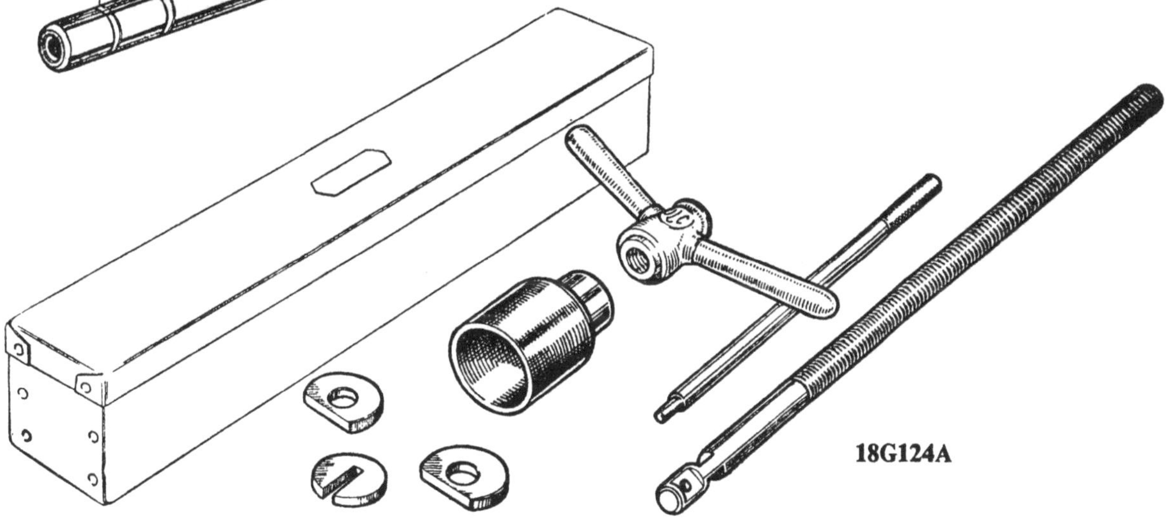

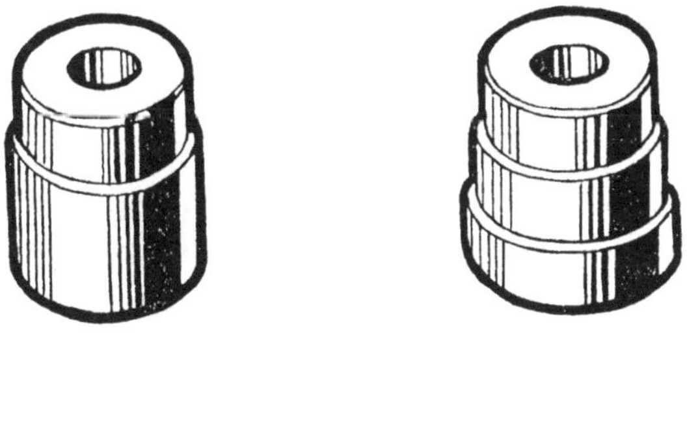

18G124A. Camshaft Liner Remover and Replacer (basic tool)

Adaptors **18G124C, 18G124D, 18G124E, 18G124F, 18G124H, 18G124L.**

This equipment consists of a basic tool 18G124A and various adaptors supplied separately. Liners can be removed and new liners pulled into the cylinder block without the damage invariably associated with the use of improvised drifts. Full operating instructions are included with each basic tool.

18G125. Track-rod Ball Joint Separator

This tool is designed on the shock and wedge principle to aid the removal of the track-rod without damage to the steering arms. The handle is short to obviate mishandling, and on no account must it be used to lever the components apart.

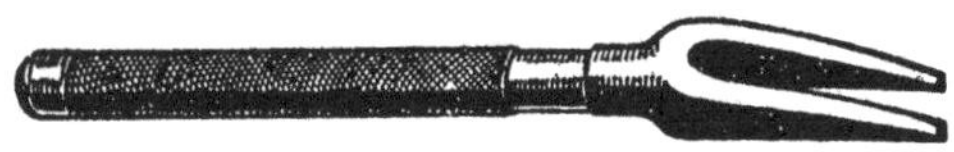

18G125

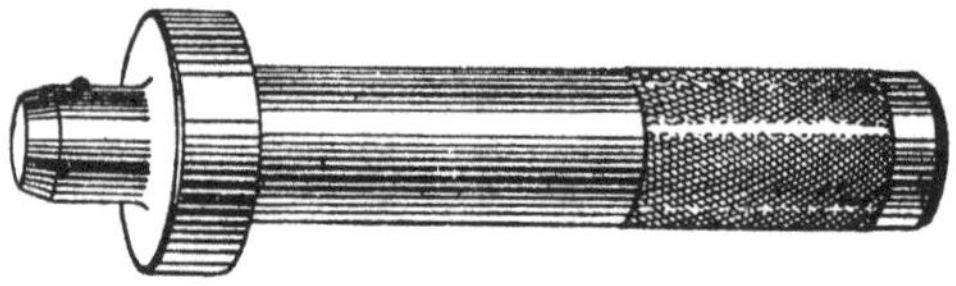

18G134

18G134. Detachable Handle

18G134K. Rear Hub Bearing Remover, Differential Bearing Replacer, and Rear Hub Assembly Replacer Adaptor

This adaptor used with the basic handle will ensure quick and easy removal of the rear hub bearings and the correct replacement of the hub assembly and the differential bearings.

The long pilot will maintain the bearing square with the axle tube or differential cage during the operation.

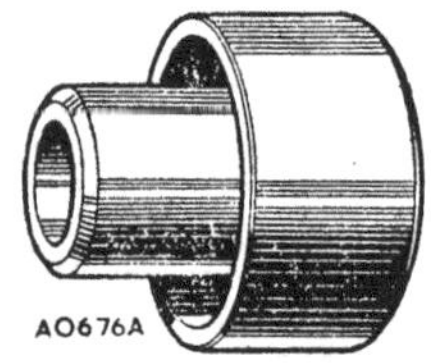

18G134K

18G134N. Oil Seal Replacer Adaptor

This tool ensures the correct fitting of a new oil seal to the gearbox rear extension.

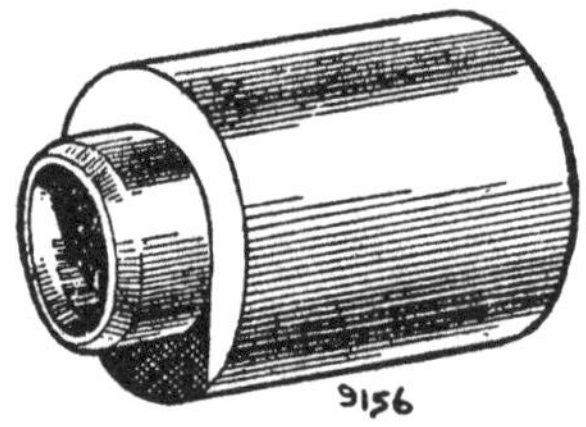

18G134N

18G134AQ. Rear Hub Oil Seal Replacer Adaptor

The adaptor fits closely to the outside rim of the oil seal and enables the seal to be driven squarely into the hub.

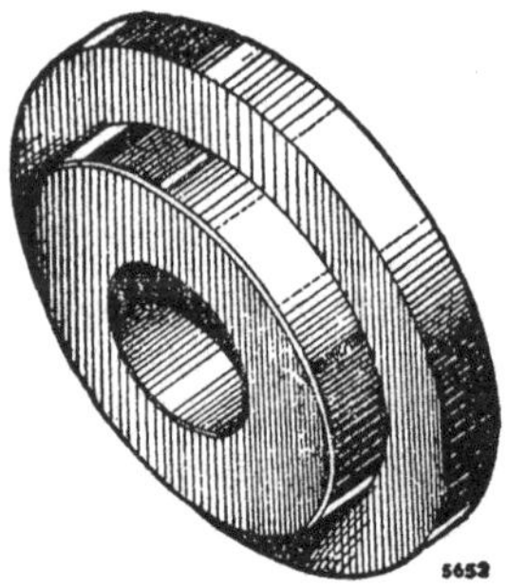

18G134AQ

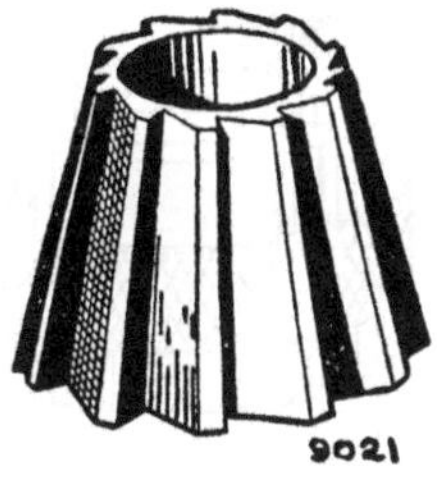

18G174C

18G174C. Valve Seat Narrowing Cutter—Bottom

For use with pilots 18G25D, 18G174D and handle 18G27.

18G174D

18G174D. Valve Seat Cutter Pilot

For use with cutters 18G28, 18G28A, 18G28B, 18G28C, 18G373, and 18G373A.

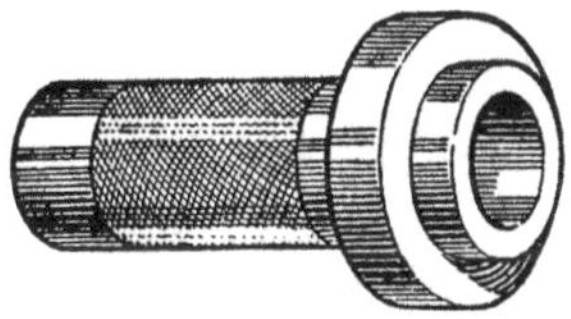

18G177

18G177. Oil Seal Replacer

Replacer designed to ensure the correct fitting of a new oil seal to the Laycock overdrive unit casing. It can be used with the gearbox and overdrive unit in position.

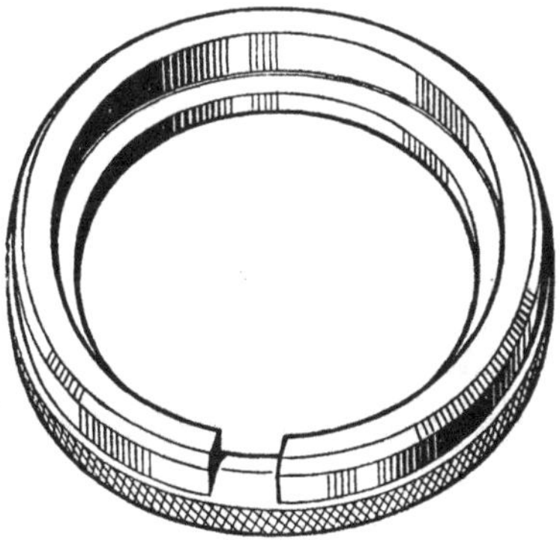

18G178

18G178. Roller Clutch Assembly Ring

This precision-made assembly ring is an essential tool for reassembling the roller clutch for the Laycock overdrive unit.

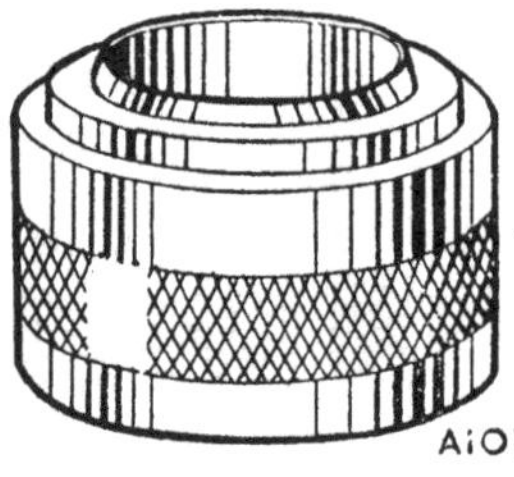

18G179

18G179. Accumulator Housing Piston Ring Compressor

A special tool for fitting the accumulator into its housing on the Laycock overdrive unit. It will ensure that the piston rings are fitted without breakage or damage to the aluminium housing.

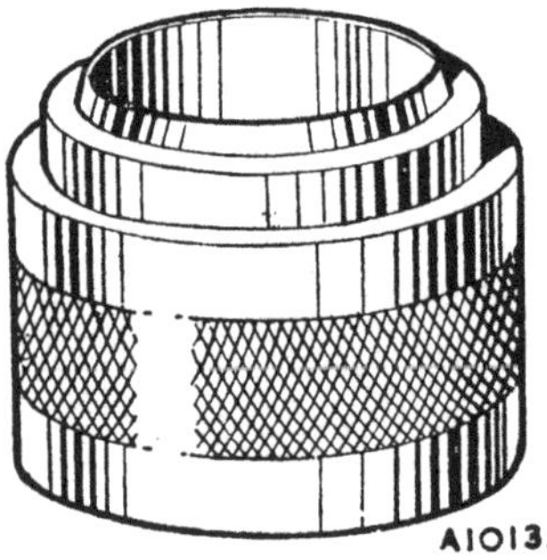

18G180

18G180. Operating Piston Ring Compresser

This tool entirely eliminates the possibility of breaking rings when fitting them to the operating piston on the Laycock overdrive unit.

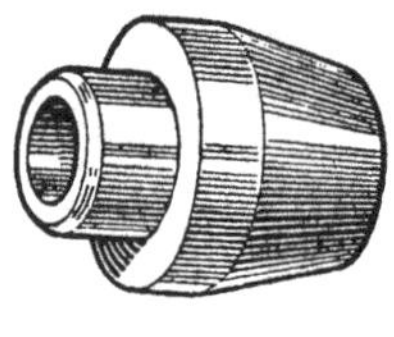

18G181

18G181. Accumulator Housing Ring Guide

This ring guide is designed to prevent permanent distortion of the lower rubber ring when it is being fitted to the accumulator housing on the Laycock overdrive unit.

18G182

18G182. Accumulator Housing Remover

The difficulty of removing the accumulator housing from its bore on the Laycock overdrive unit is completely overcome by the use of this tool.

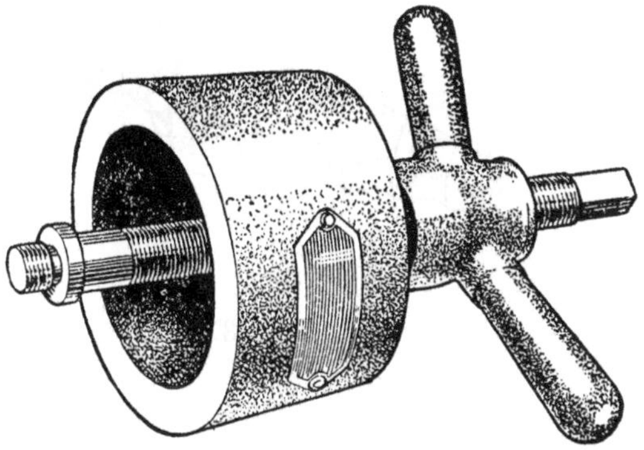

18G183

18G183. Oil Pump Body Remover

For use on the Laycock overdrive unit. The short screwed end of the extractor bolt is screwed into the oil pump body in place of the special screwed plug. With the squared end of the centre screw held firmly with a spanner the oil pump body can be withdrawn by turning the wing nut clockwise.

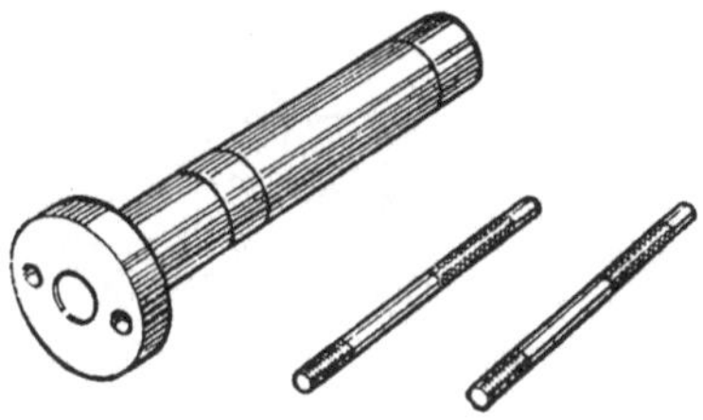

18G184

18G184. Oil Pump Body Replacer

The guide rods ensure correct lining up and easy replacement of the oil pump body to the Laycock overdrive unit.

18G185

18G185. Dummy Layshaft

This dummy mainshaft is essential when rebuilding the Laycock overdrive unit.

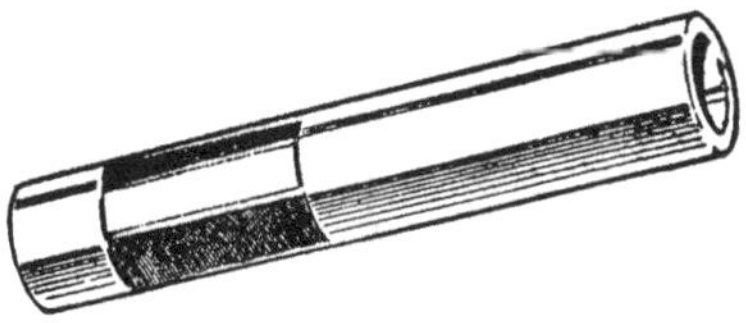

18G186

18G186. Mainshaft Bearing Replacer

This tool is designed for driving the mainshaft bearing and adaptor plate down the mainshaft and into position on the Laycock overdrive unit.

18G187

18G187. Radiator Reverse-Flush Adaptors

The adaptors should be used in pairs, one for the radiator inlet hose and one for the outlet hose. The hose connection is 1 in. (25·4 mm.) diameter.

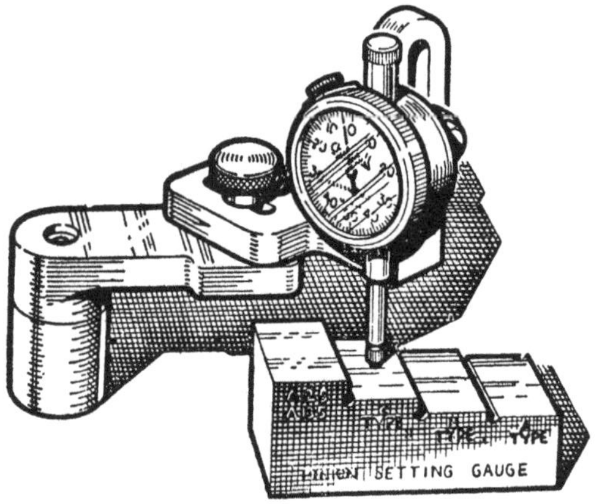

18G191

18G191. Bevel Pinion Setting Gauge

A gauge block and dial indicator is essential to obtain accurate location of the pinion in the axle case.

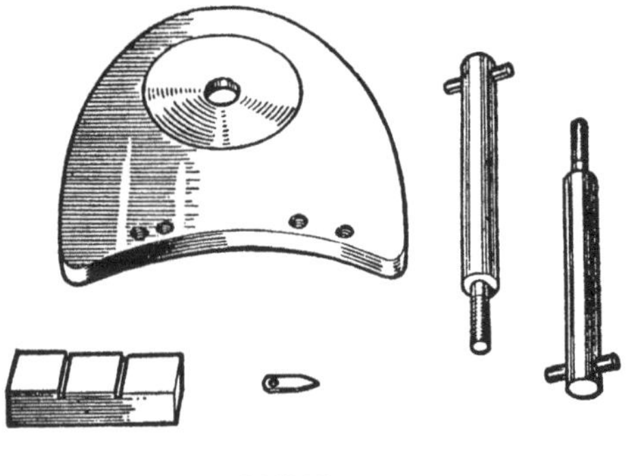

18G191A

18G191A. Differential Bearing Gauge

This gauge used with the component parts of 18G191 is designed to check the bearing width. It can also be used to mount the clock gauge on the gear carrier to check crown wheel and bevel pinion backlash.

18G207

18G207. Bevel Pinion Bearing Preload Gauge

The movable arms of the tool are located in opposite holes of the bevel pinion flange and the weight is moved along the rod to the poundage required.

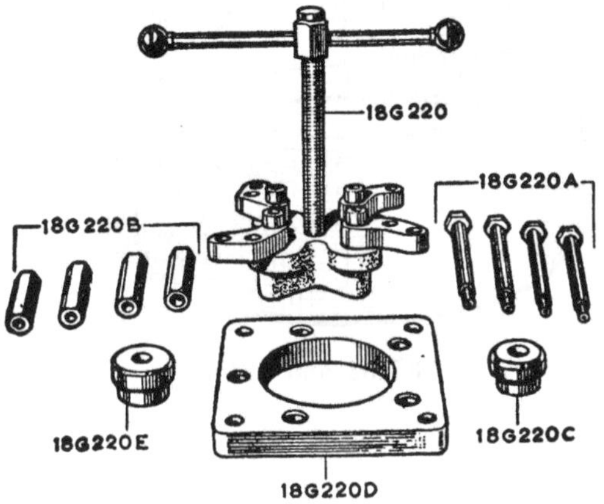

18G220
18G220A
18G220D
18G220E

18G220. Front and Rear Hub Remover
18G220A. Hub Remover Adaptor
18G220D. Hub Remover Adaptor
18G220E. Hub Remover Adaptor

A remover, designed for use with various adaptors.

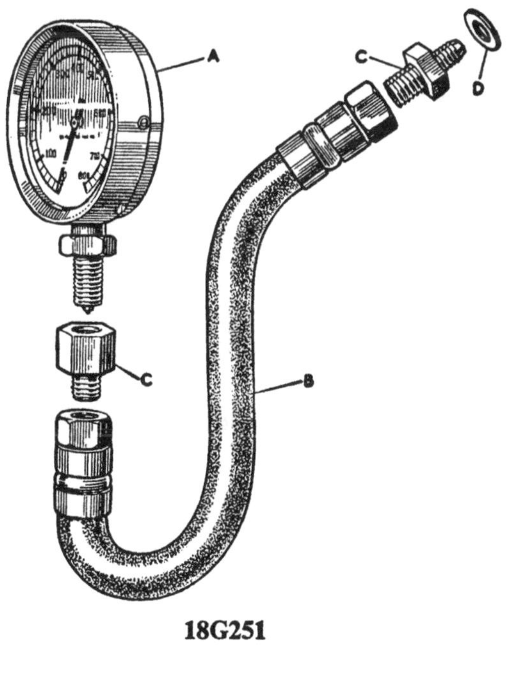

18G251

18G251. Hydraulic Pressure Gauge

When connected to the operating valve of the Laycock overdrive unit a pressure of between 430 and 460 lb./sq. in. (30·23 and 33·34 kg/cm.^{2}) should be indicated.

18G258. Rear Hub Nut Spanner

The long pilot locates in the axle tube to ensure an even pull on the thin octagonal hub nut.

18G258

18G262. Synchromesh Unit Assembly Ring (First and Second Speed)

18G263. Synchromesh Unit Assembly Ring (Third and Top Speed)

Designed to facilitate the assembly of mated synchronizer and sleeve by enabling the springs and balls to be inserted quickly and easily.

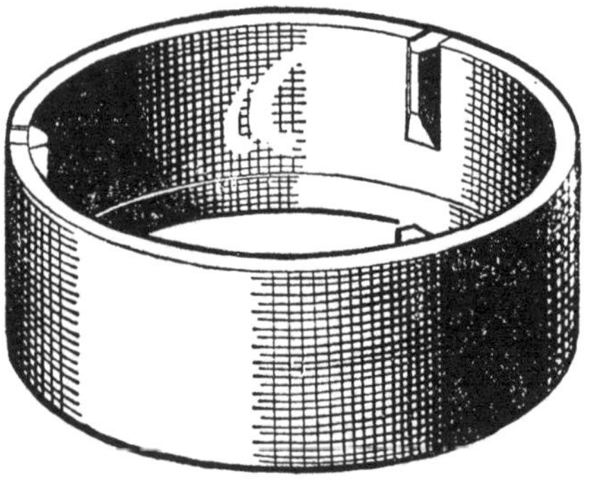

18G262 and 18G263

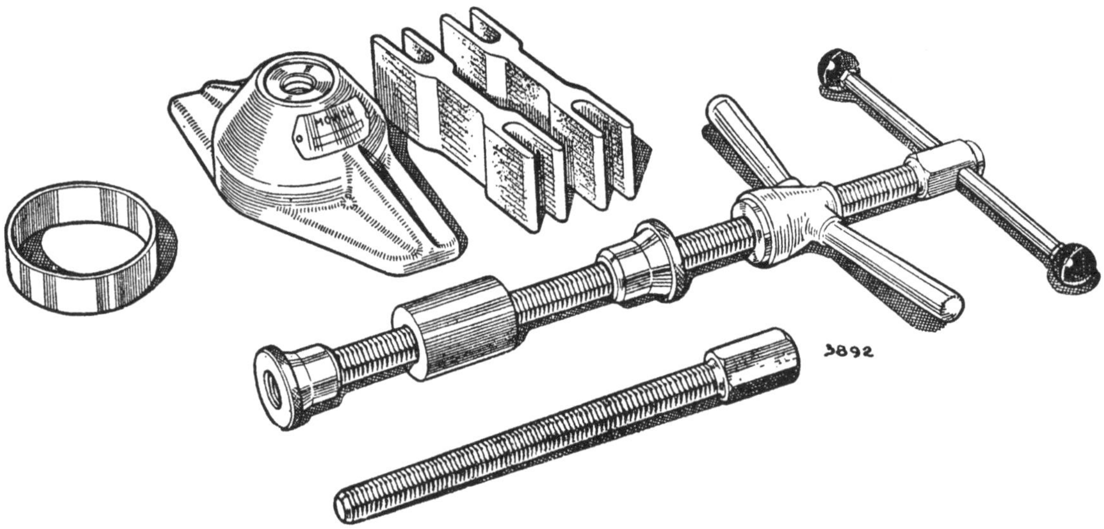

18G264

18G264. Bevel Pinion Bearing Outer Race Remover (basic tool)

Comprising a body, centre screw with extension and tommy-bar, wing nut, guide cone, and two distance pieces. A plain ring is also included to serve as a pilot when the rear bearing outer races are being replaced. Use with adaptors 18G264D and 18G264H.

18G264D and 18G264H

18G264D and 18G264H. Bevel Pinion Bearing Outer Race Remover Adaptors

Use with 18G264.

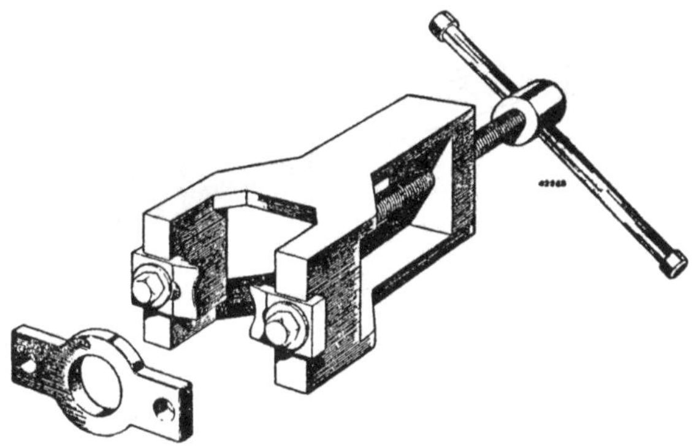

18G285

18G285. Driving Pinion Rear Bearing Inner Race and Roller Assembly Remover and Replacer

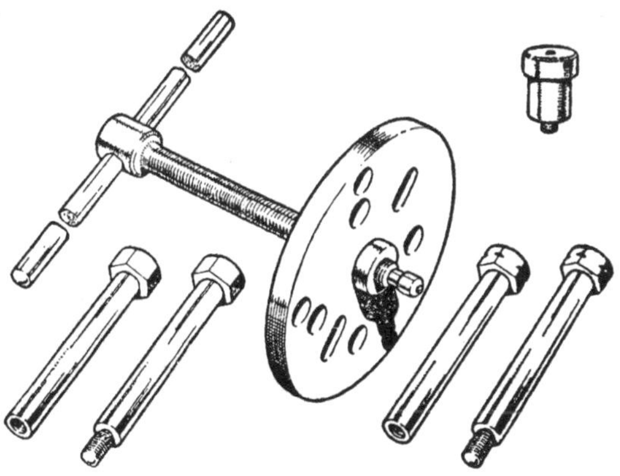

18G304
18G304B
18G304K

18G304. Hub Remover (basic tool)

18G304B. Bolt Adaptor—$\frac{7}{16}$ in. UNF.

18G304K. Thrust Pad

Screw the adaptor bolts onto the wheel studs through the holes in the plate of the basic tool. Place the thrust pad in the end of the axle tube and adjust the centre screw to engage the thrust pad. An exceptionally tight hub can be removed easily and without damage.

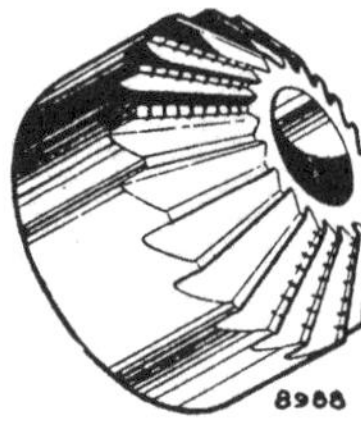

18G373. Inlet Valve Seat Finishing Cutter

Use with pilot 18G174D and handle 18G27.

18G373

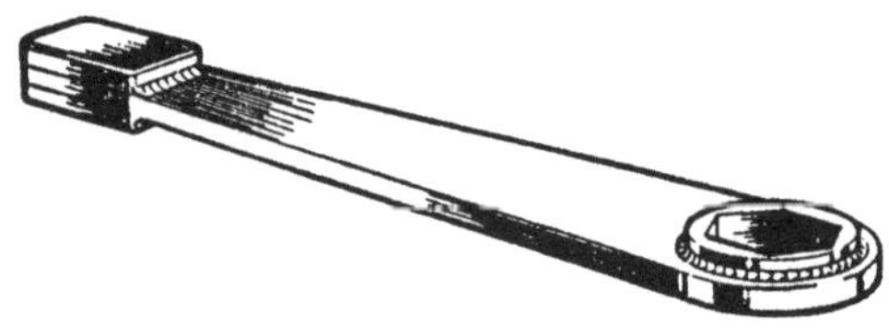

18G391. Starting Dog Nut Spanner

A "shock type" spanner designed to enable the starting dog nut to be removed without locking the crankshaft.

18G391

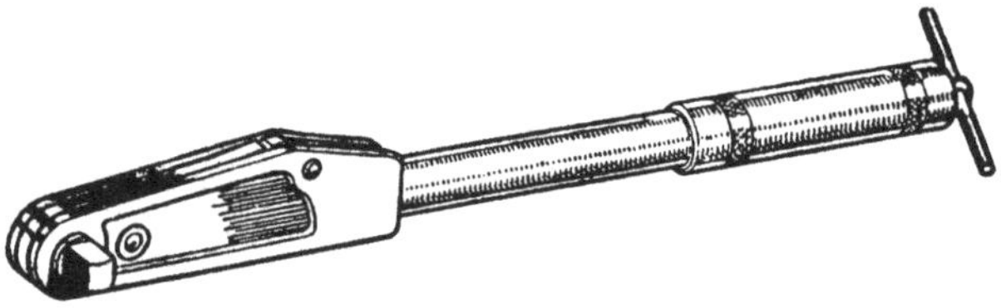

18G536. Torque Wrench 2 to 8 lb./ft. (·28 to 1·1 kg./m.)

A universal torque wrench for use with standard sockets. This tool is essential if the recommended maximum torque for various studs is not to be exceeded.

18G536

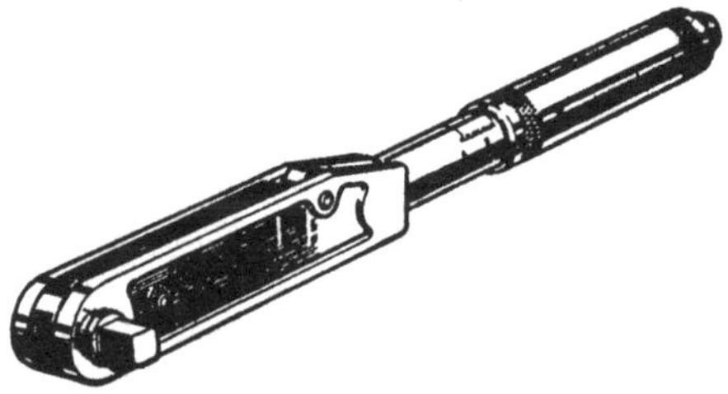

18G537. Torque Wrench 5 to 30 lb./ft. (6·9 to 4·15 kg./m.)

A universal torque wrench for use with standard sockets. This tool is essential if the recommended maximum torque for various studs is not to be exceeded.

18G537

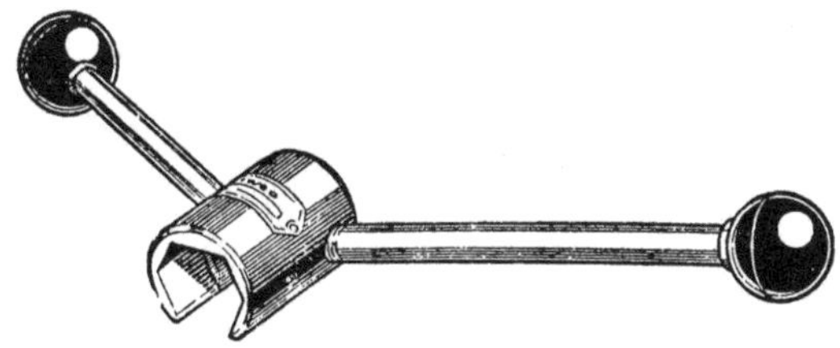

18G512

18G512. Steering Wheel Nut Spanner

This tool will enable the steering wheel nut to be loosened or tightened without completely withdrawing the stator tube.

ASJ.178

18G545

18G545. Cylinder Head Nut Spanner

This special spanner is for tightening the cylinder head nuts of "C"-type standardized engines without removing the rocker shaft.

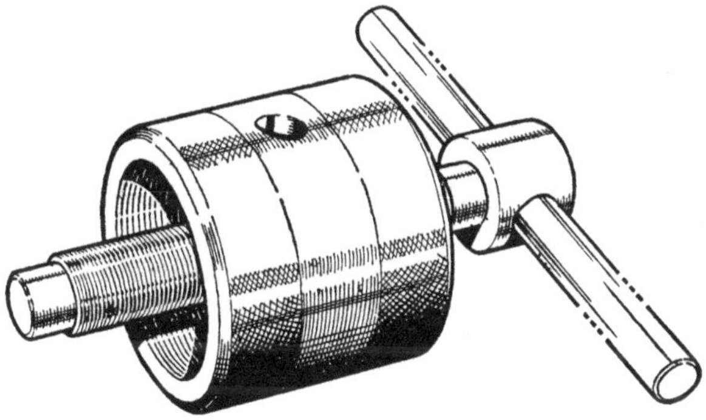

18G 363

18G 363 Hub Remover—Wire Wheels (12T.P.I.)

Designed to withdraw left-hand and right-hand 'knock-on hubs'. The body is internally threaded with a left-hand thread in one end and a right-hand thread in the other.

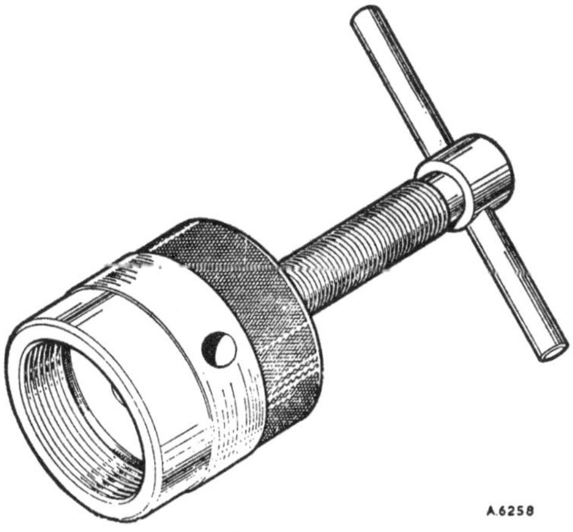

A6258

18G 1032

18G 1032 Hub Remover—Wire Wheels (8 T.P.I.)

Designed similar to 18G 363, this tool is distinguishable from 18G 363 by being knurled on the right-hand thread end of the tool.

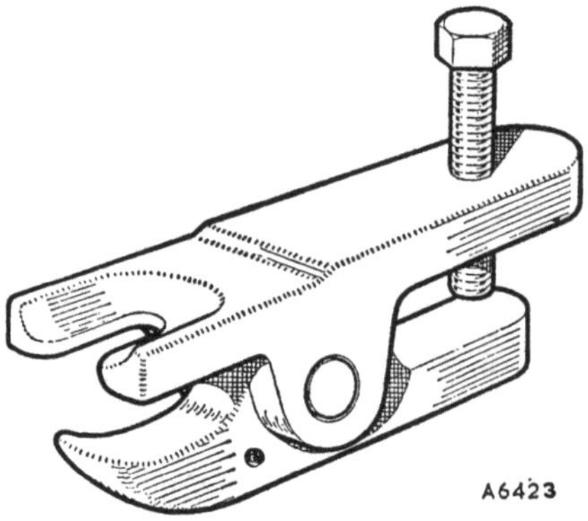

18G 1063

18G 1063 Remover—Steering Arm and Swivel Hub Ball Pins

To release the swivel hub of the steering arm ball pins.